Conference Proceedings of the Society for Experimental Mechanics Series

Series Editor

Kristin B. Zimmerman, Ph.D.
Society for Experimental Mechanics, Inc.,
Bethel, CT, USA

The Conference Proceedings of the Society for Experimental Mechanics Series presents early findings and case studies from a wide range of fundamental and applied work across the broad range of fields that comprise Experimental Mechanics. This Series volume follows the principle tracks or focus topics featured in the Society's Annual Conference & Exposition and will address critical areas of interest to researchers and design engineers working in all areas of Solid Mechanics by presenting early research findings from experimental and computational investigations related to the processing, characterization, and testing of composite, hybrid, and multifunctional materials.

More information about this series at http://www.springer.com/series/8922

Piyush R. Thakre • Raman P. Singh • Geoffrey Slipher
Editors

Mechanics of Composite, Hybrid and Multifunctional Materials, Volume 5

Proceedings of the 2018 Annual Conference on Experimental and Applied Mechanics

Springer

Editors
Piyush R. Thakre
HHB Bldg. 3B161
Dow Chemical Company
Lake Jackson, TX, USA

Raman P. Singh
College of Engineering, Architecture & Technology
Oklahoma State University
Stillwater, OK, USA

Geoffrey Slipher
United States Army Research Laboratory
Adelphi, MD, USA

ISSN 2191-5644 ISSN 2191-5652 (electronic)
Conference Proceedings of the Society for Experimental Mechanics Series
ISBN 978-3-030-07047-2 ISBN 978-3-319-95510-0 (eBook)
https://doi.org/10.1007/978-3-319-95510-0

This Springer imprint is published by the registered company Springer Nature Switzerland AG
The registered company address is: Gewerbestrasse 11, 6330 Cham, Switzerland

Preface

Mechanics of Composite, Hybrid, and Multifunctional Materials represents one of the eight volumes of technical papers presented at the 2018 SEM Annual Conference & Exposition on Experimental and Applied Mechanics organized by the Society for Experimental Mechanics and held in Greenville, SC, June 4–7, 2018. The complete proceedings also includes volumes on *Dynamic Behavior of Materials*; *Challenges in Mechanics of Time-Dependent Materials*; *Advancement of Optical Methods & Digital Image Correlation in Experimental Mechanics*; *Mechanics of Biological Systems & Micro-and Nanomechanics*; *Fracture, Fatigue, Failure and Damage Evolution*; *Residual Stress, Thermomechanics & Infrared Imaging, Hybrid Techniques and Inverse Problems*; and *Mechanics of Additive and Advanced Manufacturing.*

The commercial market for composite continues to expand with a wide range of applications from sporting equipment to aerospace vehicles. This growth has been fueled by new material developments, greater understanding of material behaviors, novel design solutions, and improved manufacturing techniques. The broad range of applications and the associated technical challenges require an increasingly multidisciplinary and collaborative approach between the mechanical, chemical, and physical sciences to sustain and enhance the positive impact of composites on the commercial and military sectors.

New materials are being developed from recycled source materials, leading to composites with unique properties and more sustainable sources. Existing materials are also being used in new and critical applications, which requires a deeper understanding of material behaviors and failure mechanisms on multiple length and time scales. In addition, the unique properties of composites present many challenges in manufacturing and in joining with other materials. New testing methods must be developed to characterize the novel composite properties, to evaluate application and product life cycle performance, and to evaluate impacts and merits of new manufacturing methods.

This volume presents early research findings from experimental and computational investigations related to the processing, characterization, and testing of composite, hybrid, and multifunctional materials.

Lake Jackson, TX, USA
Stillwater, OK, USA
Adelphi, MD, USA

Piyush R. Thakre
Raman P. Singh
Geoffrey Slipher

Contents

Chapter 1
Stimulus-Responsive Interfacial Chemistry in CNT/Polymer Nanocomposites

Frank Gardea, Zhongjie Huang, Bryan Glaz, Shashi P. Karna, Xiyuan Cheng, Zhiwei Peng, and YuHuang Wang

Abstract The enhancement of interfacial interactions in carbon nanotube (CNT)/polydimethylsiloxane (PDMS) polymer matrix composites was investigated. The approach taken was to functionalize the CNTs with the photoreactive molecule benzophenone in order to anchor the CNTs to the polymer chains on demand. The anchoring reaction was activated by the use of externally applied UV irradiation. A comparison was done on randomly dispersed and aligned CNTs in order to observe the effect of orientation on interface mechanics and overall enhancement. The effect of interfacial interaction on the mechanical response was determined through analysis of static mechanical experiments, as an increase in interfacial interaction resulted in an observable change in elastic modulus and yield stress. An increase of 22% in elastic modulus was observed in randomly oriented CNTs while an increase of 93% was observed in aligned CNT composites after exposure to UV light. In addition, alignment of CNTs lead to a more discreet yield stress which allowed for a clearer identification of the onset of interfacial failure. This work provides insight into the intelligent design of composites, starting at the nanoscale, to provide desired on-demand macroscale response.

Keywords Nanocomposites · Carbon nanotubes · Photoreactive · Interface · Adhesion

1.1 Introduction

Polymer based composites with nanoscale fillers show promise for enabling multifunctional nanocomposites with unique properties. Possibilities range from enhanced mechanical [1], electrical [2], and thermal [3] properties to functionalities such as self-healing, sensing, and power storage. However, even as advanced manufacturing methods are developed, the promise of unique materials with advanced capabilities has not been fully met. Nanoscale reinforcements have the potential to increase effective composite properties, with this ability highly dependent on the mechanics of the inclusion-polymer interface. The importance of controlling interfacial interactions between a filler and matrix to obtain desired bulk composite properties has been emphasized in literature [4–7]. Due to scale, nanoparticle composites will induce a greater effect via the interface compared to conventional microparticle composites. For example, the reinforcing effect of nanoparticles is dominated by the interfacial shear strength. The overall strength of a composite material is controlled by the filler-matrix interface and not the filler strength. The strength of the composite is directly proportional to the strength of the interface, which directly correlates to the degree of interaction between filler and matrix [1].

There has been substantial variation in reported nanocomposite properties. For example, increases in elastic modulus have been shown to improve significantly by up to 4 times at CNT loadings varying from 0.2 to 2 vol%, while other studies have shown only slight increases [1]. Furthermore, enhanced mechanical damping in CNT/polymer matrix composites has been shown to be sensitive to the interfacial shear strength [8–10]. Thermal conductivity has been shown to increase by up to 1.5 times for concentrations of 3 wt% CNT [11], while others have shown minimal improvements at similar weight fractions due to large interfacial thermal resistance [12]. The variance of reported macro-scale material property improvements shows the importance of controlled interfacial interactions.

F. Gardea (✉) · B. Glaz
Vehicle Technology Directorate, U.S. Army Research Laboratory, Aberdeen Proving Ground, Aberdeen, MD, USA
e-mail: frank.gardea4.civ@mail.mil

Z. Huang · X. Cheng · Z. Peng · Y. Wang
University of Maryland, Department of Chemistry, College Park, MD, USA

S. P. Karna
Weapons and Materials Research Directorate, U.S. Army Research Laboratory, Aberdeen Proving Ground, Aberdeen, MD, USA

P. R. Thakre et al. (eds.), *Mechanics of Composite, Hybrid and Multifunctional Materials, Volume 5*, Conference Proceedings of the Society for Experimental Mechanics Series, https://doi.org/10.1007/978-3-319-95510-0_1

Given the significant effect of the interfacial properties, controlling interfacial interactions will allow for a way to control macroscopic behavior in nanocomposites. One way of controlling the interfacial interaction is through functionalization of the filler, since an attractive interface will decrease the mobility of polymer chains surrounding it and thus increase the interfacial region [5]. Functionalization refers to the attachment of molecules to the surface of the nanoparticle to improve solubility, dispersion, and interfacial interactions [13]. Fabrication of nanocomposites with functionalized nanoparticles usually consists of a multi-step functionalization treatment of the nanoparticles [13–15] before embedding them into a polymer matrix and allowing the functionalization to naturally bond to the matrix without restriction [16–18]. Functional materials featuring tunable interfaces between components are a growing research area due to unique design tunability and resulting combinational properties. The controlled coupling and interaction at nanoscale interfaces could enable the generation of interesting and exceptional optical, electrical, thermal, and mechanical attributes. Thus, the ability to tune the interface properties during fabrication would be desirable for producing nanocomposites optimized to particular applications. Achieving in-situ control of the interface can be accomplished through stimuli-responsive functionalization. For example, dispersion of single-walled CNTs in solvent has been performed by functionalization with stimuli-responsive dispersants, whether the external stimuli is temperature, pH, CO_2, or photo-irradiation [19]. UV irradiation has been used for photopolymers, in which curing of the polymer is achieved by UV stimulus [20]. While UV light has been successful for curing of photopolymers [20], polymer grafting to thin-films [21], and polymer grafting to nanoparticles dispersed in solvent [13], the use of UV light for in-situ control of interfacial interactions in nanoparticle composites has not been reported in literature.

In this study, we investigate the chemistry of interfacial adhesion between a CNT film and a polymer matrix by the use of the photoreactive molecule benzophenone and study the feasibility of controlling the composite mechanical response by exposure to an external UV stimulus. A comparison between randomly oriented and aligned CNT composites was performed in order to determine the effect orientation has on the overall enhancement of mechanical properties. Photo-responsive materials, which can be excited by light to form chemical bonds or modify physical and chemical properties, are very promising for applications in stimuli-responsive systems with remote control.

1.2 Experimental Methods

1.2.1 Functionalization of CNT Films

The diazonium salt was synthesized based on chemistry developed previously [22, 23]. 4-aminobenzophenone (473 mg, 99%, Sigma Aldrich) was mixed with 2 mL of nanopure water and an aqueous solution of tetrafluoroboric acid (1.5 mL, 48 wt%, Sigma Aldrich). Dissolved sodium nitrite (Sigma Aldrich ≥97.0%, 331 mg in 1 mL of nanopure water) was added dropwise to the mixture and allowed to react for 15 min and the diazonium salt collected via filtration. The salt was dried and stored at 4 °C.

Drawable multiwall CNT (MWCNT) forests were synthesized using chemical vapor deposition methods [24, 25] resulting in MWCNTs with a diameter of 50 nm. The forest was drawn continuously into a film of aligned CNTs, as seen in Fig. 1.1b. The drawn films were immersed in an aqueous solution of diazonium salt (1 mg/mL) for 3 days.

HiPco SWCNTs [26] were reacted with the diazonium salt with a 1:1 molar ratio in aqueous solution with 2 wt% sodium dodecyl sulfate (SDS) surfactant for 2 h. Filtration onto a filter paper produced a randomly distributed CNT thin film. The CNT film/filter was transferred to a cured PDMS film and the filter paper dissolved in acetone.

1.2.2 Fabrication and UV Treatment of Layered Composites

PDMS monomer and crosslinker (Sylgard® 184, Dow Corning) were mixed at a 10:1 ratio. After degassing, the polymer was spin-coated at 1000 rpm for 60 s onto glass slides and cured at 90 °C for 2 h. The CNT films (either aligned or randomly dispersed) were placed on top of the polymer film and chemically treated as mentioned previously. After chemical treatment, a second layer of PDMS was spin coated on top of the CNT layer. The composites were then exposed to 365 nm UV light using an Omnicure LX500 system. The UV light was set to 500 mW/cm^2 and the duration of exposure varied between 0 and 5 min. After UV treatment, the PDMS layer was cured at 90 °C for 2 h. The final composite consisted of a PDMS-CNT-PDMS layered structure with an overall thickness of 170 μm with the CNT film positioned at the center of the composite as shown in Fig. 1.1c.

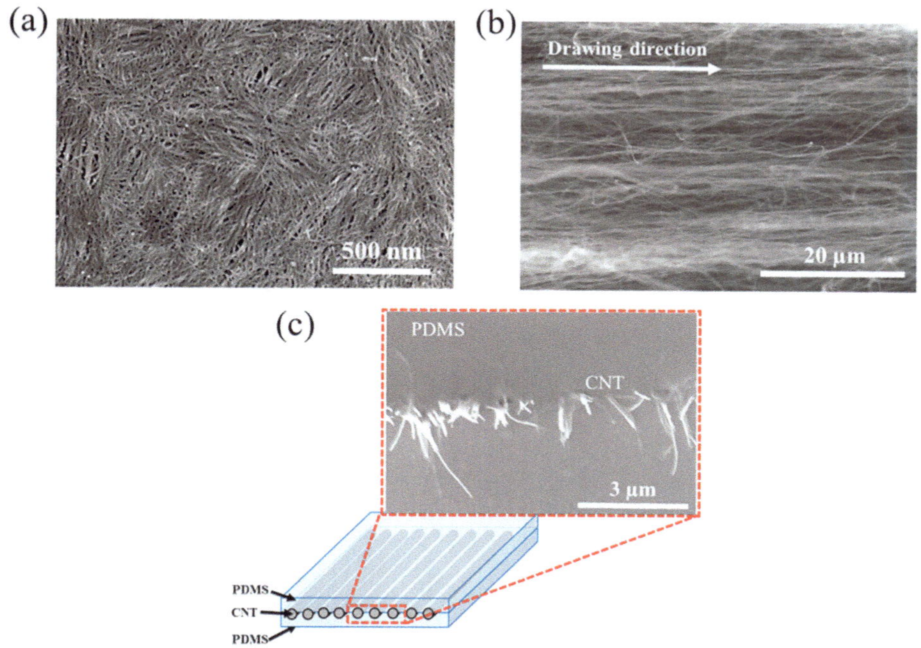

Fig. 1.1 (**a**) SEM image of randomly dispersed SWCNT film (**b**) SEM image of aligned MWCNT film (**c**) Schematic illustrating the composite layered structure. An SEM image of the composite cross-section is shown to verify the layered structure

1.2.3 Experimental Characterization

The degree of functionalization of the CNTs was characterized using Raman spectroscopy [27, 28] using a Horiba Jobin-Yvon LabRam ARAMIS Raman microscope with a 532 nm laser excitation. Absorption spectra was measured using a PerkinElmer Lambda 1050 UV-vis-NIR spectrophotometer. For the film samples, a 150 mm integrating sphere was used. Imaging and observation of CNT arrangement was performed using a Hitachi SU-70 scanning electron microscope (SEM). Static mechanical testing was performed using a Shimadzu-AGS-X series tensile tester. The elastic modulus was obtained from the initial linear portion of the response while the yield stress was taken as the stress value at the onset of nonlinearity determined using a tangent analysis method.

1.3 Results and Discussion

Inspiration from thin-film polymer-grafting technology [21] was obtained to develop a successful nanoparticle/polymer grafting method that could be controlled and activated during the nanocomposite fabrication process. The approach consisted of a "grafting to" method: functionalization of the CNT with a molecule that can bond and anchor to the polymer chain, with the anchoring reaction activated by the use of UV irradiation. UV irradiation was chosen for its simplicity and ease of intensity, frequency, wavelength, and exposure time variation. In addition, the low consumption of energy and lack of volatile organic solvents made the use of UV technology very attractive, while the short wavelength of the UV light allows for precise focusing, providing the ability for selective exposure to areas of the composite [29]. Another benefit was that UV light can be used at ambient temperature and in the presence of air, thus avoiding any thermal degradation of the polymer [20]. An advantage of most photoreactive groups is the chemical inertness in the absence of light [21]. This assures that any interfacial interaction between the filler and matrix is solely a result of the applied external stimulus.

The photoreactive molecule of interest was benzophenone (BP), with the chemical structure as shown in Fig. 1.2. The light absorbance of BP has a wide range from 310 to 390 nm wavelength. BP can be effectively excited to a diradical triplet state using UV light in this wavelength range, to form a bond with an alkyl group after hydrogen abstraction (see Fig. 1.2) [21, 30, 31]. This behavior makes BP attractive since it can attach to various polymer backbones or side groups that contain a C-H bond.

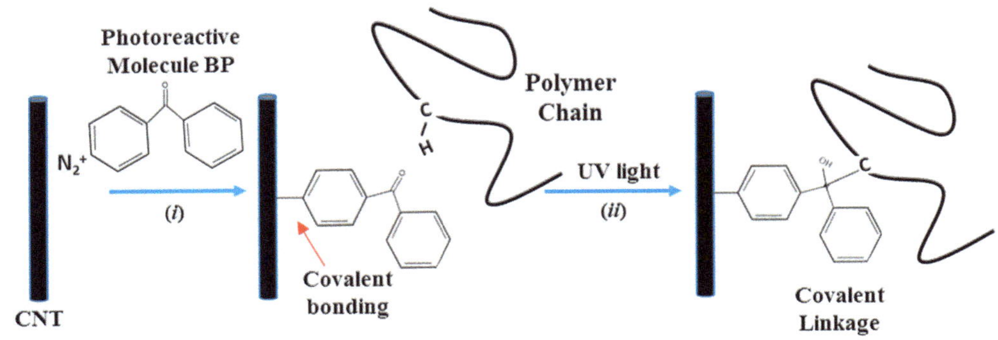

Fig. 1.2 Photoreactive benzophenone (BP) molecule bonded to a CNT via *(i)* diazonium reaction and subsequent *(ii)* photoreactive bonding to a polymer chain forming a covalent link between the CNT and polymer chain

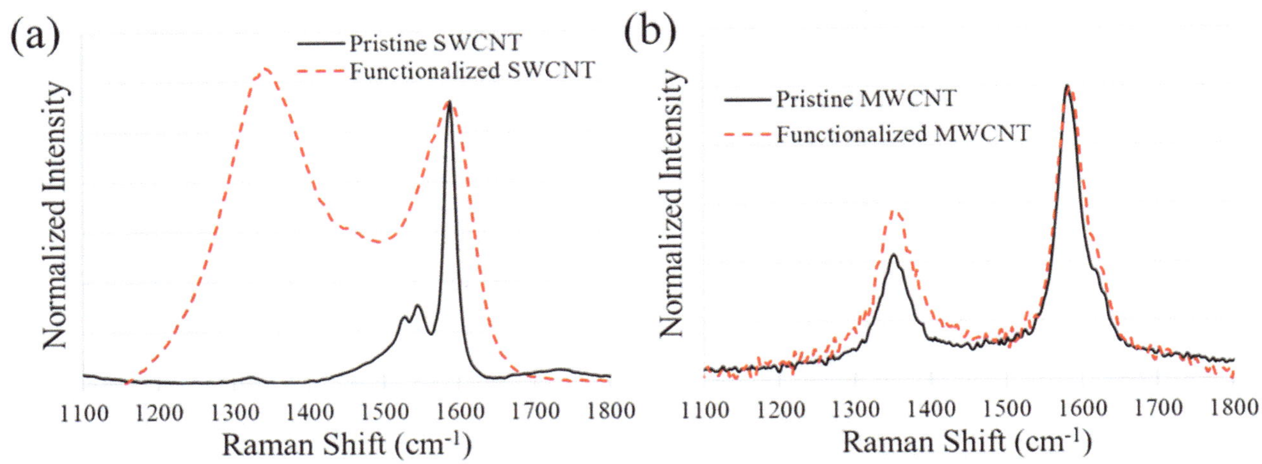

Fig. 1.3 Normalized Raman spectra of (**a**) pristine and functionalized HiPco SWCNTs and (**b**) pristine and functionalized MWCNTs

The approach was to chemically bond a BP molecule to a CNT such that when exposed to UV light, the BP would react and bond to the C-H bonds on the side groups or backbone of the polymer matrix, as schematically shown in Fig. 1.2. This photoreactive molecule would provide a controlled covalent link between the CNT and polymer chain.

The orientation of the filler is considered extremely important as the effect of this chemical bonding enhancement on the interfacial mechanics might not be observable if additional mechanisms are introduced as a result of misorientation. Studying the effect of CNT alignment provided a more theoretically grounded understanding of the interfacial load transfer mechanics.

1.3.1 Characterization of CNT Functionalization

In the Raman spectrum of a CNT, the peak ratio of the defect-derived D-band and the graphitic structure-derived G-band is a good indicator of the defect concentration in a CNT sample [27, 28, 32]. In Fig. 1.3a, the spectra of dispersed pristine HiPco SWCNTs is shown. It is observed that a high-intensity G peak (1600 cm^{-1}) and low-intensity D peak (1350 cm^{-1}) are present in the pristine sample, which indicates a low number of defect sites. After the reaction, a significantly enhanced D peak (at around 1350 cm^{-1}) is observed, resulting in a D/G peak ratio > 1, indicative of a high quantity of defects caused by grafting of the BP molecule to the CNTs. The same chemistry was effective for funtionalizing both SWCNT and MWCNT films. The Raman spectra (Fig. 1.3b) shows the enhanced D/G ratio after the reaction for aligned MWCNT films. It is worth mentioning that a D peak with considerable intensity was observed for the pristine MWCNT sample, which is due to the defects generated in the synthesis process [33]. Therefore, only a slight enhancement of D/G ratio can be observed for the functionalized sample, however the result does verify the attachment of the BP molecule to the surface of the aligned MWCNTs.

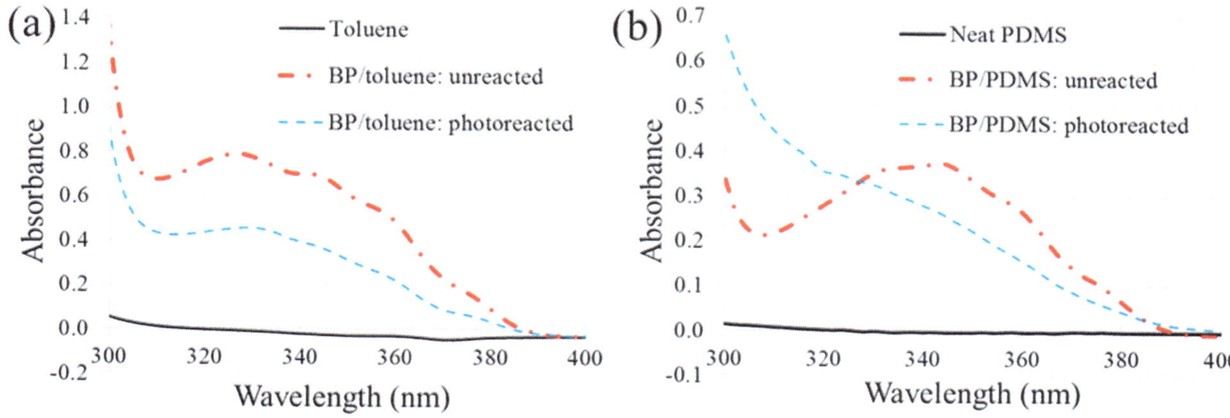

Fig. 1.4 Absorbance spectra of (**a**) BP dispersed in toluene before and after photoreaction and (**b**) BP embedded in PDMS before and after photoreaction. Results are for an irradiance of 500 mW/cm^2 and exposure time of 5 min

1.3.2 Photoreaction of Benzophenone

UV-vis spectroscopy was used to study the photoreaction of benzophenone both in a solvent and embedded in PDMS. As observed in Fig. 1.4a, the characteristic BP peak decreases with UV exposure. This result indicates that the irradiance and exposure time is sufficient to excite the BP molecule into a diradical triplet state and trigger a reaction with a C-H bond. The decrease of the characteristic peak indicates that less BP is available to absorb UV light and thus has reacted with the surrounding solvent molecules. Once embedded in PDMS, BP shows a significant decrease in the absorbance peak indicative of a reaction with the surrounding PDMS matrix.

1.3.3 Mechanical Behavior of CNT/PDMS Nanocomposites

The interfacial interaction, and thus the load transfer mechanisms, in CNT/polymer composite materials can result from a combination of the following: (1) mechanical interlocking, (2) van der Waals interactions, and (3) chemical bonding [34]. While mechanical interlocking results from surface topography on the CNT and the matrix conforming to that surface, van der Waals and chemical bonding are a function of the chemistry present at the interface. It is expected that an increase in interfacial interactions will lead to increases in elastic modulus and yield stress due to the increased efficiency of load transfer between the filler and matrix.

Evidence of improved interfacial interaction between the SWCNT film and the surrounding matrix is observed upon exposure to the UV stimulus. By triggering the photochemical covalent bonding, results in Fig. 1.5a show an increase of 22% in the elastic modulus after UV exposure compared to the functionalized case with no UV treatment.

Even though the SWCNT composites do show some slight nonlinearity, it is difficult to conclude at what point the composite yields by failure at the interface between the CNTs and matrix. This is a direct result of the random distribution of CNTs in the composite. If any interface failure occurs, it will happen over a wide range of stresses since the CNTs are not aligned in the direction of loading, thus the reason for the broad nonlinear response. Another effect of the random distribution is weak interparticle load transfer. As seen in Fig. 1.1a, there is substantial CNT-CNT interaction which could result in weak load transfer and thus would not significantly contribute to the composite modulus enhancement. Another possibility is the difficulty of PDMS penetration of the film. It is necessary for the polymer to penetrate the film in order for the polymer to act as the bridge between CNTs, as has been reported [1]. If so, only the outer CNT film surface would effectively bond with the surrounding matrix and thus the weak interparticle forces would dominate the interior of the film, resulting in weak stress transfer.

In order to address the issues observed in the SWCNT thin film results, a different approach to composite fabrication was taken which involved drawing of CNTs from a vertical forest. This procedure results in a less dense film as observed in Fig. 1.1b, allowing for the interpenetration of the PDMS chains. In addition, aligned CNTs would allow us to conclude that any observed reinforcement behavior is due solely to interfacial effects and not increased interactions between entangled CNTs since the reinforcements are in the direction of loading, similar to aligned discontinuous fiber composites. By aligning the

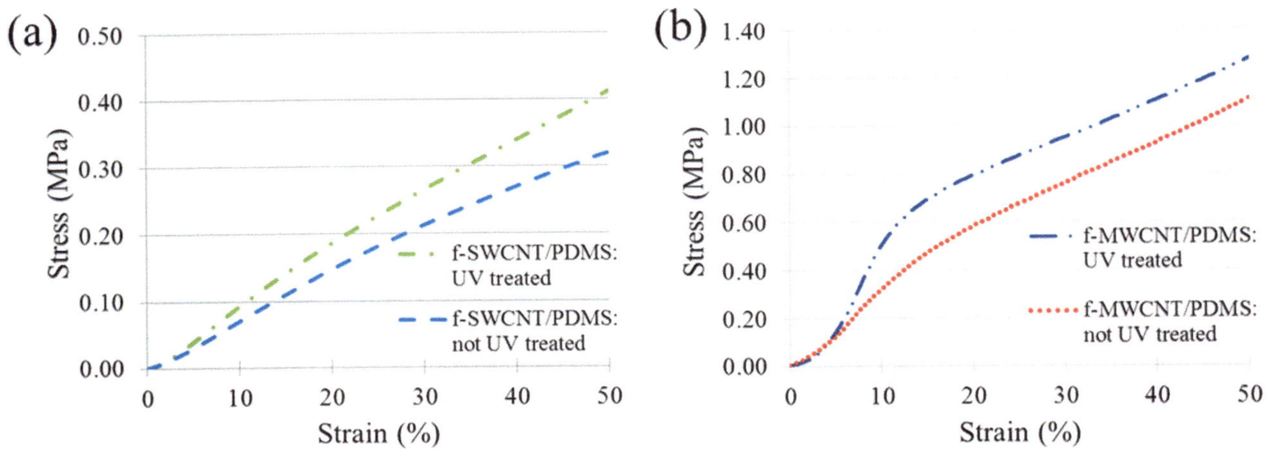

Fig. 1.5 Stress vs strain response for (**a**) randomly oriented functionalized SWCNT (f-SWCNT) film in a PDMS matrix and for (**b**) aligned functionalized MWCNT (f-MWCNT) film in a PDMS matrix, with and without exposure to a UV stimulus of 500 mW/cm^2 for 5 min

CNTs in the direction of loading, any observable change in the mechanical response would be due to altered interfacial interactions. The alignment will suppress any matrix yielding not associated with the interface and reduce CNT-CNT interaction phenomena.

The mechanical response of the aligned CNT composites is shown in Fig. 1.5b. At low strains (< 3%), all composite types show a similar elastic modulus. This behavior is as a result of the CNT films not being perfectly aligned to the loading direction but containing some initial waviness, as observed in Fig. 1.1b. As the composite material is further loaded, the waviness decreases and the CNTs further align in the direction of loading and thus show an onset in interfacial load transfer characterized by increased Young's modulus (at about 3% strain). The elastic modulus of the composite is measured past this onset point.

Results show a more pronounced yielding as compared to the SWCNT results, with a yield stress of 0.49 and 0.66 MPa for the non-UV treated and UV treated MWCNT composite samples, respectively. The UV-triggered enhanced interfacial interactions in MWCNT composites lead to a more defined and enhanced yield stress, with a total increase of 35%. In the non-UV treated samples, any interfacial interactions are due solely to mechanical interlocking and van der Waals. Upon exposure to UV light, the covalent bonding was triggered resulting in significant increase in elastic modulus from 3.88 to 7.50 MPa (93% increase) for a 5 min UV exposure time. In addition, the yielding of the composite became less broad and more defined. These results point to the significance of CNT-polymer interface engineering for enhancing and controlling the stiffness and yield stress of a composite.

In order to verify that the interface failed and that the observed enhancement was due solely to interfacial interaction, the composites were reloaded and the corresponding stress vs strain behavior captured. The enhancement in modulus due to the presence of the CNTs was lost (results not shown here) and the behavior followed that of the neat PDMS. The stress transfer at the interface deteriorated and thus the CNTs no longer carried any load, thus the reason for the response becoming similar to that of the neat matrix. This behavior was observed for every sample type. This verifies that the enhancement observed in the UV treated samples was due to enhanced interfacial interaction between the CNT and the surrounding matrix. Furthermore, this suggests a chemical bond interaction as would be expected by the covalent bond induced by the photoreactive molecule, as a broken covalent bond will not naturally re-form for another loading cycle.

In a random CNT case, it is difficult to convincingly conclude that modulus enhancement (and onset of interfacial failure) is solely due to increased interfacial interaction since many other effects (such as CNT entanglement, agglomeration, matrix tearing, curvature) are present, which affect the composite response. In addition, the stress distribution at the interface is not symmetric in off-aligned cases [35], which would result in a broad distribution of the onset of interfacial failure within the composite, making it difficult to pinpoint the onset of failure in the global mechanical response, thus the reason for the observed broadend behavior in the SWCNT composites. However, one can obtain a clearer, more theoretically grounded understanding of the interfacial load transfer mechanics by considering aligned CNT nanocomposites [8]. As our results show, interfacial effects are more pronounced in aligned composites, while in the unaligned CNT reinforcement, there are various mechanisms present that can result in not only less enhancement in mechanical properties, but also in a difficulty with isolating the salient mechanics.

1.4 Conclusion

The feasibility of using stimulus responsive chemistry to control the degree of interfacial interaction between a CNT filler and polymer matrix was studied. Both aligned and randomly oriented CNTs were functionalized with the photoreactive molecule benzophenone. Covalent bonding with the surrounding matrix was triggered by exposure to UV light at a 365 nm wavelength. Improved adhesion in randomly dispersed SWCNT composites was observed upon UV exposure, evident in the increase in elastic modulus of 22%. Aligned MWCNT film composites showed a more significant increase in elastic modulus and yield stress. The alignment of the CNTs not only allowed for the CNTs to sustain load, but also suppressed any matrix yielding not related to interfacial stresses and minimized any CNT-CNT interaction, thus resulting in an enhancement in elastic modulus of 93% when comparing UV treated and non-treated CNT composites. Repeated loading of the samples verified that the reinforcement behavior was due to solely interfacial interactions as the composites behaved like the neat PDMS material. This work shows the potential of using stimulus-responsive interfacial adhesion as a means to controlling interfacial mechanics in CNT/polymer nanocomposites.

Acknowledgement F. Gardea and Z. Huang contributed equally to this work. This work is partially supported by the Army Research Office under Cooperative Agreement No. W911NF1420024 and the Air Force Office of Scientific Research through Grant No. FA9550-16-1-0150.

References

1. Coleman, J.N., et al.: Small but strong: A review of the mechanical properties of carbon nanotube–polymer composites. Carbon. **44**(9), 1624–1652 (2006)
2. Gardea, F., Lagoudas, D.C.: Characterization of electrical and thermal properties of carbon nanotube/epoxy composites. Compos. Part B. **56**, 611–620 (2014)
3. Biercuk, M., et al.: Carbon nanotube composites for thermal management. Appl. Phys. Lett. **80**(15), 2767–2769 (2002)
4. Rittigstein, P., et al.: Model polymer nanocomposites provide an understanding of confinement effects in real nanocomposites. Nat. Mater. **6**(4), 278–282 (2007)
5. Schadler, L.: Nanocomposites: Model interfaces. Nat. Mater. **6**(4), 257–258 (2007)
6. Schadler, L., Brinson, L., Sawyer, W.: Polymer nanocomposites: A small part of the story. JOM. **59**(3), 53–60 (2007)
7. Schadler, L.S., et al.: Designed interfaces in polymer nanocomposites: A fundamental viewpoint. MRS Bull. **32**(04), 335–340 (2007)
8. Gardea, F., et al.: Energy dissipation due to interfacial slip in nanocomposites reinforced with aligned carbon nanotubes. ACS Appl. Mater. Interfaces. **7**(18), 9725–9735 (2015)
9. Koratkar, N.A., et al.: Characterizing energy dissipation in single-walled carbon nanotube polycarbonate composites. Appl. Phys. Lett. **87**(6), 063102 (2005)
10. Suhr, J., Koratkar, N.A.: Energy dissipation in carbon nanotube composites: A review. J. Mater. Sci. **43**(13), 4370–4382 (2008)
11. Wang, S., Qiu, J.: Enhancing thermal conductivity of glass fiber/polymer composites through carbon nanotubes incorporation. Compos. Part B. **41**(7), 533–536 (2010)
12. Hong, W.-T., Tai, N.-H.: Investigations on the thermal conductivity of composites reinforced with carbon nanotubes. Diamond Relat. Mater. **17**(7), 1577–1581 (2008)
13. Park, J.J., et al.: Functionalization of multi-walled carbon nanotubes by free radical graft polymerization initiated from photoinduced surface groups. Carbon. **48**(10), 2899–2905 (2010)
14. Liu, P.: Modifications of carbon nanotubes with polymers. Eur. Polym. J. **41**(11), 2693–2703 (2005)
15. Qin, S., et al.: Functionalization of single-walled carbon nanotubes with polystyrene via grafting to and grafting from methods. Macromolecules. **37**(3), 752–757 (2004)
16. Ma, P.-C., et al.: Dispersion and functionalization of carbon nanotubes for polymer-based nanocomposites: A review. Compos. A: Appl. Sci. Manuf. **41**(10), 1345–1367 (2010)
17. Ramanathan, T., et al.: Functionalized graphene sheets for polymer nanocomposites. Nat. Nanotechnol. **3**(6), 327–331 (2008)
18. Sahoo, N.G., et al.: Polymer nanocomposites based on functionalized carbon nanotubes. Prog. Polym. Sci. **35**(7), 837–867 (2010)
19. Zhao, Y.: Functionalization of single-walled carbon nanotubes (SWNTs) with stimuli-responsive dispersants. Mod Chem Appl. **1**(3), 1–2 (2013)
20. Decker, C., et al.: Synthesis of nanocomposite polymers by UV-radiation curing. Polymer. **46**(17), 6640–6648 (2005)
21. Prucker, O., et al.: Photochemical attachment of polymer films to solid surfaces via monolayers of benzophenone derivatives. J. Am. Chem. Soc. **121**(38), 8766–8770 (1999)
22. Huang, J., et al.: Covalently functionalized double-walled carbon nanotubes combine high sensitivity and selectivity in the electrical detection of small molecules. J. Am. Chem. Soc. **135**(6), 2306–2312 (2013)
23. Ng, A.L., et al.: Chemical gating of a synthetic tube-in-a-tube semiconductor. J. Am. Chem. Soc. **139**(8), 3045–3051 (2017)
24. Li, Y.L., Kinloch, I.A., Windle, A.H.: Direct spinning of carbon nanotube fibers from chemical vapor deposition synthesis. Science. **304**(5668), 276–278 (2004)
25. Huynh, C.P., Hawkins, S.C.: Understanding the synthesis of directly spinnable carbon nanotube forests. Carbon. **48**(4), 1105–1115 (2010)
26. Nikolaev, P., et al.: Gas-phase catalytic growth of single-walled carbon nanotubes from carbon monoxide. Chem. Phys. Lett. **313**(1), 91–97 (1999)

27. Kim, M., et al.: Fluorescent carbon nanotube defects manifest substantial vibrational reorganization. J. Phys. Chem. C. **120**(20), 11268–11276 (2016)
28. Piao, Y.M., et al.: Brightening of carbon nanotube photoluminescence through the incorporation of sp(3) defects. Nat. Chem. **5**(10), 840–845 (2013)
29. Srinivasan, R.: Ablation of polymers and biological tissue by ultraviolet lasers. Science. **234**(4776), 559–565 (1986)
30. Dorman, G., et al.: The life of Pi star: Exploring the exciting and forbidden worlds of the benzophenone photophore. Chem. Rev. **116**(24), 15284–15398 (2016)
31. Dorman, G., Prestwich, G.D.: Benzophenone photophores in biochemistry. Biochemistry. **33**(19), 5661–5673 (1994)
32. Deng, S., et al.: Confined propagation of covalent chemical reactions on single-walled carbon nanotubes. Nat. Commun. **2**, 382 (2011)
33. Qian, W.Z., et al.: The evaluation of the gross defects of carbon nanotubes in a continuous CVD process. Carbon. **41**(13), 2613–2617 (2003)
34. Schadler, L., Giannaris, S., Ajayan, P.: Load transfer in carbon nanotube epoxy composites. Appl. Phys. Lett. **73**(26), 3842–3844 (1998)
35. Kopp, R., et al.: Multi-fidelity modeling of interfacial micromechanics for off-aligned polymer/carbon nanotube nanocomposites. In: 58th AIAA/ASCE/AHS/ASC Structures, Structural Dynamics, and Materials Conference, 2017

Chapter 2
Devulcanized Rubber Based Composite Design Reinforced with Nano Silica, Graphene Nano Platelets (GnPs) and Epoxy for "Aircraft Wing Spar" to Withstand Bending Moment

A. B. Irez, E. Bayraktar, and I. Miskioglu

Abstract This paper aims a new composites design by using devulcanized recycled rubber (90 wt %) and epoxy 20 wt %) based composites reinforcement with nano silica and graphene nano plates (GnPs). The toughening effects of nano-silica/graphene hybrid filler at various ratios on this composite were investigated for aircraft engineering applications especially aircraft wing spar. As well known, wing spar is used two of them; one front, other rear in order to control torsional effect of the wing. Nano-silica and graphene nano plates (GnPs) have been used as the main reinforcing fillers that increase the usefulness of recycled rubber composite. As each filler retains its specific advantages, the use of nano-silica/graphene combinations should improve the mechanical and dynamic properties of recycled rubber composite.

In aircraft and aerospace applications, graphene nano plates can be used effectively new design of electrically conductive composites which can improve the electrical conductivity of these composites designed for the fuselages that it would replace copper wire which is generally used for the prevention of damage caused from lightning strikes. There are many advantages and possibilities that graphene nano plates (GnPs) can prevent water entering the wings, which adds weight to the aircraft.

In the frame of this present common research, a new devulcanized recycled rubber based composite design for aircraft wing spar has been proposed and wing load due to structure weight was calculated analytically to tolerate bending moment under the service conditions. The toughness properties and tribological behaviour indicating the reinforcement of recycled rubber based composite were evaluated. Microstructural and fractural analyses were made by Scanning Electron Microscopy (SEM).

Keywords Recycled Composites · Devulcanized Rubber · Reinforcement effect · Static 3P-Bending · Bending Moment · Analytical Model

2.1 Introduction

Rubber generally provide a good heat resistance, ease of deformation at ambient temperatures and exceptional elongation and flexibility before breaking. These properties have established rubber as excellent and relatively cheap materials for various applications in many manufacturing areas [1–9]. However, utilization of the recycled rubber for the manufacturing of the new composites is a very economical way for cost effective composite design. Last decades, the usage of the recycled rubber obtained from fresh and clean scrap rubber for different industrial applications such as automotive and aeronautical engineering has been very well developed as very useful material for the composites either as a matrix or as a reinforcement. Extensively, this rubber powder come directly manufacturing of sportive affaires, shoes etc., For this reason, it is feasible to use it after chemical (silanization) and devulcanization treatments for cost effective composite design [3, 5, 8, 9].

Reinforcement of the rubber based composites with nano graphene and/or graphene nano plates (GnPs) are very extensive applications in composite design for mainly aeronautic and aerospace applications. Graphene are known and used widely used as multi- functional reinforcement materials that can improve the electrical, piezo-resistive, dielectric, thermal, and

A. B. Irez
CentraleSupélec, Université Paris-Saclay, Gif-sur-Yvette, France

E. Bayraktar (✉)
Supmeca-Paris, School of Mechanical and Manufacturing Engineering, Saint-Ouen, Paris, France
e-mail: bayraktar@supmeca.fr

I. Miskioglu
Michigan Technology University, Engineering Mechanics Department, Houghton, MI, USA

P. R. Thakre et al. (eds.), *Mechanics of Composite, Hybrid and Multifunctional Materials, Volume 5*, Conference Proceedings of the Society for Experimental Mechanics Series, https://doi.org/10.1007/978-3-319-95510-0_2

mechanical properties of rubber based composites even at extremely lower loadings. Because of the exceptionally high surface area, as compared to other graphite derivatives, an enormous improvement in properties is observed for graphene composites. Correspondingly, they exhibit unique advantages, as compared with all other organic and inorganic fillers, and are thus useful in many applications. The interactions between the various elements of a complex multicomponent system, such as the ones found in graphene/rubber nanocomposites, play a major role on the final physico-chemical properties of the material. For this reason, in much of the literature, surface chemistry is applied to the different components of the system, in order to ensure chemical compatibility between them. In addition, it can lead to the formation of chemical bonds focused towards the improvement of the properties of the initial material and a satisfactory dispersion of the filler due to very high Young's modulus (1 TPa) and intrinsic strength of single graphene sheet 130 GPa [1].

As for the reinforcement of the rubber based composites with nano silica (SiO_2 d < 15–30 nm) particles [10–12] can also increase toughening of the rubber based composites and its distribution in matrix is very easily managed that give a multiphase hybrid-toughened composite structure. The rubber based composites reinforced with nano silica and graphene nano plates (GnPs) show higher thermal conductivity of rubber based composites containing of nano silica and GnPs in a certain ratio.

In the frame of this present common research, a new devulcanized recycled rubber based composite design for aircraft wing spar has been proposed and wing load due to structure weight was calculated analytically to tolerate bending moment under the service conditions. The toughness properties and tribological behaviour indicating the reinforcement of recycled rubber based composite were evaluated. Microstructural and fractural analyses were made by Scanning Electron Microscopy (SEM). We believe that this new composite design by using nano silica and graphene nano plates may exposed a novel interface design approach for developing multifunctional rubber based composites in order to manufacture of aircraft wing spar in safety conditions.

2.2 Experimental Conditions

2.2.1 Materials Processing

The use of recycled rubbers within epoxy matrix is a subject of interest requiring particular processes of the rubber particles. Due to the previously conducted vulcanization process, cross-linked structure of rubber restrains the movement of rubber chains and limit the interaction forces between scrap rubber particles and epoxy matrix. In consequence, a reduction in the performance of final product is observed [1–3]. In this case, to enhance materials properties of the final product, devulcanization operation is recommended. In the frame of the current work, the energy for devulcanization was provided by microwaves and it is followed by chemical surface treatments. In this study, at the beginning of the process, recycled rubber particles were blended with solid bisphenol-A type epoxy powders in pre-defined mass rates.

However, because of the insufficient free chains of rubber particles, formation of a robust and durable bonding between epoxy and recycled rubber is quite challenging. For this reason, fresh scrap rubber should be devulcanized. In fact, devulcanization is known to be an operative practice for manufacture of recycled rubbers to increase flowing capacity and also to be remoulded during manufacturing of new designed composites.

During this process, formerly created sulphur links are tried to be broken and also new other links are generated, it means that the structure of the material is modified entirely as renewable process. Thanks to this process, newly created free chains make possible have chemical bonding between rubber and epoxy [1–5, 13, 14].

A new design of devulcanized rubber based composite, reinforced with nano silica and graphene nano platelets (GnPs) are prepared in several steps:

1. Chemical treatment was applied on rubber. The principal of the chemical treatment consists in a short silanization process followed by acrylic acid followed by silanization in order to activate the surface of the rubber particles.
2. After drying of the chemically treated rubber powders, they are exposed to short microwave heating in two stages during 4 min under 900 W of power in order to avoid degradation of the main chains called devulcanization.
3. At the final stage, devulcanized rubber were mixed with a little amount of epoxy resin powders for binding of rubber matrix and reinforcements. This mixture used as a matrix after that the new designed composites are manufactured by using classical powder metallurgy methods.
4. After the mixture of the reinforcements (here nano silica and GnPs) in the matrix, a fast-rotating toothed-wheel milling process was carried out during 4 h to obtain fine rubber powder.

Table 2.1 Composition of the epoxy-rubber based composites

Devulcanized rubber based composite Rubber – Epoxy (%20 Epoxy – 80 Rubber)	CJ I	CJ II	CJ III
Reinforcements (wt %)	10 $SiO_2$4 GnPs	10 $SiO_2$6 GnPs	10 $SiO_2$8 GnPs

Table 2.2 Hardness values of CJ specimens

Hardness measurement	
Specimen	Shore D
CJ I	80.8
CJ II	77.4
CJ III	76.6

Table 2.3 Density values of CJ specimens

Hardness measurement	
Specimen	Density (g/cm^3)
CJ I	1.47
CJ II	1.45
CJ III	1.44

5. After having a homogeneous powder compound, the composite specimens were manufactured by using double uniaxial action press at a temperature of 180 °C under a pressure of 70 MPa during the heating of 30 min.
6. All of the specimens (30/50 mm in diameter) were cooled slowly in the press. All of the specimens were post-cured isothermally at 80 °C for 24 h.

This combined method consists exposing recycled rubbers to microwave heating in a short time pursued by a pre-chemical treatment that is applied in practical point of view. Thus, a good cohesion is provided at the interface between epoxy resin and devulcanized rubber powders to improve the properties of the recycled rubber with a particle size varying from 10 to 100 μm. obtained from fresh-clean scrap rubbers coming from the production of the sportive equipment. It means that they are fresh, clean and completely different from ground tire. The compositions of the nano silica and GnPs reinforced recycled rubber based composites (called as CJ I-II-III hereafter) were given in Table 2.1.

As seen the reinforcement percentages in the rubber matrix, the combination of SiO_2/GnPs for each composition gives a good adhesion of nano silica and GnPs to the rubber 80 wt% + Epoxy 20 wt%.

2.2.2 *Mechanical, Microstructure, Fracture Surface Analyses and Shore-D Hardness Measurements*

Fracture surface damage analyses and microstructural observation have been realized by means of optical (OM) and scanning electron microscopy (SEM). SEM observation was realized on fracture surface of the tested specimens with Scope/JSM-6010LA Jeol® electron microscope.

Surface hardness measurements of the specimens were performed after post curing. Shore D hardness test measurements on the polished flat surfaces of the specimens were carried out according to ASTM D 2240 using Shore D hardness tester, (type HBD-100-0). Hardness results were given in Table 2.2.

Three-point bending tests (3 PB) have been carried out according to the ASTM D790 standards. Tests were realized with the machine Zwick Proline Z050TN and during the tests crosshead speed was selected as 1 mm/min. Flexural strength and strain were obtained from the test results. At least three specimens for each composition were used and standard deviation and average values were given in results chapter with standard deviation values. In addition, fractural properties such as plain strain fracture toughness (K_{Ic}) and critical strain energy release rate (G_{Ic}) were investigated with SENB specimens and the tests were realized according to ASTM D5045 standard. Notches were introduced by tapping a fresh razor blade. Impact response of the three composites were compared with an in-house designed drop weight impact test machine. A 1.9 kg conical punch was used for the tests, punch height was 900 mm and with an impact velocity of 4 m/s.

After preparation of the specimens, densities of the compositions were measured and they were given in Table 2.3.

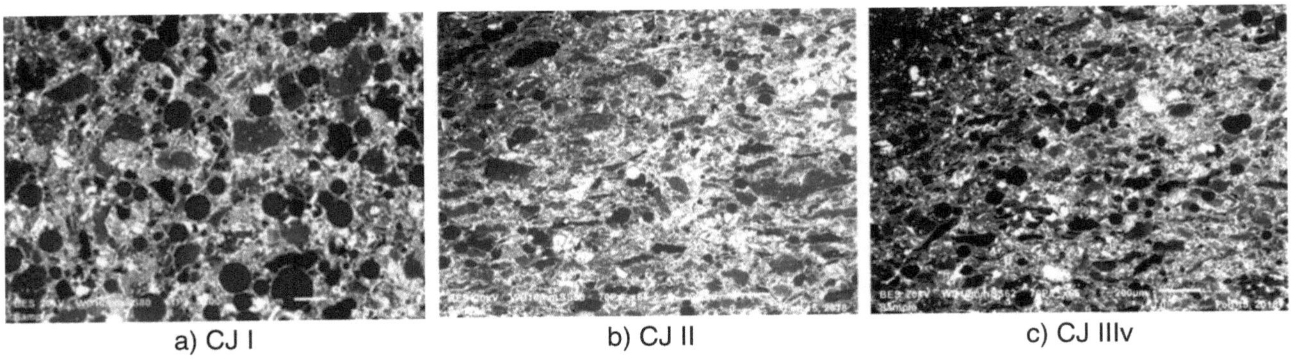

Fig. 2.1 Microstructure of the composites developed for the wings spars for this research (**a**) CJ I (**b**) CJ II (**c**) CJ III

2.2.3 *Wear Resistance (Scratch Test) and Damage Analysis Via 3D Optical Roughness Meter*

In the current research basic idea on the tribological behaviour of the epoxy and recycled elastomer based composites was evaluated performing scratch tests. A 3D optical surface scanner was utilized to assess damage zone after the scratch test in terms of scratch depth and average scratch roughness. Scratch were created over the polished specimens in two different loading conditions (10–15 N) for ten cycles on the same amplitude by using Anton Paar™ MST^3 micro scratch tester.

2.3 Results and Discussions

2.3.1 *Microstructural Evaluation of the Composites*

After hot compaction, sectioning was made then mounted specimens were polished. Then, general microstructures of the each composites were observed and they were shown in Fig. 2.1.

Then, mapping analyses was carried out for each composition were presented in Fig. 2.2 with together the general microstructures of the each composition. This type of analyze allows us to observe the distribution of the reinforcements in the matrix. It seems that a homogenous distribution of the reinforcements in the structure for each composition.

2.3.2 *Three Point Bending Tests and Fracture Surface Observation*

Three-Point Bending (3 PB) tests have been carried out for each different type of compositions with at least three specimens.

Flexural stress is calculated during three-point bending test according to the Eq. 2.1:

$$\sigma = \frac{3Pl}{2bh^2} \tag{2.1}$$

In this formula, l is the span length, P is the maximal bending load, b and h are the sample thickness and depth, respectively.

Flexural strain, $\boldsymbol{\varepsilon_f}$, was determined according to the Eq. 2.2:

$$\varepsilon_f = \frac{6Dh}{l^2} \tag{2.2}$$

D is the maximum deflection at the center of the specimen.E_B is the modulus of elasticity in bending and it is expressed with the Eq. 2.3 as follows:

$$E_B = \frac{l^3m}{4bh^3} \tag{2.3}$$

m is the tangent of the initial straight portion of the force-displacement curve.

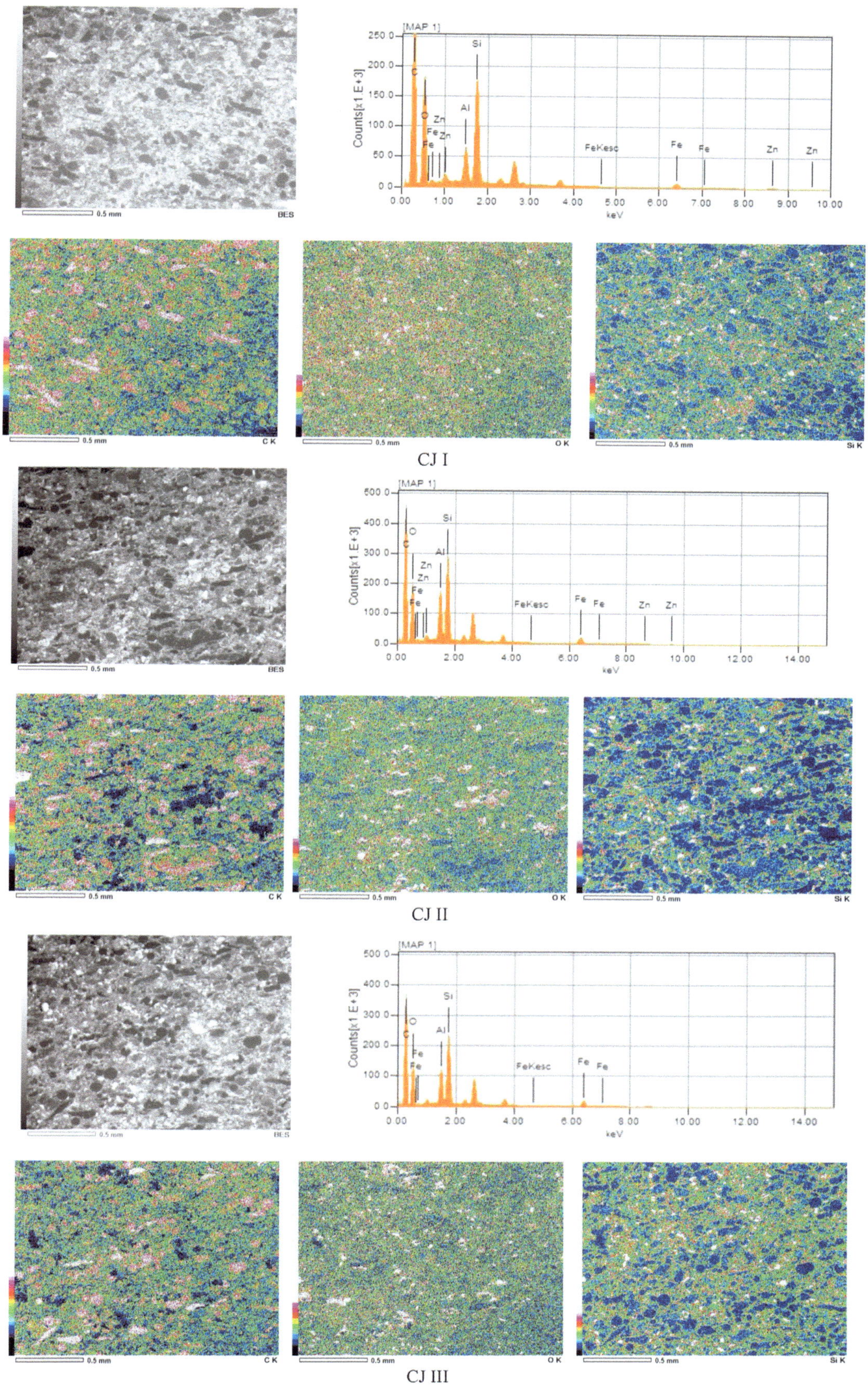

Fig. 2.2 Microstructure analysis with EDS together with Mapping for the composites CJ I CJ II and CJ III respectively

Table 2.4 Comparison of mechanical properties of CJ specimens

	Flexural stress (MPa)	Flexural modulus (MPa)	Strain in break (ε %)	K_{Ic} (MPa $m^{1/2}$)	G_{Ic} (kJ/m^2)
CJ I	15.76 ± 0.91	2.61 ± 0.28	0.348 ± 0.001	1.107 ± 0.027	0.649 ± 0.032
CJ II	11.23 ± 3.88	2.05 ± 0.93	0.359 ± 0.001	0.618 ± 0.091	0.138 ± 0.039
CJ III	15.03 ± 0.38	2.47 ± 0.066	0.363 ± 0.001	0.953 ± 0.065	0.367 ± 0.017

Table 2.5 Comparison of the experimental and analytic modelling results

Composition name	Experimental elasticity modulus	Elasticity modulus (Halpin – Tsai)	Relative difference
CJ I	2.61 ± 0.28	1.874	28%
CJ II	2.05 ± 0.93	1.937	5%
CJ III	2.47 ± 0.066	2.004	18%

The mode I fracture toughness, K_{Ic}, was determined by testing of the SENB specimens and K_{Ic} was calculated according to the Eq. 2.2:

$$K_{Ic} = \frac{F}{B\ w^{\frac{1}{2}}} f(x); \quad x = \frac{a}{W}\ , 0 < \frac{a}{W} < 1 \tag{2.4}$$

F is the maximum force from the load-elongation plot; *B* is the thickness of the specimen; *W* is the width and "a" is the total notch length.*f(x)* is the geometry correction factor and is expressed with the Eq. 2.3 as follows:

$$f(x) = 6(x)^{0.5} \left\{ \frac{\left[1.99 - x(1-x)\left(2.15 - 3.93x + 2.7x^2\right]\right.}{(1+2x)(1-x)^{1.5}} \right\} \tag{2.5}$$

Critical strain energy release rate (fracture energy) G_{Ic} was calculated using the expression Eq. 2.4:

$$G_{Ic} = \frac{K_{Ic}{}^2\left(1 - v^2\right)}{E} \tag{2.6}$$

where *E* is the elasticity modulus for plane stress approach examined for thin specimens.

Table 2.4 indicates the mechanical properties in bending mode. Fracture toughness values were also presented in the table.

Table 2.4 indicates that The hybrid rubber based composites reinforced with nano silica and graphene nano plates (GnPs) with a certain ratio have shown enhancements in the mechanical properties, and more significantly, fracture toughness (K_{IC}), which can be explained by synergistic impact coming from the intrinsic physical leftacteristics of the reinforcements it means that nano silica and graphene nano plates (GnPs). In this study, the highest K_{IC} can be obtained with addition of small amounts of graphene nano plates (GnPs). However, maximum strain values in break are the same levels.

After that, more detail analyses were given for the analytical modelling applied to three composites to compare their elasticity modulus by using experimental results. For these analyses, Halpin-Tsai model was chosen for the sake of its simplicity. Because this model is quite practical and it gives the well coherent results with the experimental methods in the case of homogeneous distribution of the fillers.

Halpin – Tsai's model was given in Eq. 2.7, where E_c, Ef and E_m are Young's modulus of the composite. V_f and V_m is the volume fraction of the AFs and the matrix, respectively. ξ is the shape factor of the fibers which is determined from the proportion of L, average length of the fibers and D, average diameter of the fibers.

$$\frac{E_c}{E_m} = \frac{1 + \xi\eta V_f}{1 - \xi\eta V_f}, \qquad \eta = \frac{\frac{E_f}{E_m} - 1}{\frac{E_f}{E_m} + \xi}, \qquad \xi = 2\,(\mathrm{L/D}) \tag{2.7}$$

According to the comparison this model with the experimental results, certain deviations are observed for the composites considered in this research. All of the results for three composites were presented in Table 2.5.

Fracture surfaces obtained from 3 PB tests have been analyzed by means of Scanning Electron Microscopy (SEM). It noticed that good adhesion of the both of the reinforcements in the rubber based matrix by creating an ideal interface for

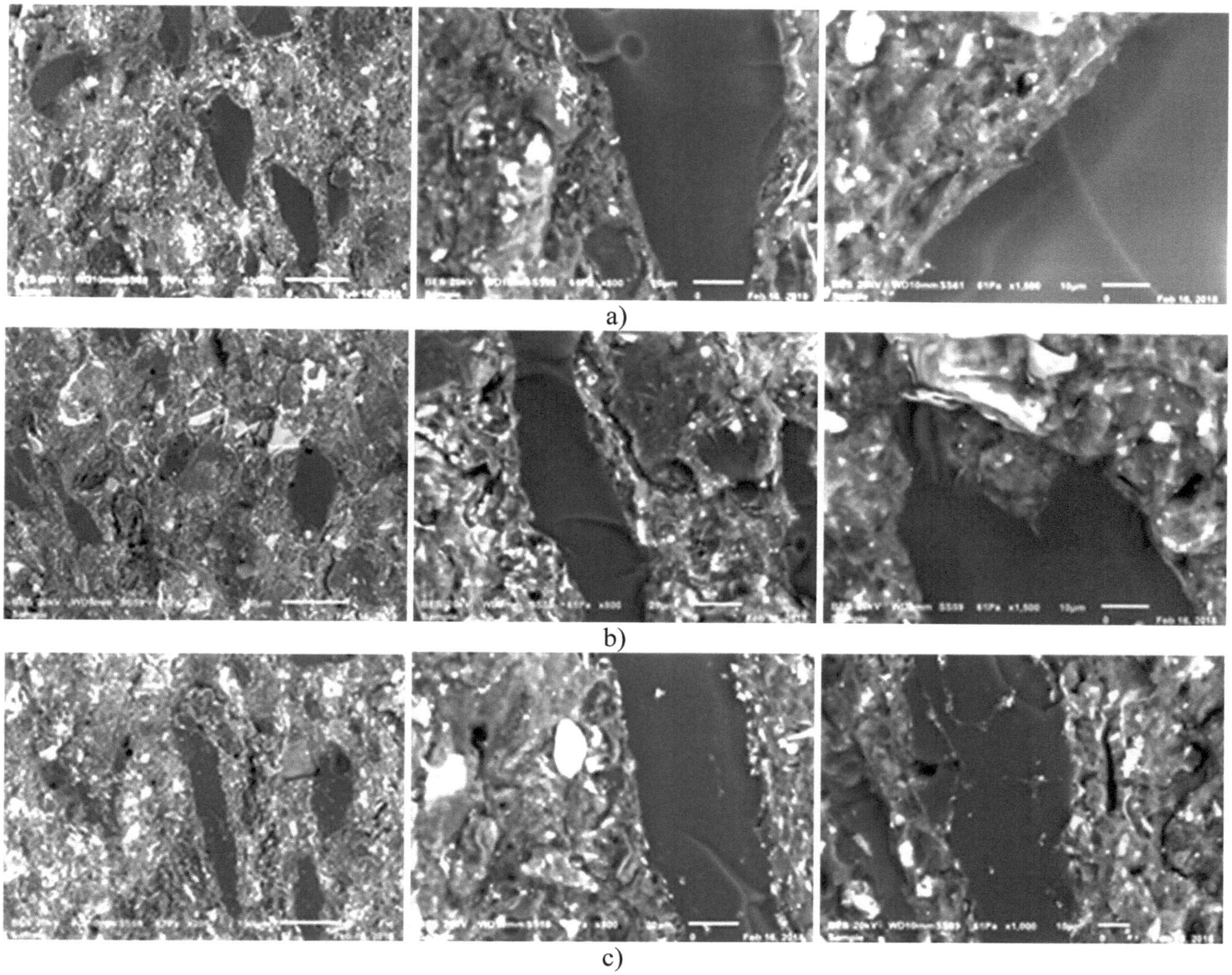

Fig. 2.3 (**a**) Fracture surfaces of the 3 different composites tested with 3P-Bending CJ I (**b**) CJ II (**c**) CJ III

each composition, as presented in Fig. 2.3 with different fracture surfaces taken by SEM. Cavitation and void formation in the rubber matrix with matrix expansion and locally, debonding of nano particles with consequent void growth have been observed in the structure as the improved toughening mechanisms. It seems also an interaction between devulcanized rubber and the epoxy is very fine and there is fusion mutually between them.

For this evolution, the mixture of devulcanized rubber (80 wt %) with epoxy (20 wt %) plays an important role. Some of the specimens with homogenous distribution of the nanoparticles have shown a typical debonding of the silica nanoparticles with GnPs that should be origins of the weakening of the matrix–particle interface. The toughness improvement in hybrid rubber based composites should be direct related to the increment of the debonding phenomena that can increase the size of the plastic zone in the structure. This case facilitate the devulcanized rubber based composites to dissipate additional fracture energy.

2.3.3 *Drop Weight Testing*

Drop weight test analyses have been carried out on the three composites for the measurement of the impact resistance and energy absorbance ability of the manufactured composites for wing spar. Four specimens were tested for each composite. Absorbed energies measured with maximum force during impact were presented in Table 2.6.

Table 2.6 Drop weight tests results on CJ specimens

Composition name	Maximum force (N)	Absorbed energy (J)
CJ I	2232.25 ± 34.68	1.27 ± 0.04
CJ II	2647.45 ± 29.94	1.49 ± 0.17
CJ III	3188.8 ± 348.74	1.66 ± 0.21

Table 2.7 Scratch test results for CJ compositions

Composition name	10 N		15 N	
	Worn surface (mm^2)	Worn volume (mm^3)	Worn surface (mm^2)	Worn volume (mm^3)
CJ I	1.32	0.0886	1.59	0.135
CJ II	1.16	0.0697	1.33	0.0972
CJ III	1.23	0.0811	1.73	0.136

From Table 2.6 it is clearly seen that GnPs improve the impact resistance of the composites. The maximum force during the impact is proportional to the absorbed energy [6, 7].Layered structure of GnPs is possibly enables to improve the damping characteristics of the composites.

Shortly, it is noticed that the absorbed energy for each specimen during the impact tests is related to the increment of the plastic zone in the structure due to the debonding of the nano particles. Higher dispersion of the values is also related to the test specimens prepared under laboratory conditions; hot compaction, cutting notch effect, etc.

2.3.4 *Damage Analysis by Means of Micro Scratch Test and 3D Optical Surface Roughness Meter*

After completing mechanical tests, tribological characterization of the composites were done by micro scratch tests. Three dimensional damages were observed three dimensionally by an optical surface scanner. The 3D damage results are presented in the Fig. 2.4.

Quantitative results related to scratch tests were given in Table 2.7. It seems that the increase of the reinforcement content, cause in the composites high resistance to wearing. In reality, because of the high shear stress at the interfaces the interfacial shear stress should probably be the main reason for damage of the matrix and reinforced filler interfaces. When the indenter is slipping, tangential tensile stress is caused on the surface behind the indenter, while in front of the indenter the tangential stress is compressive.

As expected higher force values resulted in higher damage traces. Damaged volume is significantly proportional with the force level. However, increasing force level did not manifest in the same manner for the damaged surface. In addition, increasing GnPs content generally improves the wear resistance for the composites in consideration.

2.3.5 *Numerical Approach for These Composites*

After realizing experimental procedures, mechanical test results were compared with finite element modelling (FEM). In FEM procedure, simplified approaches were utilized. In this regard, after generating 3D solid model of the tested specimens in 3 PB tests, these models were meshed with quadratic tetrahedral elements (C3D10).

After that, boundary conditions were applied according to two selected approach. In the first approach, maximum load obtained from the 3 PB tests was imposed to the material and maximum stress and the displacement in the critical region (z-direction) were obtained.

In the second approach, maximum deflection in the middle of the specimen which was obtained from 3 PB tests was imposed to the material and maximum stress in the middle of the specimen was determined. FE models were solved in linear elastic mode by using Abaqus FEM software.

First approach was executed and results were illustrated in Fig. 2.5.

The comparison of experimental and FEM results according to first approach was given in Table 2.8.

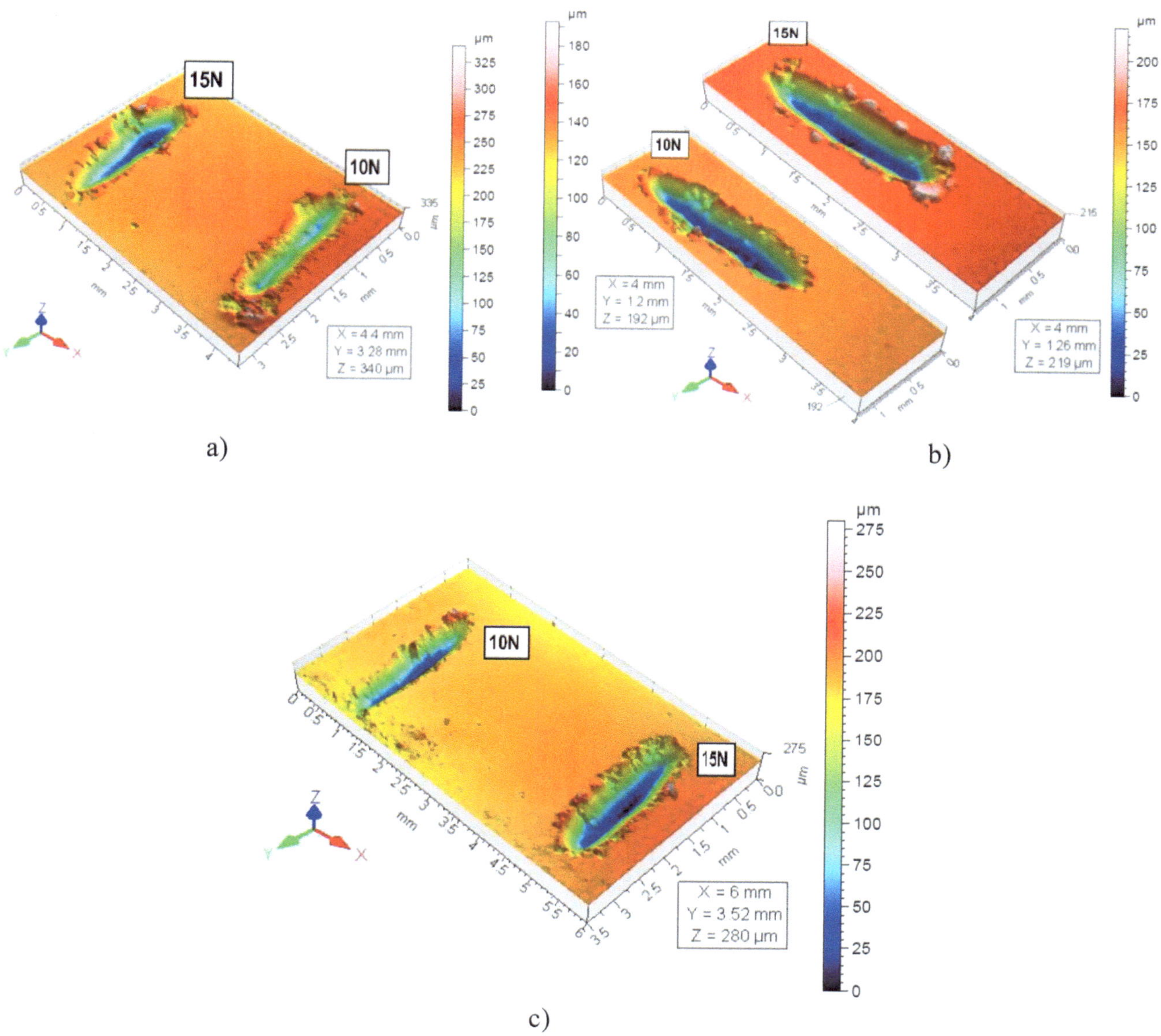

Fig. 2.4 Micro scratch test results carried out on the three compositions (**a**) CJ I (**b**) CJ II (**c**) CJ III, respectively

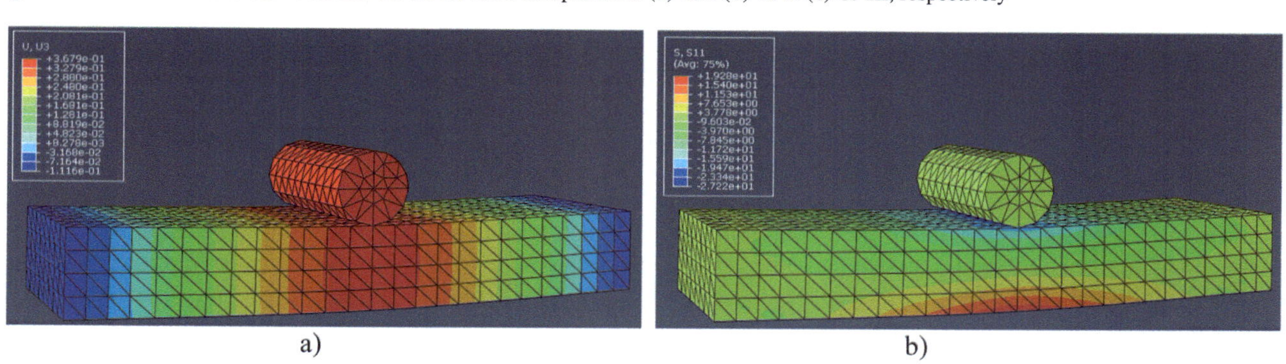

Fig. 2.5 FEM results according to maximum force approach (**a**) Displacement contour (**b**) Stress contour

According to Table 2.8, FEM gives coherent results for the stress values even if minor errors were observed. On the other hand, for the displacement, FEM does not give quite consistent results. After conducting second approach, all of the results are illustrated together in Fig. 2.6.

Table 2.8 Comparison of the modelling and experimental approaches for the imposed experimental F_{max}

1st approach: Applied load = Experimental F_{max}				
Composition name	Stress FEM (σ_{11} MPa)	Relative difference with experiments	Displacement max FEM (u_3 mm)	Relative difference with experiments
CJ I	19.28	25%	0.368	53%
CJ II	11.11	20.9%	0.214	25%
CJ III	17.33	17.33%	0.313	38%

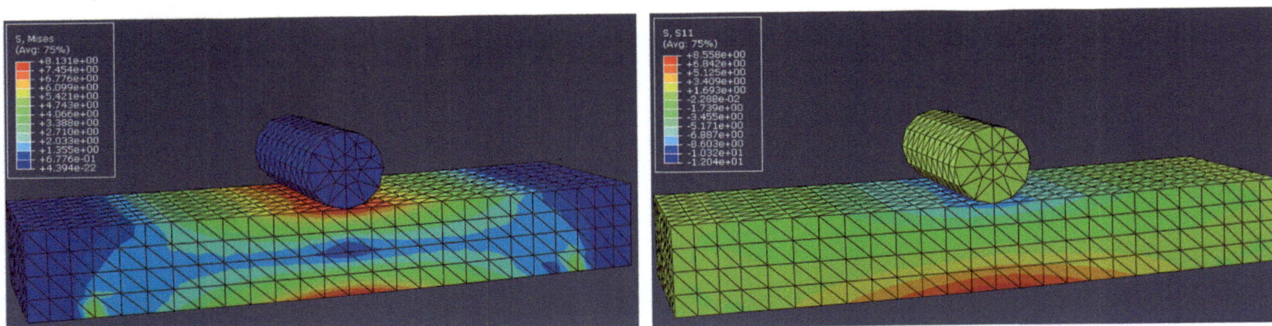

a) b)

Fig. 2.6 FEM results according to maximum displacement approach (**a**) General Stress contour (**b**) σ_{11}

Table 2.9 Comparison of the modelling and experimental approaches for the imposed experimental D_{max}

2nd approach: Applied displacement = Experimental D_{max}	
Stress FEM (σ_{11} MPa)	Relative difference with experiments
8.42	46%
12.55	27%
8.85	32%

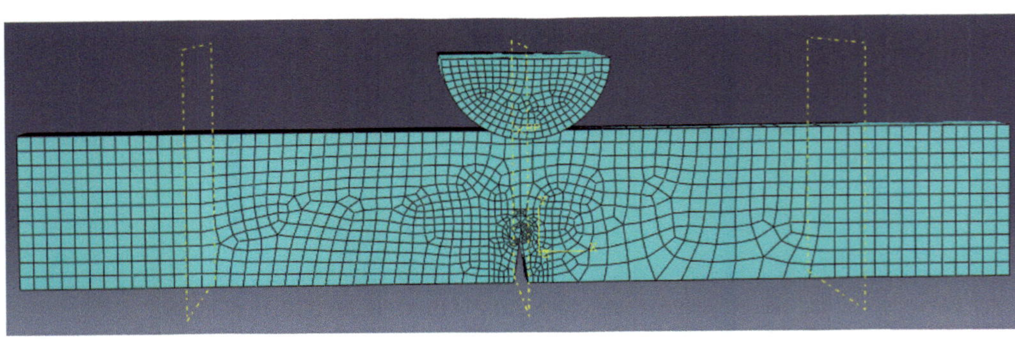

Fig. 2.7 Meshed model for SENB specimens

In this figure, the critical region was shown as the middle of the specimen as expected. Also, from these figures, the stress in the top of the sample is compressive whereas in the bottom tensile (Table 2.9).

By comparison of these two approaches, it is noticed that first approach gives more coherent results to estimate stress values. On the other hand, estimation of the displacement values requires more detailed modelling considerations.

After modelling of 3 PB results, another simplified model to estimate fracture toughness was applied to these composites. This time, solid model of the SENB specimens was generated and.

R3D4 (linear quadrilateral) and C33D8R (linear hexahedral) elements were used to mesh this model. Meshed solid model was given in Fig. 2.7. According to this model, at first maximum displacement obtained from experimental fracture toughness tests was imposed to the material as a boundary condition and elasticity modulus was selected as in the experimental results. After executing the model, maximum force was obtained.

This force was compared with the experimental value and the proportion of them was used to modify the selected elasticity modulus. After modifying the elasticity modulus on Abaqus, the model was run again. This iterative elasticity modulus

Table 2.10 Comparison of the modelling and experimental results for the fracture toughness

Composition name	K_{Ic} FEM (MPa m$^{1/2}$)	Relative difference with experiments
CJ I	0.70	37%
CJ II	0.40	34%
CJ III	0.54	43%

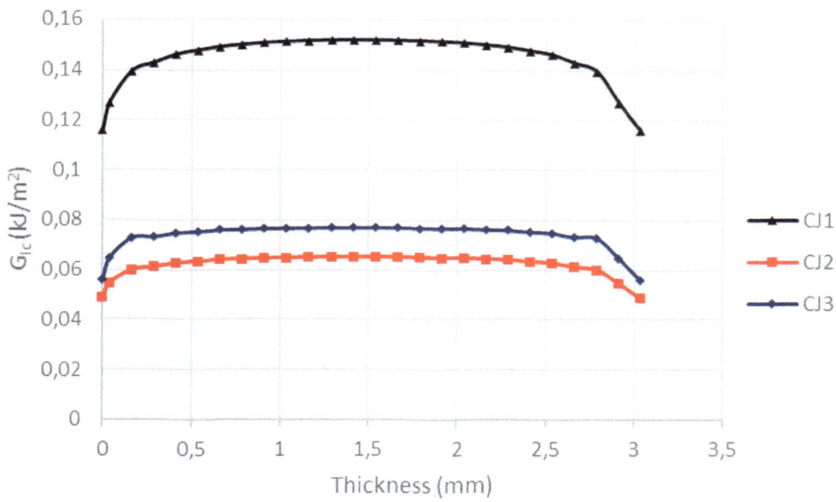

Fig. 2.8 Evolution of G_{Ic} along the specimen thickness according to FEM results

modification continued until obtaining a coherent maximum force value regarding experimental results. After converging to the maximum force, energy releasing rate (G_{IC}) of the composites were determined along the specimen thickness.

From the energy releasing rate results obtained via FEM, Eq. 2.6 can be used to calculate the K_{IC}. After completing the FEM analyses, results were given in Table 2.10.

Finite element results exhibiting significant differences with the experimental results. In fact, inhomogeneity or the existence of hard particles in the crack propagation zone can blunt the cracks. For this reason, experimental results may show a higher trend than modelling. After the comparison of FEM fracture toughness with experimental results, the evolution of the G_{Ic} obtained from FEM is given in Fig. 2.8.

According to Fig. 2.8 strain energy releasing rate have higher values in the middle of the specimen compared to the edges.

As the final step of finite element approach, the real-life conditions of these composites depending on the application area were simulated. As given in the title of this article, these composites are planned to be used in the wing spars of the aircrafts. By considering the airfoil structure and the subjected loads of the wings, the deformation and stress conditions were investigated.

To begin with, in this model, we limited our calculation to a ground situation (static, only the weight of the wing and the fuel tank). Also, the ribs inside the airfoil were not taken into consideration. This FEM approach is implemented to see the mechanical behavior of these composites under static conditions. In this study, NACA (2415) airfoil geometry was used and all parameters required for the FEM analysis were taken as real values of NACA 2415.

After modelling the "I" profile of the wing spar, a distributed load was applied according to Eq. 2.8.

$$q_t(x) = q_w(x) + q_f(x) = \begin{cases} -\frac{W_{Ws}n(C_t x - C_o x + C_o L)}{L^2(C_o+C_t)} - \frac{W_{fs}n(C_{tf}x - C_{of}x + C_{of}L)}{L_f^2(C_{of}+C_{tf})} & if\ x \le L_f \\ \frac{n(-\pi C_o L W_{Ws} + \pi C_o W_{Ws}x - \pi C_t W_{Ws}x)}{L^2\pi(C_o+C_t)} & if\ x \ge L_f \end{cases} \tag{2.8}$$

In this equation, used values are given below in Table 2.11 [7].

By using these values, a distributed load was applied on the wing spar profile as shown in Fig. 2.9. In this wing spar profile, left end of the "I" profile was taken as fixed end to the fuselage.

Table 2.11 Specifications of NACA 2415 airfoil

Total aircraft weight	$W_{to} = 4800$ kg
Weight of wing structure	$W_{Ws} = 630$ kg
Weight of fuel stored in wing	$W_f = 675$ kg
Length of wing	$L = 7$ m
Length of fuel tank within wing	$L_f = 4.8$ m
Chord length at wing root	$C_o = 1.8$ m
Chord length at wing tip	$C_t = 1.4$ m
Width of fuel tank at wing root	$C_{of} = 1.1$ m
Width of fuel tank at *Lf*	$C_{tf} = 0.85$ m
Load factor on the ground	$n = 1$

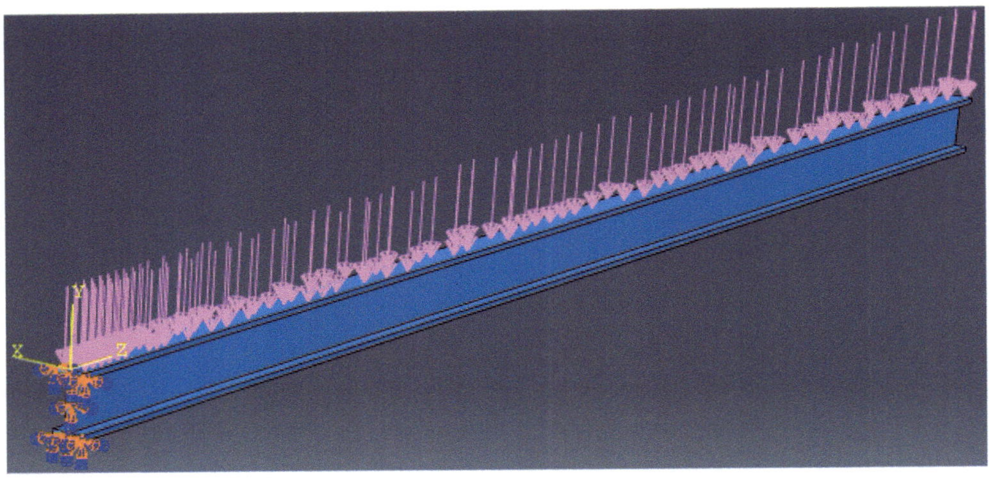

Fig. 2.9 Load profile calculated on the wing spar

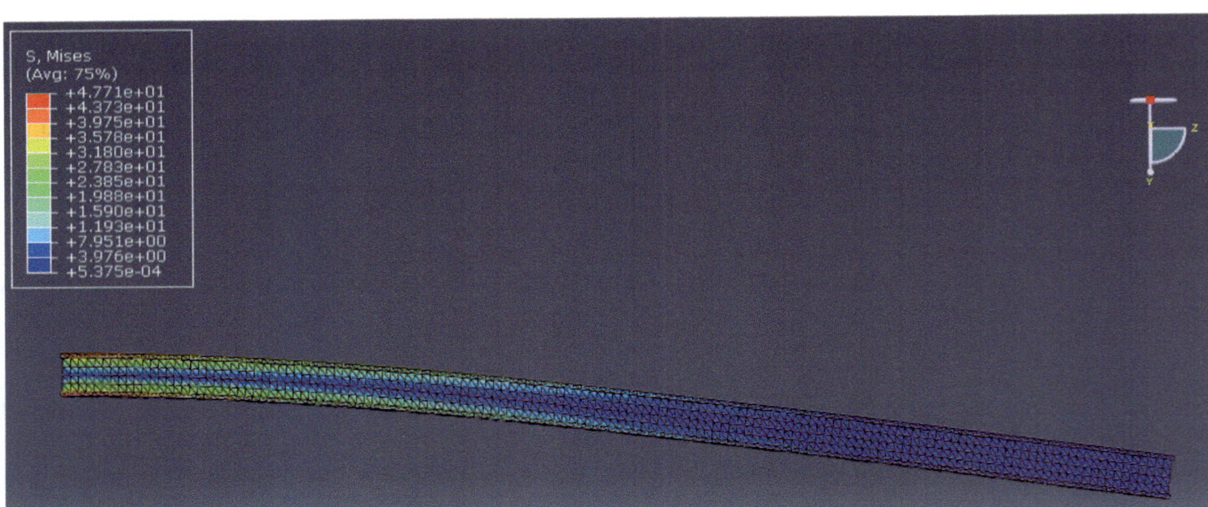

Fig. 2.10 Stress contour on wing spar manufactured by using CJ I composites

After the application of this load profile, wing spar was meshed by using C3D10 (quadratic tetrahedral) elements. Then, this model was executed on Abaqus. The stress contour for a wing spar manufactured by using CJ I composites was presented in Fig. 2.10.

In addition, the results of stress and displacement for the wing spars manufactured by using CJ composites were given in Table 2.12 in detail.

Table 2.12 Stress and displacement results of the modelled wing spar manufactured with CJ composites

Composition	Von Misses (MPa)	σ_{11} (MPa)	σ_{22} (MPa)	σ_{33} (MPa)	U2 (arrow) (m)
CJ I	47.71	21.64	20.36	59.58	1.618
CJ II	47.72	21.41	20.13	59.48	1.532
CJ III	47.72	21.21	20.01	59.17	1.716

These results may seem quite high for the composites in consideration. However, as mentioned before, the real load carrying elements in the airfoils such as aluminum based ribs and the main frame was not considered in this calculation. After integrating these elements to the model, these composites will propose a cost-efficient solution to the aircraft manufacturers.

2.4 Conclusion

In the frame of this research, a new devulcanized rubber based composite reinforced with nano silica + graphene nano plates (GnPs) was developed for the design of wing spar on the aircraft. Comparison of experimental and numerical results have been successfully carried out. Computational techniques were used for the presenting of the various loads acting on the wings of an aircraft; basically the load due to wing structure weight was calculated in Abaqus. Analytical models were also developed and the shear force and bending moments resulting from these loads are also calculated for a small aircraft design. Mechanical behaviour of the real-life application of the composites as aircraft wing spar material was numerically tested. The real load carrying elements in the airfoils such as aluminum based ribs and the main frame was not considered in this stage of the research project that is going on. Next stage of this project, the real load carrying elements in the airfoils will be included in the computational parts.

We believe that this new composite design by using nano silica and graphene nano plates may exposed a novel interface design approach for developing multifunctional rubber based composites in order to manufacture of aircraft wing spar in safety conditions.

Acknowledgments The authors thank Dr. H-A. ALHAS from Airbus-Space, London/UK for general support and also for his valuable discussion in certain stages of this project. Also, they would wish to acknowledge research foundation of Supmeca/Paris.

References

1. Lee, C., Wei, X., Kaysar, J., Hone, J.: Measurement of the elastic properties and intrinsic strength of monolayer graphene. Science. **321**(5887), 385–388 (2008)
2. Oliphant, K., Baker, W.E.: The use of cryogenically ground rubber tires as a filler in polyolefin blends. Polym. Eng. Sci. **33**(3), 166–174 (1993)
3. Naskar, A.K., Bhowmick, A.K., De, S.K.: Thermoplastic elastomeric composition based on ground rubber tire. Polym. Eng. Sci. **41**(6), 1087–1098 (2001)
4. Naskar, A.K., De, S.K., Bhowmick, A.K.: Thermoplastic elastomeric composition based on maleic anhydride–grafted ground rubber tire. J. Appl. Polym. Sci. **84**(2), 370–378 (2002)
5. De Sousa, F.D.B., et al.: Devulcanization of waste tire rubber by microwaves. Polym. Degrad. Stab. **138**, 169–181 (2017)
6. Taheri-Behrooz, F., Shokrieh, M., Abdolvand, H.: Designing and manufacturing of a drop weight impact test machine. Eng. Solid Mech. **1**, 69–76 (2013)
7. Akindapo, J.-O., Johnson Anamemena, N., Garba, D.-K.: Graphite-epoxy composite design for aircraft wing spar using computational techniques – Part I. Am. J. Mech. Eng. **5**(4), 117–127 (2017)
8. D. Zaimova, E. Bayraktar, I. Miskioglu.: Design and manufacturing of new elastomeric composites: Mechanical properties, chemical and physical analysis. Int J Composites Part B, Elsevier, USA. **1**(1), 1–12, November 2016, on line 2017
9. McCarthy, D.W., Mark, J.E., Schaeffer, D.W.: Synthesis, structure, and properties of hybrid organic-inorganic composites based on polysiloxanes. I. Poly (dimethyl siloxanes) elastomers containing silica. J. Polym. Sci. Part B Polym. Phys. **36**, 1167–1189 (1998)
10. Zaimova, D., Bayraktar, E., Miskioglu, I., Dishovsky, N.: Wear resistance of elastomeric based composites by continuous multi-cycle indentation used in manufacturing engineering. Adv. Mater. Res. **939**, 106–113 (2014), Trans Tech Publications, Switzerland, Doi:10.4028/www.scientific.net/AMR.939.106
11. Sadasivuni, K.-K., Ponnamma, D., Thomas, S., Grohens, Y.: Evolution from to graphene elastomer composites. Prog. Polym. Sci. **39**, 749–778 (2014)
12. Irez, A. B., Bayraktar, E., Miskioglu, I.: Mechanical Characterization of Epoxy – Scrap Rubber Based Composites Reinforced With Alumina Fibers, Springer link, Mechanics of Composite and Multi-functional Materials, vol 6, pp. 59–70 (2017), ISBN 978-3-319-63408-1, https://doi.org/10.1007/978-3-319-63408-1

13. Bhattacharya, M., Maiti, M., Bhowmick, A.K.: Tailoring properties of styrene butadiene rubber nano composite by various nano fillers and their dispersion. **49**(1), 81–98 (2009)
14. Irez, A. B., Miskioglu, I., Bayraktar, E.: Mechanical Characterization of Epoxy – Scrap Rubber Based Composites Reinforced With Nano Graphene, Springer link, Mechanics of Composite and Multi-functional Materials, vol 6, pp. 45–58 (2017), ISBN 978-3-319-63408-1, https://doi.org/10.1007/978-3-319-63408-1

Chapter 3
Study of Mechanical Characteristics of Banana and Jute Fiber Reinforced Polyester Composites

G. L. Easwara Prasad, B. E. Megha, and B. S. Keerthi Gowda

Abstract Composite materials are gaining utmost prominence in many fields of application, nowadays, due to their surpassing traits have acquired relevance in various spheres of Engineering. The present study consists of assessment of the mechanical characteristics of Banana and Jute Fiber reinforced Polyester Composites. Banana and Jute Fiber reinforced Polyester Composites were fabricated using banana fibers of length 10 mm and jute fibers of length 10 mm as reinforcements and polyester resin as matrix respectively. Thickness of composite panels varied from 3 to 5 mm and fiber volume fractions were adopted as 5%, 10%, 15%, 20% and 25% respectively. It is observed that Jute Fiber reinforced Polyester Composites exhibit higher values of tensile and flexural strength compared to Banana Fiber reinforced Polyester Composites. However, both Banana and Jute Fiber reinforced Polyester Composites revealed optimum tensile and flexural strength at a fiber volume fraction of 25% and 20% respectively.

Keywords Composites · Matrix · Reinforcement · Banana Fiber · Jute Fiber · Polyester

3.1 Introduction

Fiber-reinforced polymer composites have gained significance in many fields of their applications due to their remarkable qualities, such as high strength, durability and resistance to extreme atmospheric conditions [1–14]. Composites are the materials fabricated by mixing two or more than two substances, in a manner that they retain their individual properties and the resulting material has remarkable, distinguished characteristics in comparision with its ingredients. The composites, mainly consists of strong and steady component which has load sustaining potentiality and is referred to as "Reinforcement"; the other weak component which plays the role of holding the reinforcement firmly is called "Matrix".

The stiffness and strength required to sustain the loads is contributed by the Reinforcement. On the other hand, the matrix contributes towards retaining the position of the reinforcement. Considerably, although the components of the composites restrain their respective mechanical and chemical properties, the resultant composite material possesses a group of characteristics, which the constituent components cannot yield individually. The fibers performing, the role of reinforcement could be naturally obtained or artificially manufactured. Synthetic fibers vastly utilised in the fabrication of Fiber reinforced Composites are carbon, glass, boron and ceramic substances [1, 2]. In the present scenario, composites have found wide application in the field of engineering and technology.

The reinforcement in the composites typically consists of fibers, flakes or particulate matter such as carbon, boron, glass, metals, silicon carbide, and ceramic materials, which play a significant role in making the composites tough and robust [3–5]. Previous studies reveal that the fiber length and fiber volume fraction contribute notably to the strength acquired by the fiber reinforced composites [6].The matrix materials, primarily adopted in fabrication of composites are metals, polymers and ceramic materials.

Composite materials acquired by combining naturally accessible fibers with suitable matrix components are termed as Natural Fiber Reinforced Composites. Natural Fiber Reinforced Composites are gaining importance and are being utilised for a number of applications due to their availability in abundance, easy accessibility and eco-friendly nature. The counterpart

G. L. Easwara Prasad (✉)
MITE (VTU), Moodabidre, India

B. E. Megha
Department of Civil Engineering, MIT (VTU), Mysore, India

B. S. Keerthi Gowda
Centre for PG VTU Studies, Mysore, India

P. R. Thakre et al. (eds.), *Mechanics of Composite, Hybrid and Multifunctional Materials, Volume 5*, Conference Proceedings of the Society for Experimental Mechanics Series, https://doi.org/10.1007/978-3-319-95510-0_3

Fig. 3.1 Banana fibers

synthetic fibers cause risk to the environment and also involves difficult techniques of production and is costly. In the case of natural fibers, there is no urge or need of manufacturing mechanisms since they are readily, naturally available; they only need to be extracted from their respective sources and need to be processed for their usage in fabrication of composite materials. Above stated fibers are easily accessible, exhibit appreciable strength and are of course economical. The intended work portrays the formulation and the salient, characteristics of the natural fiber reinforced composites, comprising Banana, Jute fiber and polyester resin. Venkateshwaran et al. [7] carried out a study with an aim to predict the mechanical properties of woven Banana-Jute hybrid composites. Woven fabrics are arranged in three layers of different sequence. Epoxy LY556 with hardener HY951 was used and Composite specimens were prepared by hand-layup techniques. The effect of layering sequence on the mechanical properties namely tensile, flexural and impact was analysed. It was found that the tensile and flexural strength of Banana-Jute hybrid composite is higher than that of individual composites. Similarly, the impact strength of Banana-Jute hybrid composite is better than other types of composite. It is found that the moisture absorption of woven banana fiber composite is lesser than the hybrid composite. Pujari et al. [8] carried out an experimental study to inspect the mechanical properties of composites like Jute and Banana as reinforcing materials in epoxy resin based polymer matrix. These fibers were exposed to NaOH treatment before reinforcement. The reinforcing of the resin with Jute and Banana fiber was accomplished in four different orientations: 0°, 15°, 30° and 45° with reference to horizontal side of the sheet by employing optimized resin. Mechanical properties (Tensile, Impact) of both Jute fiber composites and Banana fiber composites were investigated as a function of fiber orientation. Results showed that the composite properties are strongly influenced by test direction and fabric characteristics. Comparatively, Composites tested along the Jute fabric 0° orientation obtained best overall mechanical properties.

3.2 Materials and Methodology

3.2.1 Materials

The Composites obtained from the blending of banana with polyester are Banana fiber reinforced polyester composites and those obtained from blending of jute fiber with polyester are Jute reinforced polyester composites.

Banana fiber (Fig. 3.1) is a ligno-cellulosic fiber excerpted from the pseudo stem of banana plant. Banana plant is a substantial perpetual herb with leaf envelopes that form pseudo stem. Its height ranges between 10 to 40 feet surrounding with 8–12 large leaves. It is a bast fiber with significantly exceptional mechanical properties. The chemical composition of banana fiber consists of cellulose, hemicellulose and lignin [5]. It is extremely strong fiber with light weight and biodegradable eco-friendly fiber. Density of Banana fiber used is 1.4 g/cc. It is bio- degradable and has no negative effect on environment and thus can be categorized as eco-friendly fiber. Jute (Fig. 3.2) is renowned of the substantial fibers; its primary source being Corchorus olitorious. Jute predominantly consists of cellulose and lignin. Jute is a long, soft, shiny vegetable fibre that can be spun into coarse, strong threads. It is produced primarily from plants in the genus Corchorus. It falls into the bast fibre category. The fibers are off-white to brown, and 1–4 m long [6]. The long staple fiber has high tensile strength and low extensibility and is bio-degradable. The density of Jute fiber used is 1.3 g/cc. Polyester is a thermo set polymer, whose linkage chains are strongly bonded with rigid covalent bonds after the curing process. Polyester exhibits notably increased strength, flexibility and stiffness, thus proving to be an efficient matrix component. Cobalt Napthanate and Methyl Ethyl Ketone peroxide are used as hardener and catalyst respectively.

Fig. 3.2 Jute fibers

3.2.2 *Methodology*

The Banana and Jute reinforced Polyester composites are fabricated, considering distinct thicknesses and fiber volume fractions. Banana and Jute Fiber reinforced Polyester Composites are cast to thicknesses of 3 and 5 mm, varying the fiber volume fraction from 5%, 10%, 15%, 20% and 25%.The fabrication of the various composite materials is carried out through the compression moulding technique. Banana fibers of fiber length 10 mm, Jute fibers of fiber length 10 mm, are mixed with Polyester matrix and pressed between the non-sticky sheets at the top and bottom. Cobalt Napthanate and Methyl Ethyl Ketone peroxide are used as hardener and catalyst respectively to prepare the matrix. The catalyst, hardener and Polyester Resin are mixed in the ratio 1:1:100. The mix is stirred manually to disperse the fibers in the matrix. The compression moulding machine is preheated to 82 °C. The fibers are distributed evenly within in the specimen mould (iron-frame of desired thickness).The specimen moulds of thickness 3 and 5 mm are used for the specimen fabrication. The Fiber volume fraction of 5%, 10%, 15%, 20% and 25% is adopted. After the even distribution of the fibers within the mould, it is placed on the lower plate of the Compression Moulding Machine. Then the stirred mix of Polyester resin, catalyst and hardener is poured evenly on the fibers in the specimen mould and is covered with plastic sheet and the upper plate is lowered down to the top of the specimen mould using the jack system and locked suitably. Two sheets (non-sticky) one at the top and other at the bottom of the fibers and polyester matrix mix is laid. The temperature of 82 °C is maintained for about 20 minutes. Then the specimen mould is removed from the plates and the composite specimen is un-moulded.

3.3 Results and Discussions

3.3.1 *Tensile Strength Test*

Tension Test is conducted on the test specimens as per ASTM-D3039 Standards in the Universal Testing Machine. The Tensile strength of Banana Fiber Reinforced Composites increased with the increase in fiber volume fraction and optimum value of tensile strength was obtained for fiber volume fraction of 25%. However the Tensile Strength decreased with the increase in thickness. The Tensile Strength of Jute fiber reinforced Polyester Composites increased with the increase in fiber volume fraction up to 25%. The tensile strength of Jute fiber reinforced Polyester Composites decreased with increase in thickness. Increase in thickness of composite specimens tends to decrease the tensile strength. Load required to fracture the composite specimens completely depends upon the thickness of the specimen. The extension of the specimen under the loads completely depends upon the thickness of the specimen. Load at yield point also decreases with the thickness of the specimen, which shows the weak bond of composite specimen, this may be because of increase in resin volume or improper molding of lamina.

The tensile strength values of Banana Fiber Reinforced Polyester Composites and Jute Fiber Reinforced Polyester Composites are tabulated in Tables 3.1 and 3.2 respectively.

The failure patterns of the test specimens subjected to tensile test are as shown in the Figs. 3.3 and 3.4.

Table 3.1 Tensile strength of banana fiber reinforced polyester composites

Fiber volume fraction (%)	Thickness	
	3 mm	5 mm
5	16.65	12.29
10	17.27	12.89
15	17.89	17.17
20	18.51	17.63
25	19.06	18.15

Table 3.2 Tensile strength of jute fiber reinforced polyester composites

Fiber volume fraction (%)	Thickness	
	3 mm	5 mm
5	17.09	10.68
10	22.49	14.93
15	23.45	16.59
20	35.15	24.88
25	43.94	39.28

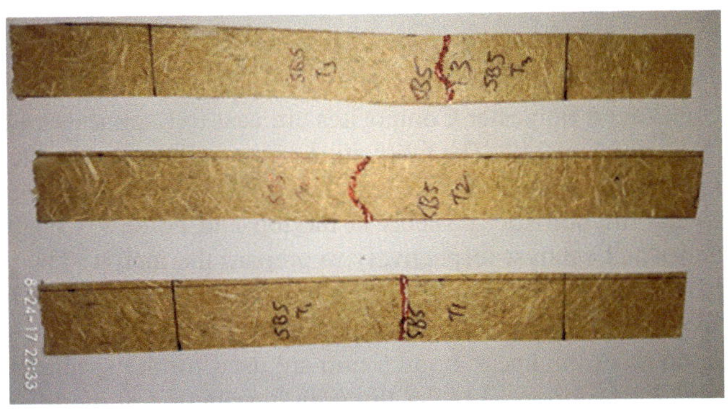

Fig. 3.3 Tensile test specimen of banana fiber reinforced composite

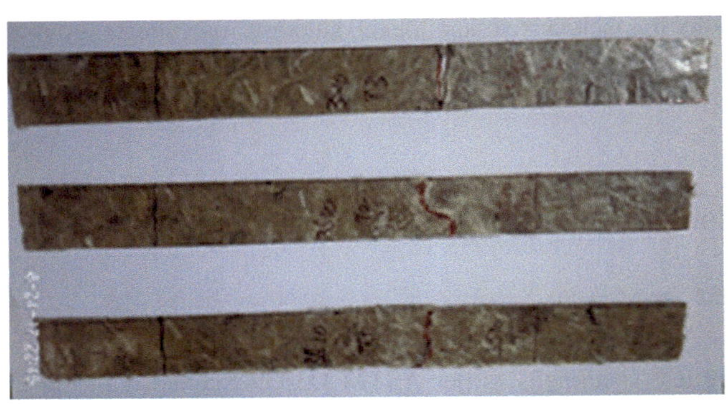

Fig. 3.4 Tensile test specimen of jute fiber reinforced composite

3.3.2 *Flexural Strength Test*

Flexural Strength test was conducted on the test specimens as per ASTM D − 7264 Standards in Universal Testing Machine. Flexural Strength of Banana fiber reinforced Polyester Composites increased up to 20% and reduced further. However the flexural strength of the Banana Fiber Reinforced Polyester Composites decreased with the increase in thickness. The Flexural Strength of Jute Fiber Reinforced Polyester Composites increased with increase in fiber volume fraction up to 20% and reduced thereafter. The flexural strength of Jute Fiber Reinforced Polyester Composites of 3 mm thickness was more than 5 mm thick composites. The flexural strength values of Banana Fiber Reinforced Polyester Composites and Jute Fiber Reinforced Polyester Composites are tabulated in Tables 3.3 and 3.4 respectively.

Table 3.3 Flexural strength of banana fiber reinforced polyester composites

Fiber volume fraction (%)	Thickness	
	3 mm	5 mm
5	28.71	16.41
10	33.99	18.05
15	36.26	19.23
20	39.66	19.77
25	38.07	18.01

Table 3.4 Flexural strength of jute fiber reinforced polyester composites

Fiber volume fraction (%)	Thickness	
	3 mm	5 mm
5	15.75	12.12
10	26.16	21.90
15	42.51	24.87
20	55.89	41.23
25	53.65	38.15

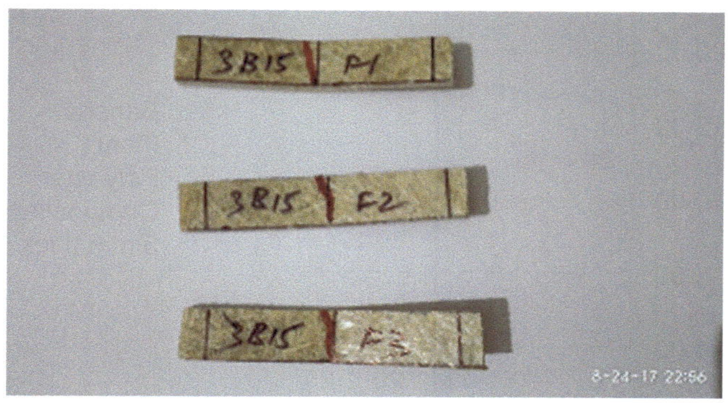

Fig. 3.5 Flexure test specimen of banana fiber reinforced composite

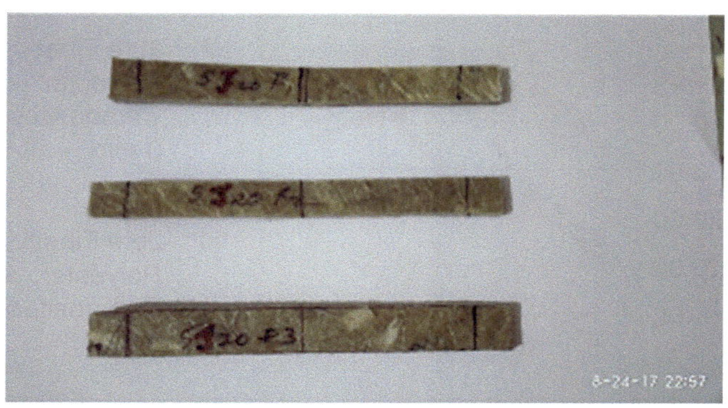

Fig. 3.6 Flexure test specimen of jute fiber reinforced composite

The failure pattern of Banana Fiber Reinforced Polyester Composites and Jute Fiber Reinforced Polyester Composites subjected to Flexural Strength are as shown in Figs. 3.5 and 3.6.

The graphical representation of the variation of tensile strength and flexural strength of Banana Fiber Polyester Composites and Jute Fiber Reinforced Polyester Composites with respect to the fiber volume fraction is shown in Figs. 3.7, 3.8, 3.9 and 3.10.

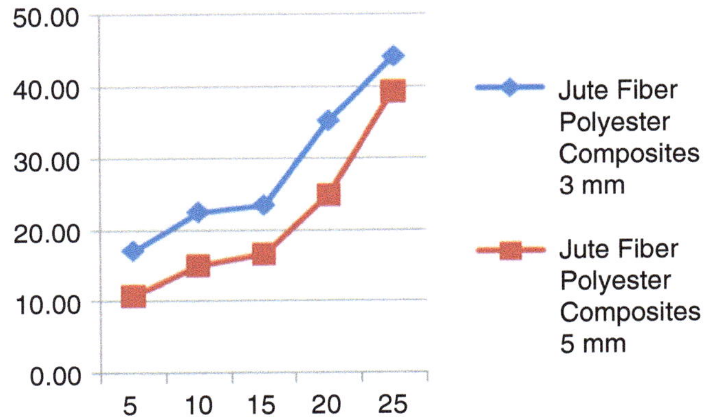

Fig. 3.7 Variation of tensile strength of jute fiber reinforced polyester composite with fiber volume fraction

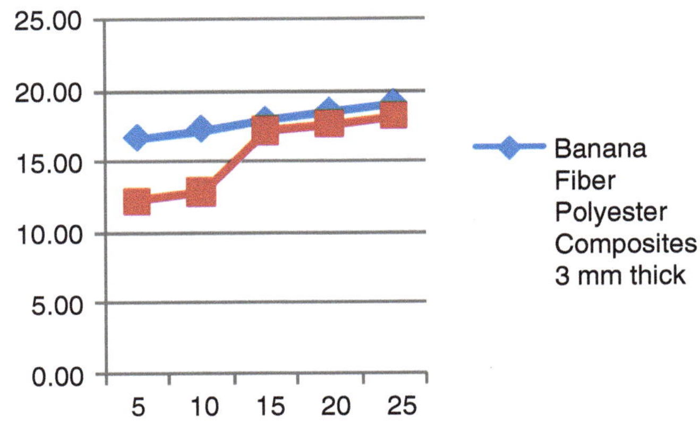

Fig. 3.8 Variation of tensile strength of banana fiber reinforced polyester composite with fiber volume fraction

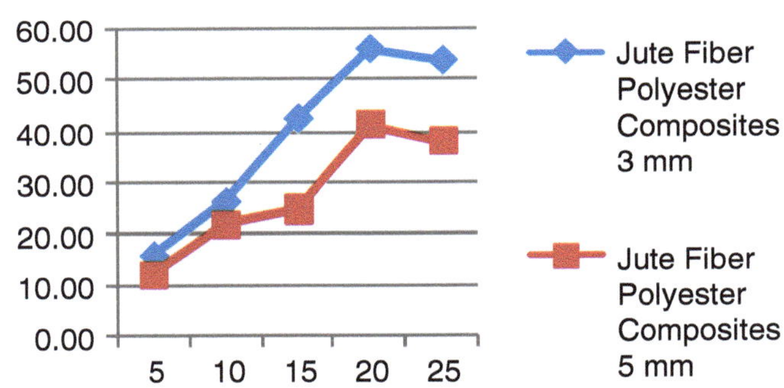

Fig. 3.9 Variation of flexure strength of jute fiber reinforced polyester composite with fiber volume fraction

3.4 Conclusions

- The Tensile strength of Banana Fiber Reinforced Polyester Composites for fiber volume fraction of 5%, 10%, 15%, 20% & 25% for 3 mm thickness specimens are 16.65 MPa, 17.27 MPa, 17.89 MPa, 18.51 MPa, 19.06 MPa and that of 5 mm thickness are 12.29 MPa, 12.89 MPa, 17.17 MPa, 17.63 MPa, 18.15 MPa respectively.
- The Tensile Strength of Jute Fiber Reinforced Polyester Composites for fiber volume fraction of 5%, 10%, 15%, 20% & 25% for 3 mm thick specimens are 17.09 MPa, 22.49 MPa, 23.45 MPa, 35.15 MPa, 43.94 MPa and that of 5 mm thickness are 10.68 MPa, 14.93 MPa, 16.59 MPa, 24.88 MPa, 39.28 MPa respectively.

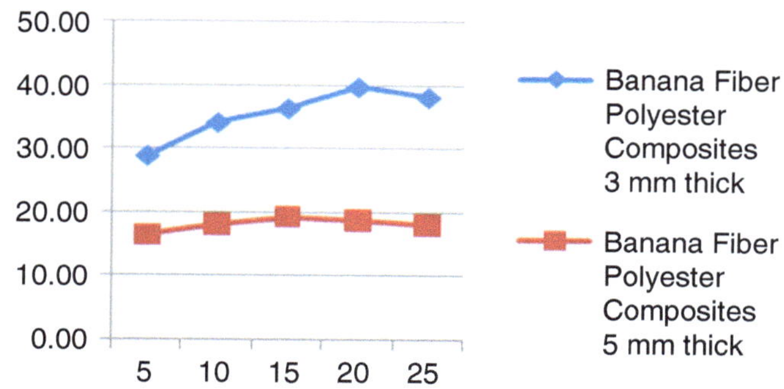

Fig. 3.10 Variation of flexure strength of banana fiber reinforced polyester composite with fiber volume fraction

- The Flexural Strength of Jute Fiber Reinforced Polyester Composites for fiber volume fraction of 5%, 10%, 15%, 20% & 25% for 3 mm thick specimens are 15.75 MPa, 26.16 MPa, 42.51 MPa, 55.89 MPa, and 53.65 MPa and for 5 mm thickness are 12.12 MPa, 21.90 MPa, 24.87 MPa, 41.23 MPa, and 38.15 MPa respectively.
- The Flexural Strength of Banana Fiber Reinforced Polyester Composites for fiber volume fraction of 5%, 10%, 15%, 20% & 25% for 3 mm thick specimens are 28.71 MPa, 33.99 MPa, 36.26 MPa, 39.66 MPa, and 38.07 MPa and for 5 mm thickness are 16.41 MPa, 18.08 MPa, 19.23 MPa, 18.77 MPa, and 18.01 MPa respectively.
- There is an increasing trend in Tensile strength of Banana Fiber Reinforced Polyester Composites and Jute Fiber Reinforced Polyester Composites with the increase in fiber volume fraction and the optimum tensile strength is achieved at a fiber volume fraction of 25%.
- There is an increase in Flexural Strength of Banana Fiber Reinforced Polyester Composites and Jute Fiber Reinforced Polyester Composites with the increase in fiber volume fraction up to 20% and then there is depreciation. Thus the contribution of banana and jute fibers to the enhancement of tensile strength and flexural strength is noteworthy. The recognizable values of tensile and flexural modulus of the banana and jute fibers are responsible for appreciation in the mechanical traits.

References

1. Wallenberger, F.T., Weston, N.: "Natural Fibers Plastics and Composites Natural", Materials Source Book. C.H.I.P.S, Texas (2004)
2. Mohanty, A.K., Misra, M.: Studies on jute composites: A literature review. Polym.-Plast. Technol. Eng. **34**(5), 729–792 (1995)
3. Boopalan, M., Niranjana, M., Umapathy, M.J.: Study on the mechanical properties and thermal properties of jute and banana fiber reinforced epoxy hybrid composites. Compos. Part B. **51**, 54–57 (2013)
4. Muthuvel, M., Ranganath, G., Janarthananand, K., Srinivasan, K.: Characterization study of jute and glass fiber reinforced hybrid composite material. Int. J. Eng. Res. Technol. **2**(4):(2013). ISSN: 2278-0181
5. Munikenche Gowda, T., Naidu, A.C.B.: Rajput Chhaya, some mechanical properties of untreated jute fabric-reinforced polyester composites, Received 10 December 1997; Accepted 1 July 1998
6. Santulli, C.: Mechanical and impact properties of untreated jute fabric reinforced polyester laminates compared with different eglass fibre reinforced laminates
7. Venkateshwaran, N., ElayaPerumal, A.: Banana fiber reinforced polymer composites – A review. J. Reinf. Plast. Compos. **29**, 2387–2396 (2010)
8. SatishPujari, A. R., Suresh Kumar, M.: Comparison of jute and banana fiber composites: A review, Accepted 10 Jan 2014, Available online 01 February 2014, Special Issue-2, 2014
9. Mohanty, S., Nayak, S.K.: Effect of mercerized banana fiber on the mechanical and morphological characteristics of organically modified fiber-reinforced polypropylene nanocomposites. Polym.-Plast. Technol. Eng. **50**(14), 14581469 (2011). https://doi.org/10.1080/03602559.2011.593079
10. Mishra, S.: Studies on electrical properties of natural fiber: HDPE composites. Polym.-Plast. Technol. Eng. **44**(4), 687–693 (2005). https://doi.org/10.1081/PTE-200057818
11. Elayaperumal, A., Sathiya, G.K.: Prediction of tensile properties of hybrid-natural fiber composites. Compos. Part B. **43**(2), 793–796 (2012)
12. Indira, K.N., Parameswaranpillai, J., Thomas, S.: Mechanical Properties & Failure Topography of Banana Fiber. Hindawi publishing corporation, ISRN Polymer science (2013)
13. Venkateshwaran, N., ElayaPerumal, A., Alavudeen, A., Thiruchitrambalam, M.: Mechanical and water absorption behaviour of banana/sisal reinforced hybrid. Composites. **32**, 4017–4021 (2011)
14. Velmurugan, R., Manikandan, V.: Mechanical properties of palmyra/glass fiber hybrid composites. Compos. Part A. **38**, 2216–2226 (2007)

Chapter 4
Toughening Mechanism in Epoxy Resin Modified Recycled Rubber Based Composites Reinforced with Gamma-Alumina, Graphene and CNT

A. B. Irez, E. Bayraktar, and I. Miskioglu

Abstract In this research, low cost devulcanized recycled rubber based composites were designed with fine gamma alumina (d < 5–10 μm) reinforcements containing minor reinforcements such as nano graphene platelets and Carbon Nano Tubes. After determination (in different wt% percentages) of the reinforcements with matrix, a special process was applied to complete successfully the manufacturing of the composites. After that, the relevant toughening mechanisms for the most suitable reinforcements were analyzed in detail for aeronautical engineering applications. For this purpose, certain mechanical and physical properties (ISO 13586:2000, K_{Ic} – Fracture toughness stress intensity factor and G_{Ic}- Critical energy release rate in mode I) have been determined by fracture toughness tests (static 3P bending test with single edge notch specimens, Charpy impact, etc.). Microstructural and fracture surfaces analysis have been carried out by means of Optical Microscopy (OM) and Scanning Electron Microscopy (SEM).

Keywords Epoxy · Recycled rubber · Wear resistance · SEM · Fracture toughness

4.1 Introduction

Rubber is generally used in different engineering applications as strategic materials and it is manufactured by using reinforcements for multifunctional composites and/or hybrid composites with higher thermal, magnetic, electric and other properties, etc. Last decades, very useful nano scale reinforcements are used extensively for the manufacturing of high performance rubber based composites that they are environment-friendly. Among these reinforcements, Graphene (GnPs) and/or graphene oxide (GO) together with Carbon Nano Tubes (CNTs) have appeared in the global market as a new alternative nano-scale materials with high static and dynamic mechanical and physical properties against Carbon blacks (CBs). They are very cheap, clean and easily produced safely by incorporating in a matrix, such as styrene-butadiene rubber (SBR) and natural rubber, etc. However it is not easy to distribute them homogenously in the matrix with traditional mixing methods such as mechanical blending [1–4].

In another aspect, utilization of the recycled rubber for the manufacturing of the new composites is a very economical way for cost effective composite design. Last decades, the usage of the recycled rubber obtained from fresh and clean scrap rubber for different industrial applications such as automotive and aeronautical engineering has been very well developed as very useful material for the composites either as a matrix or as a reinforcement. Extensively, this rubber powder come directly manufacturing of sportive affaires, shoes etc., For this reason, it is feasible to use it after chemical (silanization) and devulcanization treatments for cost effective composite design [3–9].

Reinforcement of the rubber based composites with nano graphene and/or graphene nano plates (GnPs) are very extensive applications in composite design for mainly aeronautic and aerospace applications. Graphene are known and used widely used as multi- functional reinforcement materials that can improve the electrical, piezo-resistive, dielectric, thermal and other physical and mechanical properties of rubber based composites with hybrid fillers. Because of the exceptionally high surface

A. B. Irez
CentraleSupélec, Université Paris-Saclay, Gif-sur-Yvette, France

E. Bayraktar (✉)
Supmeca-Paris, School of Mechanical and Manufacturing Engineering, Saint-Ouen, Paris, France
e-mail: bayraktar@supmeca.fr

I. Miskioglu
Michigan Technology University, Engineering Mechanics Department, Houghton, MI, USA

P. R. Thakre et al. (eds.), *Mechanics of Composite, Hybrid and Multifunctional Materials, Volume 5*, Conference Proceedings of the Society for Experimental Mechanics Series, https://doi.org/10.1007/978-3-319-95510-0_4

area, as compared to other graphite derivatives, an enormous improvement in properties is observed for graphene composites. Correspondingly, they exhibit unique advantages, as compared with all other organic and inorganic fillers, and are thus useful in many applications. Finally it may be concluded that graphene can be applied to improve the gas permeability, electrical and thermal conductivity (like carbon nanotubes) of rubber based hybrid composites.

The interactions between the various elements of a complex multicomponent system, such as the ones found in graphene/elastomer nanocomposites, play a major role on the final physico-chemical properties of the material. For this reason, in much of the literature, surface chemistry is applied to the different components of the system, in order to ensure chemical compatibility between them. In addition, it can lead to the formation of chemical bonds focused towards the improvement of the properties of the initial material and a satisfactory dispersion of the filler [10–12].

As for the reinforcement of the rubber based composites with fine gamma alumina (Al_2O_3 d < 5–10 μm) particles can increase toughening of the rubber based composites and its dispersion in the matrix is acceptable in a multiphase hybrid-toughened composite structure. The rubber based composites reinforced with nano alumina and graphene nano plates (GnPs) together with Carbon Nano Tubes (CNTs) show higher resistance capacity in a certain ratio.

For this reason, in the frame of this present work, low cost devulcanized recycled rubber based composites were designed with fine gamma alumina (d < 5–10 μm) reinforcements containing minor reinforcements such as nano graphene platelets and Carbon Nano Tubes). During this composite design, a small amount of solid epoxy powder was used as binder in the matrix called here after, epoxy modified recycled rubber based composites. The principal aim of this present work is to develop a new alternative composite design with high capacity applications in aerospace, defense industry. We believe that this new composite design may exposed a novel interface design approach for developing multifunctional rubber based composites.

4.2 Experimental Conditions

4.2.1 Materials Processing

In composite manufacturing, a good quality of bonding between matrix and reinforcements is essential to have enduring and robust final structures. However, during this study, the direct use of recycled rubbers poses a problem in terms of the final structures. Since, due their former vulcanization process, recycled rubbers do not carry any free chains to form new bonding with epoxy resin. For this reason, in order to improve the interface quality between epoxy and rubber, fresh scrap rubber should be devulcanized.

Devulcanization is known to be an efficient practice for manufacture of recycled rubbers to increase their flowing capacity and interaction with other substances. During this process, sulphur links are tried to be broken and also generated new other links, it means that the structure of the material is modified entirely as renewable process.

A new design of devulcanized rubber based composite, reinforced with carbon nano tube (CNT), graphene nano platelet (GNP) and fine γ-Al_2O_3 powders are prepared in several steps. First of all, chemical treatment was applied on rubber. After a short silanization process, a chemical bath is provided by acrylic acid and a small amount of toluene solution to activate the surface of the rubber particles [3–5, 7, 9, 10]. After drying of solutions in a conditioning oven, chemically treated rubber powders, they are exposed to short microwave heating in two stages for 3 min under 800 W in order to avoid degradation of the main chains. As the last step of matrix preparation, devulcanized rubber were mixed with epoxy resin powders. This mixture used as a matrix after that the new designed composites are manufactured by using classical powder metallurgy methods. In the second stage, the reinforcements (here CNT, GNPs and γ-alumina) were blended with the matrix, then a fast-rotating toothed-wheel milling process was carried out during 4 h to obtain fine rubber powder. After having a homogeneous powder compound, the composite specimens were manufactured by using double uniaxial action press at a temperature of 180 °C under a pressure of 70 MPa during the heating of 30 min. Finally, all of the specimens (30/50 mm in diameter) were cooled slowly in the press. All of the specimens were post-cured isothermally at 80 °C for 24 h.

The compositions of γ-alumina, GnP and CNT reinforced epoxy-recycled rubber based composites (called as GAC I-II-III hereafter) were given in Table 4.1.

Table 4.1 Composition of the epoxy-rubber based composites

Epoxy-rubber based composition Epoxy rubber (%15 Epoxy – 85 Rubber)	GAC I	GAC II	GAC III
Reinforcements (wt %)	5 γ- Al_2O_3	7,5 γ- Al_2O_3	15 γ- Al_2O_3
	1.5 GnPs	1.5 GnPs	1.5 GnPs
	1 CNT	1 CNT	1 CNT

Table 4.2 Hardness values of GAC specimens

Hardness measurement	
Specimen	Shore D
GAC I	78.6
GAC II	79.5
GAC III	81.5

Table 4.3 Density values of GAC specimens

Hardness measurement	
Specimen	Density (g/cm^3)
GAC I	1.481
GAC II	1.496
GAC III	1.502

4.2.2 Microstructure: Fracture Surface Analyses and Shore-D Hardness Measurements

Fracture surface damage analyses and microstructural observation have been realized by means of optical (OM) and scanning electron microscopy (SEM). SEM observation was realized on fracture surface of the tested specimens with Scope/JSM-6010LA Jeol® electron microscope.

Surface hardness measurements of the specimens were performed after post curing. Shore D hardness test measurements on the polished flat surfaces of the specimens were carried out according to ASTM D 2240 using Shore D hardness tester, (type HBD-100-0). Hardness results were given in Table 4.2.

Three-point bending tests (3 PB) have been carried out according to the ASTM D790 standards. Tests were realized with the machine Zwick Proline Z050TN and during the tests crosshead speed was selected as 1 mm/min and flexural strength and strain were obtained from the test results. At least three specimens for each composition were used and standard deviation and average values were given in results chapter with standard deviation values. In addition, fractural properties such as plain strain fracture toughness (K_{Ic}) and critical strain energy release rate (G_{Ic}) were investigated with SENB specimens and the tests were realized according to ASTM D5045 standard. Notches were introduced by tapping a fresh razor blade.

Charpy impact tests were realized by means of Zwick 5102 pendulum impact tester with a 1 J pendulum configuration. After preparation of specimens, densities of the compositions were measured and they were given in Table 4.3.

4.2.3 Wear Resistance (Scratch Test) and Damage Analysis Via 3D Optical Roughness Meter

In the current research basic idea on the tribological behaviour of the epoxy and recycled elastomer based composites was evaluated performing scratch tests. A 3D optical surface scanner was utilized to assess damaged zone after the scratch test in terms of scratch depth and average scratch roughness.

The contact between the sliding diamond indenter and the surface of the composite material during scratch test is analyzed. The normal and tangential forces on indenter are recorded. However, main focus is given to the damage area and volume. In the frame of the current research, the resistance to scratch deformation is evaluated in terms of scratch depth, surface and worn volume subsequent to scratching only under dry conditions and 50,000 and 100,000 number of cycles of wear.

4.3 Results and Discussions

4.3.1 *Microstructure of the Composites*

In Fig. 4.1, powder form of the composites as well as their compacted form for two different diameters and a 3 PB test specimen were shown.

For each composition, three different magnification were presented in Fig. 4.3. After hot compaction, transversal sectioning was prepared then all of the specimens for three compositions were mounted after that polished. Then, general microstructures in the transversal direction of three compositions were observed by means of OM and they were shown in Fig. 4.2. From the figures, it is said that all of the compositions have shown a considerably homogenous distribution of the reinforcements in the structure. White circular particles in different dimensions were considered as rubber. In addition, homogeneously distributed amorphous media are thought as epoxy matrix.

4.3.2 *Three Point Bending Tests and Fracture Surface Observation*

Three-Point Bending (3 PB) tests have been carried out for each different type of composites.

4.3.2.1 Flexural Testing and Fracture Toughness Determination

Flexural stress is calculated during three-point bending test according to the Eq. 4.1:

$$\sigma = \frac{3x\boldsymbol{P}\boldsymbol{x}\boldsymbol{l}}{2x\boldsymbol{b}\boldsymbol{x}\boldsymbol{h}^2} \tag{4.1}$$

In this formula, l is the span length, P is the maximal bending load, b and h are the sample width and thickness, respectively.

Flexural strain, ε_f, was determined according to the Eq. 4.2:

$$\varepsilon_f = \frac{6Dh}{l^2} \tag{4.2}$$

D is the maximum deflection at the center of the specimen.

E_B is the modulus of elasticity in bending and it is expressed with the Eq. 4.3 as follows:

$$E_B = \frac{l^3 m}{4bh^3} \tag{4.3}$$

m is the tangent of the initial straight portion of the stress-strain curve.

Fig. 4.1 Macrograph of the specimens after compacting and post curing d = 30–50 mm

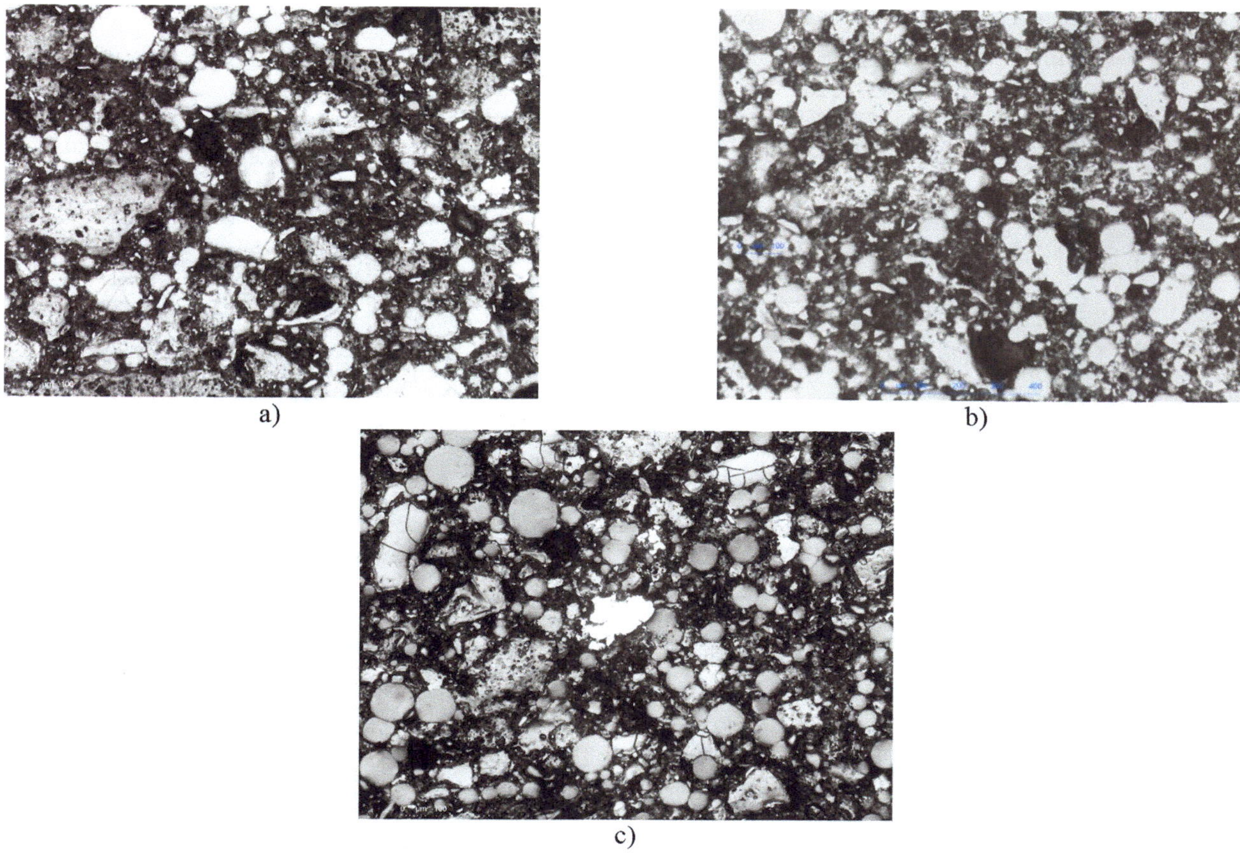

Fig. 4.2 Microstructure of GAC I-II-III after transversal sectioning

The mode I fracture toughness, K_{Ic}, was determined by testing of the SENB specimens and K_{Ic} was calculated according to the Eq. 4.2:

$$K_{Ic} = \frac{F}{B\ w^{\frac{1}{2}}} f(x); \quad x = \frac{a}{W} \quad ,0 < \frac{a}{W} < 1 \tag{4.4}$$

F is the maximum force from the load-elongation trace; *B* is the thickness of the specimen; *W* is the width and "a" is the total notch length.*f(x)* is the geometry correction factor and is expressed with the Eq. 4.3 as follows:

$$f(x) = 6(x)^{0.5} \left\{ \frac{\left[1.99 - x\,(1-x)\left(2.15 - 3.93x + 2.7x^2\right]\right.}{(1+2x)\,(1-x)^{1.5}} \right\} \tag{4.5}$$

Critical strain energy release rate (fracture energy) G_{Ic} was calculated using the expression Eq. 4.4:

$$G_{Ic} = \frac{{K_{Ic}}^2}{E} \tag{4.6}$$

where *E* is the elasticity modulus for plane stress approach examined for thin specimens.

Table 4.4 indicates the mechanical properties in bending mode. Fracture toughness values were also presented in the table.

Table 4.4 indicates that the hybrid rubber based composites reinforced with alumina and graphene nano plates (GnPs) together with carbon nano tubes (CNTs) with a certain ratio have shown improvements in the mechanical properties, and more significantly, flexural stress and fracture toughness (K_{IC}), which can be explained by synergistic impact coming from the intrinsic physical characteristics of the reinforcements.

Table 4.4 Comparison of mechanical properties of GAC specimens

	Flexural stress (MPa)	Flexural modulus (MPa)	Strain in break (ε %)	K_{Ic} (MPa m$^{1/2}$)	G_{Ic} (kJ/m^2)
GAC I	14.70 ± 1.85	933 ± 264	0.38 ± 0.076	0.38 ± 0.01	0.154 ± 0.01
GAC II	15.40 ± 3.16	972 ± 269	0.39 ± 0.06	0.39 ± 0.02	0.158 ± 0.016
GAC III	15.40 ± 4.29	957 ± 257	0.37 ± 0.02	0.54 ± 0.08	0.312 ± 0.1

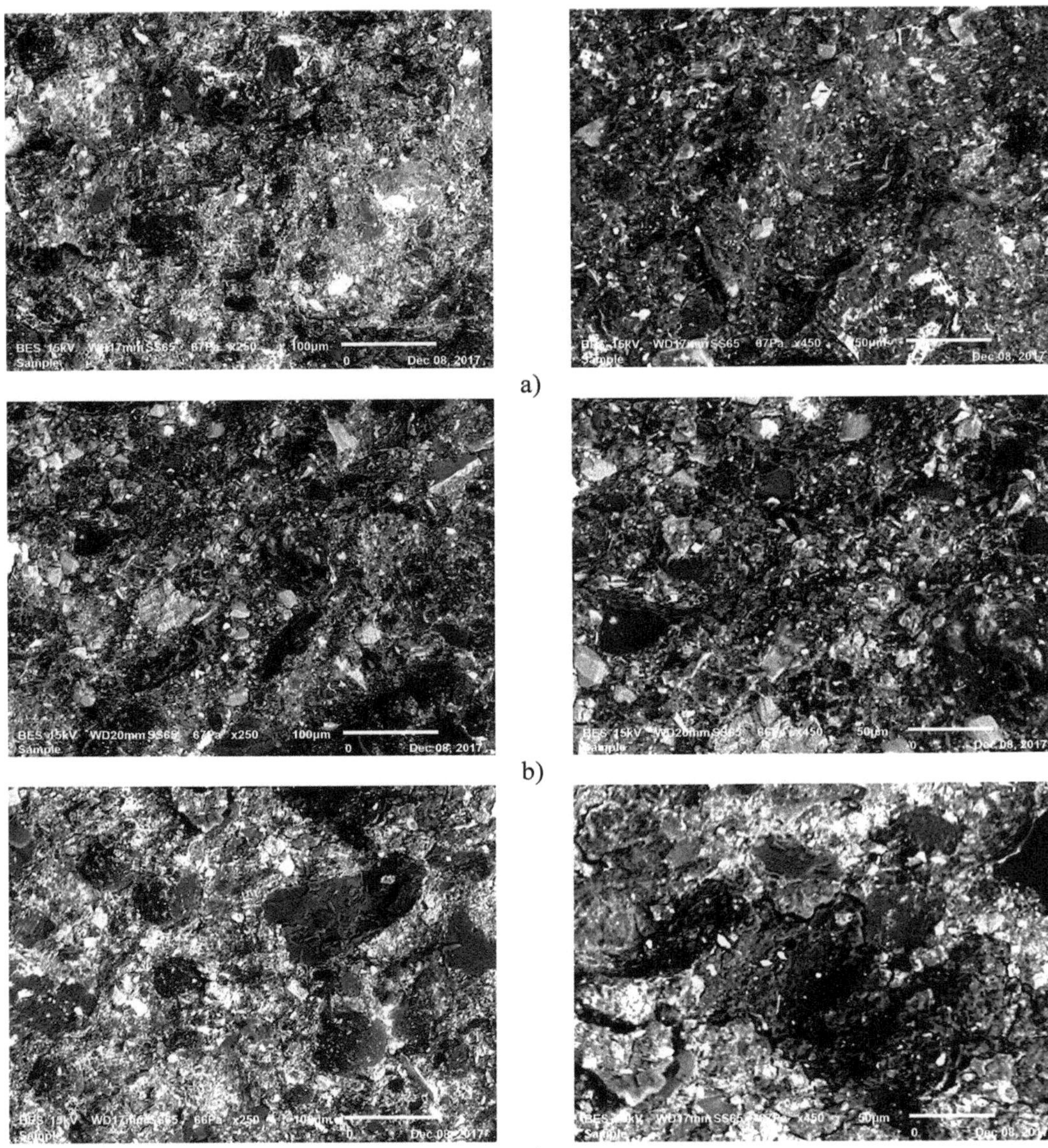

Fig. 4.3 Fracture surfaces after 3 PB testing (**a**) GAC I (**b**) GAC II (**c**) GAC III

After the realization of bending tests, fracture surfaces were observed by means of scanning electron microscopy (SEM). Fracture surfaces obtained from 3 PB tests have been analyzed by means of Scanning Electron Microscopy (SEM). It noticed that good adhesion of the both of the reinforcements in the rubber based matrix by creating an ideal interface for each composition, as presented in Fig. 4.3 with different fracture surfaces taken by SEM.

Cavitation and void formation in the rubber matrix with matrix expansion and locally, debonding of nano particles with consequent void growth can be identified in the structure as the improved toughening mechanisms. For this evolution, the mixture of devulcanized rubber (85 wt %) with epoxy (15 wt %) plays an important role. Some of the specimens with homogenous distribution of the nanoparticles have shown a typical debonding of the fine particles. This phenomena can

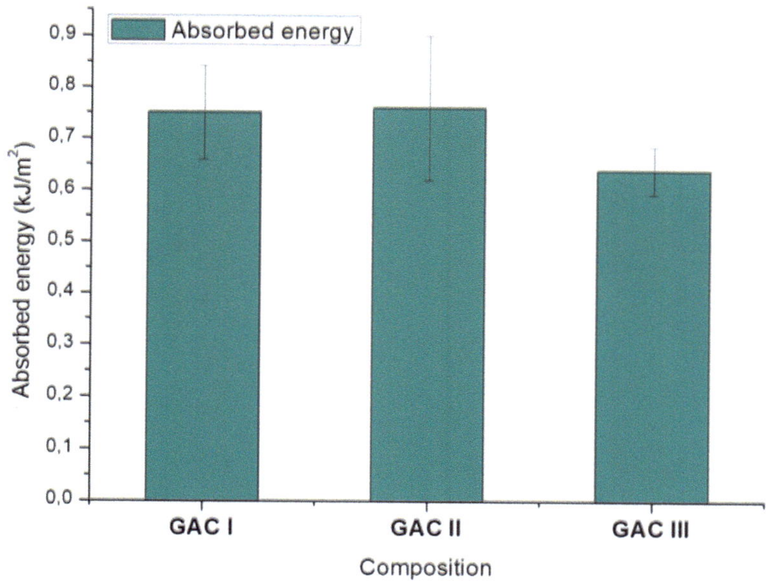

Fig. 4.4 Absorbed energy after Charpy impact tests for GAC specimens

be explained with the toughness improvement in hybrid rubber based composites that can increase the plastic zone in the structure. This case facilitate the devulcanized rubber based composites to dissipate additional fracture energy.

4.3.3 Charpy Impact Testing

Evaluation of the impact resistance and energy absorbance capability of the manufactured composites, Charpy impact test series have been carried out by using at least three standard specimens for each. Absorbed energies obtained on the Charpy impact pendulum were presented in Fig. 4.4. It seems that the absorbed energy for each specimen during the impact tests is related to the increment of the plastic zone in the structure due to the debonding of the nano particles. Higher dispersion in the values are also observed that can be explained with different parameters such as the conditions of the specimens surface roughness, notch effect, etc.

4.3.4 Damage Analysis by Means of Scratch Test and 3d Optical Roughness Meter

After completing mechanical tests, tribological characterization of the composites were done by macro scratch tests. Surface damages were observed three dimensionally by an optical surface scanner. The results are presented in the Fig. 4.5.

The volume and surface of the damage trace after scratch calculated from roughness test results are given in Fig. 4.6. It is noted that composites become more resistant to wear by making increase the percentage of the reinforcement in the structure.

In reality, the high shear stress level occurs at the interfaces that should probably be the main reason for damage of the matrix and reinforced filler interfaces. When the indenter is slipping on the specimen, tangential tensile stress is caused on the surface behind the indenter, while in front of the indenter the tangential stress is compressive.

4.4 Conclusion

The hybrid rubber based composites reinforced with fine gamma alumina and graphene nano plates (GnPs)with together small amount of the carbon nano tubes (CNTs) have shown enhancements in the mechanical properties, and more significantly, flexural stress and fracture toughness (K_{IC}), which can be explained by synergistic impact coming from the

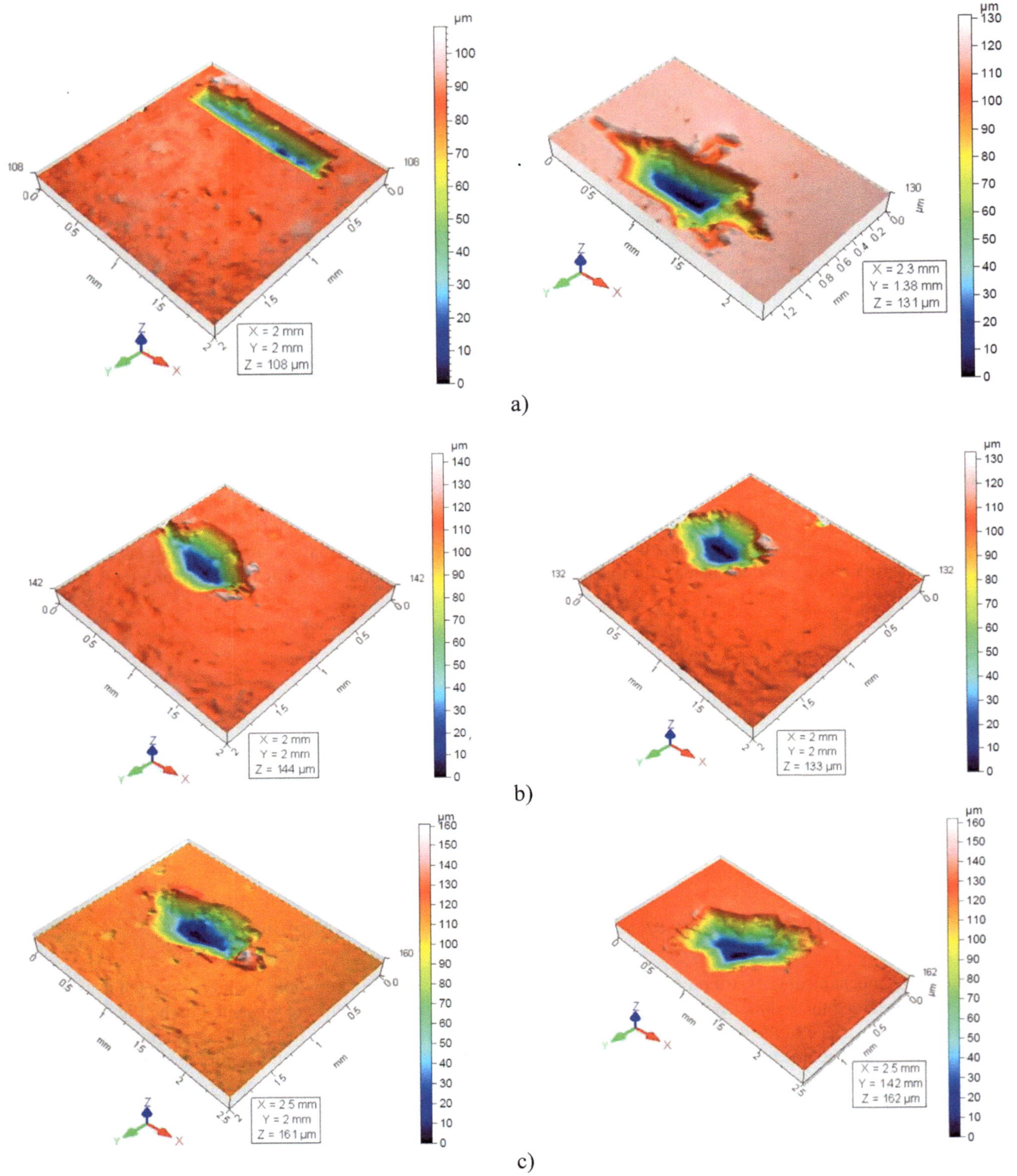

Fig. 4.5 Three dimensional damage traces obtained in the direction of width and length for (**a**) SAG I (**b**) SAG II (**c**) SAG III for 50 k cycles and for 100 k cycles

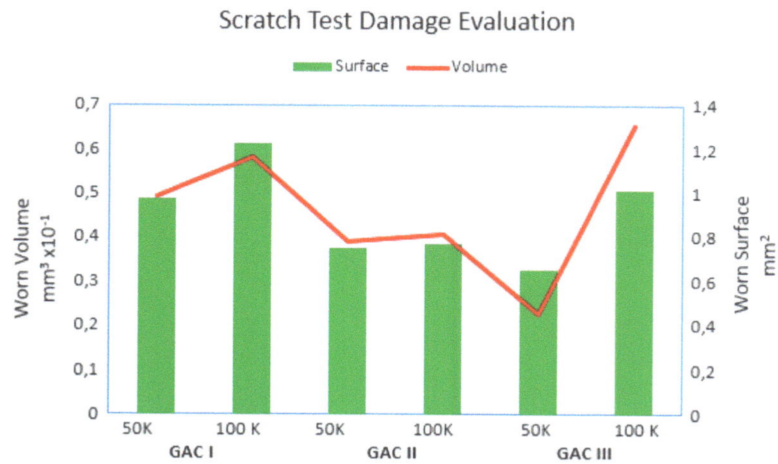

Fig. 4.6 Damage traces obtained in the direction of width and length for the specimens GAC I-II-III for 50 k and 100 k cycles

intrinsic physical characteristics of the reinforcements; the improved toughening mechanisms can be explained with the formation of the cavitation and void formation due to debonding of the fine reinforcement particles in the rubber matrix.

Probably, debonding of nano particles could create void growth with consequent locally matrix expansion like very small islands distributed everywhere in the structure. For this evolution, the mixture of devulcanized rubber (85 wt %) with solid powder epoxy (15 wt %) plays an important role. It means that the specimens with homogenous distribution of the nano particles in the structure may introduce typical "debonding islands". These small "debonding islands" should be origin of the toughness improvement in hybrid rubber based composites. This case facilitate the devulcanized rubber based composites to dissipate additional fracture energy.

Acknowledgments The authors thank Dr. H-A. Alhas from Airbus-Space, London/UK for general support and also for his valuable discussion in certain stages of this project. Also they would wish to acknowledge research foundation of Supmeca/Paris.

References

1. Wan, C., Chen, B.: Reinforcement and interphase of polymer/graphene oxide nanocomposites. J. Mater. Chem. **22**, 3637–3646 (2012)
2. Bokobza, L.: Multiwall carbon nanotube elastomeric composites: A review. Polymer. **48**, 4907–4492 (2007)
3. Zhang, G., et al.: Effect of functionalization of graphene nanoplatelets on the mechanical and thermal properties of silicone rubber composites. Materials. **9**(2), 92 (2016)
4. Shokoohi, S., Arefazar, A., Khosrokhavar, R.: Silane coupling agents in polymer-based reinforced composites: A review. J. Reinf. Plast. Compos. **27.5**, 473–485 (2008)
5. Zaimova, D., Bayraktar, E., Miskioglu, I.: Design and manufacturing of new elastomeric composites: Mechanical properties, chemical and physical analysis. Int. J. Composites Part B, Elsevier, USA **1**(1):1–12, November 2016, on line 2017
6. Sadasivuni, K.-K., Ponnamma, D., Thomas, S., Grohens, Y.: Evolution from to graphene elastomer composites. Prog. Polym. Sci. **39**, 749–778 (2014)
7. Zaimova, D., Bayraktar, E., Miskioglu, I., Dishovsky, N.: Wear resistance of elastomeric based composites by continuous multi-cycle indentation used in manufacturing engineering. Adv. Mater. Res. **939**:106–113 (2014), Trans Tech Publications, Switzerland, Doi:10.4028/www.scientific.net/AMR.939.106
8. Zhang, G., Wang, F., Dai, J., Huang, Z.: Effect of functionalization of graphene nanoplatelets on the mechanical and thermal properties of silicone rubber composites. Materials. **9**(92), 1–13 (2016)
9. Irez, A. B., Bayraktar, E., Miskioglu, I.: Mechanical Characterization of Epoxy – Scrap Rubber Based Composites Reinforced With Alumina Fibers, Springer link, Mechanics of Composite and Multi-functional Materials, vol. 6, pp. 59–70 (2017). ISBN 978-3-319-63408-1, https://doi.org/10.1007/978-3-319-63408-1
10. Ferreira, L. M. P., Miskioglu, I., Bayraktar, E., Katundi, D.: Mechanical and Tribological Properties of Scrap Rubber Reinforced With Al2O3 Fiber, Aluminium and TiO2, Springer link, Mechanics of Composite and Multi-functional Materials, vol 7, pp. 37–44 (2016). ISBN 978-3-319-41766-0, https://doi.org/10.1007/978-3-319-41766-0
11. Kinloch, A.J., Mohammed, R.D., Taylor, A.C.: The effect of silica nano-particles and rubber particles on the toughness of multiphase thermosetting epoxy polymers. J. Mater. Sci. **40**, 5083–5086 (2005)
12. Burak Irez, A., Hay, J., Miskioglu, I., Bayraktar, E.: Scrap-rubber Based Composites Reinforced With Boron and Alumina, Springer link, Mechanics of Composite and Multi-functional Materials, vol. 6, pp. 1–10

Chapter 5
AlSi10Mg Nanocomposites Prepared by DMLS Using In-Situ CVD Growth of CNTs: Process Effects and Mechanical Characterization

P. Thompson, R. Poveda, I. Bezsonov, M. Rossini, D. Orthner, K. Cobb, B. Leng, and Z. Iqbal

Abstract Carbon Nanotube (CNT) reinforced AlSi10Mg Metal Matrix Composites (MMCs) were fabricated through direct metal laser sintering (DMLS), and select mechanical properties were compared to those of virgin DMLS AlSi10Mg. The MMCs were prepared by initially depositing CNTs homogeneously on the bulk AlSi10Mg powder through the chemical vapor deposition (CVD) process. The CNT-reinforced AlSi10Mg (CNT-Al MMC) powder was then fused together through DMLS using an EOS M290 additive manufacturing printer. Virgin AlSi10Mg powder was processed as a separate build via DMLS for baseline comparison. Scanning Electron Microscopy was conducted and material property characterization was performed to observe surface morphology, microstructure and mechanical property variation of these materials. Further analysis via Scanning Electron Microscopy and Energy-Dispersive X-ray spectroscopy revealed homogeneously distributed voids within the microstructure of the CNT-Al MMC and large quantities of oxygen near these voids. The growth of CNTs directly onto powder particulates, via CVD, allows for the uniform distribution of reinforcement, which is critical to cohesive 3D printed builds. However, this process requires refinement to limit the presence of excess oxygen, which can cause microstructural defects, thereby hindering the reinforcing capabilities of CNTs.

Keywords Nanocomposites · Additive manufacturing · Carbon Nanotubes · Chemical Vapor Deposition · Mechanical Properties

5.1 Introduction

Additive Manufacturing (AM) has gained increasing attention from academia, government and industry. AM has the potential to revolutionize the automotive and aerospace industries and, therefore, numerous studies have set out to characterize the mechanical behavior of alloys produced by AM. One alloy, which is gaining considerable attention, is AlSi10Mg. Kempen et al. [1] compared the mechanical properties of AlSi10Mg produced by Selective Laser Melting (SLM) to those of AlSi10Mg produced via conventional casting. Their study revealed that the AM SLM alloy has the potential to perform as well or better than its cast counterpart. Mower et al. [2] preformed a comprehensive study on the mechanical behavior and the effects of heat treatment on Direct Metal Laser Sintered (DMLS) AlSi10Mg and compared these results to those of Al6061 produced by conventional means. Their study revealed less ductility and lower yield strength, in quasi-static loading, in AlSi10Mg than in Al6061. Additionally, the fatigue properties of AlSi10Mg were significantly lower than Al6061. In their study of AlSi10Mg across a range of strain rates ($2.77\ x\ 10^{-6}$ to $2.77\ x\ 10^{-1}S^{-1}$), Rosenthal et al. [3] observed rate dependence in the alloy. Their results indicate considerable changes in flow stress and strain hardening exponent at increased strain rates.

Since their production/ acquisition continues to become easier, Carbon Nanotubes (CNTs) have become a popular choice as reinforcement for metal matrix composites (MMCs). Numerous studies have fabricated and tested a multitude of CNT reinforced MMCs [4–7]. Kwon et al. [4] prepared Al-CNT MMCs by Spark Plasma Sintering (SPS) and hot extrusion. In this study, CNTs were dispersed onto Al particles via a nanoscale dispersion method. This lead to uniform dispersion of the reinforcement and significant improvement of mechanical properties. Others have fabricated Cu-CNT MMCs via high speed ball milling of CNTs and Cu particles followed by SPS [5]. While this study did report improved mechanical performance,

P. Thompson · R. Poveda (✉) · I. Bezsonov · M. Rossini · D. Orthner · K. Cobb
NAVAIR, Joint Base McGuire-Dix-Lakehurst, Lakehurst, NJ, USA
e-mail: ronald.poveda@navy.mil

B. Leng · Z. Iqbal
CarboMet, Morris Plains, NJ, USA

P. R. Thakre et al. (eds.), *Mechanics of Composite, Hybrid and Multifunctional Materials, Volume 5*, Conference Proceedings of the Society for Experimental Mechanics Series, https://doi.org/10.1007/978-3-319-95510-0_5

agglomeration of CNTs was also observed. Lahiri et al. [6] fabricated Al-CNT MMCs via roll bonding of Al foils coated with CNTs and studied the various strengthening mechanisms induced by CNTs. It was observed that uniform dispersion of CNTs leads to improved elastic properties, while agglomeration of CNTs yields improved strain hardening and ultimate tensile strength. In a relatively unique approach, Patel et al. [7] fabricated stainless steel-CNT MMCs by chemical vapor deposition (CVD) of CNTs directly on to stainless steel pellets. This study saw a significant increase in yield strength, elastic modulus and hardness. It also removed the step of mixing/ combining CNTs with metal particulates.

Given the rising popularity of AM and CNT-MMCs, there are surprisingly few results reported on the fabrication/ performance of CNT-MMCs prepared by AM. In one study, Wang et al. [8] fabricated Inconel 625-CNT MMCs via Selective Laser Melting (SLM). In this study, CNTs were coated on IN625 powder via molecular level mixing. This method allowed for reduced agglomeration and improved mechanical properties, like yield strength and ultimate tensile strength.

The present study examined the feasibility of preparing Al-CNT MMCs via CVD growth of CNTs directly onto AlSi10Mg powders, followed by Direct Metal Laser Sintering.

5.2 Experimental Procedure

5.2.1 CVD Growth of CNTs on AlSi10Mg Powders

Synthesis of the composite powder was carried out in a similar fashion to that found in literature [7]. This procedure is outlined, schematically in Fig. 5.1a. The process began by loading AlSi10Mg powder into a CVD quartz tube, which laid horizontally inside of a furnace, equipped with precision temperature control. After the tube was airtight-sealed, a mechanical pump brought the system to a vacuumed state, and the system was purged further by the flow of argon and carbon monoxide at constant flow rates for a designated amount of time. The vacuum valve was closed during the gas purging phase. After the system was completely purged, the temperature in the furnace was elevated at a constant rate until it reached a dwell temperature below the solidus line indicated by the AlSi10 phase diagram, which has been analyzed in previous work dealing with DMLS [9]. Once dwell temperature was reached, acetylene was introduced into the system at a constant flow rate and the system was held at dwell temperature. After the dwell period, the carbon monoxide and acetylene vents were closed, and the furnace was opened to allow cooling. Argon flow continued until the powder was ready to be removed from the CVD tube. Figure 5.1b contains an SEM image of AlSi10Mg powders with CNTs grown on their surface. In order to obtain sufficient powder with CNT content below 1 wt.%, the post-processed powder was mixed with virgin AlSi10Mg in a 1:1 ratio using a Resodyn RAM 5 acoustic mixer. This ensured homogeneous distribution and reduction of potential agglomeration. The resulting CNT-reinforced AlSi10Mg powder for use in the EOS printer contained 0.67 wt.% of CNTs after weighing all powder amounts before and after CVD.

5.2.2 DMLS of Specimens and Heat Treatment

Material testing coupons and representative parts were built by an EOS M290 system equipped with a 250 × 250 × 300 mm work envelope and an up to 400 W continuous Yb: YAG fibre laser. Two builds were performed. The first build, which served as a control build, contained specimens fabricated from virgin EOS AlSi10Mg powder (P/N: 9011-0024) from a single powder lot. The second build contained specimens fabricated from CNT-infused EOS AlSi10Mg powder from the same virgin powder lot. The targeted spot size of the machine was 80 μm in diameter. A Hatching Distance of 190 μm was utilized with a Stripe Width of 7 mm and an Overlap of 20 μm. The hatching strategy utilized was the strip scanning strategy with a 67° rotation of the laser between successive layers. The Layer thickness prior to melting was 30 μm. The builds were performed under an Argon atmosphere with maximum oxygen content of 0.1%. The build plate was kept at a temperature of 35 °C during printing. Both the Al-CNT MMC and AlSi10Mg builds, depicted in Fig. 5.2a, b, received stress relief heat treatment. This consisted of placement in a static air environment with nitrogen for 2 h at 300 °C.

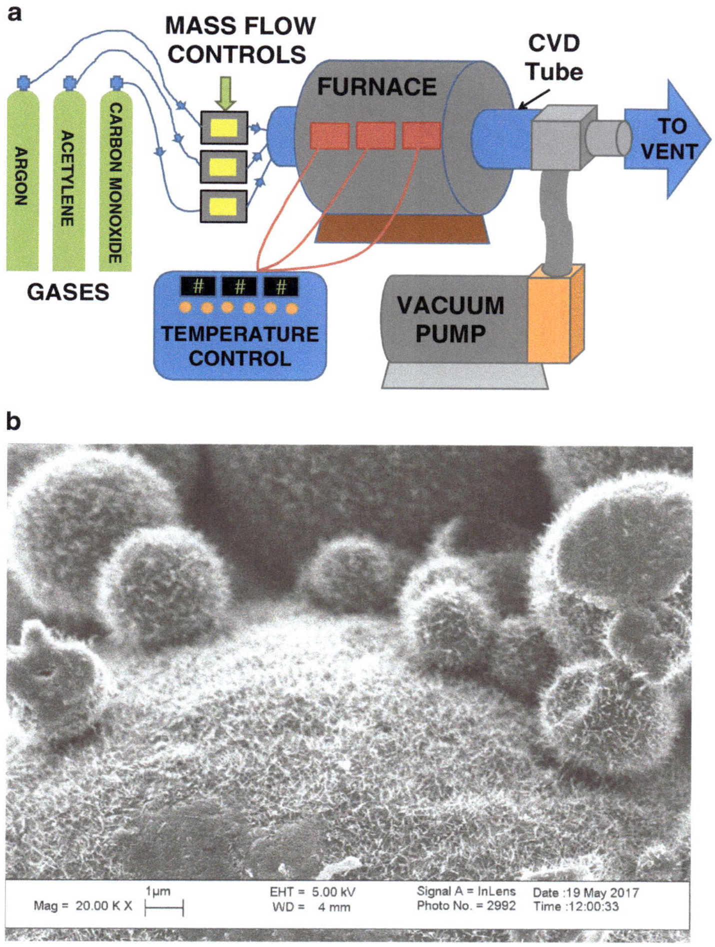

Fig. 5.1 (**a**) Schematic of CVD process (**b**) SEM Image of CNTs grown on Al Powder

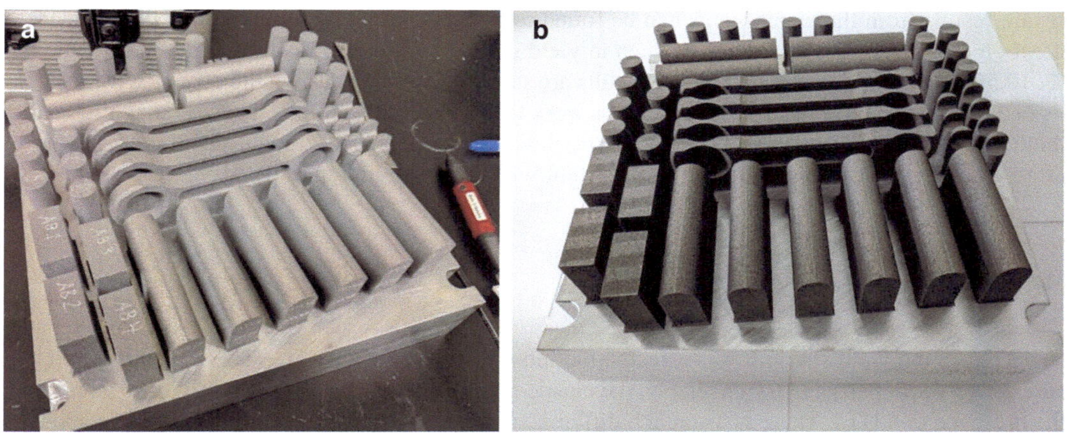

Fig. 5.2 (**a**) AlSi10Mg AM Build plate (**b**) Al-CNT MMC AM Build Plate

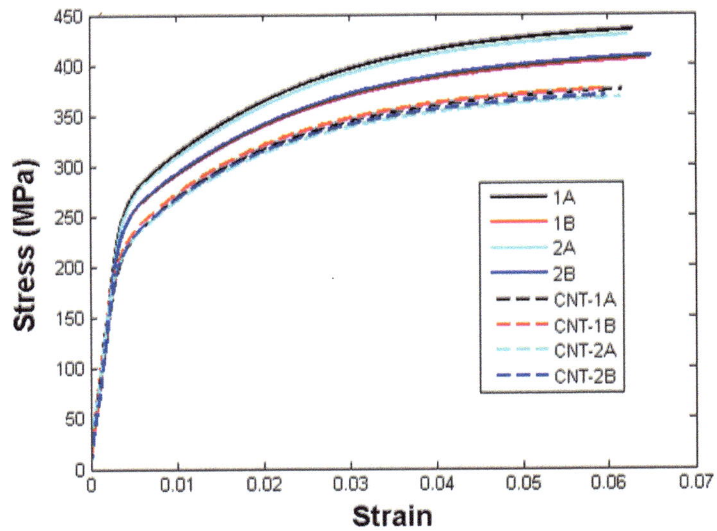

Fig. 5.3 Stress-strain Data from Tensile Experiments

5.2.3 Tensile Experiments

ASTM E8 [10] tensile specimens were machined from the Y-axis blocks of both builds via wire EDM. These specimens were placed in an MTS load frame and pulled to failure at a crosshead displacement rate of 0.003 mm/s, with elongation being measured by an extensometer. A total of 8 specimens were tested, 4 AlSi10Mg and 4 Al-CNT MMCs.

5.3 Results and Discussion

5.3.1 Tensile Test Results

True Stress-Strain results from the experiments can be found in Fig. 5.3. It can be observed that the presence of CNTs had an adverse effect on the MMC. Decreases can be seen in yield strength, ultimate tensile strength and elastic modulus to the tune of 13.3%, 10.5% and 4.9% respectively. These results are startling because, even in the presence of CNT agglomeration, improvements in performance have been seen in previous work [5, 6]. Despite their differences in magnitude, the stress-strain response of both materials was similar. In both cases, the materials experienced gradual yielding, followed by significant hardening before ultimate failure. This response is consistent with results in literature [2, 3].

5.3.2 SEM Analysis of Failure Surface

With a 13.3% decrease, the yield strength was affected the most by the presence of CNTs. In metals, yield stress is known to be a structure sensitive property [11]. Conversely, the elastic modulus is not structure sensitive and this property was least effected by the presence of CNTs. There is a staunch difference in the microstructure of the AlSi10Mg and Al-CNT MMC. Figure 5.4a, b contain SEM images of the failure surfaces of the AlSi10Mg and Al-CNT MMC respectively. One can see that the surface of the MMC is significantly more porous than that of the AlSi10Mg, with voids distributed homogeneously throughout the surface. The presence and distribution of these voids could have contributed to the decrease in strength.

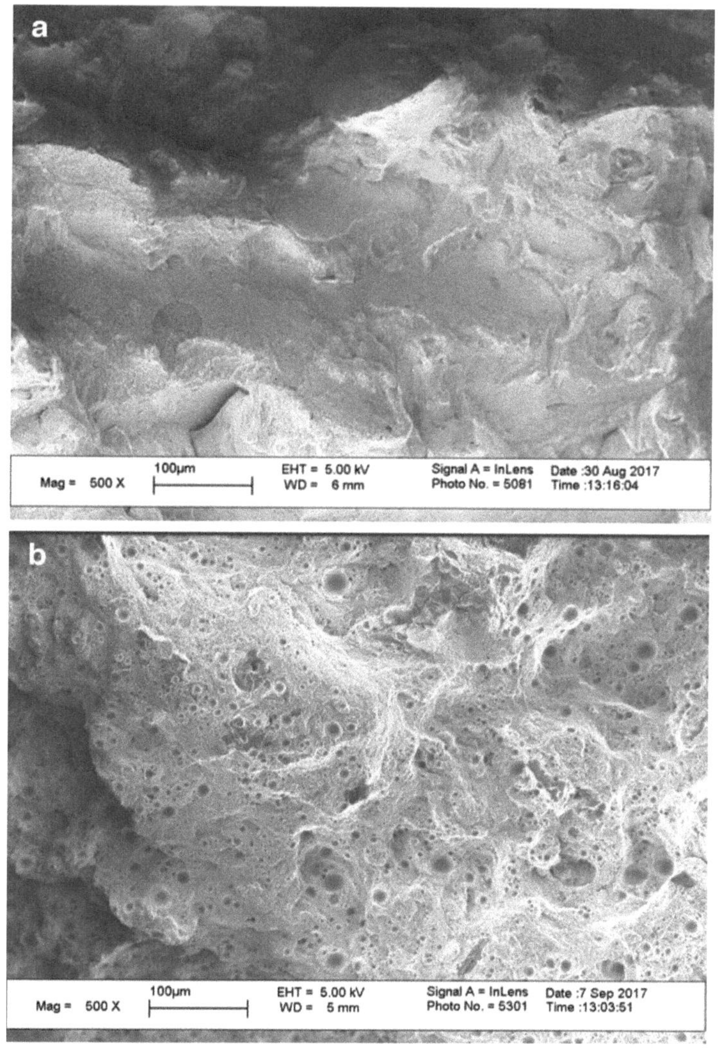

Fig. 5.4 (**a**) SEM image of AlSi10Mg fracture surface (**b**) SEM image of Al-CNT MMC fracture surface

Figure 5.5 a, b contains an SEM image, complemented by Energy Dispersive X-Ray Spectroscopy (EDX). This image reveals voids and a significant amount of oxygen near the CNT reinforcement. This phenomenon has been observed elsewhere, in literature, as an artifact of the CVD process [12]. It is speculated that the oxygen found in the MMC microstructure was released from the surface of the CNTs when they were struck by the high-powered laser.

5.4 Conclusion

The preparation of Al-CNT MMCs is feasible via AM. In situ CVD of the CNTs provides a means of homogeneously distributing CNTs, or at least CNT clusters. However, care must be taken to prevent the outgassing of any oxygen trapped on the CNT surface.

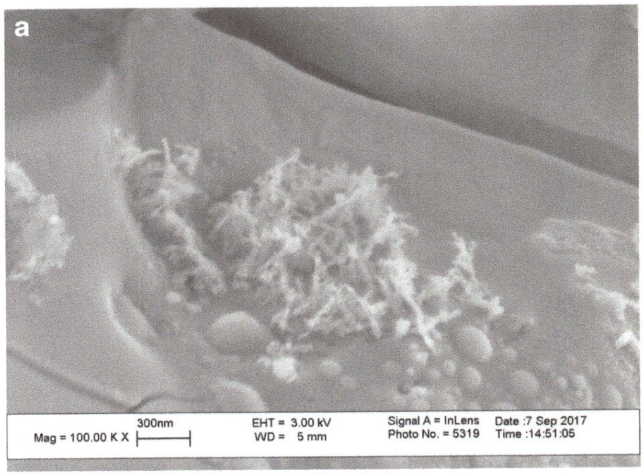

b

Element	Weight%	Atomic%
CK	7.02	12.41
O K	26.29	34.90
Mg K	2.39	2.09
Al K	64.30	50.60

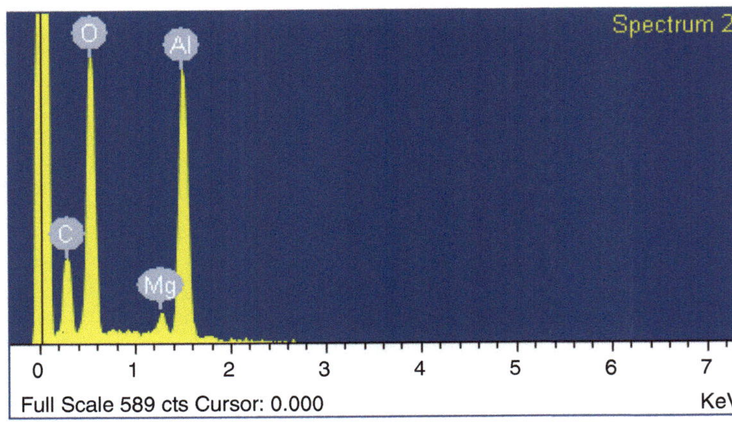

Fig. 5.5 (**a**) SEM image of CNT cluster surrounded by voids (b) EDX analysis of CNT-infiltrated specimen

References

1. Kempen, K., Thijs, L., Humbeeck, J.V., Kruth, J.P.: Mechanical properties of AlSi10Mg produced by selective laser melting. Phys. Proc. **39**, 439–446 (2012)
2. Mower, T., Long, M.: Mechanical behavior of additive manufactured, powder-bed laser-fused materials. Mater. Sci. Eng. A. **651**, 198–213 (2016)
3. Rosenthal, I., Stern, A., Frage, N.: Strain rate sensitivity and fracture mechanisms of AlSi10Mg parts produced by selective laser melting. Mater. Sci. Eng. A. **682**, 509–517 (2017)
4. Kwon, H., Estili, M.: Combination of hot extrusion and spark plasma sintering for producing carbon nanotube reinforced aluminum matrix composites. Carbon. **47**, 570–577 (2009)
5. Kim, K.T., Cha, S.I., Hong, S.H., Hong, S.H.: Microstructure and tensile behavior of carbon nanotube reinforced Cu matrix nanocomposites. Mater. Sci. Eng. A. **430**, 27–33 (2006)
6. Lahiri, D., Bakshi, S., Keshri, A., Liu, Y., Agarwal, A.: Dual strengthening mechanisms induced by carbon nanotubes in roll bonded aluminum composites. Mater. Sci. Eng. A. **523**(1–2), 263–270 (2009)
7. Patel, R.B., Liu, J., Scicolone, J.V., Roy, S., Mitra, S., Dave, R.N., Iqbal, Z.: Formation of stainless steel–carbon nanotube composites using a scalable chemical vapor infiltration process. J. Mater. Sci. **48**(3), 1387–1395 (2013)
8. Wang, P., Zhang, B., Tan, C., Raghavan, S., Lim, Y., Sun, C., Wei, J., Chi, D.: Microstructural characteristics and mechanical properties of carbon nanotube reinforced inconel 625 parts fabricated by selective laser melting. Mater. Des. **112**, 290–299 (2016)
9. Fulcher, B., Leigh, D., Watt, T.: Comparison of AlSi10Mg and Al 6061 processed through DMLS. In: Solid Freeform Fabrication Symposium, Austin (2014)
10. ASTM International: ASTM E8 / E8M – 16a – Standard Test Methods for Tension Testing of Metallic Materials. ASTM International, West Conshohocken (2016)
11. Dieter, G.E.: Mechanical Metallurgy. McGraw Hill, London (1988)
12. White, C.M., Banks, R., Hamerton, I., Watts, J.F.: Characterisation of commercially CVD grown multi-walled carbon nanotubes for paint applications. Prog. Org. Coat. **90**, 44–53 (2016)

Chapter 6
Optimization of Surface Integrity of Titanium-Aluminum Intermetallic Composite Machined by Wire EDM

S. Ezeddini, E. Bayraktar, M. Boujelbene, and S. Ben Salem

Abstract This study investigate the influence of the machining parameters on Ti-Al intermetallic composite based using a wire electrical discharge machining. This process is typically used for very hard material, which are hard to machine using a more traditional process. The aim of the work is to optimize the integrity of surface of Ti-Al intermetallic composite machined by WEDM. The first step is defined the machining parameters of the process: Start up voltage U, Pulse on time Ton or T, advanced speed S and flushing pressure P (pressure of injection of dielectric) are chosen to determine their effects on surface roughness Ra. The second step is using integrated method, which mixed Taguchi method and response surface methodology RSM, the RSM model was developed as a tool to predict the integrity of surface of Ti-Al intermetallic composite machined by WEDM. The significance of the machining parameters was obtained using analysis of variance (ANOVA) based on S/N ratio, which show that flushing pressure P, Pulse on Time T and Voltage U were the most significant parameters. Microstructure of surface and subsurface, cracks, craters, and debris and roughness surface of the samples machined at the different condition has been realized by scanning electron microscopy (SEM), and 3D-Surfscan. The integrate method of Taguchi and RSM was validated by conducting validation experiment to ensure that it can work accurately as a prediction tool.

Keywords Ti-Al Intermetallic Composite –based · WEDM · Response surface methodology RSM · Taguchi method · Surface roughness Ra · Heat Affected Zone HAZ

6.1 Introduction

The development of the metallurgical industry and the search for greater productivity have led to the use of harder materials and which have difficulties in machining and they have rendered conventional machining processes disabled. Composite intermetallic based on titanium and aluminum were discovered in the early 1950s and soon became potential materials for the aerospace, chemical and energy industries. Intermetallic TiAl-based alloys represent an important class of high temperature structural materials providing a unique set of physical and mechanical properties that can lead to substantial payoffs in aircraft engines, industrial gas turbines and automotive industry [1]. Many researchers have studied intermetallics, Paul Ervé Tchoupe Ngnekou studied in his thesis [2] the link between microstructure, oxidation and mechanical properties of intermetallic alloys based on Titanium and aluminum [3]. are studied the fatigue behaviour of two-phase ($\alpha_2 + \gamma$) TiAl alloy and showed two phases in microstructure composed of α_2-Ti_3Al and γ + TiAl and fracture mechanism is explained with different plastic incompatibilities between the two phases.

In Today's manufacturing sector, the EDM was used for machining of all types of conductive materials due to its ability to machine precise, complex, and irregular shapes of the surface electric discharge machining [4, 5]. There are researchers who have chosen to act on the parameters of the condition of machining like pulse on time, pulse off time, cutting speed, dielectric pressure flushing, duty cycle, to obtain the best surface, the most important material removal rate MRR and the smallest kerf, smallest heat affected zone HAZ and smallest residual stress [4–22]. However, the EDMed surfaces usually have defects including micro cracks and pores formed by the high temperature gradient. Other researchers have attempted to add various surfactant and powders into the dielectric fluid to further improve the EDM surface properties [6, 7]. EDM affects the surface quality, surface roughness and creation of subsurface layer as reported for various materials [8] and in detail for Ti6Al4V [9]. EDM has already been proposed as a suitable surface treatment for biomedical application due to favorable

S. Ezeddini · M. Boujelbene · S. Ben Salem
University of Tunis El Manar, ENIT, École Nationale d'Ingénieurs de Tunis, Tunis, Tunisia

E. Bayraktar (✉)
Supmeca-Paris School of Mechanical and Manufacturing Engineering, St-Ouen-France, France
e-mail: bayraktar@supmeca.fr

P. R. Thakre et al. (eds.), *Mechanics of Composite, Hybrid and Multifunctional Materials, Volume 5*, Conference Proceedings of the Society for Experimental Mechanics Series, https://doi.org/10.1007/978-3-319-95510-0_6

biocompatibility and osteointegration [10, 11]. Ilhan et al. [15] was used RSM and Taguchi orthogonal array to determine optimum experimental parameter combination to achieve the minimum surface roughness, the optimum values obtained for spindle rpm, feed rate, depth of cut and tool tip radius were respectivle, 318 rpm, 0.1 mm/rev, 0.7 mm and 0.8 mm. Pandey et al. [16] found the possibility of arcing is accurately predicted using an ANN model of suitable architecture. Amitesch et al. [17] investigate the surface integrity, material removal rate and wire wear ratio for WEDM of Nimonic 80 A alloy using grey relational analysis GRA and Taguchi method.

T_{on}	Pulse on time
P	Dielectric Pressure
WEDM	Wire electric discharge machining
U	Servo voltage
S	Speed advance
Ra	Surface Roughness
L9	Taguchi design.

6.2 Experimental Procedure

6.2.1 Equipment, Materials and Measurement

Intermetallic is known as one of the most difficult to machine. This difficulty in shaping, due to its metallurgical and mechanical characteristics. However, the Titanium has a low machinability; the surfaces are affected or damaged during the machining operation. The critical nature of the shaped parts requires thorough monitoring of the integrity of the machined surfaces.

Through this work, we first propose to explain an economical and efficient method for the manufacture of composite titanium aluminum intermetallic for the aeronautical field and to carry out tests to define the mechanical characteristics of this composite, for this purpose a combined method of powder metallurgy and thixoforming in the semi-solid state is used. In addition, we propose to recall the notion of surface and subsurface integrity. The influence of cutting conditions on surface integrity was examined after wire EDM tests. The choice of this process of machining is justified by its frequent use in the aeronautical field, the simplicity of implementation and the absence of mechanical forces during machining and the ease of observation and measurement of the integrity. The wire EDM tests were carried out under lubrication (deionized water), under machining conditions with a brass wire. Surfaces and sub-surfaces were observed and analyzed by the following means; SEM scanning electron microscope, micro-durometer, MO optical microscope, 3D surface Scan.

6.2.2 Wire Electric Discharge Machining Process and Cutting Parameters

WEDM is currently widely used in the industry for the high-precision machining of all types of conductive materials such as metals, metal alloys, graphite or even certain ceramic materials of any hardness, WEDM is used in a wide range of industries, including aerospace, biomedical, automotive, as well as molds. In this study, we are interested in the intermetallic based on titanium and aluminum, which are known to be among the most difficult materials to machine, so tests of cutting by the machine of the wire EDM were carried out on the composite material intermetallic based on titanium aluminum. In order to optimize this method, an integrated method was used which brings together the Taguchi method and the RSM response surface methodology.

The material studied in this work is the titanium aluminum composite which has been used with the chemical specification given in Table 6.1. A plate with a size of 100 × 8 × 4 mm was cut to 9 samples (Fig. 6.1). The experiments were carried out on a wire rod erosion machine Robofil 190. The fixed process parameters during the experiment were:

- Work piece: composite intermetallic TiAl-based.
- Electrode (tool): 250 μm diameter of brass wire (Fig. 6.2) and (Table 6.2).
- Dielectric: deionized water.

Table 6.1 Chemical composition

Composition	Ti	Al	Nb_2Al	Nb	Yttrium-Doped-Zirconia(Y-ZrO2)	Mo	B	Zn-St
%	53	27	4	15	1	0.1	0.15	1

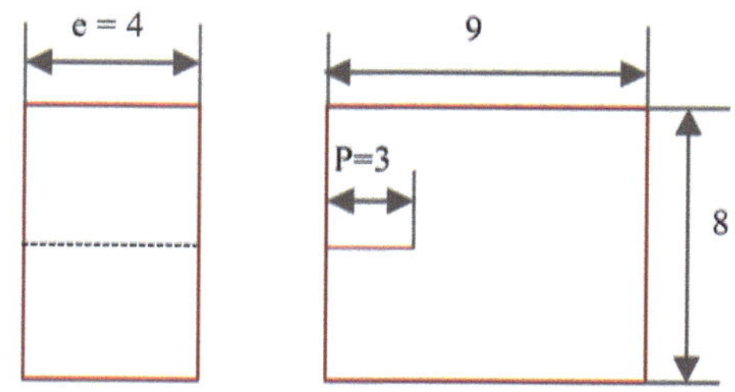

Fig. 6.1 Experimental specimen

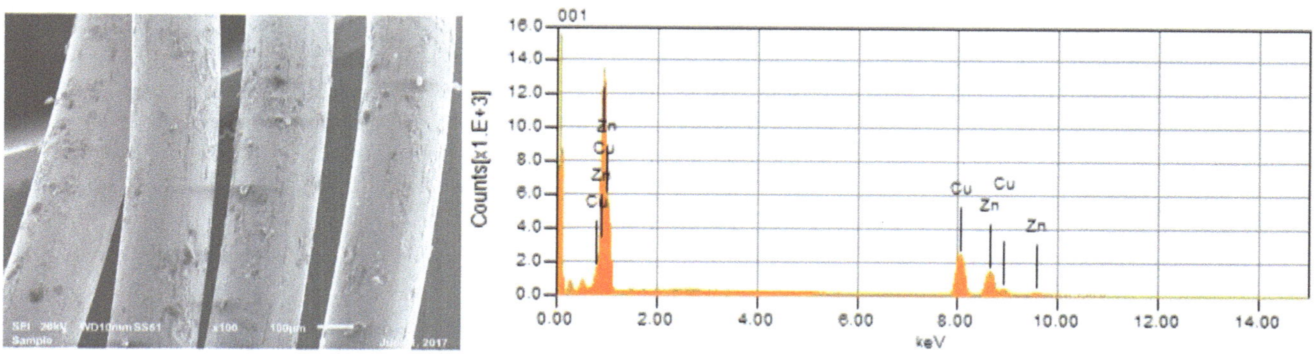

Fig. 6.2 SEM picture of wire tool and composition of wire tool from SEM analysis

Table 6.2 Chemical composition of wire tool

Composition of wire	Mass%	Atom%
Cuivre-Cu	57.59	58.28
Zinc-Zn	42.41	41.72

The chemical composition of the pieces studied was presented in Table 6.1.

The roughness of the surface was measured using surface roughness (3D-SurfaScan), the area examined was 2 × 1 mm. The level of the ranges of cut parameters and the values of the initial parameters have been chosen in the manufacturer's manual recommended for the tested material. The control factors, their symbols and their designated range are given in Table 6.5. Figure 6.4 shows the configuration of the experiment carried out during this study.

Two basic tools in the Taguchi technique are the orthogonal network (OA) and the signal-to-noise ratio (S/N). OA is used in the Taguchi method to save time and cost experiments. The S/N ratio is used to measure the deviation of the quality characteristics from the desired values, including the highest Higher-The-Better (HTB), Nominal-The-Better (NTB) and Lower-The-Better LTB). In [20] they studied the variation of kerf and rate of removal of MRD WEDM materials by using ANOVA and SN ration. In this study we aim to optimize the surface roughness Ra, so that the most objective type of objective function is Lower-The-Better (STB) were used. The exact relationship between the S / N ratio and the signal is given by the following Eq. (6.1):

LTB: Eq. (6.1)

$$\frac{S}{N} = -10 * \log\left(\sum_{i=0}^{n}\left(1/Yi^2\right)/n\right) \tag{6.1}$$

Where n is the number of experiments, and y_i is the value of the Ra

The Response Surface Method (RSM) proposed by Box and Wilson (1951) in the early 1950s received considerable attention because of its good empirical performance in modeling. It is a set of mathematical and statistical techniques that provide adapted models between input parameters and responses to develop, improve and optimize a process. Several researches have used the RSM method to improve and optimize the performance of machining processes such as surface integrity (roughness, heat affected zone, residual stress), MRR material removal rate, thread wear TWR, kerf [13–15, 18,

Table 6.3 The basic Taguchi L9 (3^4) orthogonal array

	Control factors and levels				Coding				Results	
Run	U(V)	Ton(μs)	S(mm/min)	P(bar)	X1	X2	X3	X4	Ra(μm)	HAZ (μm)
1	80	0.8	29	60	−1	−1	−1	−1	2.92	9.12
2	80	0.9	36	80	−1	0	0	0	1.93	11.1
3	80	1	43	100	−1	1	1	1	1.74	7.85
4	100	0.8	36	100	0	−1	0	1	1.96	8.99
5	100	0.9	43	60	0	0	1	−1	2.48	12.13
6	100	1	29	80	0	1	−1	0	2.01	12.24
7	120	0.8	43	80	1	−1	1	0	2.48	12.62
8	120	0.9	29	100	1	0	−1	1	1.8	10
9	120	1	36	60	1	1	0	−1	2.84	12.4

21]. The most extensive application of RSM is found in the industrial world, especially in the case where several input variables potentially influence the performance measures or the product or process quality characteristics [21]. This study seeks an appropriate approximation method for analyzing the surface roughness relationship Ra with respect to independent input parameters. A mathematical equation of the second order polynomial response surface is used as indicated in the equation; the coefficients of the function can be obtained by the least squares method

$$Y = \beta_0 + \sum\nolimits_{i}^{n} \beta_i X_i + \sum\nolimits_{i<j}^{n} \sum \beta_{ij} X_i X_j + \sum \beta_{ij} X_i^2 + \varepsilon \tag{6.2}$$

Where y denotes Ra, x represents the parameters of WEDM (T_{on}: pulse time, U: start up voltage or servo voltage, S: feed rate or speed advance and P: flushing pressure or dielectric injection pressure), B_{ij} is the coefficients of each term, ε is a residual error.

Table 6.4 shows the basic Taguchi L9 orthogonal array (Table 6.3).

6.2.3 *Influence of Cutting Parameters on the Surface Roughness*

Increasing the discharge energy and increasing pulse duration removing more melted material and generating deeper and larger discharge craters, increasing the roughness of the surface.

The term surface defect encompasses a number of forms of alterations in the surface produced as a result of machining of the material [18–22]: advance marks, traces, tears, tearing, micro and macro cracks, glued particles, Sliding zones, micro-chip debris, deformed grains, cavities, texturing of the surface These alterations generated by the machining operation can be minimized by optimizing the cutting conditions [22]. Figure 6.3 show the surface quality of two WEDM-cut samples with different cutting conditions. The images of the figure illustrate what has been concluded, the surfaces which are machined with the lowest pressure P = 60 bar are the roughest compared with the surfaces machined with the high pressure P = 100 bar in a fix value of feed rate 36 S = mm/min. Figures 6.4, and 6.5 show the topographies of the 3D surfaces realized on the Mountains software of samples 4 and 9. The images taken by the SEM electron scanning microscope of samples 4 and 9, then the surface machined with a lower pressure P = 60 bar comprises larger craters And more numerous compared with the surface machined with a higher pressure P = 100 bar which comprises craters of very small and smaller sizes.

The surface is characterized by juxtaposed craters, resulting from the evaporation of the material during the machining, according to the images in Fig. 6.3 which are taken by the SEM it is seen that the craters are deeper and the irregularities are larger in the surface (b) machined with a voltage U = 120 V, pulse duration Ton = 1 μs and a dielectric injection pressure P = 60 bar on the same image, the presence of deep cracks and very little debris in the same Fig. 6.3 on the surface (a) which has machined with a smaller voltage U = 100 V, and pulse on time Ton = 0.8 μs, and a greater pressure P = 100 bar, on this surface Less debris, less cracks with shallower craters because the high dielectric injection pressure helps to evacuate as much fused material as possible and thereafter to minimize the number of debris which gives the irregularity of the surface and the results of roughness of the two surface showed in Fig. 6.6, Ra = 1.56 μm for the surface (a) and Ra = 2.84 μm for the surface (b).

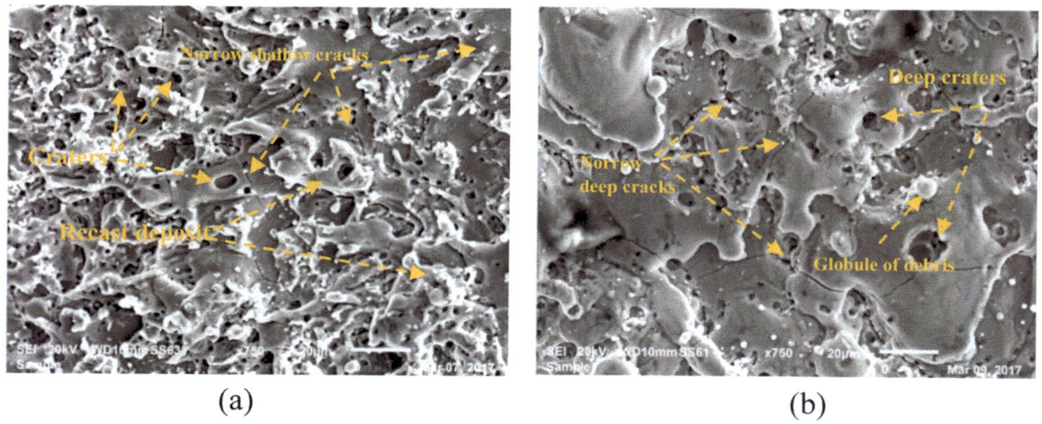

Fig. 6.3 SEM observation of surface after electric discharge machining at machining parameters: (**a**):U = 100V Ton = 0.8 µs, S = 36 mm/min, P = 100 bar; (**b**):U = 120V Ton = 1 µs, S = 36 mm/min, P = 60 bar

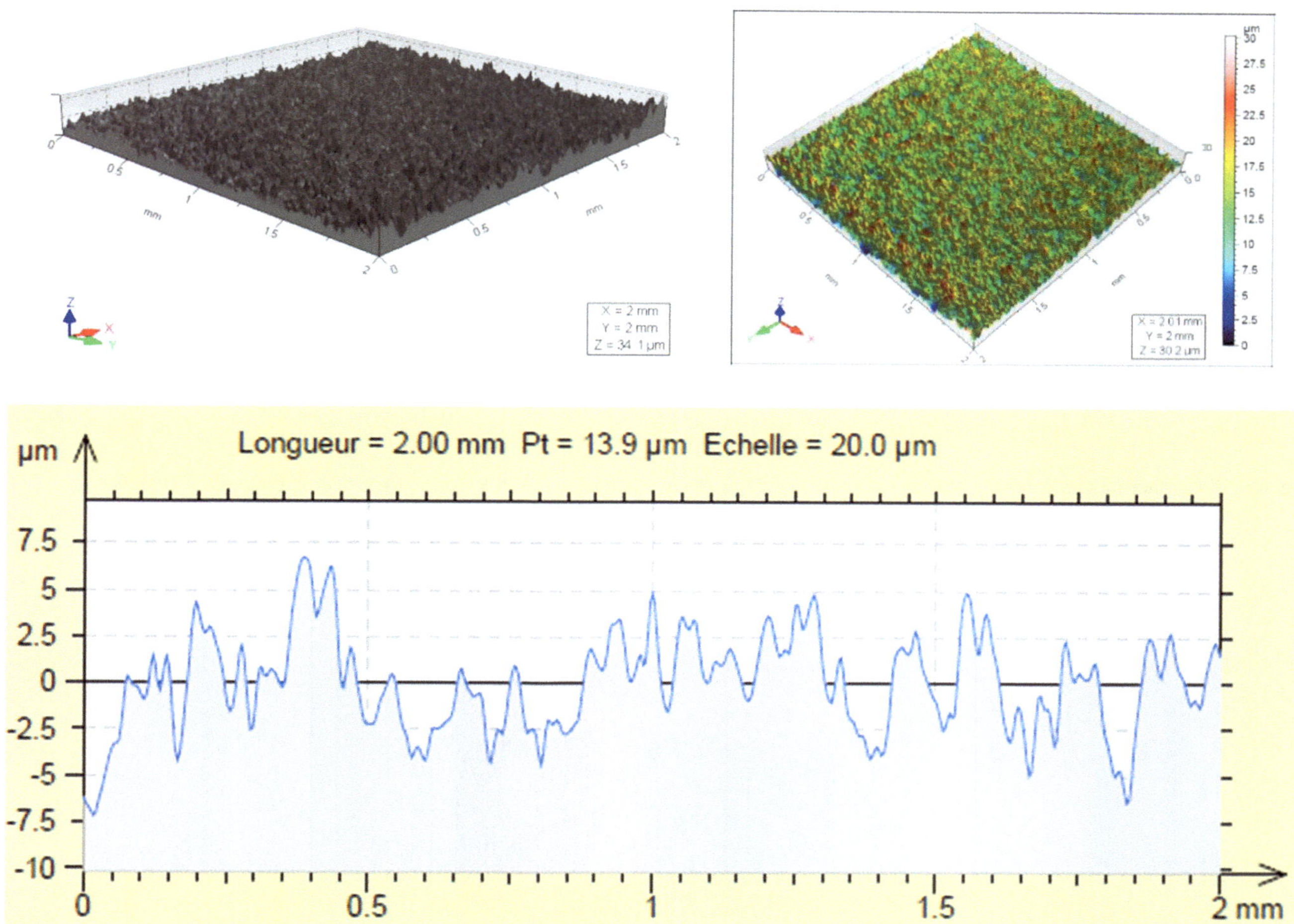

Fig. 6.4 3D surface topographies for work pieces profile of surface cutting edge at U = 100 V Ton = 0.8 µs, S = 36 mm/min, P = 100 bar

Figure 6.6 showed the difference between the value of roughness surface Ra for the surface at machining condition different, at U = 100 V Ton = 0.8 µs, S = 36 mm/min, P = 100 bar, Ra = 1.56 µm and Ra = 2.84 µm at U = 120 V Ton = 1 µs, S = 36 mm/min, P = 60 bar

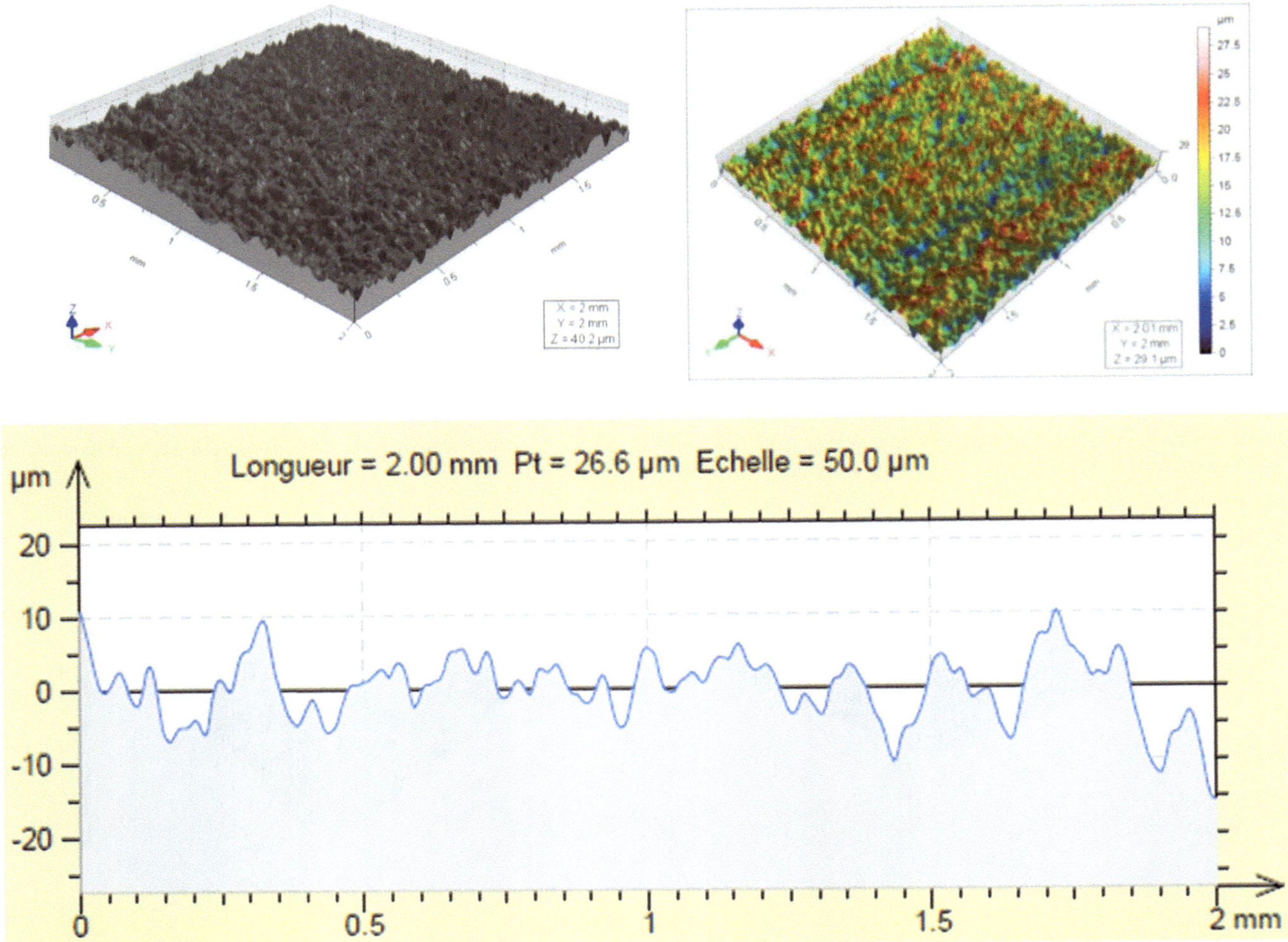

Fig. 6.5 3D surface topographies for workpiece profile of surface cutting edge at U = 120 V Ton = 1 μs, S = 36 mm/min, P = 60 bar

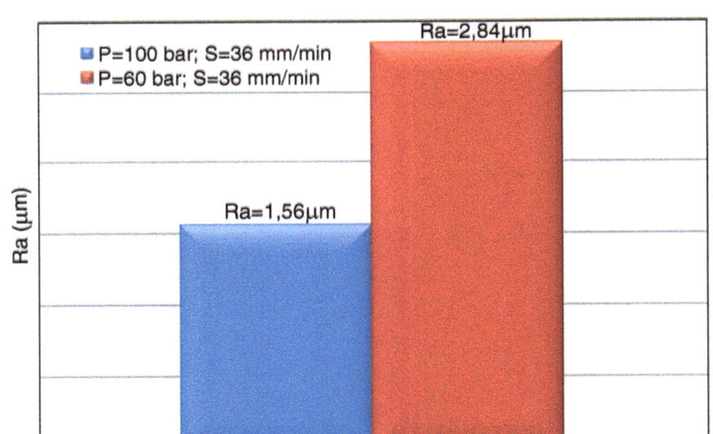

Fig. 6.6 Value of roughness at U = 100 V Ton = 0.8 μs, S = 36 mm/min, P = 100 bar and U = 120 V Ton = 1 μs, S = 36 mm/min, P = 60 bar

Figure 6.7 shows the pictures taken by the SEM electron scanning microscope of samples 2, 6 and 9, present the surfaces which are machined with the lowest speed advance S = 29 mm/min are the roughest compared with the surfaces machined with the middle value of speed advance S = 36 mm/min and the surface machined with the high value of speed advance S = 42 mm/min and we fixed the value of P = 80 bar to view the influence of the feed rate on the roughness surface.

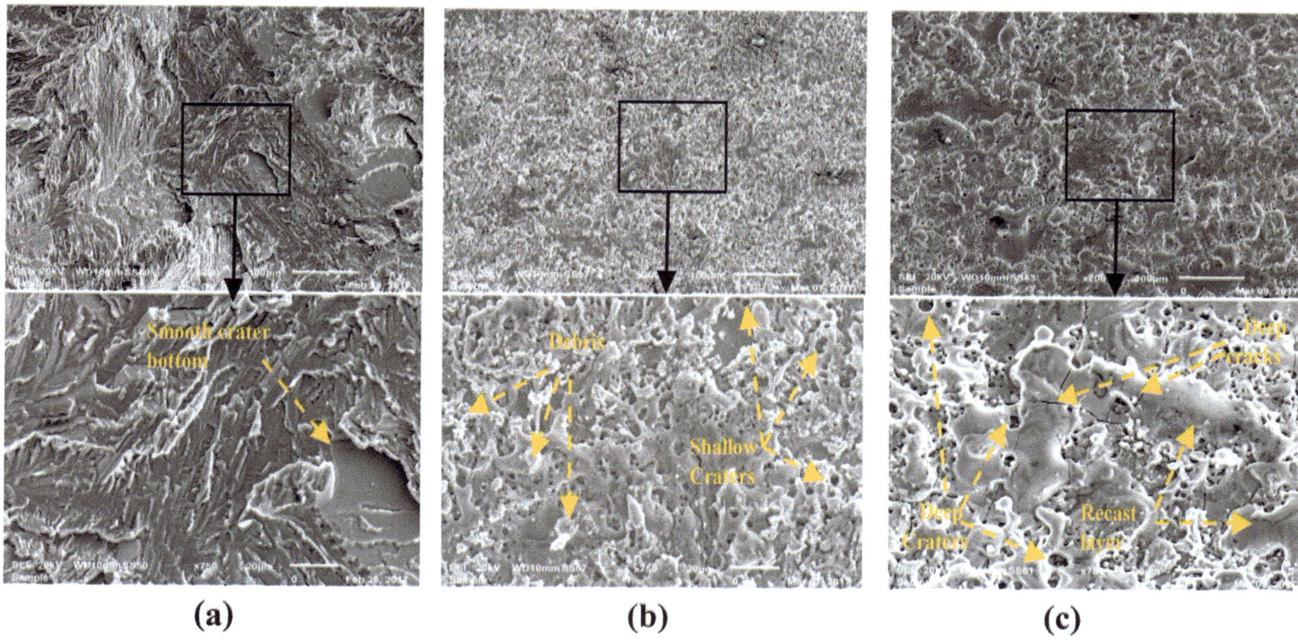

Fig. 6.7 SEM observation of surface after electric discharge machining at machining parameters (**a**):U = 80V Ton = 0.9 μs, S = 36 mm/min, P = 80 bar, P2; (**b**):U = 100V Ton = 1 μs, S = 29 mm/min, P = 80 bar, P6; (**c**):U = 120VTon = 0.8 μs, S = 43 mm/min, P = 80 bar, P7

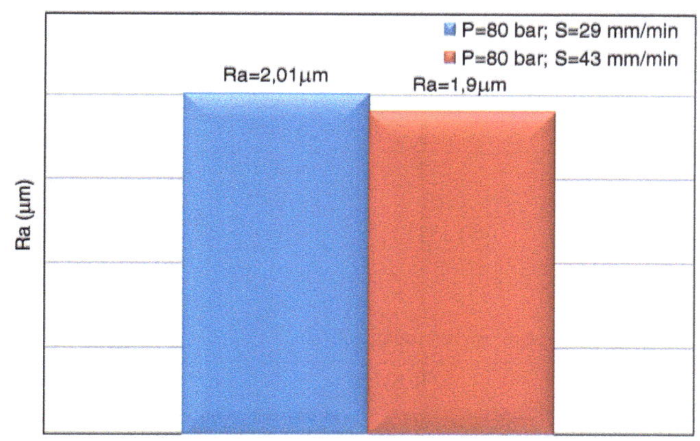

Fig. 6.8 Value of roughness at U = 100 V, Ton = 1 μs, S = 29 mm /min, P = 80 bar and at U = 120 V, Ton = 0, 8 μs, S = 43 mm/min, P = 80 bar

Figure 6.8 showed the difference between the value of roughness surface Ra for the surface at machining condition different, at S = 29 mm/min, P = 80 bar, Ra = 2.01 μm and Ra = 1.9 μm at S = 43 mm/min, P = 80 bar.

6.3 Statistical Analysis

The S/N ratio was used to determine the optimum parameters for a good surface roughness in WEDM machining of Ti-Al intermetallic compositeand according to Fig. 6.9.

Table 6.4 shows response table for means and signal to noise ratio for Ra and HAZ of intermetallic based of Ti-Al material successive. This response table represents the effects of various input factors on Ra. Higher the slope in the main effects plot corresponding values of delta is higher in the response table. The rank represents directly the level of effect of input based on the values of delta. Here according to ranks, the effects of various input factors on Ra in sequence of its effect are power and Speed. That means power affects the Ra at highest level and speed at lowest level.

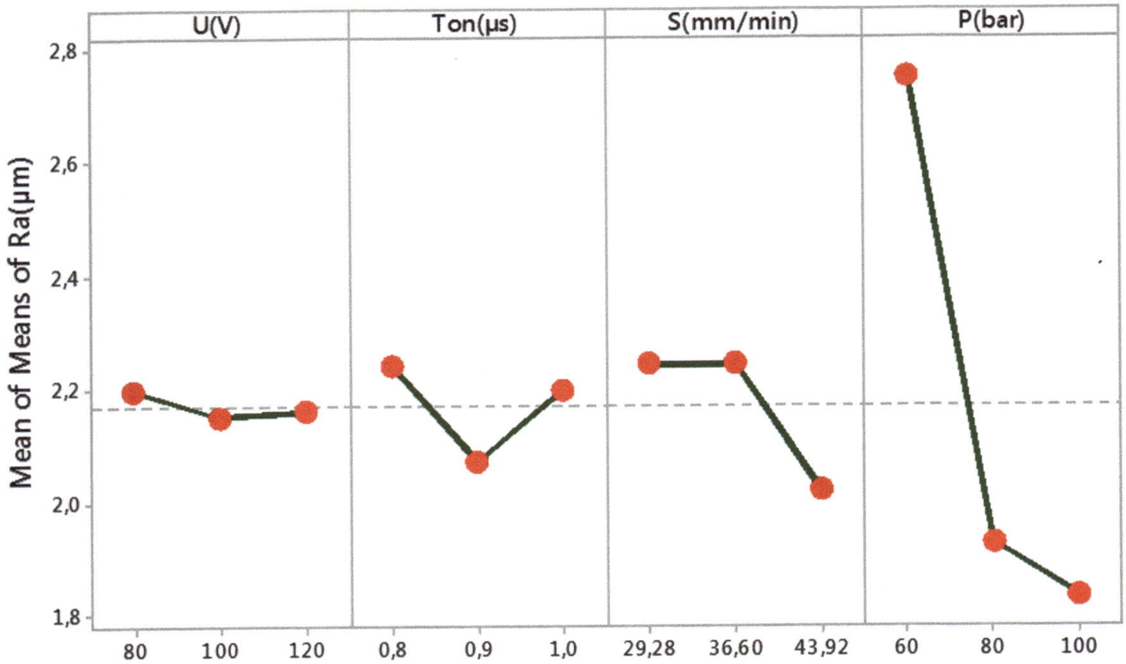

Fig. 6.9 The effect of machining parameters on surface roughness (Ra)

Table 6.4 Response table for signal to noise ratios smaller is better and for means of (Ra)

		Means				S/N Ratio			
Process Parameters	Level	U(V)	Ton(µs)	S(mm/min)	P(bar)	U(V)	Ton(µs)	S(mm/min)	P(bar)
Average value	L1	2.197	2.240	2.243	2.747	−6.610	−6.816	−6.826	−8.754
	L2	2.150	2.070	2.243	1.927	−6.599	−6.235	−6.874	−5.690
	L3	2.160	2.197	2.020	1.833	−6.489	−6.647	−5.999	−5.254
	Delta	0.047	0.170	0.223	0.913	0.121	0.581	0.875	3.501
	Rank	4	3	2	1	3	2	4	1

Table 6.5 S/N ratio means and predicted S/N ratio and predicted means of (Ra)

	Control factors and levels				Results					
Run	U(V)	Ton(µs)	S(mm/min)	P(bar)	Ra	Pred Ra	SNRA	PSNRA	Mean	P Mean
1	80	0.8	29	60	2.92	2.9288	−9.30847	−9.30847	2.92	2.92
2	80	0.9	36	80	1.93	1.9475	−5.71301	−5.71301	1.93	1.93
3	80	1	43	100	1.74	1.7400	−4.81328	−4.81328	1.74	1.74
4	100	0.8	36	100	1.96	1.9425	−5.84523	−5.84523	1.96	1.96
5	100	0.9	43	60	2.48	2.4713	−7.88967	−7.88967	2.48	2.48
6	100	1	29	80	2.01	1.9838	−6.06435	−6.06435	2.01	2.01
7	120	0.8	43	80	2.48	1.8487	−5.29649	−5.29649	1.84	1.84
8	120	0.9	29	100	1.8	1.8175	−5.10666	−5.10666	1.80	1.80
9	120	1	36	60	2.84	2.8400	−9.06723	−9.06723	2.84	2.84

Table 6.5 shows the values of signal to noise ratio (SNRA) and Predicted signal to noise ratio (PSNRA) for Ra and HAZ of intermetallic-based Ti-Al material successive. The values of predicted signal to noise is very much close to the calculated signal to noise values hence the analysis of Taguchi for signal to noise ratio is correct, it's perfect. The representation of effects of various parameters on Ra and optimize condition is very much nearby.

Figure 6.9 shows the influence of each cutting parameter: the pulse on time, the starting voltage, the feed rate and the injection pressure. However, according to the curves of the effects, the pressure has the greatest influence on the roughness values and the value P3 = 120 bar corresponds to the smallest roughness value compared with P1 and P2. In addition, the speed S3 corresponds to the smallest roughness value with compared with S1 and S2, the pulse time $T_{on}2$ corresponds to the smallest roughness value compared with $T_{on}1$ and $T_{on}3$.

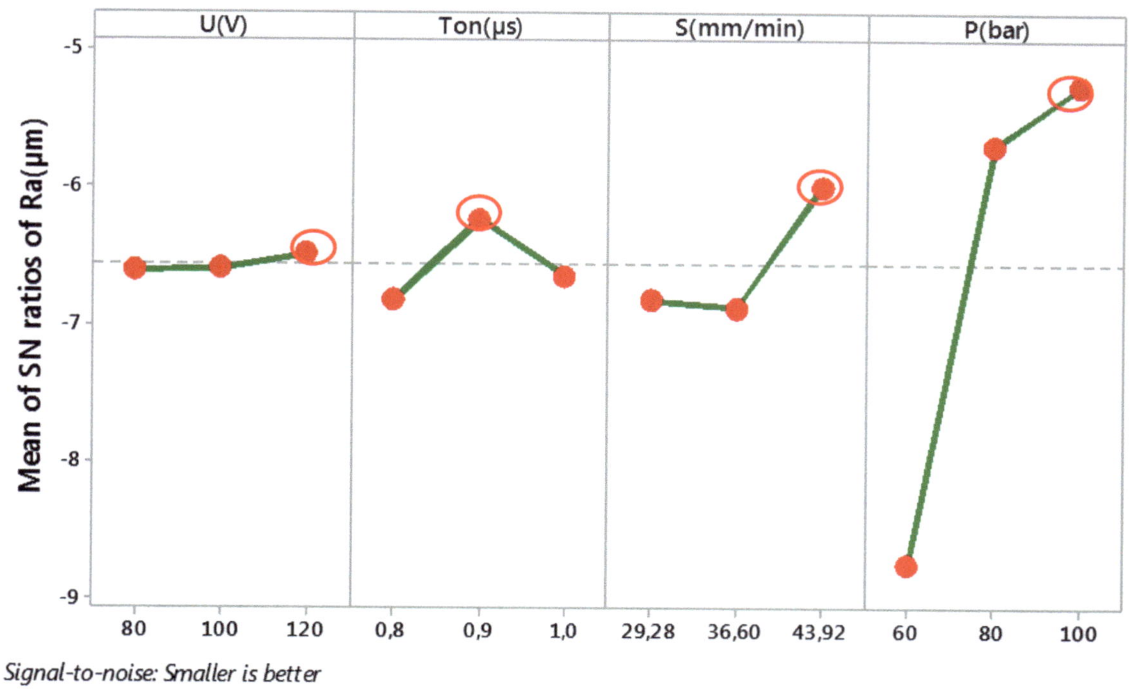

Fig. 6.10 Main effects plot for SN ratios for Ra

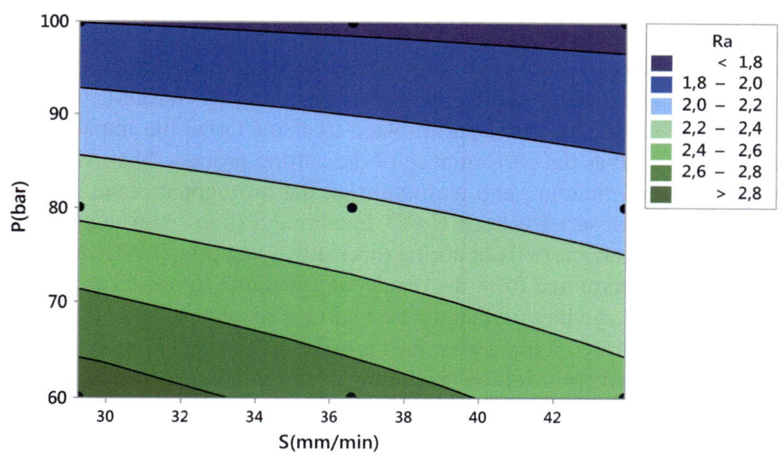

Fig. 6.11 Contour plots for roughness surface Ra

The S/N ratio was used to determine the optimum parameters for a smaller value of Ra in WEDM machined surface of Ti-Al intermetallic compositeand according to Fig. 6.10 and Table 6.4 the optimal level of the machining parameters is the level with the greatest S/N ratio.

Regression Equation

$$Ra\ (\mu m) = 20.30 - 0.0151 \times U - 26.92T_{on} + 0.1373 \times S - 0.16817 \times P$$
$$+ 0.00071 \times U^2 + 14.83 \times T_{on}{}^2 - 0.002084 \times S^2 + 0.000908\times \quad (6.3)$$

Figure 6.11 shows the effect of speed S and flushing pressure P on Ra. As is clear, smaller values of Ra can be obtained by selecting greater speed and higher values of flushing pressure.

The RSM models are carried out with 9 experimental tests. For each combination of the input factors, the prediction value of the response $Y_{j,pred}$ is compared with the experimental value of the response $Y_{j,exp}$. Figure 6.12 illustrates the comparison

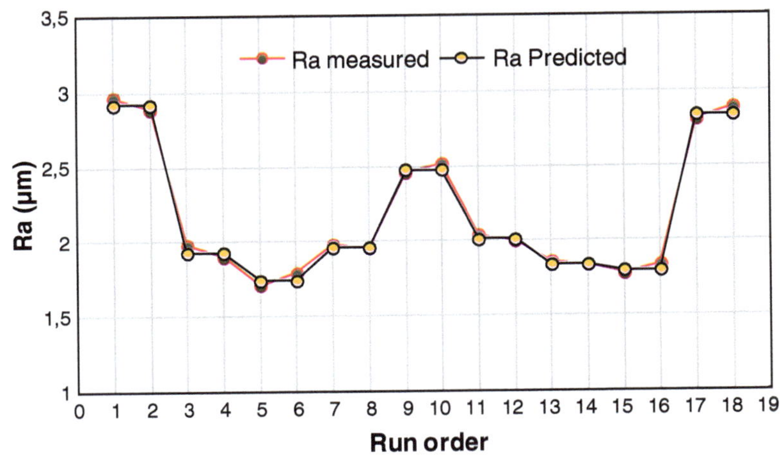

Fig. 6.12 Comparison between measured and predicted values for surface roughness Ra

between the experimental and predictive response values, comparing predictive and experimental surface roughness to test the interpolation of the prediction of the models developed by RSM. We conclude that the predictive results with RSM are very close to the experimental results.

6.4 Conclusion

For obvious safety reasons, components of the aluminum-based composite aeronautics are sensitive parts for which all precautions must be taken when shaping them, especially when machining by WEDM. Good surface integrity are to be sought. Through this work, a set of means has been put in place to characterize the impact of WEDM wire EDM on the integrity of the surfaces produced as well as the performance of the cutting process. Nine WEDM cutting tests were carried out on the composite material based on titanium and aluminum on the orthogonal plane of Taguchi L9, by changing the cutting parameters (pulse time TON (µs), servo voltage U (V), Feed rate S (mm / min) and flushing pressure P (bar)) three tests for each factor, so 9 combinations were carried out during this study. The optimal WEDM cutting conditions of titanium-aluminum intermetallic composite is determined from the results of the signal-to-noise ratio in a Taguchi plane of different inputs on each performance. However, it can be said that the best surface roughness obtained is for the following conditions: a voltage of $U = 120$ V, a pulse time $Ton = 0.9$ µs, a wire feed speed $S = 43$ mm / min and a pressure $P = 100$ bar. Finally, the integrated method used in this paper to optimize the machining of intermetallic composite by WEDM brings together the Taguchi method and the RSM method, which has given us interesting models Can use it to predict desired performance such as; The surface roughness.

References

1. Paul Ervé TCHOUPE NGNEKOU Le 29 Avril 2010.thése : Microstructure, oxydation et propriétés mécaniques d'alliages intermétalliques à base de TiAl, Université de Toulouse
2. Lapin, J.: TiAl-based alloys: Present status and future perspectives, Hradec nad Moravici (2009)
3. Bayraktar, E., Bathias, C., Xue, H., Hao, T.: On the giga cycle fatigue behavior of two-phase(alpha 2+gama) TiAl alloy. Int. J. Fatigue. **26**, 1263–1275
4. Baraskar, S.S., Banwait, S.S., Laroiya, S.C.: Multiobejctive optimization of electrical discharge machining process using hybrid method. Mater. Manuf. **28**, 348–354 (2013)
5. Saedon, J.B., Jaafer, N., Jaafar, R., Saad, N., Kasim, M.S.: Modeling and multi-response optimization on WEDM Ti6Al4V. Appl. Mech. Mater. **510**, 123–129 (2014)
6. Sidhu, S.S., Batish, A., Kumar, S.: Study of surface properties in particulate-reinforced metal matrix composites (MMCs) using powder-mixed electrical discharge coating (EDM). Mater. Manuf. Process. **29**, 46–52 (2014)
7. Janmanee, P., Muttamara, A.: Surface modification of tungsten carbide by electrical discharge coating (EDC) using a titanium powder suspension. Appl. Surf. Sci. **258**, 7255–7265 (2012)

8. Kumar, S., Singh, R., Singh, T.P., Sethi, B.L.: Surface modification by electrical discharge machining/ a review. J. Mater. Process. Technol. **209**, 3675–1208 (2009)
9. Reissig, L., Volkl, R., Mills, M.J., Glatzel, U.: Investigation of near surface structure in order to determine process-temperature during different machining process of Ti-6Al-4V. Scr. Mater. **50**, 121–126 (2004)
10. Peng, P.W., Ou, K.L., Lin, H.C., Pan, Y.N., Wang, C.H.: Effect of electrical discharging on formation of nanoporous biocompatible layer on titanium. J. Alloys Compd. **492**, 625–630 (2010)
11. Bigerelle, M., Anselme, K., Noel, B., Ruderman, I., Hardouin, P., Lost, A.: Improvement in the morphology of Ti-based surfaces: A new process to increase in vitro human osteoblast response. Biomaterials. **23**, 1563–1577 (2002)
12. Oktem, H.: An integrated study of surface roughness for modeling and optimization of cutting parameters during end milling operation. Int. J. Adv. Technol. **43**, 852–861 (2009)
13. Jeyakumar, S., Marimuthu, K., Ramachandran, T.: Optimization of machining parameters of Al6061 composite to minimize the surface roughness- modelling using RSM and ANN. Indian J. Eng. Mater. Sci. **22**, 29–37 (2015)
14. Kanagarajan, D., Karthikeyan, R., Palanikumar, K., Sivaraj, P.: Influence of process parameters on electric discharge machining of WC/30%Co composite. Proc. IMechE B J. Eng. Manuf. **222**, 807–815 (2009)
15. Asilturk, I., Neseli, S., Ince, M.A.: Optimisation of parameters affecting surface roughness of Co28Cr6Mo medical material during CNC lathe machining by using the Taguchi and RSM methods. Measurement. **78**, 120–128 (2016)
16. Pandey, A.B., Brahmankar, P.K.: A method to predict possibility of arcing in EDM of TiB_{2p} reinforced ferrous matrix composite. Int. J. Adv. Manuf. Technol. **86**, 2837–2849 (2016)., GrossMark.
17. Goswami, A., Kumar, J.: Investigation of surface integrity, material removal rate and wire wear ratio for WEDM of Nimonic 80A alloy using GRA and taguchi method. Eng. Sci. Technol. Int. J. **17**, 173 184 (2014)
18. Assarzadeh, S., Ghoreishi, M.: Statistical modeling and optimization of process parameters in electro-discharge machining of cobalt-bonded tungsten carbide composite (WC/6%Co). Procedia CIRP. **6**(2013), 463–468 (2013)
19. Farnaz Nourbakhsh, K.P., Rajurkar, A.P.M., Cao, J.: Wire electro-discharge machining of titanium alloy. Procedia CIRP. **5**, 13–18 (2013)
20. Tosun, N., Cogun, C., Tosun, G.: A study on kerf and material removal rate in wire electrical discharge machining based on taguchi method. J. Mater. Process. Technol. **152**, 316–322 (2004)
21. Rajesh, R., Dev Anand, M.: The optimization of electro-discharge machining process using response surface methodology and genetic algorithms. Procedia Eng. **38**, 3941–3950 (2012)
22. Chalisgaonkar, R., Kumar, J.: Multi-response optimization and modeling of trim cut WEDM operation of commercially pure titanium (CPTi) considering multiple user's preferences. Eng. Sci. Technol. Int. J. **18**, 125–134 (2015)

Chapter 7
Design of Cost Effective Epoxy + Scrap Rubber Based Composites Reinforced with Titanium Dioxide and Alumina Fibers

A. B. Irez, I. Miskioglu, and E. Bayraktar

Abstract In last decades, aerospace and automotive industries are in search of multi-functional high performance, low cost materials due to certain environmental regulations. Epoxy-recycled rubber based structural composites (ERCs) are used in these type of engineering applications thanks to their favorable properties such as corrosion resistance, low cost and light weight. In addition, the use of recycled materials gives an economic and environmental aspect to the manufacturers. The data for basic material parameters of these composites is essential in order to realize an efficient engineering development process. For this reason, this paper is focused on the design of ERCs reinforced with ceramic powders in different ratios in a matrix of epoxy-fresh scrap rubber. The mechanical and some physical properties of these composite systems were studied in this research. Titanium dioxide (titania-TiO_2) and alumina fibers (Al_2O_3) are used as reinforcements in pre-defined weight percentages. During this study, mechanical and wear properties of these composite systems are studied. Three-point bending tests and nanoindentation were conducted to evaluate mechanical properties. After that, wear resistance is examined by means of nano-scratch tests. As the final step, fracture surfaces were observed with scanning electron microscopy (SEM) to identify damage mechanisms of these composites.

Keywords Nano indentation · Epoxy – Rubber · Alumina fibers · Titanium oxide · SEM · Recycling

7.1 Introduction

In the last decades, industry is in search of high performance thermosetting and thermoplastic polymer composites. In this regard, epoxy and its matrix composites gained ground in marine, automotive and aeronautical industries thanks to their decent properties as high specific tensile strength and stiffness, high fatigue endurance, high damping and low thermal coefficient [1]. However, ecological and economical concerns brings some restrictions to the manufacture of these composites. In this regard, recycling is thought as a good option to answer these constraints.

The modification of epoxy resins is generally done in different modes. First mode is that the addition of hard particles such as glass beads and fibers. Second mode is the inclusion of thermoplastics or elastomeric materials [2]. These modifiers can change the mechanical properties of epoxy matrices by modifying the molecular architecture and structure. For instance, a highly-crosslinked epoxy resin exhibits high brittleness and because of the restriction of the plastic deformation it has poor resistance to crack initiation and growth. Modifiers less rigid than the matrix can serve as unique tougheners which enhance the ductility. They can activate yielding processes because of the reduction of the local yield stress. Therefore, a significant amount of energy is dissipated within the plastic zone near the crack tip [3]. In this regard, in order to improve the fracture toughness of epoxies, adding a rubber phase such as commercially available styrene butadiene rubber with different functional terminal groups has been implemented by different researchers [4]. The volume fraction of the toughening rubber phase generally varies from 5% to 20%. Nevertheless, the incorporation of rubber phase increases the viscosity of the epoxy resin mixture whereas the crosslink density, elasticity modulus and tensile strength diminishes. Last but not least,

A. B. Irez
CentraleSupélec, Université Paris-Saclay, Gif-sur-Yvette, France

I. Miskioglu (✉)
Michigan Technology University, Engineering Mechanics Department, Houghton, MI, USA
e-mail: imiski@mtu.edu

E. Bayraktar
Supmeca-Paris, School of Mechanical and Manufacturing Engineering, Saint-Ouen, Paris, France

P. R. Thakre et al. (eds.), *Mechanics of Composite, Hybrid and Multifunctional Materials, Volume 5*, Conference Proceedings of the Society for Experimental Mechanics Series, https://doi.org/10.1007/978-3-319-95510-0_7

the incorporation of some inorganic – particulate fillers such as alumina [5–8] and titania [9] has some positive effects on resin stiffness [10] and wear resistance [5].

In addition to previously mentioned properties of the fillers, TiO_2 has specifically attracted attention due to its good optical, dielectric, electronic, structural, mechanical and anticorrosion properties [11]. TiO_2 is easily found in the market and it can effectively improve the wear resistance [12] of a matrix material. Also, as the second main reinforcement, alumina has good thermal conductivity, inertness to most acids and alkalis, high adsorption capacity, thermal stability and electrical insulation, and so on. Also, it is inexpensive, non-toxic and highly abrasive [13–15].

The present work presents processing of recycled rubbers blended with epoxy resin to create novel composites in an economic way. Main objective of this research is to determine the mechanical properties of these composites. During this study, mechanical properties were determined by bending tests with notched and smooth specimens. After that, nanoindentation tests were examined to see the time dependent behaviour and wear characteristics. In addition, surface hardness was determined by means of Shore D hardness measurement. At the end, fracture surfaces were observed by means of scanning electron microscopy (SEM) to study the toughening and damage mechanisms.

7.2 Experimental Conditions

7.2.1 Materials Processing

In the frame of the present work, a new combined method was used to devulcanize fresh scrap rubber (SBR) with a particle size varying from 30 to 130 μm. This combined method contains a pre-chemical treatment of rubbers followed by exposing them to microwave heating for a short time that is applied in two stages. By this way, a good cohesion is achieved at the interface between epoxy resin and rubber powders to improve the properties of the recycled rubber coming from the manufacture of sportive equipment which are free of contaminants.

The principal of the chemical treatment consists of a short silanization process followed by acrylic acid and a small amount of toluene solution to activate the surface of the rubber particles. After drying of the chemical treated rubber powders, they are exposed to short microwave heating in two stages for 4 min in order to avoid degradation of the main chains.

In the final stage, devulcanized rubber was mixed with epoxy resin powder in different ratios as a matrix after that the new designed composites are manufactured using classical powder metallurgy methods. The reinforcements TiO_2 and Al_2O_3 fibers were added to the matrix in pre-defined ratios and mixed to obtain a homogeneous compound, then the composite specimens were manufactured by double uniaxial action press at a temperature of 180 °C under a load of 50 kN. The dwell time at this step was 30 min. All of the specimens (30/50 mm in diameter) were cooled slowly in the press. Post-curing of the specimens were done isothermally at 80 °C for 24 h (Fig. 7.1).

The compositions designed for two types of the composites in this work (hereafter called as ERAT I-II) are given in Table 7.1.

General compositions of all the composites manufactured are given in the Table1 with specified weight ratios for each constituent.

Fig. 7.1 Picture of the specimens after compacting and post curing (TiO_2 + Al_2O_3+ epoxy-rubber) d = 30 and 50 mm

Table 7.1 Composition of the epoxy-rubber based composites

Epoxy-Rubber based composition Epoxy – SBR rubber (10 phr)	ERAT I	ERAT II
Reinforcements (wt %)	5 Alumina fiber (Al_2O_3)10 Titanium dioxide	10 Alumina fiber (Al_2O_3)10 Titanium dioxide

Table 7.2 Hardness values of ERAL specimens

Hardness measurement	
Specimen	Shore D
ERAT I	86
ERAT II	91

7.2.2 *Experimental Procedure*

Microstructure and fracture surface damage analyses have been carried out by means of optical (OM) and scanning electron microscopy (SEM). SEM observation was realized on fracture surface of the tested specimens with Scope/JSM-6010LA Jeol® electron microscope.

After post curing, hardness measurements on the specimens were performed. Shore D hardness test measurements on the polished flat surfaces of the specimens were carried out according to ASTM D 2240 using Shore D hardness tester, (type HBD-100-0). Hardness results were given in Table 7.2. Increased hard particles content resulted in harder surface.

There point bending tests (3 PB) have been carried out according to ASTM D790 standards. Deflection of the specimen was measured by the crosshead position and crosshead speed was selected as 1 mm/min. Flexural strength and strain were obtained from the test results. The test specimen was placed in the Instron 5569 bending test fixture and force was applied until fracture of the specimen occurred.

Creep tests using a nanoindenter were performed on the two compositions manufactured to see the time dependent behaviour. On each sample 25 indents were performed on a 5x4 grid with a Berkovich indenter. The indents were spaced 50 μm along the 5-indent side and 75 μm along the 4-indent side. The load was increased at a rate of 1 mN/s to the max load and kept at the maximum load for 500 s then unloaded.

Scratch testing capability of a nanoindenter was utilized to perform relatively fast wear tests to compare the wear behavior of the different samples. Wear tests were conducted by a conical tip with a 90° cone angle. Tests were run under a normal load of 20 mN applied over a linear track of 500 nm for 50 cycles.

7.3 Results and Discussions

7.3.1 *Microstructure of the Composites*

General microstructure is shown in the Fig. 7.2 after sectioning of the specimens. Final microstructure showed a considerably homogenous distribution of the reinforcements. White circular particles in Fig. 7.2 show the alumina fibers in different sizes. Some small local agglomerations are observed in the structure due to the mixture process. Since, homogenous distributions of the reinforcements need longer mixing time.

7.3.2 *Three Point Bending Tests and Fracture Surface Observation*

Three-Point Bending (3 PB) tests have been carried out for each different type of composites.

Flexural stress is calculated during three-point bending test according to Eq. 7.1:

In this formula, l is the span length, P is the maximal bending load, b and h are the specimen thickness and depth, respectively.

$$\sigma = \frac{3Pl}{2bh^2} \tag{7.1}$$

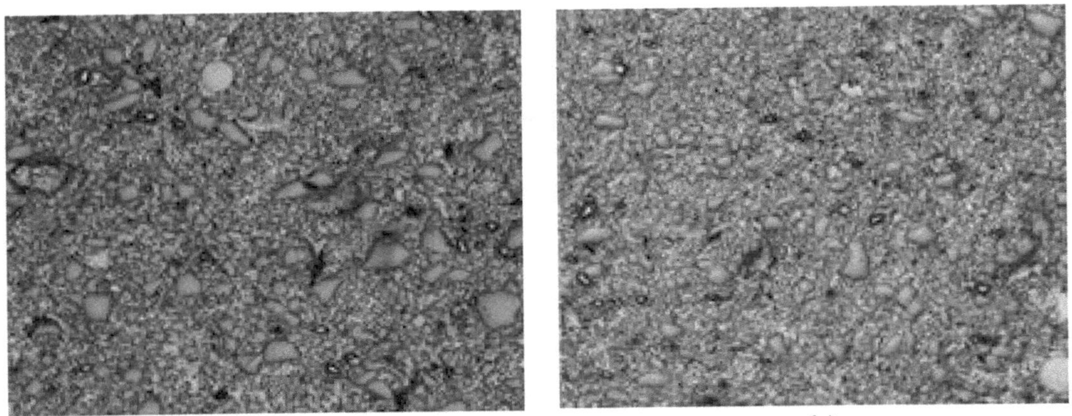

a) b)

Fig. 7.2 Microstructure of the two different composition for (**a**) ERAT I and (**b**) ERAT II, respectively

Table 7.3 Comparison of mechanical properties of ERAL specimens

	Flexural stress (MPa)	Flexural modulus (GPa)	Strain at break (ε %)
Matrix	42.13 ± 6.01	7.33 ± 1.41	0.66 ± 0.11
ERAT I	39.76 ± 3.20	5.05 ± 0.33	0.76 ± 0.07
ERAT II	36.15 ± 2.11	3.85 ± 0.24	1.01 ± 0.08

In this formula, l is the span length, P is the maximal bending load, b and h are the sample width and thickness, respectively.

Flexural strain, ε_f, was determined according to Eq. 7.2:

$$\varepsilon_f = \frac{6Dh}{l^2} \tag{7.2}$$

D is the maximum deflection at the center of the specimen.

E_B is the modulus of elasticity in bending and it is expressed with Eq. 7.3 as follows:

$$E_B = \frac{l^3 m}{4bh^3} \tag{7.3}$$

m is the tangent of the initial straight portion of the load-deflection curve.

From Table 7.3, one can comment that flexural strength and modulus of the composites were decreased while their strain at break was improved. This situation is commented as the existence of agglomerated hard ceramic particles behaves as stress concentrators which facilitates crack opening. Also, new composites show less rigid characteristics which is seen from the modulus. This situation brings out as improved flexibility seen from the strain at break.

After 3 PB tests, SEM observation was done on fracture surfaces. Figure 7.3 shows the fracture surfaces of ERAT I-II composites in different magnifications. As first impression, it can be commented that rough fracture surfaces give a sign about the good adhesion between matrix and reinforcements. Also, the fibers which are perpendicular to fracture surface created a link (crack bridging) between two sides of the crack and it has a positive effect against to crack opening. Besides, existence of holes refers to pull-out of fibers under loading. Last, damage on fibers was also observed in their extremities. In recycled rubber modified epoxies the main toughening mechanism is thought as crack deflection with the secondary contributions of shear deformations [16] but here important mechanism is attributed to bridging effect of fibers.

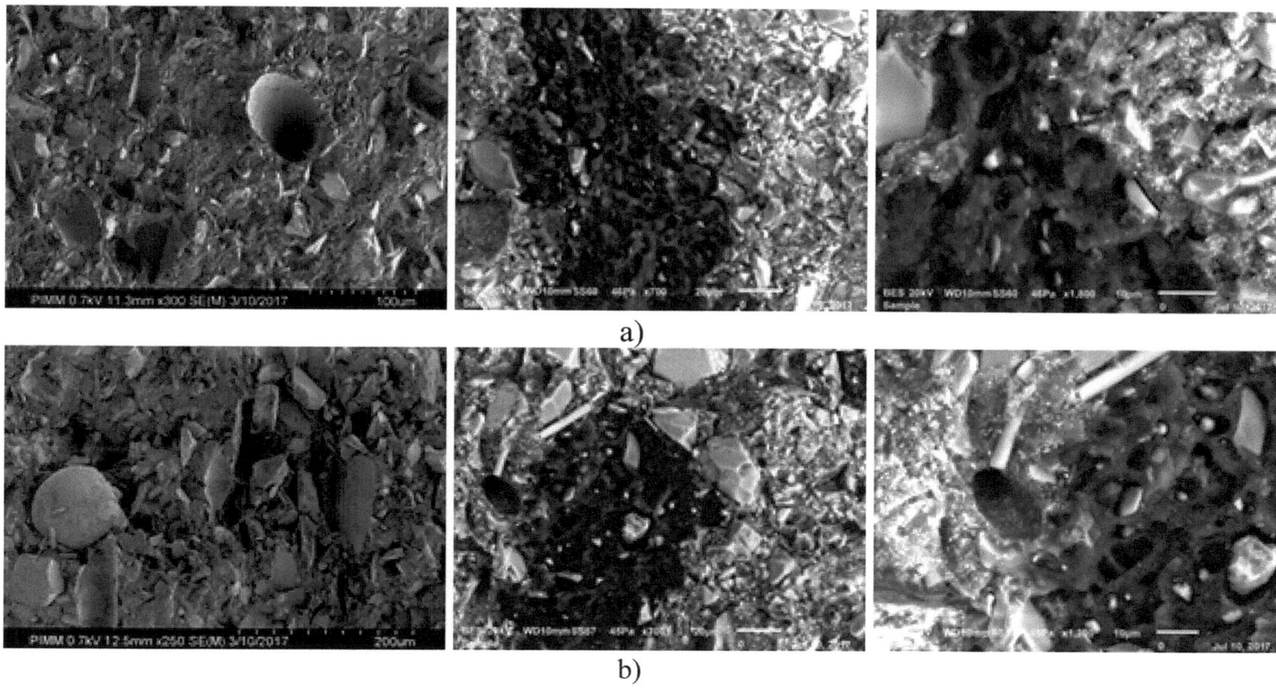

Fig. 7.3 (a) Fracture surfaces after 3 PB testing ERAT I (b) ERAT II

7.3.3 Time Dependent Behaviour by Means of NanoIndentation

During each test, data collected was used to calculate the creep compliance and the stress exponent defined in Eq. 7.4 [17]:

$$\varepsilon(t) = \sigma_0 J(t) \tag{7.4}$$

where σ_0 is the constant stress applied and *J (t) is* calculated using Eq. 7.5

$$J(t) = A(t)/\,(1-\nu)\,P_0 \tan\theta \tag{7.5}$$

In Eq. 7.5 *A(t)* is the contact area, P_0 constant applied load, θ is the effective cone angle which is 70.3° for a Berkovich indenter and the Poisson's ratio ν is assumed to be 0.3. This approach takes into account how the contact area under the Berkovich tip alters while displacement into the surface changes.

The strain versus time behavior during creep is characterized by a high strain rate $\dot{\varepsilon} = d\varepsilon/dt$ in the primary stage of creep and then in the secondary, steady state stage of creep, the strain rate is given in Eq. 7.6 can be written as

$$\dot{\varepsilon} = K\sigma^n \tag{7.6}$$

where *K* is a constant and *n* is the stress exponent. The strain rate is calculated in the software and in turn *n* is obtained from the log-log plot of strain rate versus stress in the secondary stage of creep.

The materials under consideration are heterogeneous in nature and the fact that the nano-indentation test is carried out over a small area/volume, a large scatter in the data is observed and to overcome this sampling number was taken as large as possible then Fig. 7.4 was plotted from the average values.

According to Fig. 7.4 it is seen that creep compliance values are not quite dependent to the alumina fiber content of the compositions. For instance, creep compliance is not very sensitive to fiber content at 20 mN of loading.

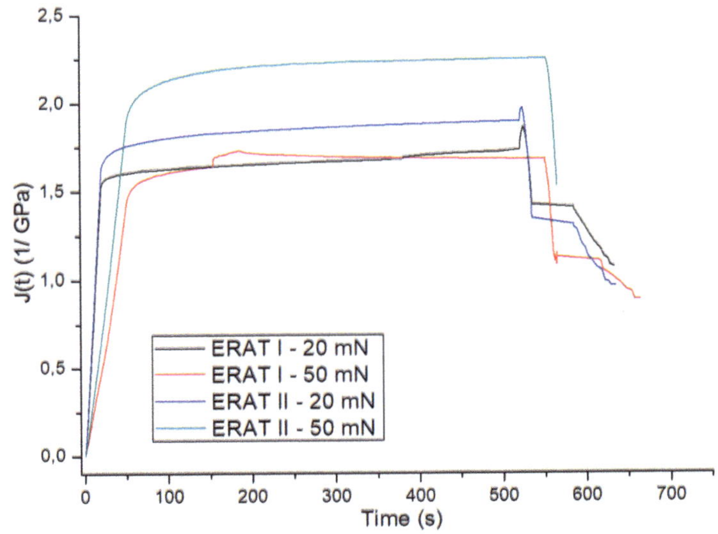

Fig. 7.4 Creep Compliance curves for specimens ERAT I-II, under 20 mN and 50 mN load

According to Fig. 7.4 it is seen that creep compliance is improving with the increasing amount of alumina fibers which means that ERAT II has a higher capacity to flow in return to an abruptly applied stress.

Besides, the trends in the compliance values with time seem alike for different force levels examined with Berkovich tip, and this implies difference of the time-dependent behaviour from stress-dependent behaviour [18].

Results may be compared with linear viscoelastic models. However, these models do not consider the case of tip-specimen adhesion which may differ from experimental results.

The average indentation modulus and stress exponent values were obtained from the 20 indentations performed under 20 and 50 mN constant loads and they are presented in Fig. 7.5a, b with error bars showing ± one standard deviation.

From Fig. 7.5b it is seen that, increasing Al_2O_3 fiber amount results in a drop in the stress exponent which shows more viscoelastic behaviour of ERAT-I. Whether or not indenter tip comes across an alumina fiber or titanium rich zone or a matrix rich area on the specimen, it will introduce error into this method. Different researchers have indicated for polymers and polymer-based composites that modulus as determined by nanoindentation is higher than that determined by macroscopic tensile tests. This inequality is thought because of pile up of material around the contact impression and viscoelasticity of the polymer and polymer based composites which are not taken into consideration during the modulus determination by the Oliver-Pharr method [19–21].

7.3.4 Wear Testing by Nanoindentation

Total of 10 wear tests for each specimen were performed under the two normal loads. The wear in a track is characterized as the area between the initial profile and the residual profile of the wear track.

In the same way, the averages of the wear track deformation are shown in Fig. 7.6.

There is not a significant change was observed in the wear track deformations for the 20 mN force value. However, for more considerable loading values increasing rate of alumina fibers resulted in less resistant structure in wear.

7.4 Conclusions

In this study, the epoxy-rubber based composites were fabricated by powder metallurgy. Preparation processes may introduce inhomogeneity in the material. Favorable effects of surface activation were observed compared to our previous studies. Microstructure observation indicated very small amount of agglomerations of rubber particles which is thought as a consequence of short milling time; increase in the milling time should improve the distribution of particles in the matrix

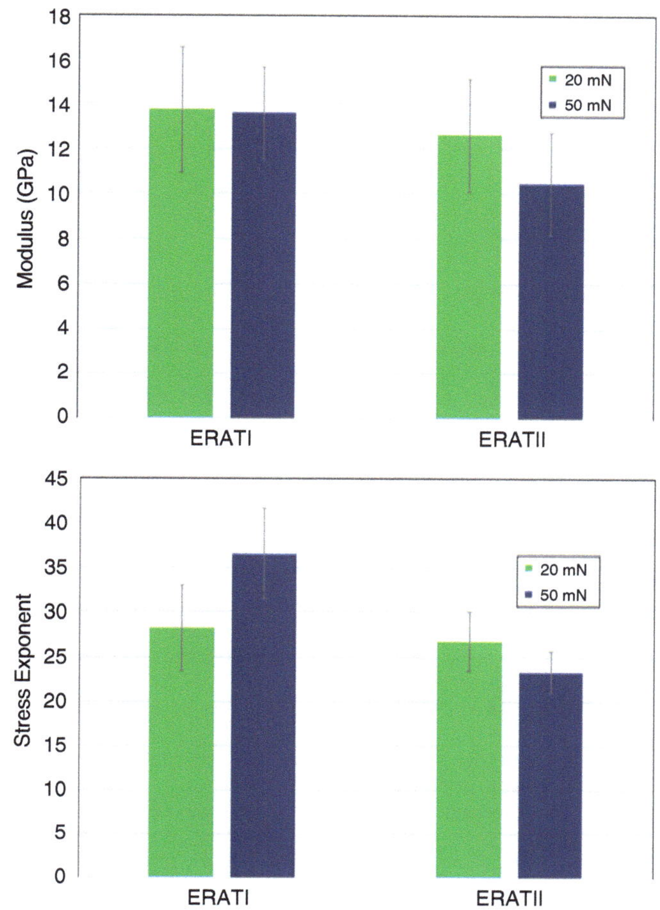

Fig. 7.5 (**a**) Indentation modulus (**b**) Stress exponent for the two composites under 20 mN and 50 mN

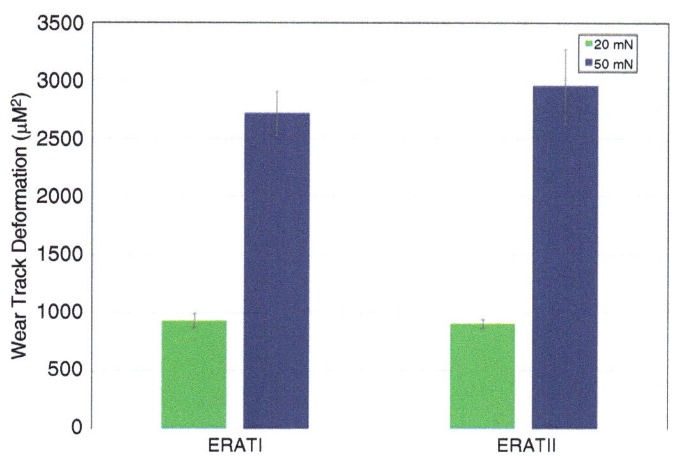

Fig. 7.6 Comparison of wear track deformation for ERAT composites

homogeneously. However, long processing time may result in over heating of the matrix and the binding characteristics of the matrix can be damaged.

Mechanical characterization demonstrated that reinforcements increase the fracture toughness of these composites a certain amount. However, excessive increase of alumina fibers leads to a descending tendency. After mechanical tests, fracture surfaces were observed by means of SEM to identify toughening mechanisms. Crack deflection and crack blunting with the contribution of fiber bridging were thought as the main toughening mechanisms. Then, creep compliance curves obtained from nanoindentation revealed the time dependent nonlinear behaviour with different responses to alumina content. At the end, according to scratch capability of the nanoindenter, increase in the alumina content makes the material less resistant to wear.

References

1. Pendhari, S.S., Kant, T., Desai, Y.M.: Application of polymer composites in civil construction: a general review. Compos. Struct. **84**(2), 114–124 (2008)
2. Lowe, A., Hyok Kwon, O., Wing Mai, Y.: Fatigue and fracture behavior of novel rubber modified epoxy resins. Polymer. **37**(4), 565–572 (1996)
3. Wetzel, B., Rosso, P., Haupert, F., Friedrich, K.: Epoxy nanocomposites – fracture and toughening mechanisms. Eng. Fract. Mech. **73**, 2375–2398 (2006). https://doi.org/10.1016/j.engfracmech.2006.05.018
4. Zaimova, D., Bayraktar, E., Katundi, D., Dishovsky, N.: Design of new elastomeric composites used in manufacturing engineering. In: 14th International Materials Symposium–IMSP, pp. 10–12 (2012)
5. Wetzel, B., Haupert, F., Qiu Zhang, M.: Epoxy nanocomposites with high mechanical and tribological performance. Compos. Sci. Technol. **63**, 2055–2067 (2003). https://doi.org/10.1016/S0266-3538(03)00115-5
6. Zee, R.H., Huang, Y.H., Chen, J.J., Jang, B.Z.: Properties and processing characteristics of dielectric-filled epoxy resins. Polym. Compos. **10**, 205–214 (1989). https://doi.org/10.1002/pc.750100402
7. Jin Kim, D., Hyun Kang, P., Chang Nho, Y.: Characterization of mechanical properties of γAl2O3 dispersed epoxy resin cured by γ-ray radiation. J. Appl. Polym. Sci. **91**, 1898–1903 (2004). https://doi.org/10.1002/app.13250
8. Arayasantiparb, D., Mcknight, S., Libera, M.: Compositional variation within the epoxy/adherend interphase. J. Adhes. Sci. Technol. **15**, 1463–1484 (2001). https://doi.org/10.1163/156856101753213312
9. Bittmann, B., Haupert, F., Schlarb, A.K.: Preparation of TiO2 epoxy nanocomposites by ultrasonic dispersion and resulting properties. J. Appl. Polym. Sci. **124**(3), 1906–1911 (2012)
10. Rothon, R.N.: Mineral Fillers in Thermoplastics: Filler Manufacture and Characterisation. Mineral Fillers in Thermoplastics I, pp. 67–107. Springer, Berlin/Heidelberg (1999). https://doi.org/10.1007/3-540-69220-7_2
11. Pinto, D., Bernardo, L., Amaro, A., Lopes, S.: Mechanical properties of epoxy nanocomposites using titanium dioxide as reinforcement–a review. Constr. Build. Mater. **95**, 506–524 (2015)
12. Wetzel, B., Haupert, F., Friedrich, K., Zhang, M.Q., Rong, M.Z.: Polym. Eng. Sci. **42**, 1919 (2002)
13. Mirjalili, F., Mohamad, H., Chuah, L.: Preparation of nano-scale α-Al2O3 powder by the sol–gel method. Ceram-Silikáty. **55**, 378–383 (2011)
14. Irez, A.B., Bayraktar, E., Miskioglu, I.: Mechanical characterization of epoxy–scrap rubber based composites reinforced with alumina fibers. In: Mechanics of Composite and Multi-Functional Materials, Volume 6, pp. 59–70. Springer, Cham (2018)
15. Zhang, Z., Lei, H.: Preparation of α-alumina/polymethacrylic acid composite abrasive and its CMP performance on glass substrate. Microelectron. Eng. **85**, 714–720 (2008)
16. Kaynak, C., Sipahi-Saglam, E., Akovali, G.: A fractographic study on toughening of epoxy resin using ground tyre rubber. Polymer. **42**(9), 4393–4399 (2001)
17. Irez, A.B., Miskioglu, I., Bayraktar, E.: Mechanical characterization of epoxy–scrap rubber based composites reinforced with nano graphene. In: Mechanics of Composite and Multi-Functional Materials, Volume 6, pp. 45–57. Springer, Cham (2018)
18. Ting, T.C.T.: The contact stresses between a rigid indentor and a viscoelastic half-space. J. Appl. Mech. **88**, 845 (1966)
19. Tranchida, D., et al.: Mechanical characterization of polymers on a nanometer scale through nanoindentation. A study on pile-up and viscoelasticity. Macromolecules. **40**(4), 1259–1267 (2007)
20. Lagoudas, D.C., Thakre, P.R., Amine Benzerga, A.: Nanoindentation of cnt reinforced epoxy nanocomposites. In: Fracture of Nano and Engineering Materials and Structures, pp. 649–650. Springer, Dordrecht (2006)
21. Tranchida, D., et al.: Accurately evaluating Young's modulus of polymers through nanoindentations: a phenomenological correction factor to the Oliver and Pharr procedure. Appl. Phys. Lett. **89**(17), 171905 (2006)

Chapter 8
Reinforcement of Recycled Rubber Based Composite with Nano-Silica and Graphene Hybrid Fillers

A. B. Irez, E. Bayraktar, and I. Miskioglu

Abstract Nano-silica and Graphene have been used as the main reinforcing fillers that increase the usefulness of recycled rubber composite. As each filler retains its specific advantages, the use of nano-silica/graphene combinations should improve the mechanical and dynamic properties of recycled rubber composite. But, the optimum nano-silica/graphene ratio giving rise to the optimum properties requests to be explained. In this work, reinforcement of recycled rubber composite with nano-silica/graphene hybrid filler at various ratios was studied in order to determine the optimum nano-silica/graphene combinations. The toughness properties and tribological behaviour indicating the reinforcement of recycled rubber based composite were evaluated. Microstructural and fractural analyses were made by Scanning Electron Microscopy (SEM).

Keywords Epoxy · Recycled rubber · Wear resistance · SEM · Fracture toughness

8.1 Introduction

Elastomers are very well classified in the polymer family that exhibit rubber-like elasticity [1–8]. Their general characteristics involve viscoelastic behaviour, a low modulus of elasticity, a high failure strain along with very weak inter-molecular forces. Among other properties, rubber provide good heat resistance, ease of deformation at ambient temperatures and exceptional elongation and flexibility before breaking. These properties have established rubber as excellent and relatively cheap materials for various applications in many sectors including automotive, industrial, packaging, healthcare and many others. Currently the rubber industry is vast developing because of the wide commercial penetration of specific materials accompanied by an increased industrial and academic interest, resulting in a steady rise in global annual revenues, predicted to be US56$ billion by 2020 [1, 3, 4, 8]. However, utilization of the recycled rubber for the manufacturing of the new composites is a very economical way for cost effective composite design. Last decades, the usage of the recycled rubber obtained from fresh and clean scrap rubber for different industrial applications such as automotive and aeronautical engineering has been very well developed as very useful material for the composites either as a matrix or as a reinforcement. Extensively, this rubber powder come directly manufacturing of sportive affaires, shoes etc., For this reason, it is feasible to use it after chemical (silanization) and devulcanization treatments for cost effective composite design [3, 5, 8, 9].

Reinforcement of the rubber based composites with nano graphene and/or graphene nano plates (GnPs) are very extensive applications in composite design for mainly aeronautic and aerospace applications. Graphene are known and used widely used as multi- functional reinforcement materials that can improve the electrical, piezo-resistive, dielectric, thermal, mechanical and gas barrier properties of elastomer based composites even at extremely lower loadings. Because of the exceptionally high surface area, as compared to other graphite derivatives, an enormous improvement in properties is observed for graphene composites. Correspondingly, they exhibit unique advantages, as compared with all other organic and inorganic fillers, and are thus useful in many applications. Finally it may be concluded that graphene can be applied to improve the gas permeability (like layered silicates), electrical and thermal conductivity (like carbon nanotubes) of rubber based hybrid composites.

A. B. Irez
CentraleSupélec, Université Paris-Saclay, Gif-sur-Yvette, France

E. Bayraktar (✉)
Supmeca-Paris, School of Mechanical and Manufacturing Engineering, Saint-Ouen, Paris, France
e-mail: bayraktar@supmeca.fr

I. Miskioglu
Michigan Technology University, Engineering Mechanics Department, Houghton, MI, USA

P. R. Thakre et al. (eds.), *Mechanics of Composite, Hybrid and Multifunctional Materials, Volume 5*, Conference Proceedings of the Society for Experimental Mechanics Series, https://doi.org/10.1007/978-3-319-95510-0_8

The interactions between the various elements of a complex multicomponent system, such as the ones found in graphene/elastomer nanocomposites, play a major role on the final physico-chemical properties of the material. For this reason, in much of the literature, surface chemistry is applied to the different components of the system, in order to ensure chemical compatibility between them. In addition, it can lead to the formation of chemical bonds focused towards the improvement of the properties of the initial material and a satisfactory dispersion of the filler.

As for the reinforcement of the rubber based composites with nano silica (SiO_2 d < 15–30 nm) particles [10–12] can increase toughening of the rubber based composites and its distribution in matrix is very easily managed that give a multiphase hybrid-toughened composite structure. The rubber based composites reinforced with nano silica and graphene nano plates (GnPs) show higher thermal conductivity of rubber based composites containing of nano silica and GnPs in a certain ratio. According to the former experience obtained in the same research project, these composites have shown higher thermal conductivities (variable between 25–30 $\pm$ 5.5 W/m K) at a fixed ratio of the reinforcement as considered in this work (nano Silica/GnPs = 2.5) [3].

In the frame of the common research project that is going on, a hybrid rubber based composite reinforced with nano silica+ graphene nano-plates (GnPs) has been designed for electronic applications, mainly in the manufacturing of the electronic devices for the control panels in aerospace, defense industry due to the high capacity of thermal conduction, etc. We believe that this new composite design by using nano silica and graphene nano plates may exposed a novel interface design approach for developing multifunctional rubber based composites.

8.2 Experimental Conditions

8.2.1 Materials Processing

In this study, at the beginning recycled rubber particles were blended with solid bisphenol-A type epoxy powders in pre-defined mass rates. However, because of the insufficient free chains of rubber particles, formation of a robust and durable bonding between epoxy and recycled rubber is quite challenging. For this reason, fresh scrap rubber should be devulcanized. In fact, devulcanization is known to be an operative practice for manufacture of recycled rubbers to increase flowing capacity and also to be remoulded during manufacturing of new designed composites. During this process, formerly created sulphur links are tried to be broken and also new other links are generated, it means that the structure of the material is modified entirely as renewable process. Thanks to this process, newly created free chains make possible have chemical bonding between rubber and epoxy [3–5, 13, 14].

A new design of devulcanized rubber based composite, reinforced with nano silica and graphene nano platelets (GnPs) are prepared in several steps:

1. Chemical treatment was applied on rubber. The principal of the chemical treatment consists in a short silanization process followed by acrylic acid followed by silanization in order to activate the surface of the rubber particles.
2. After drying of the chemically treated rubber powders, they are exposed to short microwave heating in two stages during 4 min under 900 W of power in order to avoid degradation of the main chains called devulcanization.
3. At the final stage, devulcanized rubber were mixed with a little amount of epoxy resin powders for binding of rubber matrix and reinforcements. This mixture used as a matrix after that the new designed composites are manufactured by using classical powder metallurgy methods.
4. After the mixture of the reinforcements (here nano silica and GnPs) in the matrix, a fast-rotating toothed-wheel milling process was carried out during 4 h to obtain fine rubber powder.
5. After having a homogeneous powder compound, the composite specimens were manufactured by using double uniaxial action press at a temperature of 180 °C under a pressure of 70 MPa during the heating of 30 min.
6. All of the specimens (30/50 mm in diameter) were cooled slowly in the press. All of the specimens were post-cured isothermally at 80 °C for 24 h.

The compositions of silica and GnPs reinforced epoxy-recycled rubber based composites (called as SG I-II-III hereafter) were given in Table 8.1.

As seen the reinforcement percentage in the rubber matrix, the ratio SiO_2/GnPs = 2.5 was kept as constant to calculate molar ratio in the mixture for adding to the rubber matrix. By this way, adhesion of nano silica and GnPs to the rubber 90 wt% + Epoxy 10 wt% are carried out at very high levels.

Table 8.1 Composition of the epoxy-rubber based composites

Rubber based composition Epoxy – SBR rubber (10 wt% Epoxy – 90 wt% Rubber)	SG I	SG II	SG III
Reinforcements (wt %)	5 nano $SiO_2$2 GnPs	7,5 nano $SiO_2$3 GnPs	10 nano $SiO_2$4 GnPs

Table 8.2 Hardness values of SG specimens

Hardness measurement	
Specimen	Shore D
SG I	77.2
SG II	78.3
SG III	79

Table 8.3 Density values of SG specimens

Hardness measurement	
Specimen	Density (g/cm^3)
SG I	1438
SG II	1455
SG III	1467

8.2.2 *Microstructure: Fracture Surface Analyses and Shore-D Hardness Measurements*

Fracture surface damage analyses and microstructural observation have been realized by means of optical (OM) and scanning electron microscopy (SEM). SEM observation was realized on fracture surface of the tested specimens with Scope/JSM-6010LA Jeol® electron microscope.

Surface hardness measurements of the specimens were performed after post curing. Shore D hardness test measurements on the polished flat surfaces of the specimens were carried out according to ASTM D 2240 using Shore D hardness tester, (type HBD-100-0). Hardness results were given in Table 8.2.

Three-point bending tests (3 PB) have been carried out according to the ASTM D790 standards. Tests were realized with the machine Zwick Proline Z050TN and during the tests crosshead speed was selected as 1 mm/min. Flexural strength and strain were obtained from the test results. At least three specimens for each composition were used and standard deviation and average values were given in results chapter with standard deviation values. In addition, fractural properties such as plain strain fracture toughness (K_{Ic}) and critical strain energy release rate (G_{Ic}) were investigated with SENB specimens and the tests were realized according to ASTM D5045 standard. Notches were introduced by tapping a fresh razor blade.

Charpy impact tests have been carried out by means of Zwick 5102 pendulum impact tester with a 1 J pendulum configuration for measuring the resilience behaviour of the composites. Relative density for each composition was also measured and they were summarized in Table 8.3.

8.2.3 *Wear Resistance (Scratch Test) and Damage Analysis via 3D Optical Roughness Meter*

In the current research basic idea on the tribological behaviour of the epoxy and recycled elastomer based composites was evaluated performing scratch tests. A 3D optical surface scanner was utilized to assess damage zone after the scratch test in terms of scratch depth and average scratch roughness.

The contact between the sliding diamond indenter and the surface of the composite material during scratch test was analyzed. The normal and tangential forces on indenter were recorded. However, main focus was given to the damage area and volume. In the frame of the current research, the resistance to scratch deformation was evaluated in terms of scratch depth, worn surface and volume subsequent to scratching only under dry conditions and 50,000 and 100,000 number of cycles of wear.

8.3 Results and Discussions

8.3.1 *Microstructure of the Composites*

In Fig. 8.1 two different specimens were shown in different diameters for different characterizations. They were manufactured by means of hot compaction technics.

After hot compaction, transversal sectioning was made then mounted specimens were polished. Then, general microstructures in the transversal direction of three compositions were observed by means of OM and they were shown in Fig. 8.2. From the figures, it is said that all of the compositions have shown a considerably homogenous distribution of the reinforcements in the structure. White circular particles in different dimensions were considered as rubber. In addition, homogeneously distributed amorphous media are thought as epoxy matrix. However, more precise identification requires EDS analysis. On the other hand, silica and GnPs cannot be detected via OM due to their nano-scale dimensions.

8.3.2 *Three Point Bending Tests and Fracture Surface Observation*

Three-Point Bending (3 PB) tests have been carried out for each type of compositions. These tests have been repeated with 4–5 specimens to optimize the results for the all of three composition.

8.3.2.1 Flexural Testing and Fracture Toughness Determination

Flexural stress is calculated during three-point bending test according to the Eq. 8.1:

$$\sigma = \frac{3x\,P\,x\,l}{2x\,b\,x\,h^2} \tag{8.1}$$

In this formula, l is the span length, P is the maximal bending load, b and h are the sample width and thickness, respectively.

Flexural strain, ε_f, was determined according to the Eq. 8.2:

$$\varepsilon_f = \frac{6Dh}{l^2} \tag{8.2}$$

where, D is the maximum deflection at the center of the specimen.

Fig. 8.1 Macrograph of the specimens after compacting and post curing d = 30–50 mm

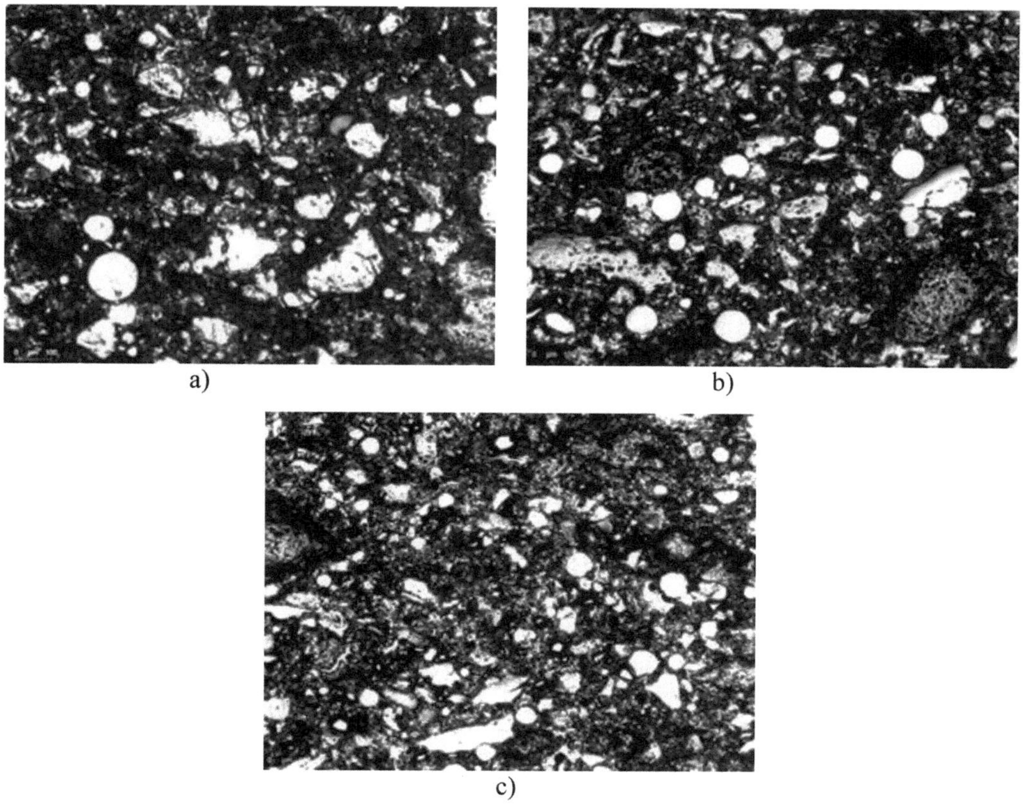

Fig. 8.2 Microstructure of the composites called here as (a) SGI, (b) SGII and (c) SGIII respectively

Therefore, E_B is the modulus of elasticity in bending and it is expressed with the Eq. 8.3 as follows:

$$E_B = \frac{l^3 m}{4bh^3} \tag{8.3}$$

where, m is the tangent of the initial straight portion of the stress-strain curve.

The mode I fracture toughness, K_{Ic}, was determined by testing of the SENB specimens and K_{Ic} was calculated according to the Eq. 8.2:

$$K_{Ic} = \frac{F}{B\ w^{1/2}} f(x); \quad x = \frac{a}{W}\ , 0 < \frac{a}{W} < 1 \tag{8.4}$$

where F is the maximum force from the load-elongation plot; B is the thickness of the specimen; W is the width and "a" is the total notch length.

f(x) is the geometry correction factor and is expressed with the Eq. 8.3 as follows:

$$f(x) = 6(x)^{0.5} \left\{ \frac{\left[1.99 - x(1-x)\left(2.15 - 3.93x + 2.7x^2\right]\right.}{(1+2x)(1-x)^{1.5}} \right\} \tag{8.5}$$

Critical strain energy release rate (fracture energy) G_{Ic} was calculated using the expression Eq.8.4:

$$G_{Ic} = \frac{{K_{Ic}}^2}{E} \tag{8.6}$$

Table 8.4 Comparison of mechanical properties of SG specimens

	Flexural stress (MPa)	Flexural modulus (MPa)	Strain in break (ε %)	K_{Ic} (MPa m$^{1/2}$)	G_{Ic} (kJ/m^2)
SG I	10.41 ± 7.9	929.1 ± 696	0.,63 ± 0.16	0.66 ± 0.12	0.48 ± 0.18
SG II	14.53 ± 4.8	884.9 ± 380	0.40 ± 0.05	0.62 ± 0.02	0.43 ± 0.03
SG III	17.84 ± 0.3	1092.2 ± 304	0.38 ± 0.06	0.90 ± 0.09	0.75 ± 0.15

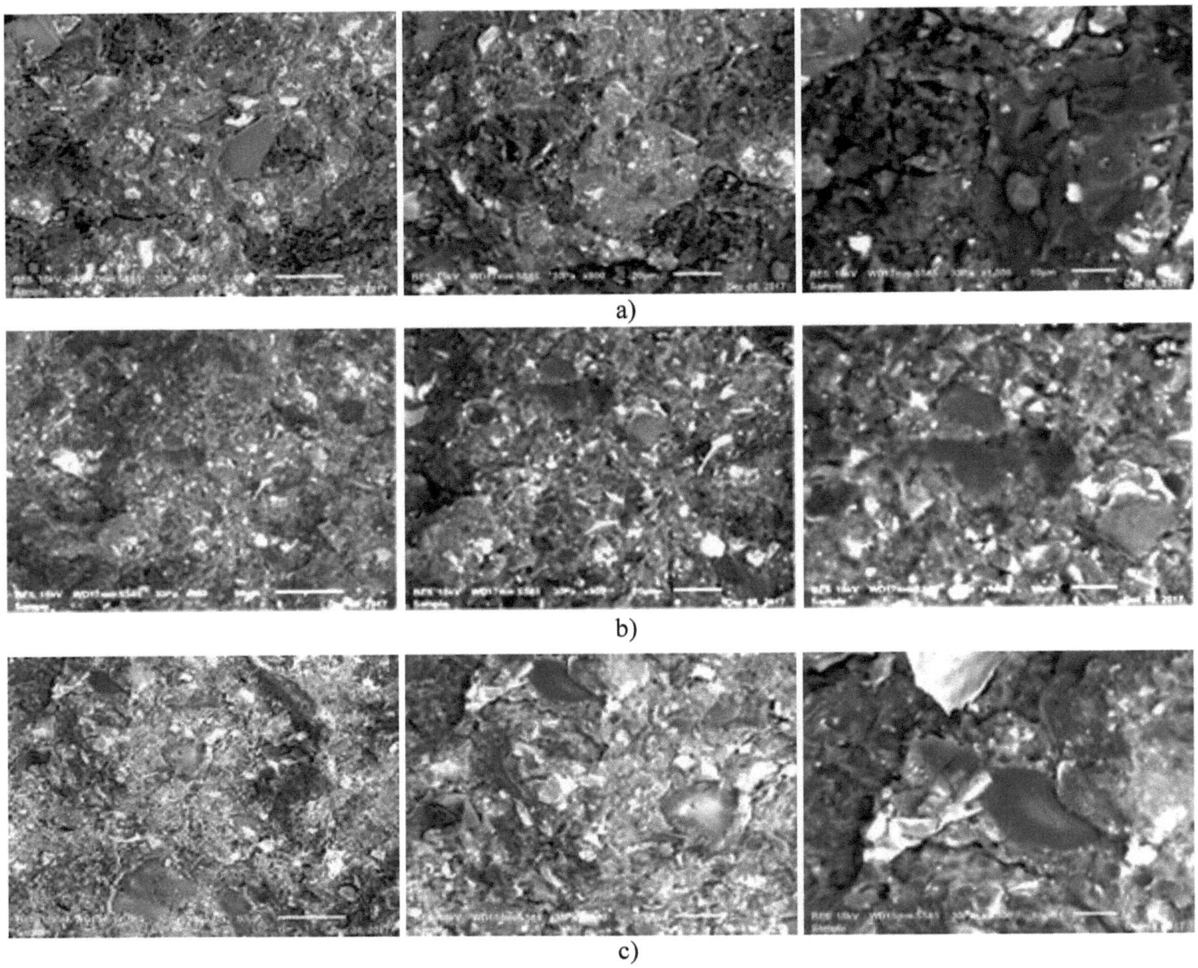

Fig. 8.3 (**a**) Fracture surfaces after 3 PB testing SG I (b) SG II (**c**) SG III

where E is the elasticity modulus for plane stress approach examined for thin specimens.

Table 8.4 indicates the mechanical properties in bending mode. Fracture toughness values were also presented in the table.

Table 8.4 indicates that The hybrid rubber based composites reinforced with nano silica and graphene nano plates (GnPs) with a certain ratio have shown enhancements in the mechanical properties, and more significantly, fracture toughness (K_{IC}), which can be explained by synergistic impact coming from the intrinsic physical characteristics of the reinforcements it means that nano silica and graphene nano plates (GnPs). In this study, the highest K_{IC} can be obtained with addition of small amounts of nano-silica particles (5, 7.5 and 10 wt %) with a constant ratio of 2.5 with graphene nano plates (GnPs).

Fracture surfaces obtained from 3 PB tests have been analyzed by means of Scanning Electron Microscopy (SEM). It noticed that good adhesion of the both of the reinforcements in the rubber based matrix by creating an ideal interface for each composition, as presented in Fig. 8.3 with different fracture surfaces taken by SEM.

Cavitation and void formation in the rubber matrix with matrix expansion and locally, debonding of nano particles with consequent void growth have been observed in the structure as the improved toughening mechanisms. For this evolution, the mixture of devulcanized rubber (90 wt %) with epoxy (10 wt %) plays an important role. Some of the specimens with homogenous distribution of the nanoparticles have shown a typical debonding of the silica nanoparticles with GnPs that

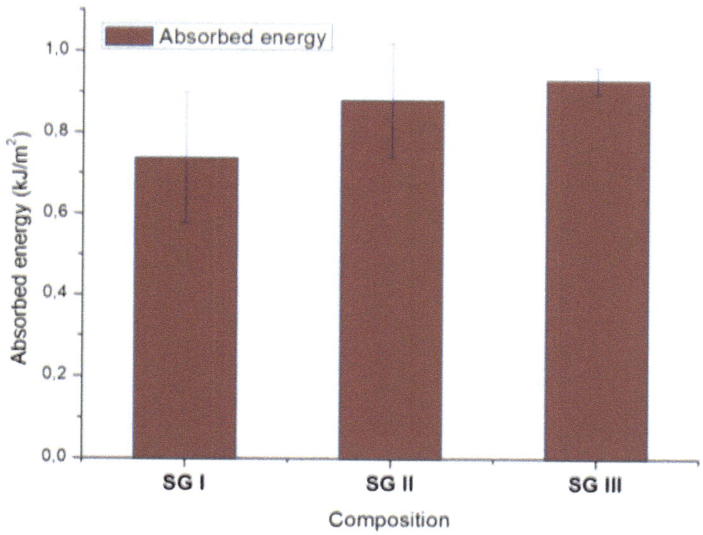

Fig. 8.4 Absorbed energy during charpy impact tests for SG specimens

should be origins of the weakening of the matrix–particle interface. The toughness improvement in hybrid rubber based composites should be direct related to the increment of the debonding phenomena that can increase the size of the plastic zone in the structure. This case facilitate the devulcanized rubber based composites to dissipate additional fracture energy.

8.3.3 Charpy Impact Testing

Last mechanical characterization was realized to see the impact resistance and energy absorbance capability of the manufactured composites. Each composition group was tested with at least three specimens. Absorbed energies during impact testing of the composites were presented in Fig. 8.4. It seems that the absorbed energy for each specimen during the impact tests is related to the increment of the plastic zone in the structure due to the debonding of the nano particles. Higher dispersion of the values is also related to the test specimens prepared under laboratory conditions; hot compaction, cutting notch effect, etc.

8.3.4 Damage Analysis by Means of Scratch Test and 3d Optical Roughness Meter

After completing mechanical tests, tribological characterization of the composites were done by macro scratch tests. Surface damages were observed three dimensionally by an optical surface scanner. The results are presented in the Fig. 8.5.

In Fig. 8.6 the volume and surface of the damage trace after scratch calculated from roughness test results are given. It can be said that, by the increase of the reinforcement content, composites become more resistant to wear.

In reality, because of the high shear stress at the interfaces the interfacial shear stress should probably be the main reason for damage of the matrix and reinforced filler interfaces. When the indenter is slipping, tangential tensile stress is caused on the surface behind the indenter, while in front of the indenter the tangential stress is compressive.

8.4 Conclusion

The hybrid rubber based composites reinforced with nano silica and graphene nano plates (GnPs) have shown enhancements in the mechanical properties, and more significantly, fracture toughness (K_{IC}), which can be explained by synergistic impact coming from the intrinsic physical characteristics of the reinforcements; it means that nano silica and graphene nano plates

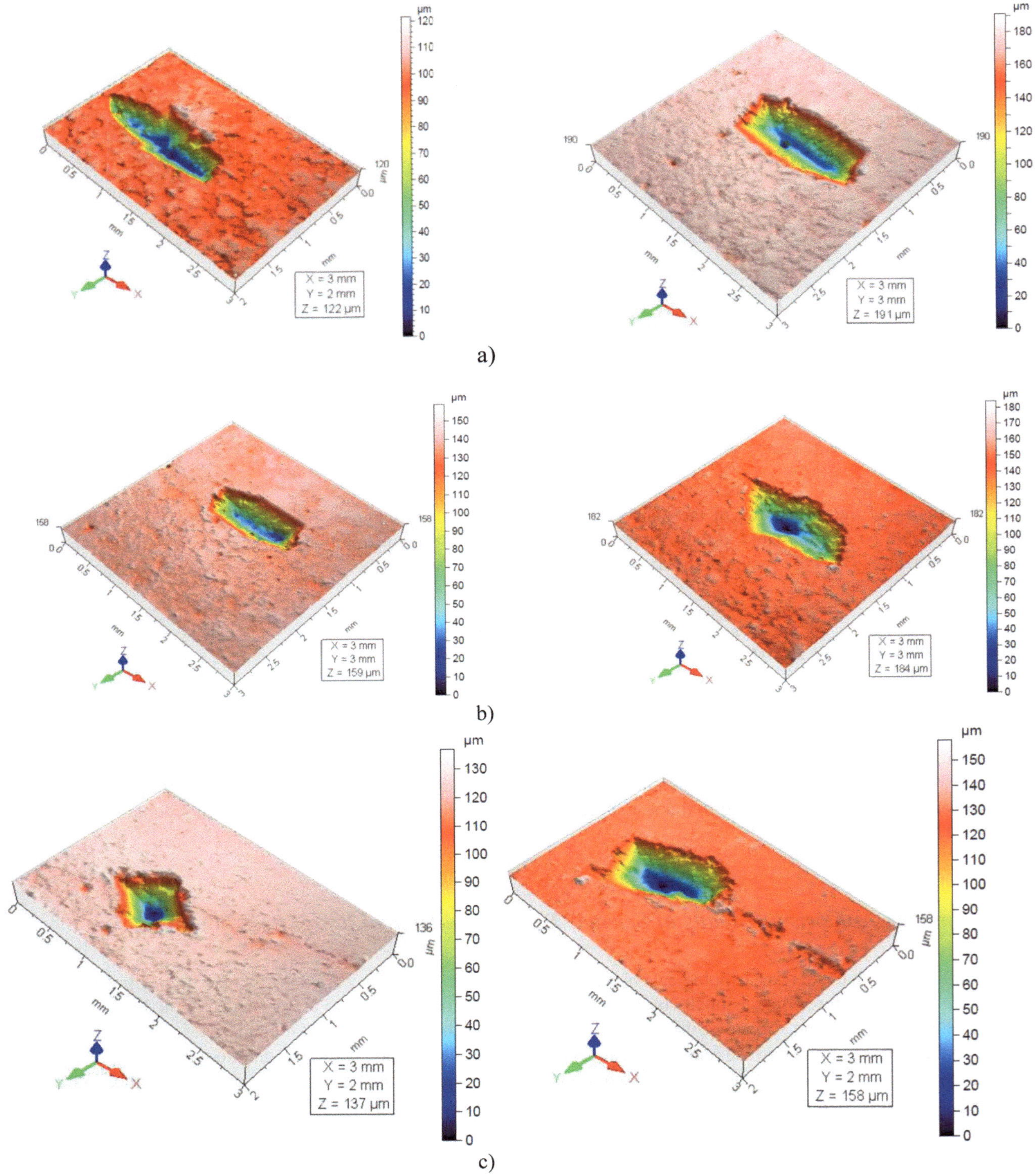

Fig. 8.5 Three dimensional damage traces obtained in the direction of width and length for (**a**) SG I (**b**) SG II (**c**) SG III for 50 k cycles and for 100 k cycles

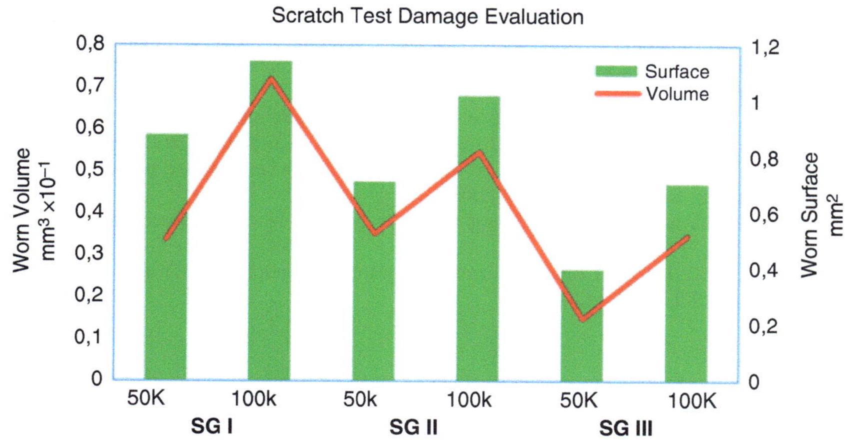

Fig. 8.6 Comparison of scratch damage traces presented as volume lost and damage surfaces obtained by wear-software for the specimens SG I-II-III for 50 and 100 k cycles, respectively

(GnPs) with certain ratio. In this study, the highest K_{IC} can be obtained with addition of small amounts of nano-silica particles (5, 7.5 and 10 wt %) with a constant ratio of 2.5 with graphene nano plates (GnPs).

Cavitation and void formation in the rubber matrix with matrix expansion and locally, debonding of nano particles with consequent void growth have been observed in the structure as the improved toughening mechanisms. For this evolution, the mixture of devulcanized rubber (90 wt %) with epoxy (10 wt %) plays an important role. Some of the specimens with homogenous distribution of the nanoparticles have shown a typical debonding of the silica nanoparticles with GnPs that should be origins of the weakening of the matrix–particle interface. The toughness improvement in hybrid rubber based composites should be direct related to the increment of the debonding phenomena that can increase the size of the plastic zone in the structure. This case facilitate the devulcanized rubber based composites to dissipate additional fracture energy.

In another aspect of these composites is related to the higher thermal conductivity of rubber based composites containing of GnPs graphene in varying ratios. According to the former experience obtained in the same research project, these composites have shown higher thermal conductivities (variable between 25–30 $\pm$ 5.5 W/m K) at a fixed ratio of the reinforcement as considered in this work (nano Silica/GnPs = 2.5).

Acknowledgments The authors thank Dr. H-A. Alhas from Airbus-Space, London/UK for general support and also for his valuable discussion in certain stages of this project. Also they would wish to acknowledge research foundation of Supmeca/Paris.

References

1. Bokobza, L.: Multiwall carbon nanotube elastomeric composites: a review. Polymer. **48**, 4907–4492 (2007)
2. Irez, A.B., Miskioglu, I., Bayraktar, E.: Mechanical characterization of epoxy: scrap rubber based composites reinforced with nanoparticles, Springer link. Mechanics Composite Multi-funct. Mater. **6**, 33–44 (2017., ISBN 978-3-319-63408-1). https://doi.org/10.1007/978-3-319-63408-1
3. Zaimova, D., Bayraktar, E., Miskioglu, I.: Design and manufacturing of new elastomeric composites: mechanical properties, chemical and physical analysis. Int. J. Composites, Part B, Elsevier, USA. **1**(1), 1–12 (2016)., on line 2017
4. McCarthy, D.W., Mark, J.E., Schaeffer, D.W.: Synthesis, structure, and properties of hybrid organic-inorganic composites based on polysiloxanes. I. Poly (dimethyl siloxanes) elastomers containing silica. J. Polym. Sci., Part B Polym. Phys. **36**, 1167–1189 (1998)
5. Zaimova, D., Bayraktar, E., Miskioglu, I., Dishovsky, N.: Wear resistance of elastomeric based composites by continuous multi-cycle indentation used in manufacturing engineering. Adv. Mater. Res. **939**, 106–113 (2014., Trans Tech Publications, Switzerland). https://doi.org/10.4028/www.scientific.net/AMR.939.106
6. Sadasivuni, K.-K., Ponnamma, D., Thomas, S., Grohens, Y.: Evolution from to graphene elastomer composites. Prog. Polym. Sci. **39**, 749–778 (2014)
7. Irez, A.B., Bayraktar, E., Miskioglu, I.: Mechanical characterization of epoxy – scrap rubber based composites reinforced with alumina fibers, Springer link. Mech. Composite Multi-funct. Mater. **6**, 59–70 (2017., ISBN 978-3-319-63408-1). https://doi.org/10.1007/978-3-319-63408-1
8. Ferreira, L.M.P., Miskioglu, I., Bayraktar, E., Katundi, D.: Mechanical and tribological properties of scrap rubber reinforced with Al2O3 fiber, aluminium and TiO2, Springer link. Mech. Composite Multi-funct. Mater. **7**, 37–44 (2016., ISBN 978-3-319-41766-0). https://doi.org/10.1007/978-3-319-41766-0

9. Irez, A.B., Miskioglu, I., Bayraktar, E.: Mechanical characterization of epoxy – scrap rubber based composites reinforced with nano graphene, Springer link. Mech. Composite Multi-funct. Mater. **6**, 45–58 (2017., ISBN 978-3-319-63408-1). https://doi.org/10.1007/978-3-319-63408-1
10. Bhattacharya, M., Maiti, M., Bhowmick, A.K.: Tailoring properties of styrene butadiene rubber nano composite by various nano fillers and their dispersion. **49**(1), 81–98 (2009)
11. Burak Irez, A., Hay, J., Miskioglu, I., Bayraktar, E.: Scrap-rubber based composites reinforced with boron and alumina, Springer link. Mech. Composite Multi-funct. Mater. **6**, 1–10 (2017., ISBN 978-3-319-63408-1). https://doi.org/10.1007/978-3-319-63408-1
12. Kinloch, A.J., Mohammed, R.D., Taylor, A.C.: The effect of silica nano-particles and rubber particles on the toughness of multiphase thermosetting epoxy polymers. J. Mater. Sci. **40**, 5083–5086 (2005)
13. Shokoohi, S., Arefazar, A., Khosrokhavar, R.: Silane coupling agents in polymer-based reinforced composites: a review. J. Reinf. Plast. Compos. **27**(5), 473–485 (2008)
14. Zhang, G., Wang, F., Dai, J., Huang, Z.: Effect of functionalization of graphene nanoplatelets on the mechanical and thermal properties of silicone rubber composites. Materials. **9**(92), 1–13 (2016)

Chapter 9
Testing the 2-3 Shear Strength of Unidirectional Composite

Joel S. Fenner and Isaac M. Daniel

Abstract In this study, the objective was to measure the "out-of-plane" 2-3 shear strength of unidirectional composite, working within constraints in supplied material geometry. Unidirectional carbon/epoxy composite material was tested using a sandwich-type beam specimen under 3-point bending with a low span-to-thickness ratio to achieve failure under 2-3 shear. Specimens were carefully designed to deliberately cause shear failure near the midplane, avoiding other possible failure mechanisms. A photoelastic coating and post-mortem microscopy were used to verify failures. Results were compared with a simple analytical description of failure and found to have good agreement. Notably, this approach was able to accommodate the limitations of the supplied material (thin sheets) while still providing an accurate means of obtaining the F_{23} shear strength of the material. The results also imply the possibility of testing the transverse tensile strength (F_{2t}) in lieu of performing a shear test, which is far simpler, and inferring the out-of-plane shear strength F_{23}.

Keywords Composites · Interlaminar · Test methods · Shear strength · Photoelasticity

9.1 Introduction

High-performance continuous-fiber composites have grown steadily in their widespread adoption since their commercial introduction decades ago. In the course of time, these materials have presented a vast array of challenges for the engineer, from fabrication to characterization. They traditionally exhibit highly anisotropic behavior, requiring a broader range of testing techniques than more conventional alternatives (metals, polymers) to achieve thorough characterization, which has been chief among the impediments to widespread adoption. With the advent of modern computational platforms for structural design, some of the inherent complexity in working with such anisotropic materials has been reduced. However, concurrent with these developments has come the need for expeditious and convenient methods of obtaining complete sets of characterization data. Certain material parameters historically seen as having lesser importance, such as the "out-of-plane" or "interlaminar" shear strengths F_{13} and F_{23}, are gradually attaining greater significance, especially as inputs to increasingly stringent and complicated computational models.

Out-of-plane properties tend to be matrix-dominated in unidirectional composites, rendering them much less impressive, and often embodying an encumbrance to be grappled with from a testing perspective, as it can be difficult to obtain an accurate result from conventional tests [1, 2]. For instance, there exist several well-established forms of shear strength tests (Arcan, Iosipescu, double notch lap shear) [3, 4]; however, the majority of these tests rely on the availability of thick sections of material to produce viable specimens. As high-performance composites are often manufactured in thin sections to take advantage of their high strength and stiffness (along fiber directions), this can make application of these tests infeasible. Furthermore, many such specimen designs involve the use of notches to generate a region of locally-uniform shear stress, which may not manifest itself properly in the 2-3 plane because of the way in which the elastic constants of the material may present themselves [5].

An approach sometimes used for testing the 1-3 interlaminar shear strength of unidirectional composites is to prepare a beam in bending with a low span-to-thickness ratio [6, 7]. Such a sample, if properly designed, is likely to fail in 1-3 shear at the midplane rather than by compression or tension at its surfaces. By itself, such a specimen does not work for bending in the 2-3 plane of most composites, simply because the transverse tensile strength of the composite (F_{2t}) is far too low. However, by carefully constructing a laminated sandwich-type specimen [8] composed only of unidirectional composite, the material anisotropy may be exploited to cause the desired shear failure in the 2-3 plane.

J. S. Fenner (✉) · I. M. Daniel
Center for Intelligent Processing of Composites, Northwestern University, Evanston, IL, USA
e-mail: JoelFenner2012@u.northwestern.edu

P. R. Thakre et al. (eds.), *Mechanics of Composite, Hybrid and Multifunctional Materials, Volume 5*, Conference Proceedings of the Society for Experimental Mechanics Series, https://doi.org/10.1007/978-3-319-95510-0_9

9.2 Experimental Methodology

Successful testing requires careful specimen design to assure the desired failure mode. The goal is to cause shear failure in the "core" of the sample, comprised of material in the 2-3 plane, by providing support from the "facesheets" of the sample, loaded within the 1-3 plane (Fig. 9.1). For a sandwich-beam under three-point bending, the normal-stress and shear-stress profiles show discontinuities or "kinks", owing to the difference in elastic constants between the two layers. As the elastic modulus of the facesheets is increased above that of the core, the normal stress distributions of core and facesheet become more disparate, with the facesheets taking up more of the load (as in Fig. 9.1). This in turn causes the shear stress distribution to become more uniform within the core.

Likewise as the ratio of core to facesheet thickness is increased, the shear stress distribution in the core becomes more "peaked" or pronounced (as in Fig. 9.2), lowering the shear stress at the interface between core and facesheet, though following a more complicated mathematical relationship than the contribution of elastic moduli. In conjunction, these two effects may be exploited to:

1. Attempt to minimize the shear stress at the interface between facesheet and core (a "weak point" dominated by an adhesive bond)
2. Attempt to maximize the "sharpness" of the shear stress distribution within the core so as to assure a shear failure at the sample midplane (neutral axis)
3. Avoid tensile or compressive failures of the facesheets.

For purposes of specimen design, the significant sample stresses are calculated readily from ordinary beam theory. The maximum normal stresses at the outermost surfaces of the beam (the "facesheet" layers) are

$$\sigma_{\mathrm{max,FS}} = \frac{PL}{4I}\left(\frac{E_f}{E_c}\right)\left(h_c + h_f\right) \tag{9.1}$$

and the maximum normal stress in the core (at the facesheet-core interface) is.

$$\sigma_{\mathrm{max,core}} = \frac{PL}{4I} h_c \tag{9.2}$$

where

P is the applied load at the central roller
L is the total span of the sample
b is the sample width

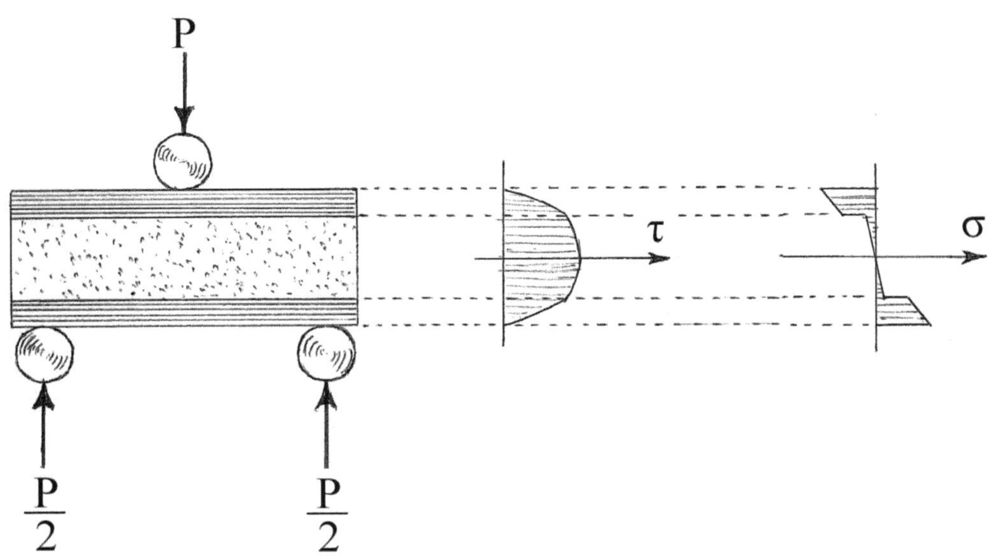

Fig. 9.1 Illustration of short sandwich beam and related shear and normal stress profiles

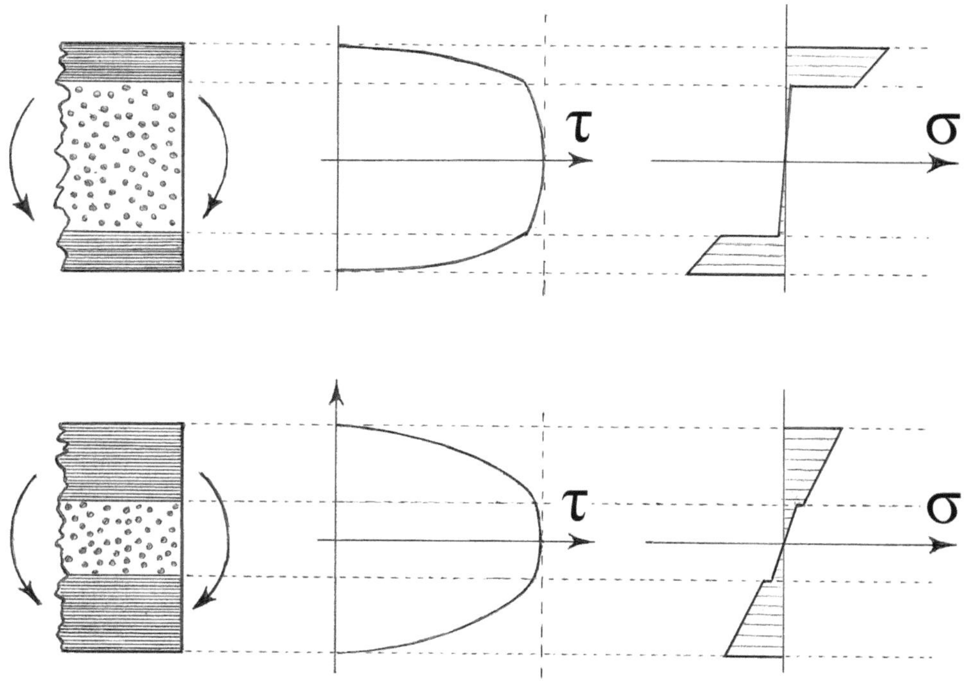

Fig. 9.2 Illustration of effects on stress profiles arising from geometry changes. **Top**: Thick core, thin facesheets – low normal stresses in core, with more nonuniform shear stress through core thickness. **Bottom**: Thick facesheets, thin core – higher normal stresses in core, but with much flatter profile to shear stress in core

h_f is the thickness of a single facesheet layer
$2h_c$ is the total thickness of the core
E_f and E_c are the elastic moduli of the facesheet and core, respectively, in the plane of bending
I is the effective second area moment of inertia of the specimen given as

$$I = \frac{2b}{3}\left[\frac{E_f}{E_c}\left(h_f^3 + 3h_c h_f^2 + 3h_c^2 h_f\right) + h_c^3\right] \tag{9.3}$$

The critical shear stresses, at the interface between facesheet and core and within the core itself, are obtained from integration of the normal-stress gradient within the specimen as

$$\tau_{\text{interface}} = \frac{P}{4I}\left(\frac{E_f}{E_c}\right) h_f\left(h_f + 2h_c\right) \tag{9.4}$$

and

$$\tau_{\text{nax,core}} = \frac{P}{4I}\left[\frac{E_f}{E_c}\left(h_f^2 + 2h_f h_c\right) + h_c^2\right] \tag{9.5}$$

For the unidirectional carbon/epoxy composite in this study, material was supplied primarily in the form of 2.4 mm and 5.0 mm sheets – the manufacture of thicker sheets being impractical due to difficulties with exotherm in epoxy curing. The composite was comprised of fibers similar to AS4 in their behavior (Dow-Aksa A42) at a nominal volume fraction of approximately 50%. Basic material properties obtained from prior uniaxial tension and compression testing (Table 9.1) were utilized in determining suitable specimen geometry.

A limiting factor in such a sample tends to be the shear strength of the adhesive used to join the layers in the sandwich-beam. Hence, a nominal value of 50 MPa was set as a design limit (Loctite Hysol 9430) for the interfacial shear stress between layers of the sandwich beam (Eq. 9.4) in order to avoid adhesive failure, allowing a slightly higher value due to the

Table 9.1 Basic unidirectional composite lamina properties

Property	Value
Longitudinal elastic modulus, E_1	134 GPa
Transverse elastic modulus, E_2	8.89 GPa
Longitudinal tensile strength, F_{1t}	2020 MPa
Longitudinal compressive strength, F_{1c}	1070 MPa
Transverse tensile strength, F_{2t}	61.7 MPa
Transverse compressive strength, F_{2c}	220 MPa

presence of a compressive load at the interface (from the shear force). Use of Eqs. 9.1, 9.2, 9.3, 9.4, and 9.5 gave several viable sample geometries, of which two were selected.

For all tests, the total span was set at $L = 31$ mm. Samples with $2h_c = 5.0$ mm (total core thickness) and $h_f = 2.5$ mm (facesheet thickness) were selected as a suitable candidate, with predicted valid failures within the core. Alternate samples, with "double facesheets" (and hence 5 total layers) were built up from the available 2.5 mm composite sheets, with $2h_c = 2.5$ mm and $h_f = 5.0$ mm (by bonding 2 layers), under the rationale that the intermediary adhesive joints would preferentially alter the shear stress profile. A metallic pad (of aluminum shims) was placed between the central roller and the specimen to "blunt" or distribute the local contact stresses from loading to help avoid compressive failure of the sample at the contact point.

To verify the behavior of the specimen through failure, and to assess failure location, a photoelastic coating was employed (Vishay PS-1). A rudimentary setup with a white-light source and circular polarizer was employed to take stress-fringe images of the coating during loading. The observed stress fringes had the effect of highlighting

1. The uniformity of the shear stresses within the layers
2. The effect of shear stress concentrations at the interfaces between layers in the sandwich-beam specimen
3. The locations of sample failure cracks (to assess valid sample failure)

After failure, sample surfaces were re-polished for the purpose of taking optical micrographs of cracks. Samples showing no "core" failures (i.e. interfacial adhesive failures only) were rejected from consideration.

9.3 Results

Tests were conducted in a servo-hydraulic load frame (Instron 1331) at a displacement rate of 0.025 mm/min (machine minimum) to achieve quasi-static loading in the sample. Tests gave essentially linear load-displacement curves (as in Fig. 9.3) up to failure, concluding with a marked load drop, audible sample cracking, and a rapid change in the fringe pattern in the photoelastic coating indicative of sample failure.

The photoelastic fringe pattern was quite helpful in interpreting the results of the tests. In the configuration employed in this study, the photoelastic coating largely served as a qualitative indicator of shear strain gradients in different portions of the sample (Fig. 9.4). Fringe patterns at low loads were largely limited to the stress concentrations arising from rollers (Fig. 9.4a). At higher loads, a clear region of higher shear strain was visible within the 2-3 plane sample "core" (Fig. 9.4b, c) due to the strong difference in elastic moduli between the layers ($E_c << E_f$). The fringe pattern within the "core" became muddled near failure due to the emergence of large numbers of polychromatic fringes. At failure (Fig. 9.4f), the locations of failure cracks could be ascertained from "singularities" in the photoelastic fringe pattern.

Post-mortem micrographs were used to verify valid sample failures. Samples were imaged prior to test to verify the absence of manufacturing defects or extant cracks (Fig. 9.5). Samples showing angular post-motem cracks within the 2-3 "core", as in Fig. 9.6 (left), were indicative of acceptable sample failures. Most samples show clear initiation at a random location within the core, with the crack propagating outward toward one of the facesheets, typically with the crack deviating into the adhesive interface and remaining within it (Fig. 9.6, right).

Test results showed the out-of-plane shear strength F_{23} to be 52.7 MPa with 12.0% CoV (7 tests). The observed angle of crack propagation (with respect to the neutral axis) in the samples also ranged from 43° to 57° (mean 52°, 16% CoV).

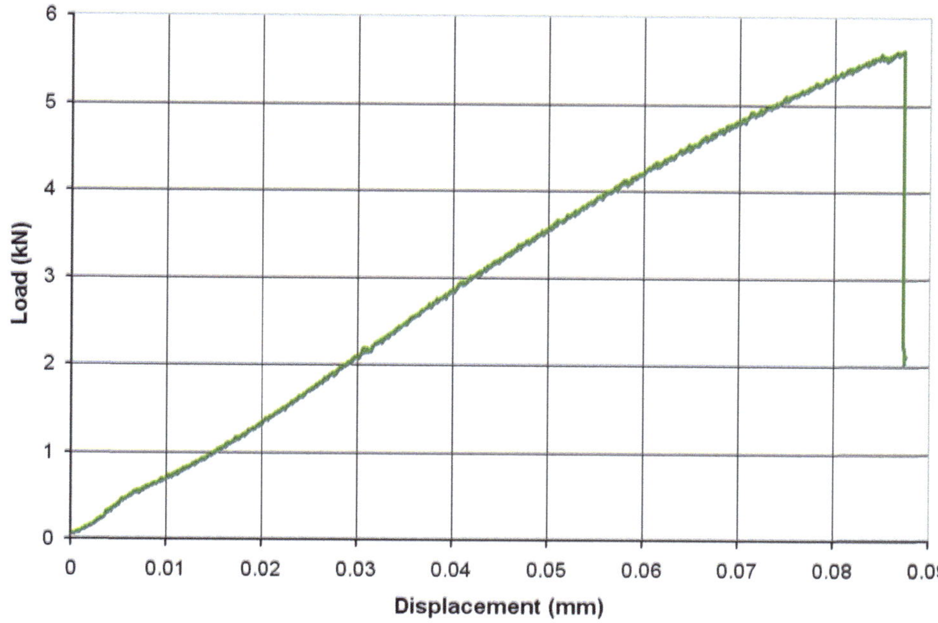

Fig. 9.3 Representative load-displacement curve for a sandwich-type short-beam shear test in carbon/epoxy composite

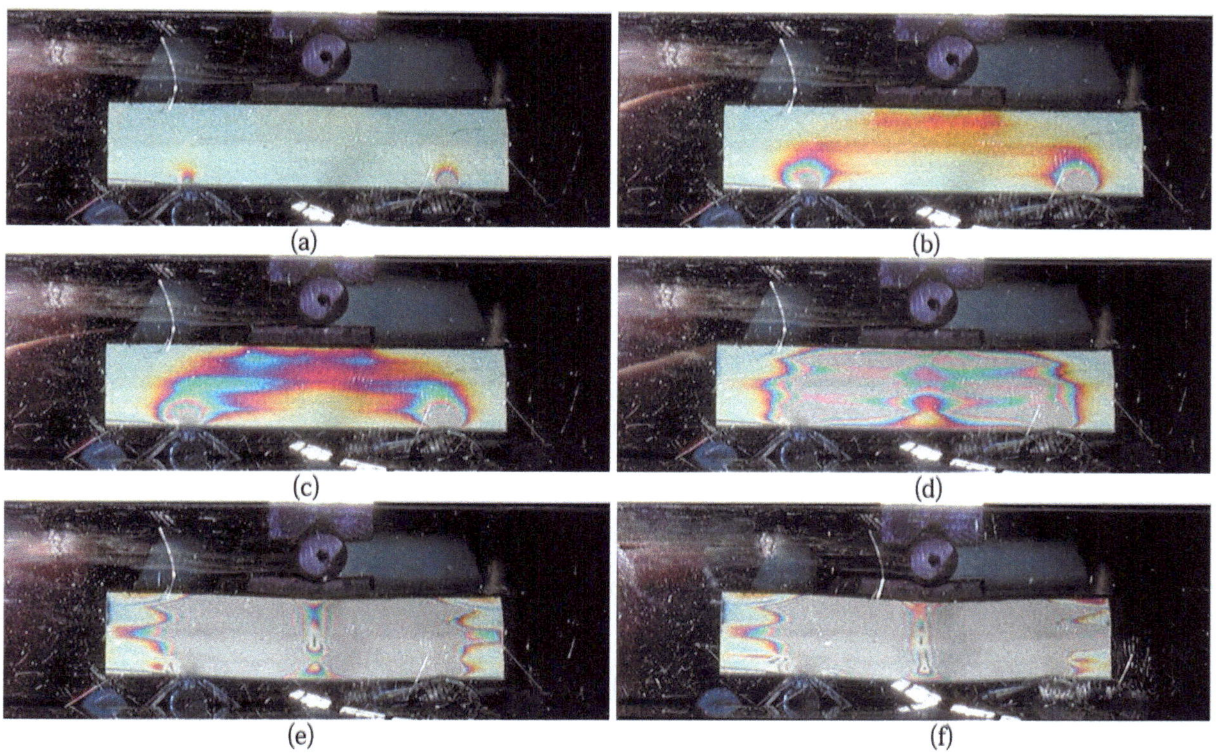

Fig. 9.4 Photoelastic fringe patterns observed during test (5 layer sample, center layer shows 2-3 plane, other layers show 1-3 plane) (**a**) fringe pattern at minimal sample preload, (**b**) fringe pattern at 5% of failure load, (**c**) fringe pattern at 10% of failure load, (**d**) fringe pattern at 40% of failure load, (**e**) fringe pattern at 90% of failure load, (**f**) fringe pattern at failure showing localization around crack at center of sample

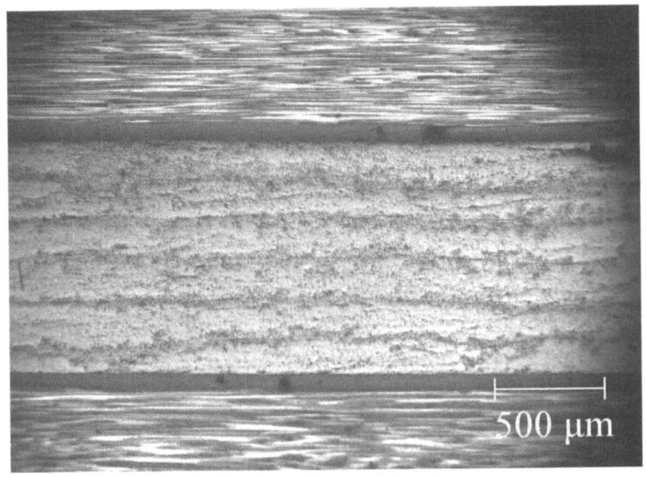

Fig. 9.5 Micrograph of representative unidirectional composite sample prior to testing, showing 2-3 plane "core", adhesive interfaces, and 1-3 plane "facesheets". Visibility of individual laminae within core is a result of the manufacturing process

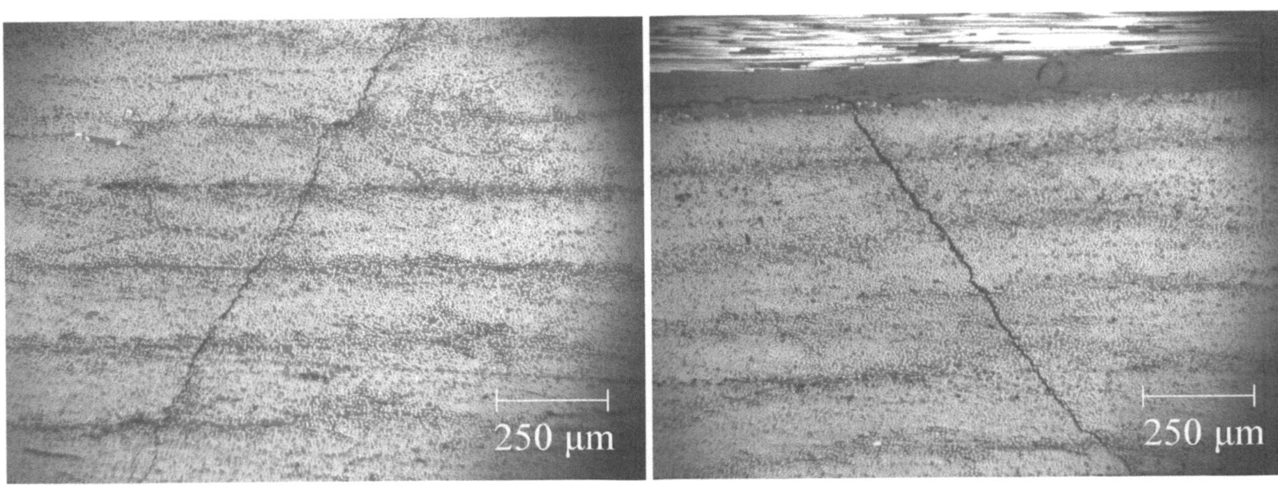

Fig. 9.6 Post-mortem micrographs showing crack formation in "core" layers of specimens. **Left**: Representative angular crack through sample core, appearing to initiate at an internal defect. **Right**: Crack propagation from core to adhesive interface, with crack deviating to follow interface

9.4 Analysis & Discussion

In the case of unidirectional composites, it is often assumed that there is great symmetry between the 2- and 3- directions of the material because of random distribution of fibers in the 2-3 plane. Hence, from an analytical standpoint, one may consider a unidirectional composite as orthotropic with 2-3 symmetry.

If the 2-3 plane of a unidirectional composite is then subjected to a pure shear stress (Fig. 9.7, left), the stress may be transformed, by rotation, to any arbitrary angle within that plane without concern for material orientation. One may then contemplate failure in terms of maximum principal stresses regardless of orientation with respect to the material coordinate system. For such a case, the maximum principal normal stresses exist at 45° to the case of the original applied shear stress (Fig. 9.7, middle). In this case,

$$\tau_{23} = \sigma_{45^\circ t} = -\sigma_{45^\circ c} \tag{9.6}$$

or because of symmetry between the 2- and 3- axes (Fig. 9.7, right),

$$\tau_{23} = \sigma_2 = -\sigma_3 \tag{9.7}$$

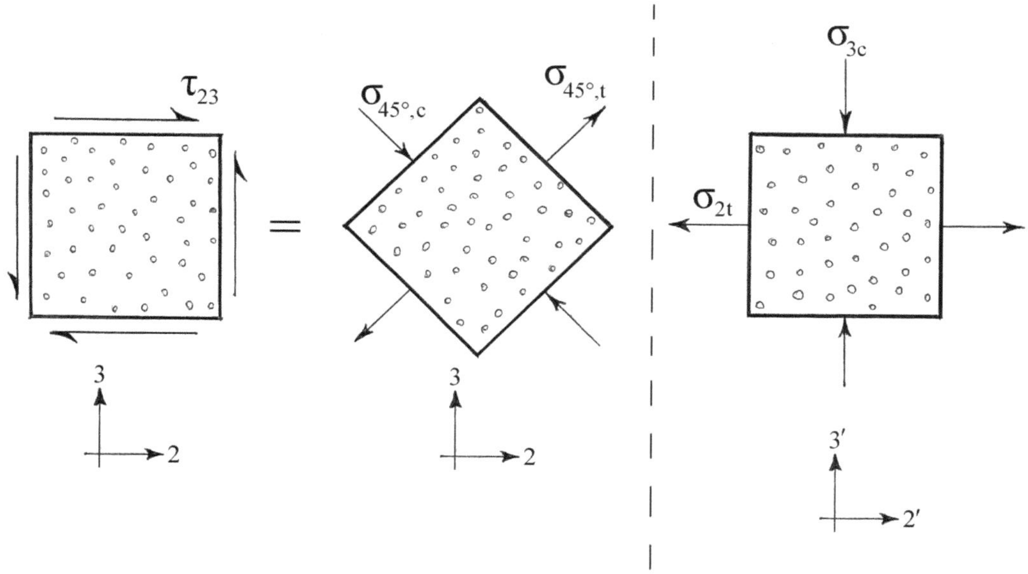

Fig. 9.7 Illustration of shear stress transformation within the 2-3 plane of the composite. **Left**: Original pure 2-3 shear stress. **Middle:** Equivalent tension-compression stress pair at 45°. **Right**: Equivalent normal stress state due to material symmetry in 2-3 plane

Material behavior on the 2-3 plane tends to be matrix-dominated, as there are no continuous fibers in either of these directions to carry load. Matrix-dominated tensile failure (either F_{2t} or F_{3t}) also tends to be the lowest out of all material strengths for a unidirectional continuous-fiber composite, and is likely to be the cause of material failure. By a simple maximum-stress failure criterion, one may then expect that.

$$F_{23} = F_{2t} \tag{9.8}$$

For the material in this study, the transverse tensile strength F_{2t} was determined to be 61.7 MPa (Table 9.1), which is reasonably close to the measured value of 52.7 MPa for F_{23}. Likewise, the crack propagation angle (mean 52°) is also fairly close to the theoretical prediction of 45°. This provides a reasonable justification for concluding that the out-of-plane shear strength F_{23} is closely related to the transverse tensile strength F_{2t}. Because the short sandwich beam specimen unavoidably imposes some small normal stress on its "core" as one moves away from the neutral axis, there does not exist a truly *pure* state of shear stress in much of the volume of the samples tested. It is unlikely, however, that the "principal" normal compressive stress contributes much to the failure, as the transverse compressive strength F_{2c} is 220 MPa, more than four times higher than the magnitude of the applied compressive stress. It is plausible, then, that the discrepancy may be a result of imperfections in the specimen.

Perhaps the most useful consequence of this study is to demonstrate that one need not, in principle, directly test the out-of-plane shear strength F_{23} of a composite in order to obtain its value with some reasonable accuracy – one may simply rely on a test to determine the transverse tensile strength F_{2t}, which is profoundly easier to carry out. This is, however, predicated on the assumption that material manufacture is of sufficiently high quality to justify an expectation of orthotropy (symmetry between the 2- and 3- axes of the material).

9.5 Conclusions

From this study, it was demonstrated that the out-of-plane shear strength F_{23} of unidirectional continuous-fiber composites may be successfully tested by means of a cross-ply sandwich-beam made entirely of the composite under test. Careful design of the specimen geometry is needed to ensure a valid shear failure in the intended mode, and multiple techniques are required to assess and evaluate acceptable failures. However, the specimen does permit a fairly straightforward means of testing the

2-3 shear strength when the supplied form of material is not well suited to other test methods. Furthermore, there exists the possibility that the 2-3 shear strength is closely linked to the transverse tensile strength F_{2t}, which if more thoroughly established could obviate direct testing of F_{23}.

Acknowledgment This study was supported by the Department of Energy under grant DE-EE0006867, with the Ford Motor Company as a principal contractor.

References

1. Lodeiro, M.J., Broughton, W.R., Sims, G.D.: Understanding the limitations of through-thickness test methods. In: Proc. 4th European Con$ on composites: testing and standardisation, pp. 80–90. IOM Communications Ltd., London (1998)
2. Abot, J.L., Daniel, I.M.: Through-thickness mechanical characterization of woven fabric composites. J. Comp. Mater. **38**(7), 543–553 (2004)
3. Arcan, M., Hashin, Z., Voloshin, A.: A method to produce uniform plane-stress states with applications to Fiber-reinforced materials. Exp. Mech. **18**, 141–146 (1978)
4. Walrath, D.E., Adams, D.F.: The Iosipescu shear test as applied to composite materials. Exp. Mech. **23**(1), 105–110 (1983)
5. Gipple, K., Hoyns, D.: Measurement of the out-of-plane shear response of thick section composite materials using the V-notched beam specimen. In: US Navy Technical Report CRDKNSWC-SSM-64-92/22. Naval Surface Warfare Center, Bethesda (1993)
6. Berg, C.A., Tirosh, J., Israeli, M.: Analysis of short beam bending of Fiber reinforced composites. In: Composite Materials: Testing and Design (Second Conference), ASTM STP 497, pp. 206–218. American Society for Testing and Materials, West Conshohocken (1972)
7. Whitney, J.M., Browning, C.E.: On short-beam shear tests for composite materials. Exp. Mech. **25**(3), 294–300 (1985). (critical)
8. Short, S.R.: Characterization of interlaminar shear failures of graphite/epoxy composite materials. Composites. **26**(6), 431–449 (1995)

Chapter 10
Nondestructive Damage Detection of a Magentostricive Composite Structure

Michael Coatney, Asha Hall, Mulugeta Haile, Natasha Bradley, Jin Hyeong Yoo, Brandon Williams, and Oliver Myers

Abstract The integrity of composite structures gradually degrades due to the onset of damage such as matrix cracking, fiber/matrix debonding, and delamination. Over the last two decades, great strides have been made in structural health monitoring (SHM) community using various sensing techniques such as acoustic emission, eddy current, strain gages, etc., to diagnose damage in aerospace, mechanical and civil infrastructures. Embedded sensing offers the prospects of proving for real-time, in-service monitoring of damage were weight savings is a major factor in Aerospace Industry. It also provides for a new nondestructive indication of early stage damage. Defect detection and monitoring of fatigue in structural materials can be captured through local indicators of a change of the magnetic properties within the damaged sites. In this present work, magnetostrictive particles such as Terfenol-D were embedded in a composite structure, along with acoustic emissions technique, to validate the damage in a composite system undergoing qausi static and fatigue loading. As the applied load and fatigue cycles increased, the change in the magnetization flux density was captured using a non-contact magnetic field sensor. It was confirmed through numerous tests that a change in the magnetic properties of the composite served as an indicator of early stage damage detection.

Keywords Embedded sensing · Structural health monitoring · Magnetostriction · Terfenol-D · Damage detection

10.1 Introduction

The integrity of composite structures gradually degrades due to the onset of damage such as matrix cracking, fiber/matrix debonding, and delamination. Over the last two decades, great strides have been made in structural health monitoring (SHM) community using various sensing techniques such as acoustic emission, eddy current, strain gages, etc., to diagnose damage in aerospace, mechanical and civil infrastructures. While various conventional non-destructive evaluation (NDE) techniques are used to detect, monitor, and visualize damage of composite structures under fatigue loading, they all have a common limitation in that continuous damage assessment cannot be made in real-time while in service. Our current research examines the possibilities of embedding nanomaterials into composite laminates as a minimally invasive, non-contacting, nondestructive evaluation and sensing technique. The embedded sensing composite effectively becomes a smart structure with the capability of self-sensing damage by evaluating the integrity of the material in real-time.

Ferromagnetics possess unpaired electronic spins that line up parallel with each other in a region called a domain. The domains are randomly oriented when the ferromagnetic material is not under an external magnetic field. Once an external magnetic field is applied the electronic spins line with each other and become magnetized. Magnetostriction is a property of ferromagnetic materials such that when an external magnetic field is applied the domains align to the external field causing their shape or dimensions to change during the process of magnetization. For example, TerFenol-D, ($TbxDy_1 - xFe_2$) strains to about 2000 microstrain in a field of 2 kOe when under a mechanical-bias. Due to their high magnetostrictive coefficients

M. Coatney (✉) · A. Hall · M. Haile · N. Bradley
US Army Research Laboratory, Aberdeen, MD, USA
e-mail: michael.d.coatney.civ@mail.mil

J. H. Yoo
US Naval Surface Warfare Center (NSWC) Carderock Division, West, MD, USA

B. Williams · O. Myers
Mechanical Engineering Department, Clemson University, Clemson, SC, USA

P. R. Thakre et al. (eds.), *Mechanics of Composite, Hybrid and Multifunctional Materials, Volume 5*, Conference Proceedings of the Society for Experimental Mechanics Series, https://doi.org/10.1007/978-3-319-95510-0_10

TerFenol-D serves as an excellent candidate for in-situ sensing of damage in various kinds of polymer composites under different mechanical loading situations.

In this study, magnetostrictive particles such as Terfenol-D were embedded in a composite structure, along with acoustic emissions technique, to validate the damage in a composite system undergoing qausi static and fatigue loading. As the applied load and fatigue cycles increased, the change in the magnetization flux density was captured using a non-contact magnetic field sensor. It was confirmed through numerous tests that a change in the magnetic properties of the composite served as an indicator of early stage damage detection.

10.2 Preliminary Results

Figure 10.1 depicts a stress – strain curve of a magnetostrictive CFRP sample. The elastic modulus of the Non Magnetostrictive CFRP sample was 110 GPa, while the Elastic Modulus of the Magnetostrictive sample is 94 GPa, which is a 7% decrease in the failure stress. This indicates a slight decrease in strength as the Terfenol-d particles are added.

The results from experimental tensile testing of unidirectional carbon fiber reinforced polymer composite specimens with and without magnetostrictive particles showed that the particle layer was slightly intrusive on the quasi-static tension properties of the beam. Based on the number of samples tested, the amount of change seen in both the tensile strength and the modulus was statistically negligible. Future tests will be conducted incorporating acoustic emission to further validate the correlation between change in magnetic flux intensity to the increase in microcrack density.

A non-contact external magnetic sensor comprised of a driving coil and a pick up coil was engineered to send a magnetic field and then pick up the magnetic flux density response of the magnetostrictive composite specimens while under an applied load. The change in magnetic flux density, depicted by delta RMS, is an indicator of the degradation of the material properties due to damage (e.g matrix cracking, fiber matric de-bonding, and/or delamination).

Initial tests were measured to find the ultimate tensile strength (UTS) of the 15 wt% specimens. Once the UTS was known, the next set of samples were loaded at 10% intervals, up to 70% of the UTS. At each load interval, the load was held while a signal was recorded for approximately 3–4 s at a 5000 Hz sampling rate. Figure 10.2 depicts the peak voltage of the sample as each load intervals were applied. At 50% applied load there is a decrease in voltage RMS (i.e. magnetic flux density) of the specimen as.

Well as a slight frequency shift. There is a suspicion of damage such as a micro cracking may have occurred at the 50% UTS applied load which would have shifted resonance frequency. On the contrary, as the UTS percentage increases to the 60% and 70% UTS the voltage RMS returns to the initial voltage RMS as recorded earlier in the loading regime. The presence of internal stresses in a structure could influence the domain boundary movement of magnetostrictive particles in the structure which can be detected by an external magnetic sensor thereby revealing the early stage of the initiation of damage.

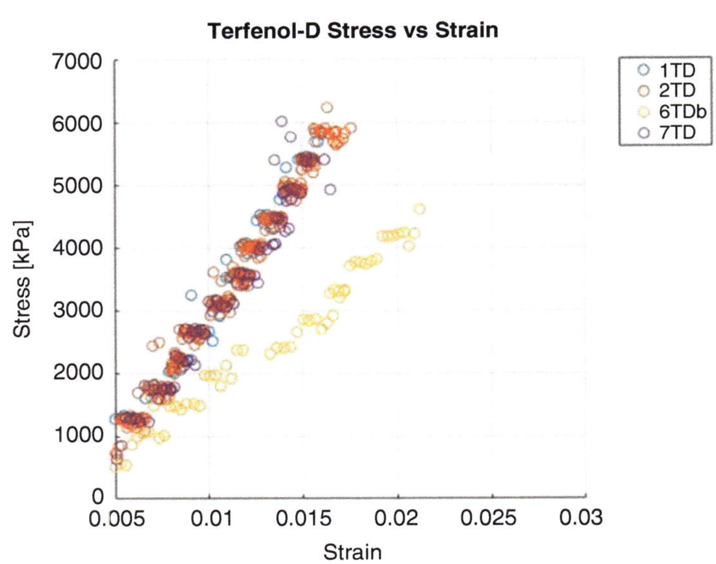

Fig. 10.1 Stress-strain curve of magnetostrictive specimens

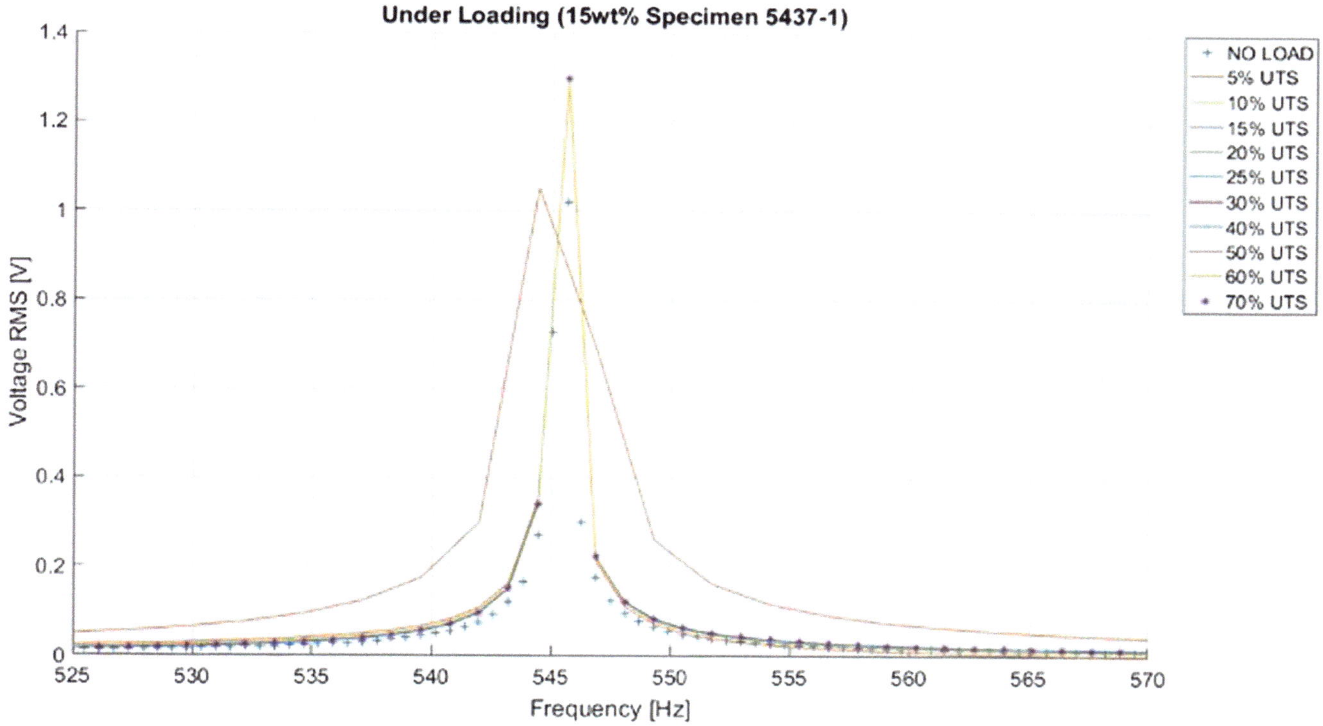

Fig. 10.2 Voltage RMS recorded with driving coil and pick-up coil sensor

10.3 Conclusion

In this study, magnetostrictive Terfenol-D powders were embedded in Terfenol-D composite specimens. This technique was performed to examine the changes in magnetostriction from tensile fatigue loading. The results from experimental tensile testing of (0/90) carbon fiber reinforced polymer composite specimens with and without magnetostrictive particles showed that the particle layer was minimally intrusive on the quasi-static tension properties of the beam. Based on the number of samples tested, the amount of change seen in both the tensile strength and the modulus was statistically negligible. Future tests will be conducted incorporating acoustic emission to further validate the correlation between change in magnetic flux intensity to the increase in microcrack density.

References

1. Wang, S., Shui, X., Fu, X., Chung, D.D.L.: Early fatigue damage in carbon-fibre composites observed by electric resistance measurements. J. Mater. Sci. **33**, 3875–3884 (1998)
2. A.J. Hall, R. E. Brennan, A. Ghoshal, K. C. Liu, M. Coatney, R. Haynes, N. Bradley, V. Weiss, J Tzeng, " Damage precursor investigation of fiber-reinforced composite materials under fatigue loads, ARL-TR-6622, 2013
3. ASTM Standard D3039, 2007: Standard test method for tensile properties of polymer matrix composite materials. ASTM International, West Conshohocken (2007). https://doi.org/10.1520/D3039_D3039M-08
4. Ramsteiner, F., Ambrust, T.: Fatigue crack growth in polymer. Polym. Test. **20**(3), 321–327 (2001)
5. Mudivarthi, C., Datta, S., Atulaimha, J., Flatau, A.B.: A bidirectionally coupled magnetoelastic model and its validation using a Galfenol unimorph sensor. Smart Mater Struct. **17**(3), 035005 (2008)
6. Calkins, F.T., Flatau, A.B., Dapino, M.J.: Overview of magnetostrictive sensor technology. J. Intell. Mater. Syst. Struct. **18**(10), 1057–1066 (2007)
7. Purekar, A., Ragunath, G., Flatau, A.B., Yoo, J.-H.: Development of galfenol based non-contact torque sensor. In: 53rd AIAA/ASME/ASCE/AHS/ASC structures, structural dynamics and materials conference (2012)
8. Clark, A.E., Abbundi, R., Gillmor, W.R.: Magnetization and magnetic anisotropy of TbFe2, DyFe2, Tb0. 27Dy0.73Fe2 and TmFe2. IEEE Transact. Magnet. **14**(5), 542–544 (1978)

Chapter 11
Thermo-Mechanical Properties of Thermoset Polymers and Composites Fabricated by Frontal Polymerization

M. Yourdkhani, B. Koohbor, C. Lamuta, L. M. Dean, P. Centellas, D. G. Ivanoff, I. D. Robertson, S. R. White, and N. R. Sottos

Abstract Thermoset polymers are commonly used as the matrix material in fiber-reinforced polymer composites (FRPCs) due to their good mechanical properties, chemical stabilities, and ease of manufacturing. Conventional curing of thermosets and their composites requires heating the matrix monomers at elevated temperatures during long cure cycles for producing fully crosslinked polymers, resulting in high manufacturing cost in terms of time, energy, and capital investment. Frontal polymerization (FP) is a promising approach for rapid, energy-efficient fabrication of high-performance thermosets and FRPCs. In FP, a thermal stimulus (trigger) causes a self-propagating exothermic reaction wave that transforms liquid monomers to fully cured polymers, eliminating the need for external energy input by large ovens or autoclaves. We have used the FP of dicyclopentadiene (DCPD) to successfully fabricate thermoset polymers and composite parts. In this novel curing strategy, the final degree-of-cure of the polymer, and thereby its mechanical performance, is governed by the heat transfer phenomenon that occur at the polymerization front. During the fabrication of FRPCs some fraction of the generated heat is absorbed by continuous fibers or lost through the tooling. In this work, we will discuss the characterization of the thermo-mechanical properties of DCPD polymer manufactured by FP curing.

Keywords Frontal polymerization · Polymer composites · Digital image correlation · Residual stress · Advanced manufacturing

Thermoset polymers are generated via the formation of irreversible 3D crosslinking networks between monomers often facilitated by applying external energy in the form of heat or light. The resulting crosslinked network yields the polymer thermally stable with superior mechanical performance and high glass transition temperature (T_g) compared to thermoplastic counterparts. As a result, thermosets have been widely used in high-performance, light-weighting applications in aerospace, automotive, marine, and energy industries. In particular, the matrix polymer in fiber-reinforced polymer composites is typically composed of a thermoset such as epoxy. Current technologies for the manufacturing of high-performance polymeric and composite parts rely on curing in large autoclaves or ovens, leading to high manufacturing time, energy, and cost. It is highly desirable to cure these polymers with less energy to reduce their cost and carbon footprint and further their applications.

M. Yourdkhani (✉) · C. Lamuta
Beckman Institute for Advanced Science and Technology, University of Illinois at Urbana-Champaign, Urbana, IL, USA
e-mail: yourd@illinois.edu

B. Koohbor · L. M. Dean · D. G. Ivanoff · N. R. Sottos
Beckman Institute for Advanced Science and Technology, University of Illinois at Urbana-Champaign, Urbana, IL, USA

Department of Materials Science and Engineering, University of Illinois at Urbana-Champaign, Urbana, IL, USA

P. Centellas · S. R. White
Beckman Institute for Advanced Science and Technology, University of Illinois at Urbana-Champaign, Urbana, IL, USA

Department of Aerospace Engineering, University of Illinois at Urbana-Champaign, Urbana, IL, USA

I. D. Robertson
Beckman Institute for Advanced Science and Technology, University of Illinois at Urbana-Champaign, Urbana, IL, USA

Department of Chemistry, University of Illinois at Urbana-Champaign, Urbana, IL, USA

P. R. Thakre et al. (eds.), *Mechanics of Composite, Hybrid and Multifunctional Materials, Volume 5*, Conference Proceedings of the Society for Experimental Mechanics Series, https://doi.org/10.1007/978-3-319-95510-0_11

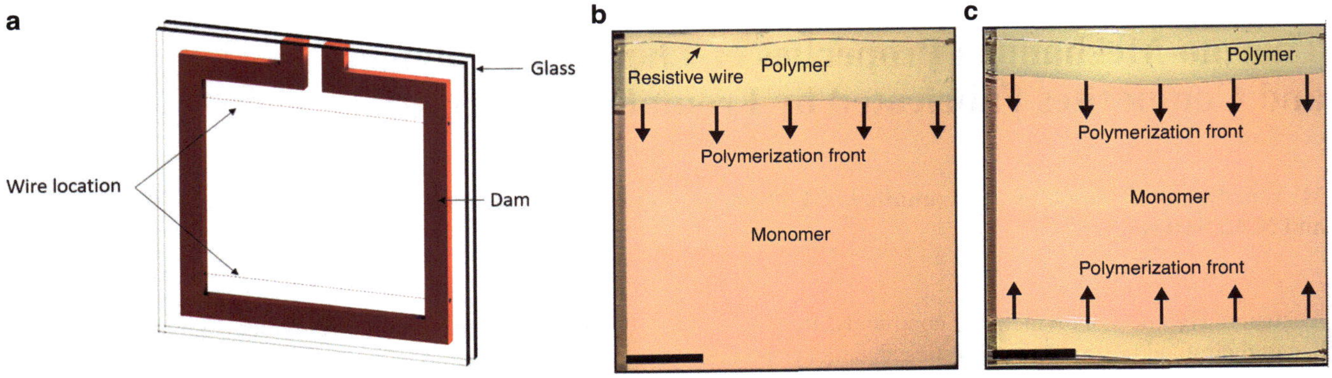

Fig. 11.1 Triggering modes of frontal polymerization. (**a**) Schematic of the open glass mold used to manufacture polymer panels. (**b**) One resistive wire embedded in the mold and connected to a power source initiates the reaction that propagates from top to bottom. (**c**) Two resistive wires on the opposite ends of the molds trigger the reaction simultaneously, where the reaction fronts propagate in opposite directions and merge at the middle of the mold. The scale bar is 2.5 cm

One promising approach to substantially reduce the manufacturing time and energy is to employ the enthalpy of polymerization to produce the necessary thermal energy for curing the polymer rather than using external energy. Frontal polymerization (FP) is a propagating reaction wave based on the heat release of an exothermic polymerization reaction [1]. A solution of monomer and latent initiator is heated locally until the initiator is activated, producing heat from polymerization, and further accelerating the reaction. The result is a propagating wave of polymerization, which rapidly polymerizes the available monomers. The manufacturing time of a product is thereby directly dependent on the front speed as well as the traveling distance of the front. Using more than one thermal trigger point at different locations allows for reducing the travel distance, and consequently, the manufacturing time, which is of great importance in manufacturing of large components.

Here, we have used the FP of dicyclopentadiene (DCPD) [2] to produce high-performance thermoset and FRPC parts. The FP reaction is initiated by applying an electrical current through a resistive wire embedded in an open glass mold (Fig. 11.1a, b) or composite layup for manufacturing neat resin and FRPC panels, respectively. The frontal speed in the neat resin and FRPC is 7.5 and 10 cm/min, respectively. Using two resistive wires in opposite ends of the glass mold (Fig. 11.1c) or composite layup reduces the manufacturing time by half compared to a single triggering mode. However, as the two fronts merge, the heat produced by the monomers in opposite fronts cannot be removed and will lead to the formation of a thermal spike that may result in undesirable volume changes, variation in material properties, or creation of residual stresses. The present study aims at understating the effect of the thermal overshoot that occurs during the merging of two fronts on the mechanical properties of produced polymer parts.

Local mechanical response of neat polymer samples is evaluated through full-field strain measurements facilitated by Digital Image Correlation (DIC). DIC area of interest is selected such that merged front area would remain within the field of view throughout the entire experiment. Time-lapse images are acquired from the speckled area of interest at a rate of 1 Hz, synchronized with load-cell data sampling rate. Images are post-processed in the DIC software Vic-2D to obtain the full-field distribution of in-plane strain components. Figure 11.2 shows representative contour maps obtained at different global stress levels. Strain localization is evident at near-to-failure stress levels for double front polymerized samples. Possible mechanisms giving rise to such localized strain response will be discussed in the presentation.

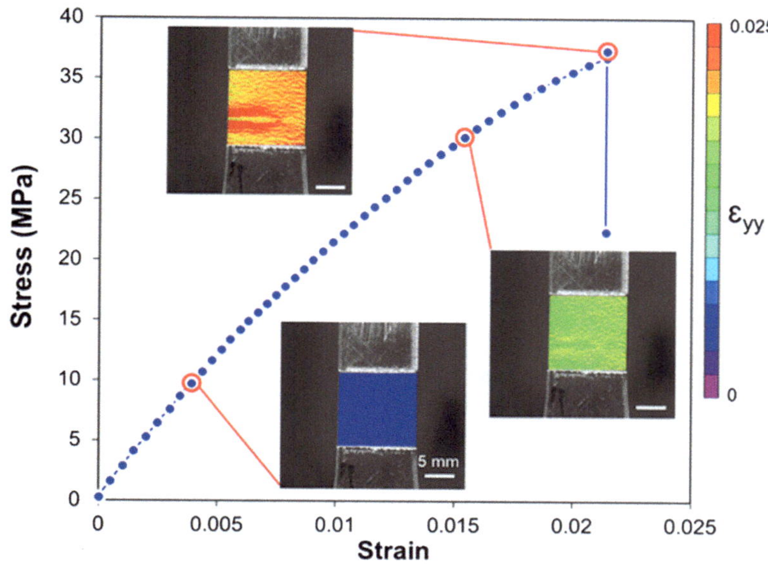

Fig. 11.2 Contour maps showing the distribution of vertical in-plane strain component, ε_{yy}, at various stress levels. Tensile load is applied in *y*-direction

Acknowledgement This work was supported by the United States Air Force Office of Scientific Research through award FA9550-16-1-0017.

References

1. Pojman, J.A.: Frontal Polymerization. In: Matyjaszewski, K., Möller, M. (eds.) Polymer Science: A Comprehensive Reference, pp. 957–980. Elsevier (2012)
2. Robertson, I.D., et al.: Alkyl phosphite inhibitors for frontal ring-opening metathesis polymerization greatly increase pot life. ACS Macro Lett. **6**(6), 609–612 (2017)

Chapter 12
Design of Magnetic Aluminium (AA356) Composites (AMCs) Reinforced with Nano Fe_3O_4, and Recycled Nickel: Copper Particles

L-M. P. Ferreira, E. Bayraktar, I. Miskioglu, and M-H. Robert

Abstract New classes of aluminium matrix composites (AMCs – 356) were designed by three different manufacturing techniques; only sintering, through combined method called here after "Sinter + Forging" and/or "sinter + thixoforming". Main reinforcement was magnetic iron oxide, Fe_3O_4 (10, 20 and 30 wt %) and two recycled reinforcements, nickel, Ni and pure electrolytic copper, Cu, given by French Aeronautical Society were also used and preceded under the constant process parameters such as hot compaction, sinter-forging, sintering time, Forging temperature and Force, etc. As auxiliary element, hybrid graphene nano-platelets, GNPs, was added in the structure. Microstructural analyses (by using SEM), magnetic, mechanical and physical properties of the composites were compared with three different manufacturing processes. Static compression tests, Micro hardness tests, measurement of magnetic permeability and also electrical conductivity, have shown that the mechanical and physical properties of these composites can be improved with the optimization of process parameters. In the present work, an alternative and a low cost manufacturing process were proposed for these composites.

Keywords Aluminium Matrix Composites · Graphene · Fe_3O_4 · AMCs · Magnetic Permeability · Electrical Conductivity · Mechanical Properties · Wear-SEM

12.1 Introduction

Aluminium matrix composite materials (AMCs) are the important class of advanced composite materials giving engineers the opportunity to modify the material properties according to the needs of new technology. These composites are actually used in high-tech structural and functional applications including aerospace, defence, automotive, and thermal management areas. New class of "AMCs" reinforced with nano iron oxide (Fe_3O_4), hybrid graphene nano-platelets, GNPs ($d < 10$ nm), Nickel, etc., show good thermal conductivity and also low thermal expansion. For this reason, these composites are categorized as very high multifunctional lightweight materials.

The present paper is based on the low cost manufacturing light and efficient materials and using them in engineering applications. For this reason, a comprehensive study has been carried out on the magnetic properties of the AMCs (AA356) reinforced basically with nano Fe_3O_4-iron oxide and other alloying elements such as hybrid graphene nano-platelets, GNPs, Nickel, etc.

In the frame of the research project that is going on. Three different manufacturing techniques were used; only sintering, through combined method called here after "Sinter + Forging" and/or "sinter + thixoforming" with low cost manufacturing of light and efficient multifunctional materials or much more interesting one is multiferroïques composite for aeronautical applications that is one of the aims of the present work.

L.-M. P. Ferreira
Federal University of Southern and South-eastern, Marabá, Pará, Brazil

Supmeca-Paris, School of Mechanical and Manufacturing Engineering, Saint-Ouen, France

E. Bayraktar (✉)
Supmeca-Paris, School of Mechanical and Manufacturing Engineering, Saint-Ouen, Paris, France
e-mail: bayraktar@supmeca.fr

I. Miskioglu
Michigan Technology University, Engineering Mechanics Department, Houghton, MI, USA

M.-H. Robert
University of Campinas, Mechanical Engineering Faculty, Campinas, Brazil

P. R. Thakre et al. (eds.), *Mechanics of Composite, Hybrid and Multifunctional Materials, Volume 5*, Conference Proceedings of the Society for Experimental Mechanics Series, https://doi.org/10.1007/978-3-319-95510-0_12

Table 12.1 Chemical composition of scrap A356 powder (wt %)

Element	Al	Cu	Fe	Mg	Mn	Si	Ti	Zn
Wt %	Balance	≤0.20	≤0.20	≤0.25	≤0.10	6.5–7.5	≤0.20	≤0.10

Table 12.2 Compositions of the composites prepared in two groups (wt %)

Composition name	Fe_3O_4	Pure Al	Ni	Cu	GNPs	Zn- Stearate
A356-I	10	5	0	1	1	1
A356-II	10	5	4	1	1	1
A356-III	20	5	4	1	1	1
A356-IV	30	5	4	1	1	1

Furthermore, Multiferroïques are an attractive class of materials that are instantaneously ferromagnetic and ferroelectric, and they have potential applications in multifunctional devices, transducers, actuators and sensors [6–9]. There are two types of multiferroïques used in engineering applications: intrinsic magneto electrics that exist as a single-phase state, a few ones such as MnO_3, $BiMnO_3$ and $BiFeO_3$, and extrinsic multiferroïques that are composites or solid solutions of ferroelectric and ferromagnetic crystals. Since the number of intrinsic multiferroïques is not so high [4], there have been many researches to manufacture multiferroïques composites [5–9].

For the composites of ferromagnetic and ferroelectric crystals, the connection between the ferromagnetic and ferroelectric order parameters is through the strain order parameter. In other words, the magnetostrictive and electrostrictive (or piezoelectric) effects of the ferromagnetic and ferroelectric crystals lead to a combination between magnetization and polarization through the elastic interactions.

In the last few decades, high performance AMCs have been widely developed with high strength, high stiffness, low density, and good wear resistance capacity [1–5]. Among them, A356 aluminium is widely used as matrix. A356 is an interesting option for aeronautical, military and automobile applications as the alloy having important properties of high strength, light weight and good capacity for foundry (fluidity). This alloy can also be successfully used as matrix from fresh scrap powder to produce high quality metal matrix composites in an economic way [6–11]. In the present work, Fe_3O_4, GNPs and Ni were chosen as the basic reinforcements added to the AMCs structures to improve mechanical behaviour and wear resistance of these novel composites [7–13].

Morphology of the composites was evaluated by Scanning Electron Microscopy (SEM). Scratch tests were conducted to evaluate wear properties.

12.2 Experimental Conditions

In the present work, Aluminium Matrix Composites (AMCs with scrap A356 powder given by French Aeronautical Society) were designed by using different reinforcements such as magnetic iron oxide (Fe_3O_4), hybrid graphene nano-platelets, GNPs and also scrap Nickel and copper, Cu. These composited have been manufactured by three different manufacturing techniques; only sintering, through combined method called here after "Sinter + Forging" and/or "sinter + thixoforming". Main reinforcement were magnetic iron oxide, Fe_3O_4 (10, 20 and 30 wt %) and recycled reinforcements, nickel, Ni (4 wt %). For easy wettability of the reinforcements to the matrix, certain amount of pure nano aluminium was added in the mixture. These composites have a good advantages regarding to conventional manufacturing processes such as low cost, capability of producing products with complex shapes, and processing simplicity, etc.

Effectively, this work aims to improve the quality and mechanical and wear behaviour of the composites After atomization of scrap A356 powder, a planetary ball milling was used during 2 h, these powders was successfully combined with particulate reinforcements. After that, this main structure was mixed with ball milling during 4 h to homogenize the mixture. The general composition of scrap A356 powder was given in the Table 12.1 and also, the compositions of two types of compositions were presented in two groups (Table 12.2).

Microstructural characterization was done by means of scanning electron microscope (SEM). The dispersion of reinforcement particles in the matrix and interface at matrix/reinforcements was evaluated. Micro hardness tests ($HV_{0.3}$) have been carried out on the polished and etched specimens.

All the density measurements of the specimens were carried out by using *Archimedes* method. These values change between 2.45 and 2.62, 2.79 and 2.83 respectively. And then, these results were compared. 4–5 cylindrical specimens

(H/D $\geq$ 1.5) for each composition. Quasi-static compression tests have been carried out in a servo-hydraulic MTS Universal test system (model: 5500R) at an initial rate of 10 mm/min and second rate of 5 mm/min. Maximum load endpoint was 5000 N. These test were made for only the series of A356-IV where magnetic iron oxide is 30 wt %.

Wear resistance was measured by scratch wear tests at a frequency of 15 Hz for only the specimens of A356-IV for only two different numbers of cycles ($50*10^3$, $100*10^3$). After scratch test, damaged zone was investigated by 3D optical roughness meter. Damage surfaces and Volume loss/time and maximum depth were evaluated. Electrical and magnetic measurements have been carried out by the laboratory of the French aeronautical society on the four specimens for each composite and evaluated in this work.

12.3 Results and Discussion

12.3.1 Microstructural Evaluation of the Composites

A novel composite is aimed with an optimized magnetic permeability and good electrical conductivity. For this reason, a magnetic iron oxide (Fe_3O_4) was used as primary reinforcement in these composites and GNPs, Ni, etc., were added as secondary reinforcements for obtaining an improved mechanical behaviour and wear resistance. As indicated in the papers of the same authors of this work, nano magnetic iron oxide was produced easily at the university laboratory [5, 10, 11, 12, 14]. Figure 12.1 shows general microstructure of the composite of AA356 [SEM2017] and crystal structure of the magnetite iron oxide (Fe_3O_4 cubic, Fd3m, the black ball present Fe^{2+}, the green ball present Fe^{3+} and the red ball present O^{2-} ions), at the right picture and also XRD analyses for the final structure of the composites after sinter+forging process on the local zone, spectrum 1 and 2 for only giving an information about the structure [6, 7].

12.3.2 Evaluation of Magnetic Properties for A356-I, II, III, IV

Magnetic measurements have been carried out by the laboratory of the French aeronautical society. Two test specimens were used for each composite and evaluated by drawing the magnetic saturation, remanence values and compared certain hysteresis parameters and presented in the Figs. 12.2, 12.3, 12.4 and 12.5 for the compositions respectively.

Magnetic properties of the composite structures can be improved with only small grain size distribution well in the matrix especially in the nanoscale [17, 21, 23].

As shown in these figures, the percentage of the Fe_3O_4 particles and Nickel increase considerably the magnetic properties of the composites studied. The field dependence of magnetization measured is presented in the same figures; the magnetic saturation (Ms) and remanence values, (Mr) have been improves in regular way Ms. values. As indicated in the former papers [5, 11, 17, 21, 23], a basic and important parameter in the characterization of soft magnetic materials is the power loss; this kind of power gives a measure of the energy density available in the material for a specific application. If data obtained from the analysis of the hysteresis curve as the saturation magnetization (M_s), remanence magnetization (M_r), coercive field (H_c), are compared for all of the composites, the highest magnetic permeability (Ms = 9. 98 emu/g) was obtained on the specimens of A356-IV regarding to the other composite series.

It is noted that the value of M_s must be maximized as much as possible to confirm a better response under the application of the pieces. These results are the first time published and original for these composites and other measurements for different composites are going on in the frame of the common research project for aeronautical applications. At this stage, we cannot give more detail about this part due to the actual research project that is going on.

After choosing the composite, AA356-IV containing 4 wt% Nickel and 30 wt % Fe_3O_4, further characterizations such as static compression and wear (scratch) tests have been carried out on this composite.

12.3.3 Static Compression Test Results and Micro Hardness Measurements

Actually, sinter + forging and/or sinter+ thixoforming processes are a novel processes regarding to the conventional processes mainly called as near-net shape process for the manufacturing of the pieces [13–15]. Mainly, this process is used for bulk materials in industrial applications. In fact, low-cost sinter+forging approach to processing of particle-reinforced

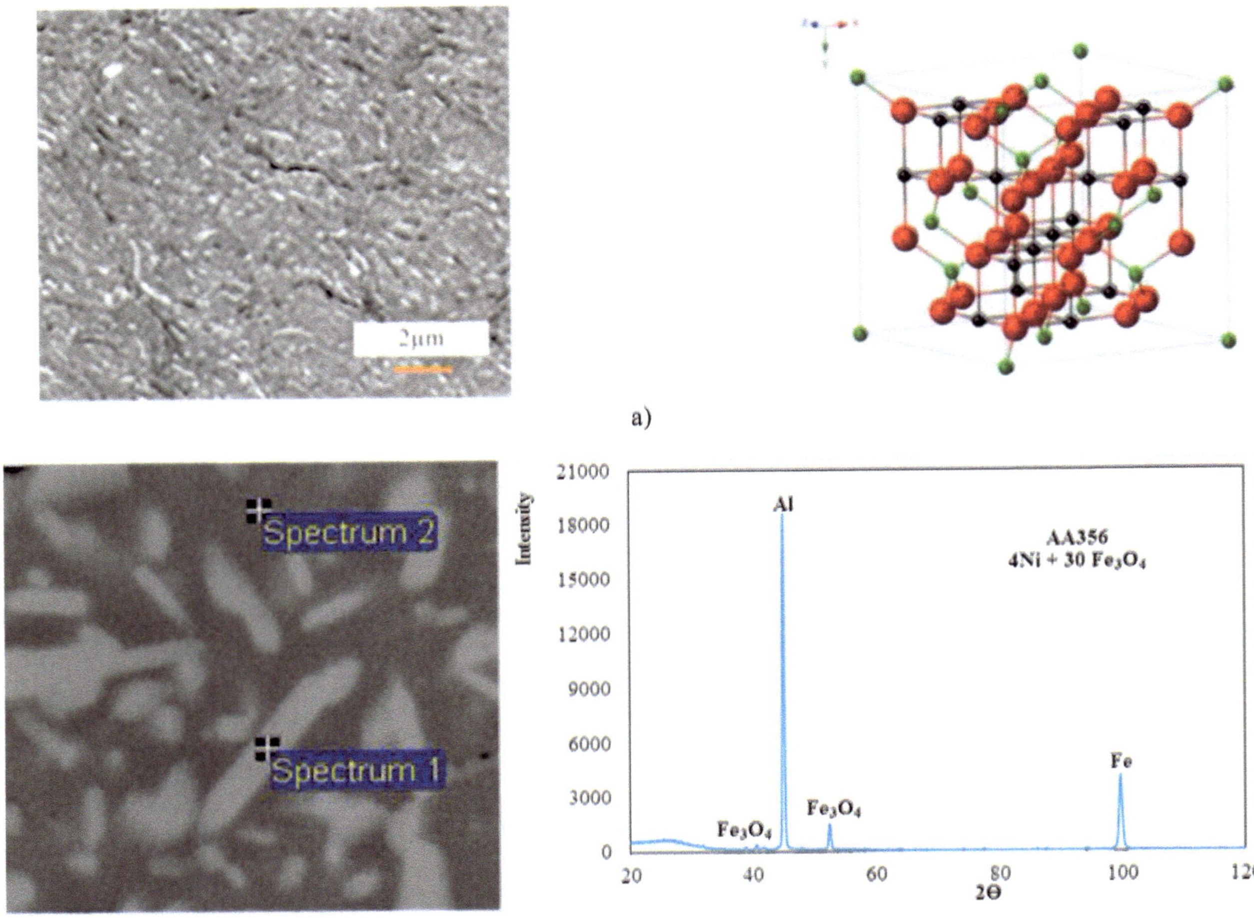

Spectrum N°	In stats.	O	Al	Fe	Ni	Cu	Total
Spectrum 1	Yes	1.62	63.19	26.09	7.22	1.88	100.00
Spectrum 2	Yes		68.05	19.03	11.55	1.38	100.00

b)

Fig. 12.1 (**a**) General microstructure of the composite AA356 [SEM2017] and crystal structure of the magnetite iron oxide (Fe_3O_4 cubic, Fd3m, the black ball present Fe^{2+}, the green ball present Fe^{3+} and the red ball present O^{2-} ions), at the right picture and (**b**) XRD analyses for the final structure of the composites after sinter+forging process on the local zone, spectrum 1 and 2 [6, 7]

metal matrix composites gives always high performance applications of the industrial pieces (fatigue-creep, static and impact compression behaviours, etc.).

In the frame of this present work, only a simple static compression tests have been carried on the chosen compositeAA356-IV among the other series by using of three different manufacturing processes. As mentioned in the former session, this composite show very satisfied results in order to optimize the magnetic permeability and electrical conductivity. This idea should be developed very well on the many other composites in industrial scale. It means that very tough and strong pieces can be obtained with a microstructure by this manufacturing process: combined method sinter+forging.

Figure 12.6 gives static compression test results carried out on the series of A356-IV with three manufacturing processes. These values are mean values obtained from four tests for each composition. Evidently standard deviation is variable around ±15–20 MPa for all of three manufacturing processes due to internal (micro) defects – porosity in the microstructure [17, 19, 23]. The same confirmation was given by micro hardness measurements taken from the specimens produced for A356-IV by three different manufacturing processes; HV = 107 ± 10 (sintering), HV = 119 ± 5 (sinter + thixoforming) and HV = 129 ± 15 (sinter+forging) respectively. One may observe that the specimens produced by the sinter+ thixoforming process have tougher than the others. These values are only experimental results that should be improved with the specimens improved without defect in the microstructure.

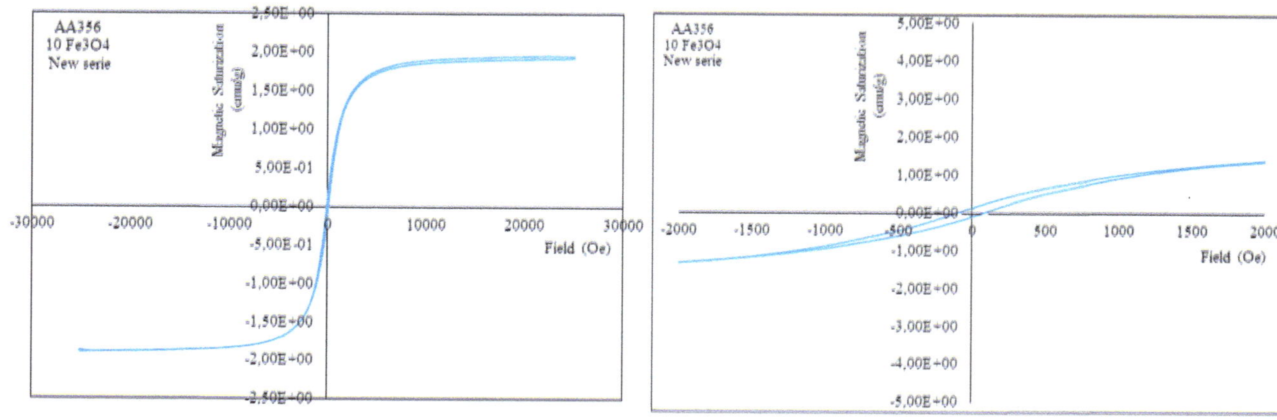

Fig. 12.2 Magnetic properties; magnetic saturation curves at the left side and the hysteresis curves at the right side, for the composite, AA356 containing only 10 wt % Fe_3O_4 respectively

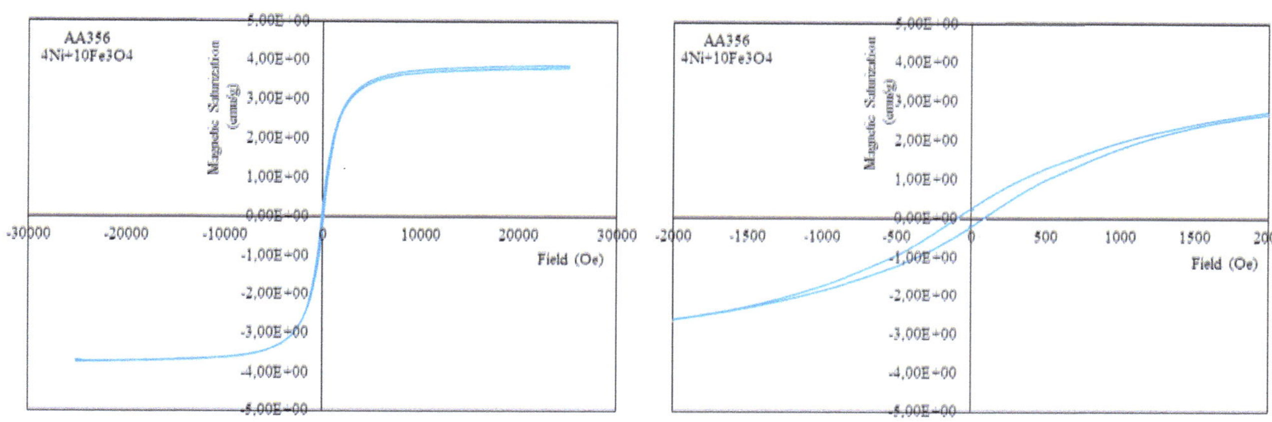

Fig. 12.3 Magnetic properties; magnetic saturation curves at the left side and the hysteresis curves at the right side, for the composite, AA356 containing 4 wt% Nickel and 10 wt % Fe_3O_4 respectively

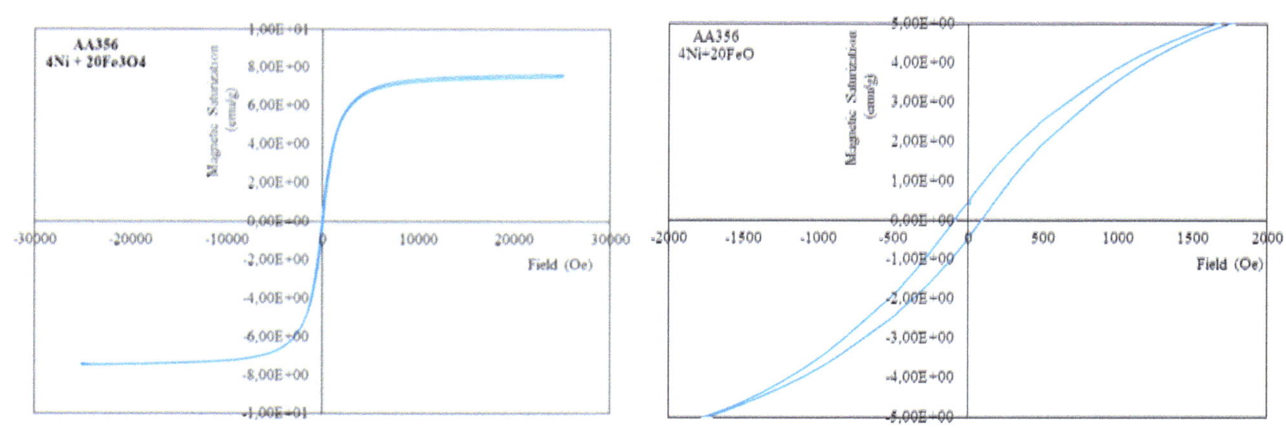

Fig. 12.4 Magnetic properties; magnetic saturation curves at the left side and the hysteresis curves at the right side, for the composite, AA356 containing 4 wt% Nickel and 20 wt % Fe_3O_4 respectively

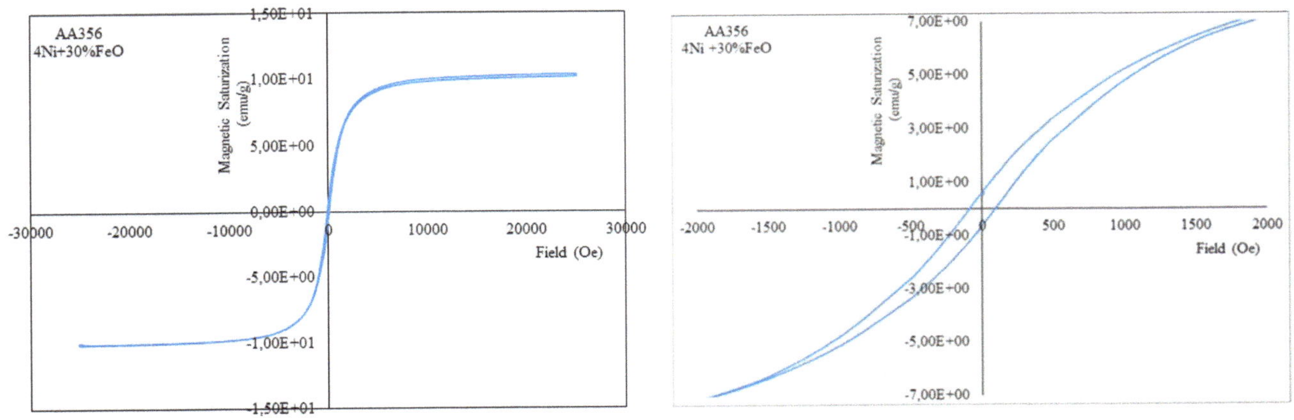

Fig. 12.5 Magnetic properties; magnetic saturation curves at the left side and the hysteresis curves at the right side, for the composite, AA356 containing 4 wt% Nickel and 30 wt % Fe_3O_4 respectively

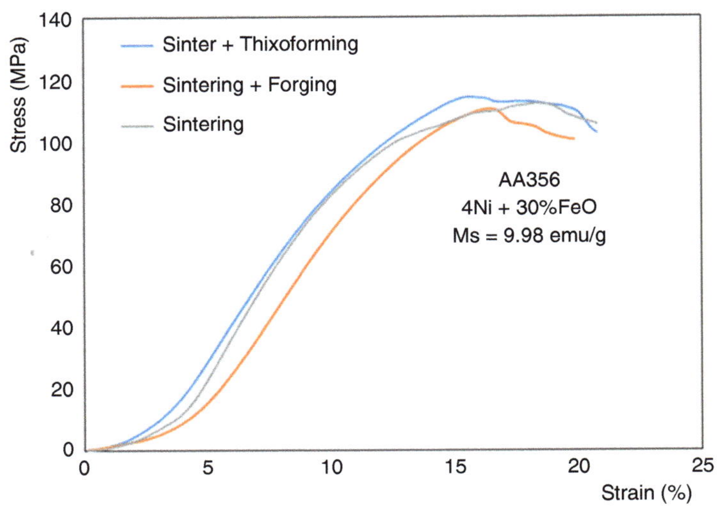

Fig. 12.6 Static compression test results for three manufacturing process of A356 (4 wt % Ni + 30 wt % Fe_3O_4); with only sintering, sinter+ forging and with sinter + thixoforming

12.3.4 *Wear Resistance by Scratch Test*

Reduction of friction rests the main challenge for wear-resistant composites. Evaluation of wear resistance has been carried out on the same composite A356-IV having high magnetic values. This time, only one manufacturing process was used for evaluation of the wear resistance. Two different numbers of cycles, ($50*10^3$, $100*10^3$ cycles) have been used for this evaluation. These results were presented in the Fig. 12.7 for two different wear tests conditions.

The surface, volume and the values of the damage surfaces are presented in the same figures for each test condition. These tests are going on actually for the new series under different parameters. Considering the test conditions, it seems that not so much differences in overall wear results even if wear resistance of GNPs indicated in literature [24].

12.4 Conclusions

In the frame of this common project, the new design of magnetic AMCs composites have been carried out from fresh scrap aluminium A356 reinforced with magnetic nano iron oxide (Fe_3O_4) and nickel and graphene nano particulates (GnPs). Iron oxide was doped with other auxiliary elements by means of a novel thermo mechanical treatment in an economic way.

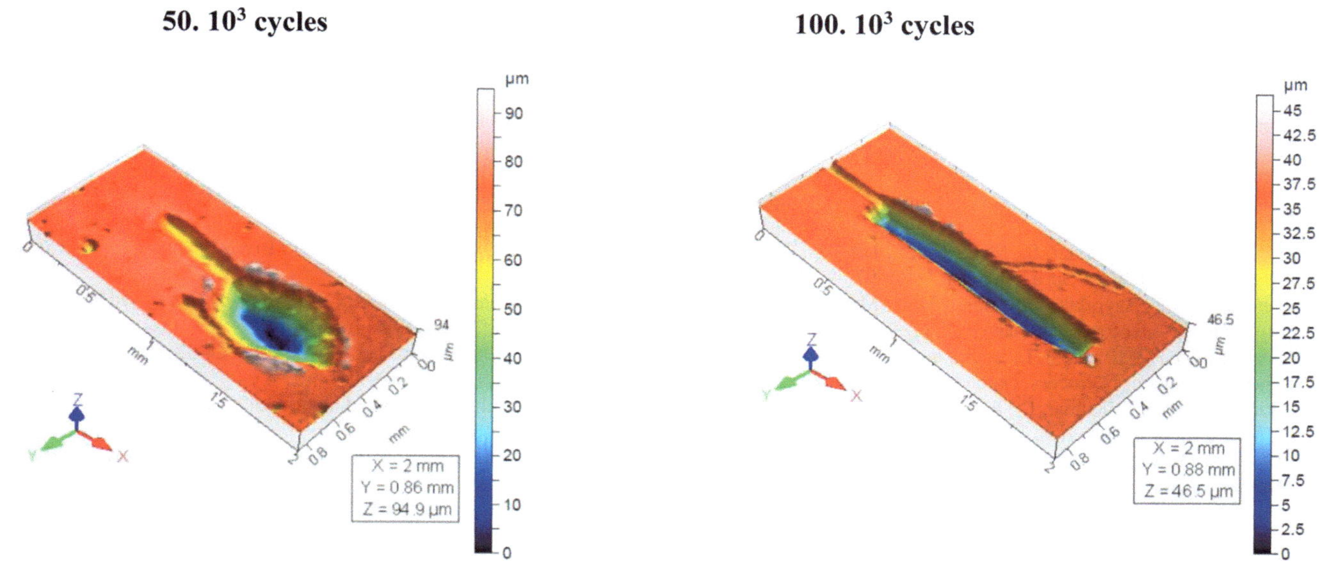

Composite	50. 10^3 cycles		100. 10^3 cycles	
Composition AA356 4Ni+30Fe_3O_4	Worn Surface mm^2	Worn Volume μm^3	Worn Surface mm^2	Worn Volume μm^3
	0,269	3997979	0,155	1629717

Fig. 12.7 Images of surface damage zones produced by scratch (wear) test for A356-IV specimens under the cycles of 50. 10^3 and 100.10^3 cycles respectively

Low cost manufacturing of these composites have been successfully managed in this work by combined method of sinter + forging processes. These composites are potentially be a useful material for aeronautic and automotive applications.

Microstructural analysis has shown that a good diffusion bonding at interface of matrix-reinforcement, a tough and sound microstructure was obtained without porosity. Wear resistance and ductility should be improved with doping process and good powder mixture preparation conditions; ball milling in longer time is needed for helping the fine and homogeneous distribution of the particles in the matrix.

Acknowledgments Some of the magnetic measurements in the laboratory of French aeronautical industry. We appreciate so much for their kind help. We acknowledge to Dr. H-A. Alhas from Airbus-space-London/UK for valuable technical helps and discussions.

References

1. N. Chawla, industrial paper, MMCs in automotive applications, Advanced Materials and Processes, pp. 29–31, (2006)
2. Din, T., Campbell, J.: High strength aerospace aluminum casting alloys a comparative study. Mater Sci. Technol. **12**, 644–650 (1996)
3. Choi, D.-H., et al.: Microstructure and mechanical property of A356 based composite by friction stir processing. Trans. Nonferrous Met. Soc. China. **23**, 335––340 (2013)
4. Bayraktar, E., Katundi, D.: New aluminum matrix composites reinforced with iron oxide. J. Achiev. Mater. Manufact. Eng., JAMME. **38**(1), 7–14 (2010)
5. Wu, W., Wu, Z., Yu, T., Jiang, C., Kim, W.S.: Review: recent progress on magnetic iron oxide nanoparticles: synthesis, surface functional strategies and biomedical applications. Sci. Technol. Adv. Mater. **16**, 43p (2015)
6. F. Ayari, D. Katundi, and E. Bayraktar, Damage of Aluminium Matrix Composite reinforced with Iron Oxide (Fe3O4): Experimental and Numerical Study AIP Conference Proceedings 1315–1328, 27 (2011)
7. Ferreira, L.-M.-P., Bayraktar, E., Robert, M.-H.: Magnetic and electrical properties of aluminium matrix composite reinforced with magnetic nano iron oxide. JAMPT, J. Adv. Mater. Process. Technol. **2**(1), 165–173., Taylor & Francis-USA (2016)
8. Bayraktar, E., Ayari, F., Tan, M.-J., Tosun Bayraktar, A., Katundi, D.: Manufacturing of aluminum matrix composites reinforced with iron-oxide nanoparticles: microstructural and mechanical properties. Metall. Mater. Trans. B. **45B**(26), 352–362., ed. David E. Laughlin, ASM-TMS/USA (2014)

9. Katundi, D., Ferreira, L.P., Bayraktar, E., Miskioglu, I., Robert, M.H.: Design of magnetic aluminium (A356) based composites through combined method of sinter + forging. SEM, Mech. Composite Multi-funct. Mater. **6**, 89–101 (2017). https://doi.org/10.1007/978-3-319-63408-1
10. Ferreira, L.M.P., Robert, M.H., Bayraktar, E., Zaimova, D.: New design of aluminium based composites through combined method of powder metallurgy and thixoforming. Adv. Mater. Res. AMR. **939**(1), 68–75 (2014)
11. Ferreira, L.-M.-P., Bayraktar, E., Robert, M.-H., Miskioglu, I.: Optimization of magnetic and electrical properties of new aluminium matrix composite reinforced with magnetic nano iron oxide (Fe_3O_4). Mech. Composite Multifunct. Mater., SEM-Springer-USA. **7**(1), 11–18 (2015)
12. Ferreira, L.F.P., Bayraktar, E., Miskioglu, I., Robert, M.H.: Recycle of aluminium (A356) for processing of new composites reinforced with magnetic nano iron oxide and molybdenum. Mech. Composites Multifunct. Mater., Chapter 18. **7**, 153–161 (2016). SEM-Springer-USA, ISBN 978-3-319-41766-0
13. Ferreira, L.F.P., Bayraktar, E., Miskioglu, I., Katundi, D.: Design of Hybrid Composites from Scrap Aluminum Bronze Chips, chapter 15, pp. 131–138 (2016). SEM-Springer-USA, ISBN 978-3-319-41766-0
14. Ferreira, L.F.P., Bayraktar, E., Robert, M.H., Miskioglu, I.: Particles Reinforced Scrap Aluminum Based Composites by Combined Processing Sintering + Thixoforming, chapter 17, pp. 145–152 (2016). SEM-Springer-USA, ISBN 978-3-319-41766-0
15. Ferreira, L.F.P., Miskioglu, I., Bayraktar, E., Robert, M.H.: Aluminium Matrix Composites Reinforced by Nano Fe3O4 Doped with TiO_2 by Thermo mechanical Process, chapter 30, pp. 251–259 (2016). SEM-Springer-USA, ISBN 978-3-319-41766-0
16. Garg, P., Gupta, P., Kumar, D., Parkash, O.: Structural and mechanical properties of graphene reinforced aluminum matrix composites. JMES, J. Mater. Environ. Sci. **7**(5), 1461–1473 (2016). ISSN: 2028-2508
17. F. Ayari, D. Katundi, and E. Bayraktar, Damage of Aluminium Matrix Composite reinforced with Iron Oxide (Fe3O4): Experimental and Numerical Study AIP Conference Proceedings 1315–1328, 27 (2011)
18. Ferreira, L.-M.-P., Bayraktar, E., Robert, M.-H.: Magnetic and electrical properties of aluminium matrix composite reinforced with magnetic nano iron oxide. JAMPT, J. Adv. Mater. Process. Technol. **2**(1), 165–173., Taylor & Francis-USA (2016)
19. Bayraktar, E., Ayari, F., Tan, M.-J., Tosun Bayraktar, A., Katundi, D.: Manufacturing of aluminum matrix composites reinforced with iron-oxide nanoparticles: microstructural and mechanical properties. Metall. Mater. Trans. B. **45B**(26), 352–362., ed. David E. Laughlin, ASM-TMS/USA (2014)
20. Ferreira, L.M.P., Robert, M.H., Bayraktar, E.: Diana Zaimova new design of aluminium based composites through combined method of powder metallurgy and thixoforming. Adv. Mater. Res., AMR. **939**(1), 68–75 (2014)
21. Ferreira, L.-M.-P., Bayraktar, E., Robert, M.-H., Miskioglu, I.: Optimization of magnetic and electrical properties of new aluminium matrix composite reinforced with magnetic nano iron oxide (Fe_3O_4). Mech. Composite Multifunct. Mater., SEM-Springer-USA. **7**(1), 11–18 (2015)
22. Ferreira, L.F.P., Bayraktar, E., Miskioglu, I., Robert, M.H.: Recycle of aluminium (A356) for processing of new composites reinforced with magnetic nano iron oxide and molybdenum. Mech. Composites Multifunct. Mater., Chapter 18. **7**, 153–161 (2016). SEM-Springer-USA, ISBN 978-3-319-41766-0
23. Ferreira, L.F.P., Bayraktar, E., Miskioglu, I., Katundi, D.: Design of Hybrid Composites from Scrap Aluminum Bronze Chips, chapter 15, pp. 131–138 (2016). SEM-Springer-USA, ISBN 978-3-319-41766-0
24. Ferreira, L.F.P., Bayraktar, E., Robert, M.H., Miskioglu, I.: Particles Reinforced Scrap Aluminum Based Composites by Combined Processing Sintering + Thixoforming, chapter 17, pp. 145–152 (2016). SEM-Springer-USA, ISBN 978-3-319-41766-0
25. Ferreira, L.F.P., Miskioglu, I., Bayraktar, E., Robert, M.H.: Aluminium Matrix Composites Reinforced by Nano Fe3O4 Doped with TiO_2 by Thermomechanical Process, chapter 30, pp. 251–259 (2016). SEM-Springer-USA, ISBN 978-3-319-41766-0
26. Garg, P., Gupta, P., Kumar, D., Parkash, O.: Structural and mechanical properties of graphene reinforced aluminum matrix composites. JMES, J Mater Environ Sci. **7**(5), 1461–1473 (2016). ISSN: 2028-2508
27. Ferreira, L.-M.-P., Bayraktar, E., Miskioglu, I., Robert, M.-H.: New magnetic aluminium matrix composites (Al-Zn-Si) reinforced with nano-magnetic Fe3O4 for aeronautical applications. J Adv Mater Process Technol. **4**, 1–12 (2018). https://doi.org/10.1080/2374068X.2018.1432940

Chapter 13
Reinforcement Effect of Nano Fe_3O_4 and Nb_2Al on the Mechanical and Physical Properties of Cu-Al Based Composites

L.-M. P. Ferreira, I. Miskioglu, E. Bayraktar, and D. Katundi

Abstract In the present work, Copper-Aluminium based composites(CAMCs) were designed through 3 different manufacturing processes; "Sintering, "Sinter + Forging" and also "sinter + thixoforming" was made a pre alloy from pure electrolytic copper that was doped with atomized recycled aluminium alloy chips (AA7075) that was given by French Aeronautical Society. After that, a typical composite was created by adding nano magnetic iron oxide (Fe_3O_4) with a special treatment. Nb_2Al intermetallic was also used as reinforcements to increase the wear resistance of the final structure. Graphene Nano Plates (GNPs) and nickel were also added to the CAMCs structures to improve mechanical behaviour and wear resistance of these novel composites. The addition of hard ceramic and/or intermetallic particles to soft copper matrix can significantly improve the mechanical properties and wear resistance, without any serious deteriorating of both magnetic, thermal and electrical conductivities of copper based composites.

Briefly, Cu-Al based composite can be successfully used to produce as high quality metal matrix composites in an economic way in the electronic field. Microstructural evaluation was performed by Scanning Electron Microscopy (SEM) and EDS analyses to optimize influence of the major reinforcements distributed in the matrix.

Mechanical and physical properties of the composites designed here can be improved with the processing method and reinforcement volume fractions. Macro scratch, nano wear and micro hardness tests were also made for these composites.

Keywords AMMCs · Compression test · Nano wear · Magnetic permeability · SEM analyze

13.1 Introduction

Copper-Aluminium based composite materials are actually used in the manufacturing of the electronic devices for aeronautic and automotive applications. Normally copper is known as low strength materials. Strength of the copper matrix was Alloying with Ag, Nb, Cr and Zr the Copper –Aluminium (Cu-Al) matrix Composites (CAMCs) reinforced with nano iron oxide (Fe_3O_4) exhibit good physical and mechanical behaviour (electrical conductivity and magnetic permeability), which makes it an excellent multifunctional lightweight material.

Novel metal matrix composites containing magnetic particulate reinforcements with metallic matrix have been progressively used in aerospace and automotive industries due to their desirable properties [1–7]. They are very attractive if they can be produced at low cost for such engineering applications. During the last two decades, they became very popular as multifunctional materials mainly known multi ferroic composites [3]. Naturally, there are optimistic views for innovative applications, especially if the properties of cost-effective magnetic parts can be manufactured to new requirements such as corrosive and wear stability and/or high temperature applications, etc. Today, permanent magnets are unique in their capability to deliver magnetic flux into the air gap of a magnetic circuit without any continuous expenditure of energy essentially in the computer technologies and also other electronic device industry.

L.-M. P. Ferreira
Federal University of Southern and South-Eastern, Marabá, Pará, Brazil

Supmeca/Paris, School of Mechanical and Manufacturing Engineering, Saint-Ouen, France

I. Miskioglu
Michigan Technology University, Engineering Mechanics Department, Houghton, MI, USA

E. Bayraktar (✉) · D. Katundi
Supmeca/Paris, School of Mechanical and Manufacturing Engineering, Saint-Ouen, France
e-mail: bayraktar@supmeca.fr

P. R. Thakre et al. (eds.), *Mechanics of Composite, Hybrid and Multifunctional Materials, Volume 5*, Conference Proceedings of the Society for Experimental Mechanics Series, https://doi.org/10.1007/978-3-319-95510-0_13

As for magnetic iron oxide, it is very easy to produce as nanoscale particles that were presented in the former papers. Magnetic iron oxide nanoparticles (Fe_3O_4) with a lattice parameter 0.8397 nm are very adaptable for new electromagnetic applications with good magnetic and electrical properties. For this reason, magnetite Fe_3O_4 is one of the favourite and mostly characterized filler materials [9–14]. Cost reduction can be obtained by reducing the total raw material cost as well as more efficient manufacturing and assembly.

In the present work, First of all, Cu-Al based composites was made a pre alloy from pure electrolytic copper that was doped with atomized recycled AA7075 chips. After that, a typical composite was created by adding nano magnetic iron oxide with a special treatment. Nb_2Al intermetallic was also used as reinforcements to increase the wear resistance of the final structure. Graphene Nano Plates (GNPs) and were also added to the CAMCs structures to improve mechanical behaviour and wear resistance of these novel composites.

Briefly, Cu-Al composite can be successfully used to produce as high quality metal matrix composites in an economic way in electronic field [6–13]. Microstructural evaluation was performed by Scanning Electron Microscopy (SEM) and EDS analyses to optimize influence of the major reinforcements distributed in the matrix.

13.2 Experimental Conditions

In the present work, Copper-Aluminium Matrix Composites (CAMCs) have been produced with three sintering techniques; "Sintering, "Sinter + Forging" and also "sinter + thixoforming". Copper was doped with AA7075 (recycled chip) after atomization. Main reinforcements were magnetic nano iron oxide, Fe_3O_4 and Nb_2Al, intermetallic. Ni, GnPs and other auxiliary elements are also used in the composites. For easy wettability of the reinforcements to the matrix, certain amount of pure nano aluminium was added in the mixture. A planetary ball milling was used during 2 h, these powders was successfully combined with particulate reinforcements. These composites have a good advantages regarding to conventional manufacturing processes such as low cost, capability of manufacturing products with complex shapes, and simple processing, etc. After that, this main structure was mixed with ball milling during 4 h to homogenize the mixture. Final composition was given in the Table 13.1 with the chemical composition of AA7075 together (Table 13.2).

Microstructural characterization was done by means of scanning electron microscope (SEM). The dispersion of reinforcement particles in the matrix and interface at matrix/reinforcements was evaluated. Micro hardness tests ($HV_{0.1}$) have been carried out on the polished and etched specimens. The micro hardness values obtained for three different processing techniques are found for Sintering, "Sinter + Forging" and "sinter + thixoforming as 64, 82 and 129 with ±10–15% accuracy for all of the manufacturing processing techniques respectively.

All the density measurements of the specimens were carried out by using *Archimedes* method. These values change between 4.21 and 5.65 ± 20% accuracy respectively. Quasi-static compression tests have been carried out in a servo-hydraulic MTS Universal test system (model: 5500R) at an initial rate of 10 mm/min and second rate of 5 mm/min. Maximum load endpoint was 5000 N. 4–5 cylindrical specimens ($H/D \geq 1.5$) was used for each manufacturing processing.

Wear resistance was measured by macro scratch wear tests at a frequency of 15 Hz for only the specimens produced by sinter + Forging with different numbers of cycles ($50*10^3$ and $100*10^3$ cycles). After scratch test, damaged zone was investigated by 3D optical roughness meter. Damage surfaces and volume loss/time, etc. have been evaluated. Electrical and magnetic measurements have been carried out by the laboratory of the French aeronautical society on the specimens produced by sinter + forging specimens and evaluated in this work. Finally, nano wear tests have been carried out by nano indentation under two different normal loads (20 and 50mN) applied over a linear track of 500 nm for 50 cycles. A conical tip with a 90° cone angle was used for this damage. One cycle is defined as a pass and return over the track; the total distance for one test was 050 μm. The speed of the tip during wear tests was 50 μm/s. A total of 10 wear tests were performed for each sample.

Table 13.1 Composition of the composites prepared in two groups (wt. %)

Composition name	Matrix copper alloyed with AA7075	Nb_2Al	Fe_3O_4	Ni	GNPs	Zn-St
CAF2	Balance	10	40	<2	<1	<2

Table 13.2 Chemical composition of scrap AA7075 powder (wt. %)

Element	Al	Cu	Fe	Mg	Mn	Si	Ni	Zn	Cr	Zr
wt. %	Balance	1.48	0.23	2.11	0.07	0.10	0.01	5.29	0.22	0.02

Nano wear testing was carried out by nano indentation test under two different normal loads (20 and 50mN) applied over a linear track of 500 μm for 50 cycles. One cycle is defined as a pass and return over the track; the total distance for one test was 0.050 m. The speed of the tip during wear tests was 50 μm/s. A total of 10 wear tests were performed for each sample.

13.3 Results and Discussion

13.3.1 Microstructure and Mapping Analyses of the Compositions Produced by "Sinter+ Forging Process

A novel composite is aimed with an optimized magnetic permeability and good electrical conductivity. For this reason, a magnetic iron oxide (Fe_3O_4) in nano scale and Nb_2Al intermetallics particles were used as primary reinforcement in these composites. Additionally, Graphene nanoparticles (GNPs) and Ni were also added as secondary reinforcements for obtaining an improved mechanical behaviour and wear resistance. As indicated in the papers of the same authors of this work, nano magnetic iron oxide was produced easily at the university laboratory [1–8]. Figure 13.1 shows general microstructure of the composite of (CAF2) produced by "sinter+ forging" process [6–11]. We have chosen this process among the others because the mechanical behaviour of this process higher than the others, and also the manufacturing price for a piece is cheaper than the others [3, 8, 10–14]. These results related to the economic aspect were not considered here in the frame of this present work. it is noted that the reinforcements in the microstructure and interface relation with matrix can be observed from these pictures (Fig. 13.2). All of the microstructure of the composites were taken from the specimens produced by sinter + forging process [6–8].

13.3.2 Macro Wear (Scratch Test) Results

As specified in the former session, production of the wear-resistant composites are requested for many industrial applications. Eliminating of the damage due to the friction phenomena is the main task for the soft matrix composites. For this reason, wear resistance of the novel composite reinforced with certain amount of hard particles is a main part of this work. Therefore, evaluation of the wear resistance of the composite "CAF2" has been carried out on the same polished specimens that have made for the microstructural analyses. For the same reason, the specimens produced only by "sinter+ forging" process was chosen for evaluation of the wear resistance. Two different numbers of cycles, ($50*10^3$, $100*10^3$ cycles) have been used for this evaluation. These results were presented in the Fig. 13.3 for two different wear tests conditions. The surface, volume and the values of the damage surfaces are presented in the same figures for each test condition. These tests are going on actually for the new series under different parameters. Considering the test conditions, it seems that not so much differences in overall wear results even if wear resistance of GNPs indicated in literature [14].

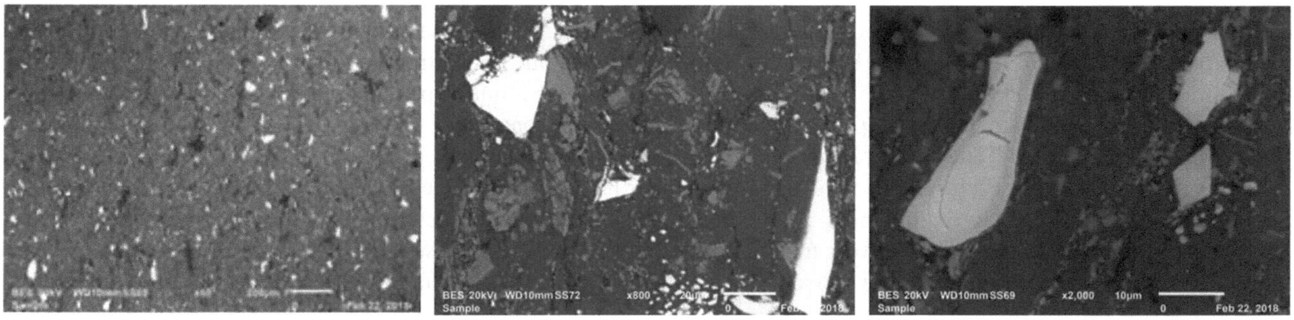

Fig. 13.1 General microstructure of the composite CAF2 produced by "sinter+ forging" process

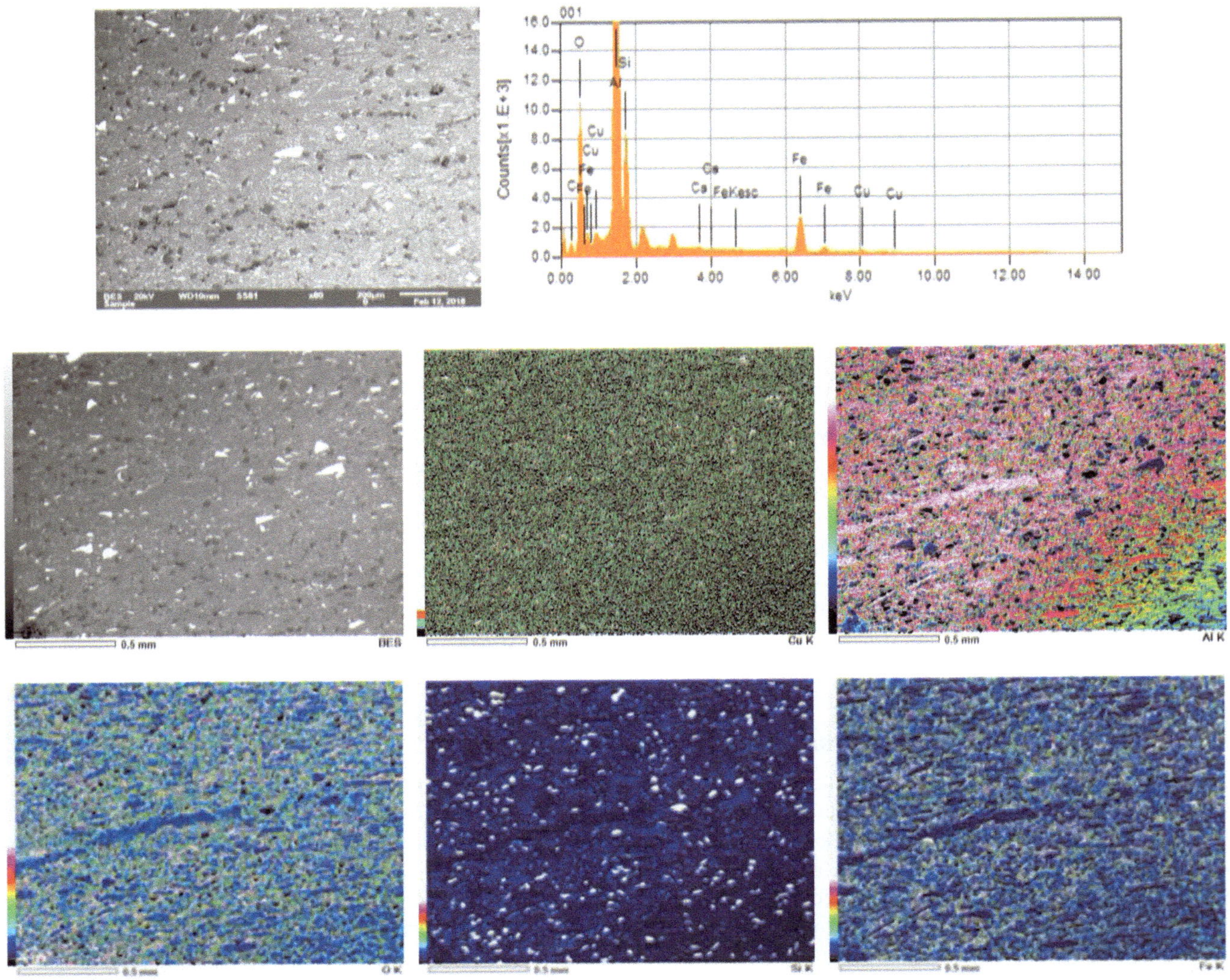

Fig. 13.2 Mapping analyses of the composite for CAF2 produced by combined process; Sinter + Forging and EDS analyses in the selected zone

13.3.3 Nano Wear Testing Results Obtained by Nanoindentation

Nano wear track deformation are very useful parameters because nano wear tests obtained by using a standard nano indentation give much more details in the hybrid composite containing fine microstructure mainly detail of interface relation of the reinforcements with the matrix that is an important evaluation for the time-dependent materials. During each nano wear test, as usually, one cycle is defined as a pass and return of the nano indenter over the track, so the total distance measured for one wear test was 50 μm. The speed of the tip during wear tests was 50 μm/s.

Total of 10 wear tests for each sample were performed under the two normal loads of 20 and 50 mN. The wear in a track is characterized as the area between the initial profile and the residual profile of the wear track that are obtained.

In the same way, the averages of the wear track deformation as "μm^2" are shown in the Fig. 13.4. The values presented in these figures are experimental results and evidently show high percentage of accuracy at the levels of ±20–30% accuracy.

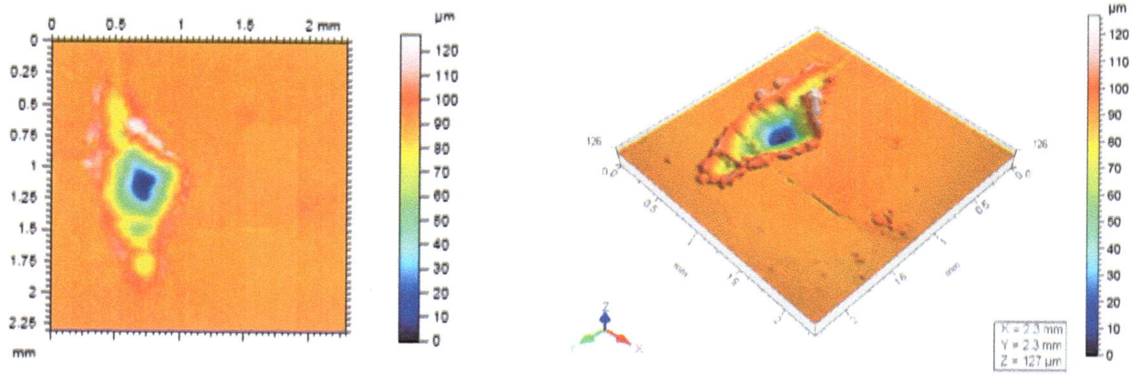

N= 50. 10³ cycles

Surface: 0.340 mm²
Volume: 10931931 µm³

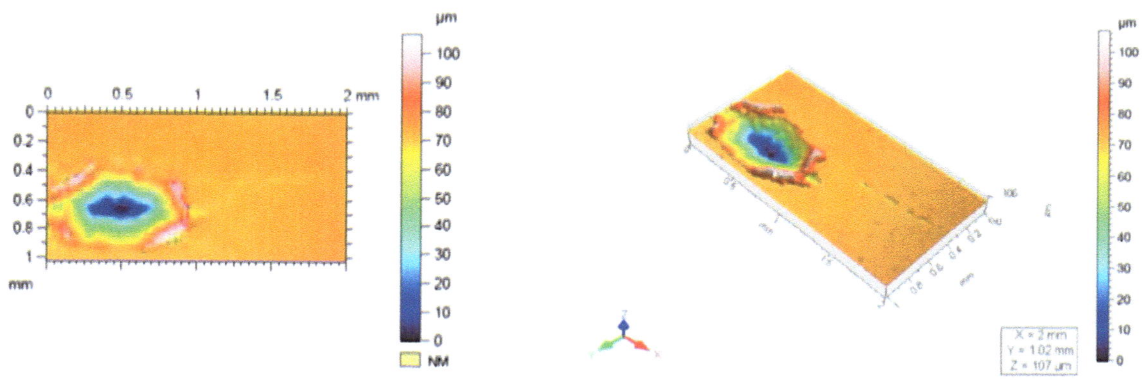

N= 100. 10³ cycles

Surface: 0.556 mm² and **Volume:** 16180432 µm³

Fig. 13.3 Images of surface damage zones produced by scratch (wear) test for "CAF2" specimens produced by "sinter+ forging" under the cycles of 50. 10^3 and 100.10^3 cycles respectively

13.3.4 Static Compression Test Results and Micro Hardness Measurements

Actually, sinter + forging and/or sinter+ thixoforming processes are always indicated that they are novel processes regarding to the conventional sintering processes mainly called as near-net shape process for the manufacturing of the pieces [13–15]. Mainly, these processes are used for the bulk materials in industrial applications. In fact, low-cost sinter+forging approach to processing of hard particle-reinforced metal matrix composites gives always high performance values of the industrial pieces (fatigue-creep, static and impact compression behaviours, etc.).

In the frame of this present work, quasi static compression tests have been carried out on the specimens of CAF2 for three manufacturing processes in order to compare all of three processes. As mentioned in the former session, this composite show very satisfied results in order to optimize the toughening capacity, magnetic permeability and electrical conductivity, etc. It means that adding of hard particles does not influence these values. This idea should be developed very well on the many other composites in industrial scale. It means that very tough and strong pieces can be obtained with a microstructure by this manufacturing process: combined method "sinter + forging".

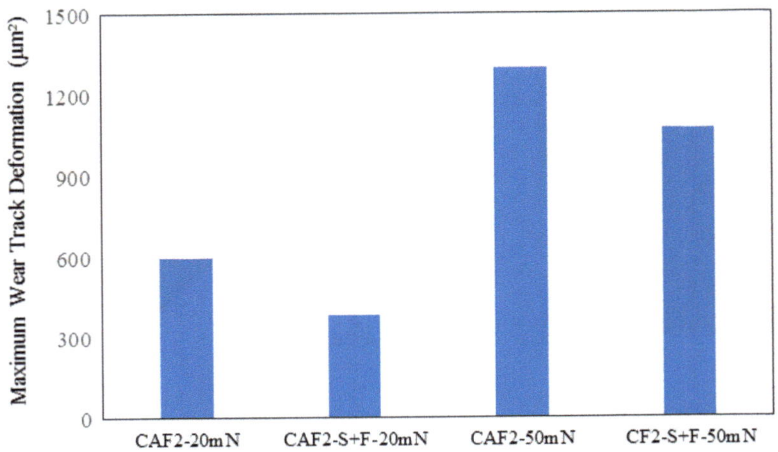

Fig. 13.4 Maximum wear track deformation obtained by two applied forces, 20 and 50 mN

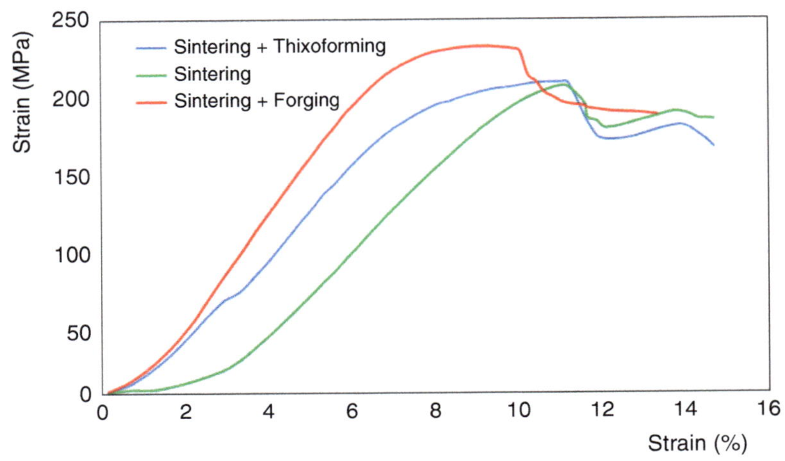

Fig. 13.5 Static compression test results for three manufacturing process of CAF2, with produced by three process; "only sintering", "sinter+ forging" and with "sinter + thixoforming

Figure 13.5 gives comparison of the static compression test results of the specimens produced with different manufacturing processes carried out on the CAF2. These values are mean values obtained from four tests results for each group. Evidently standard deviation is variable around ±15–20 MPa for all of three manufacturing processes due to internal (micro) defects - porosity in the microstructure coming from the laboratory conditions [2–9]. The same confirmation was given by micro hardness measurements taken from the specimens produced for CAF2 for three different manufacturing processes; HV = 117 ± 10 (sintering), HV = 215 ± 5 (sinter + thixoforming) and HV = 229 ± 15 (sinter + forging) respectively. One may observe that the specimens produced by the sinter+ forging process have tougher than the others. These values are only experimental results that should be improved with the specimens without defect in the microstructure.

13.4 Evaluation of Magnetic Properties for CAF2 Produced with "Sinter + Forging" Process

Magnetic measurements have been carried out by the laboratory of the French aeronautical society. As higher strength values were found for the specimens produced by sinter+forging, evaluation of the magnetic permeability have been made on these specimens. Three test specimens were used for the measurements that were evaluated by drawing the magnetic saturation, remanence values and compared certain hysteresis parameters and presented in the Fig. 13.6. One may conclude that magnetic permeability of the composite structures can be improved with only with a microstructure well prepared containing small grain size distribution well with the nano scale reinforcements found in the microstructure [1, 4, 8, 15].

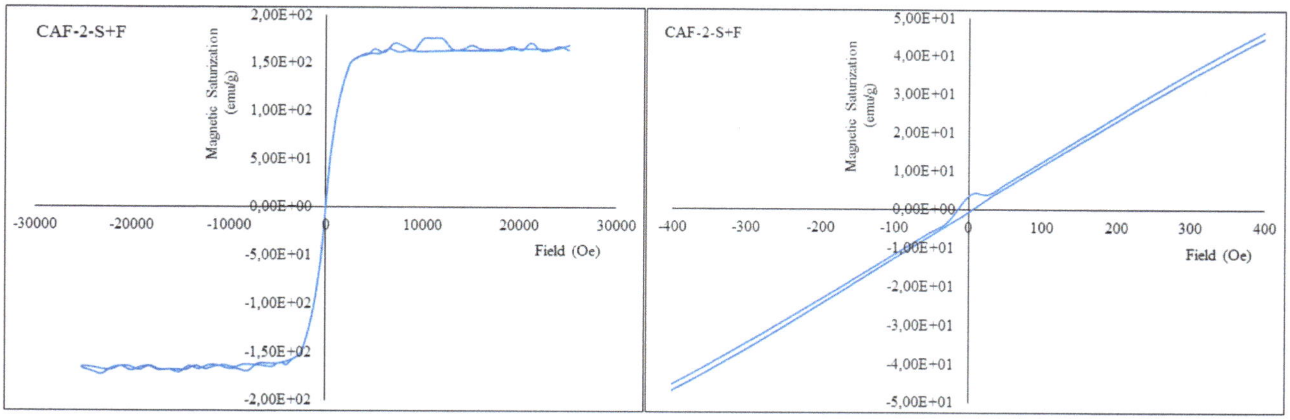

Fig. 13.6 Magnetic properties; magnetic saturation curves at the left side and the hysteresis curves at the right side, for the composite, CAF2 produced by "sinter + forging"

As shown in these figures, the manufacturing process, the percentage of the reinforcements, here basically the nano Fe_3O_4 and nickel particles play significantly a role on the magnetic properties of the composite studied here. The field dependence of magnetization measured is presented in the same figure; the magnetic saturation (Ms) and remanence values, (Mr) have been improved in regular way the Ms. values. As indicated in the former papers [5, 11, 15], a basic and important parameter in the characterization of soft magnetic materials is the power loss; this kind of power gives a measure of the energy density available in the material for a specific application. Finally, if the data obtained from the analysis of the hysteresis curve as the saturation magnetization (M_s), remanence magnetization (M_r), coercive field (H_c), etc. are compared for all of the sintering processes, the highest magnetic permeability (Ms $\geq$150 emu/g) was obtained on the specimens of CAF2 produced by sinter+forging regarding to the other manufacturing processes. One may give a reasonable results that sintering followed forging process gives always fine microstructure that ideal and efficient this type of the composites.

13.5 Conclusions

In the frame of this common project, the new design of magnetic copper-aluminium based composites CAMCs have been carried out with the reinforcements of the fine hard particles such as Nb_2Al intermetallics, magnetic nano iron oxide (Fe_3O_4) and fine nickel and graphene nano particulates (GnPs), etc.

Nano iron oxide was doped at the first stage of the process with other auxiliary elements by means of a special thermo mechanical treatment in an economic way. Low cost manufacturing of these composites have been successfully managed in this work and the highest mechanical properties and magnetic permeability were obtained by combined method of "sinter + forging" process.

These composites are potentially be a useful material for aeronautic and electronic applications recently similar composites were used for the electrical brushes, etc. Microstructural analysis has shown that a good diffusion bonding at interface of matrix-reinforcement, a tough and sound microstructure was obtained without porosity as quasi net shape composite pieces. Wear resistance and ductility should be improved with doping process magnetic vs and good powder mixture preparation conditions; plenary ball milling in shorter time is needed for helping the fine and homogeneous distribution of the fine, micro and nano scale particles in the matrix.

Acknowledgments Some of the magnetic measurements in the laboratory of French aeronautical industry. We thank so much for their kind help. We acknowledge to Dr. H-A. ALHAS from the Airbus-Space-London/UK for valuable technical helps and discussions.

References

1. Bayraktar, E., Katundi, D.: New aluminium matrix composites reinforced with iron oxide. J. Achieve. Mater. Manufact. Eng., JAMME. **38**(1), 7–14 (2010)
2. Wu, W., Wu, Z., Yu, T., Jiang, C., Kim, W.S.: Review: recent progress on magnetic iron oxide nanoparticles: synthesis, surface functional strategies and biomedical applications. Sci. Technol. Adv. Mater. **16**, 43p (2015)

3. Ferreira, L.-M.-P., Bayraktar, E., Robert, M.-H.: Magnetic and electrical properties of aluminium matrix composite reinforced with magnetic nano iron oxide. JAMPT, J. Adv. Mater. Process. Technol. **2**(1), 165–173., Taylor & Francis-USA (2016)
4. Ayari, F., Katundi, D., Bayraktar, E.: Damage of aluminium matrix composite reinforced with iron oxide (Fe3O4): experimental and numerical study AIP conference proceedings. **27**, 1315–1328 (2011)
5. Rajkovic, V., Bozic, D., Stasic, J., vWang, H., Jovanovic, M.-T.: Processing, characterization and properties of copper-based composites strengthened by low amount of alumina particles. Powder Technol. **268**, 392–400 (2014)
6. Bayraktar, E., Ayari, F., Tan, M.-J., Tosun Bayraktar, A., Katundi, D.: Manufacturing of aluminium matrix composites reinforced with iron-oxide nanoparticles: microstructural and mechanical properties. Metall. Mater. Trans. B. **45B**(26), 352–362., ed. David E. Laughlin, ASM-TMS/USA (2014)
7. Akbarpour, M.R., Salahi, E., Alikhani Hesari, F., Yoon, E.Y., Kim, H.S., Simchi, A.: Microstructural development and mechanical properties of nano-structured copper reinforced with SiC nano particles. Mater. Sci. Eng. A. **568**, 33–39 (2013)
8. Katundi, D., Ferreira, L.P., Bayraktar, E., Miskioglu, I., Robert, M.H.: Design of magnetic aluminium (A356) based composites through combined method of sinter + forging. SEM, Mech. Composite Multi-funct. Mater. **6**, 89–101 (2017). https://doi.org/10.1007/978-3-319-63408-1
9. Li, Y.J., Zeng, X.H., Blum, W.: Transition from strengthening to softening by grain boundaries in ultrafine-grained cu. Acta Materiala. **52**, 5009–5018 (2004)
10. Ferreira, L.M.P., Bayraktar, E., Robert, M.-H., Miskioglu, I.: Optimization of magnetic and electrical properties of new aluminium matrix composite reinforced with magnetic nano iron oxide (Fe_3O_4). Mech. Composite Multifunct. Mater. SEM-Springer-USA. **7**(1), 11–18 (2015)
11. Need, R.F., Alexander, D.J., Field, R.D., Livescu, V., Papin, P., Swenson, C.A., Mutnick, D.B.: The effects of equal channel angular extrusion on the mechanical and electrical properties of alumina dispersion-strengthened copper alloys. J. Mater. Sci. Eng. A. **565**, 450–458 (2013)
12. Ferreira, L.-M.-P., Bayraktar, E., Miskioglu, I., Robert, M.H.: Recycle of aluminium (A356) for processing of new composites reinforced with magnetic nano iron oxide and molybdenum. Mech. Composites Multifunct. Mater., Chapter 18. **7**, 153–161 (2016). SEM-Springer-USA, ISBN 978-3-319-41766-0
13. L.F.P. Ferreira, I. Miskioglu, E. Bayraktar, and M.H. Robert, Aluminium matrix composites reinforced by nano Fe3O4 doped with TiO_2 by thermo mechanical process, chapter 30, pp. 251–259, 2016., SEM-Springer-USA, ISBN 978-3-319-41766-0
14. Garg, P., Gupta, P., Kumar, D., Parkash, O.: Structural and mechanical properties of graphene reinforced aluminium matrix composites. JMES, J. Mater. Environ. Sci. **7**(5), 1461–1473 (2016). ISSN: 2028-2508
15. Ferreira, L.-M.-P., Bayraktar, E., Miskioglu, I., Robert, M.-H.: New magnetic aluminium matrix composites (Al-Zn-Si) reinforced with nanomagnetic Fe3O4 for aeronautical applications. J. Adv. Mater. Process Technol. **4**, 1–12 (2018). https://doi.org/10.1080/2374068X.2018.1432940
16. Ferkel, H.: Properties of copper reinforced by laser-generated Al_2O_3 nanoparticles. Nano Struct. Mater. **11**, 595–602 (1999)

Chapter 14
Recycled Ti-17 Based Composite Design; Optimization Process Parameters in Wire Cut Electrical Discharge Machining (WEDM)

Sonia Ezeddini, Mohamed Boujelbene, Emin Bayraktar, and Sahbi Ben Salem

Abstract This work present a comprehensive study on the effect and optimization of machining parameters on the kerf (cutting width) and material removal rate (MRR) in wire electrical discharge machining (WEDM) process by using the response surface methodology (RSM) and Taguchi method. The experimental studies were conducted under varying parameters. The main input parameters on this model are the cutting parameters such us pulse on time (T_{on}), servo voltage (U), Speed of advance or feed rate (S) and injection pressure or flushing pressure (P). Recycled Titanium based composite (an alloy Ti17) was used for machining operations and the combined effects of cutting parameters on the material removal MRR rate and kerf were investigated while using the analysis of variance ANOVA. Mathematical models were used for the objective of minimum kerf and maximum MRR, Cut edge surface analysis was carried out using an optic microscope and Scanning Electron Microscope (SEM) to evaluate the kerf.

Keywords WEDM · ANOVA · Taguchi method · MRR · Kerf (cutting width) · Recycled composites · Ti17

14.1 Introduction

Electrical discharge machining (EDM) an important non-conventional manufacturing method used for hard to cut conductive material and has been accepted worldwide as a standard process in the manufacture of forming tools to produce plastics mouldings, forming dies, die casting, since it does not require cutting tools and allows machining involving hard, brittle, thin and complex geometry. Wire electrical discharge machining (WEDM) is now used in automobile, aérospace and medical industries, as good as in practically all areas of conductive material machining.

WEDM is based on electric discharge machining process and it is an unconventional method of machining that includes electrically conductive materials, and involves the removal of material by the action of energy dissipated between an electrode and a workpiece. This process has kneed important progress that made it a lead of machining processes of precision and many several researchers are studied these process and their different aspects. It is the machining process during which material is eroded by series of controlled sparks between electrode and the workpiece. Workpiece and the electrode are immersed in the dielectric fluid. The area around the spark is heated to 10,000–20,000 °C and the dielectric fluid around this area is vaporized, leading to increase in the pressure. Also small amount of workpiece and electrode material melts and vaporizes, which creates small craters on the surface. A layer of melted and resolidified material called recast can be found on the surface after EDM [1–7]. I was observed from the intensive literature review that most of the studies have target the optimization of multiple quality characteristics in WEDM process such as cutting rate, wire wear rate and wire failure frequency, kerf size, Material removal rate MRR and most important quality of machined surface the surface roughness SR. Those characteristics are influenced by machining parameters. Optimization of these parameters for each material or groupe of materials is necessary [1, 9].

S. Ezeddini · M. Boujelbene · S. Ben Salem
University of Tunis El Manar, ENIT, École Nationale d'Ingénieurs de Tunis, Tunis, Tunisia

E. Bayraktar (✉)
Supmeca-Paris, School of Mechanical and Manufacturing Engineering, Saint-Ouen, Paris, France
e-mail: bayraktar@supmeca.fr

P. R. Thakre et al. (eds.), *Mechanics of Composite, Hybrid and Multifunctional Materials, Volume 5*, Conference Proceedings of the Society for Experimental Mechanics Series, https://doi.org/10.1007/978-3-319-95510-0_14

A large number of researchers have reported the optimization of multiple correlated responses of WEDM process using traditional industry. W. Tebni et al. [10] studied surface integrity by varying the parameters of EDM machining. Ozkul et al. [11] studied the influence of the machining parameters such as the pulse time On, the current I, the voltage U and the dielectric pressure and set the pause time Off. Khan and saifuddin [12] examined the wear for two copper and aluminum tools for two types of material (stainless steel and carbide), as a result of the study show that aluminum electrode was left to smother surface than copper electrode on stailess stell and carbide. Guu [13] has been worked on imaging surface on implemented electro discharge machining AISI D2 tool steel by atomic force microscopy. Prabhu and Vinayagam [14] was researched AISI D2 tool steel hardware surface features implemented electric dis-charge machining process with single wall caebon nanontubes. Observation that the surface roughness and half-cro-fractures increased proportionally with the power. Ay and Aydogdu [15] have been worked out by investigating the effects of technical parameterization on the parameters of technical parameters and their predictive values.

Present study is targeted at investigation of machining characteristics of material removal rate and kerf width in WEDM of Recycled Titanium based composite (an alloy Ti17). The outcome of this study would add to the database of the machinability of Recycled Titanium based composite (an alloy Ti17) and also would be very beneficial for the machinist as the technology charts for WEDM of this material would be updated.

14.2 Experimental Procedure

14.2.1 Equipment, Materials and Measurement

The material studied in this work is Recycled Titanium based composite (an alloy Ti17) which has been used with the chemical specification given in Table 14.1 and the chemical specification of electrode (wire) given in Fig. 14.3 and Table 14.2, a circular workpiece with a diameter of 50 mm and thikness of 3 mm was cut to 9 samples (Fig. 14.2). The experiments were carried out on a wire rod erosion machine Robofil 190. The fixed process parameters during the experiment given in Table 14.3.

Ti17 or TA5CD4 (AFNOR) is a titanium-based alloy to which the alloy elements are added (17% in all, hence its name). Like pure titanium, above a so-called β transus temperature, Ti17 has a cubic crystallographic structure centered at 100%. In contrast, unlike pure titanium hexagonal structure (h.c.) at 100% at room temperature, Ti17 is two-phase with 30% phase β (c.c.) and 70% phase α (h.c.). The structure of our material machined by WEDM is classified in the acircular structure then this structure requires only a heat treatment. It is obtained during a cooling started in the β domain and is characterized by the presence of needle in the thickness is according to the cooling rate. The structure of Widmanstatten (or basketry) produced by a slower cooling allowing the transformation of β into α. The alpha phase develops as lamellae all the thicker as the cooling rate is slower. Figure 14.1 is an example which shows the structures of Ti17 after machining with WEDM and which shows the relationship between the cooling rate and the sizes of the alpha phase lamellae. in our case we consider EDM machining as a heat treatment and the cutting speed and the dielectric pressure are proportional to the cooling rate.

14.2.2 Wire Electric Discharge Machining Process and Cutting Parameters

WEDM is currently widely used in the industry for the high-precision machining of all types of conductive materials such as metals, metal alloys, graphite or even certain ceramic materials of any hardness, WEDM is used in a wide range of industries, including aerospace, biomedical, automotive, as well as molds. In this study, we are interested in the recycled Titanium based

Table 14.1 Chemical composition

Element	Ti	Al	Mo	Cr	Zr	Sn	O
%	83	5	4	4	1.9	2	0.1

Table 14.2 Chemical composition of wire tool

Composition of wire	Mass%	Atom%
Cuivre-Cu	57.59	58.28
Zinc-Zn	42.41	41.72

Table 14.3 Fixed parameters

Wire	Brass wire of diameter 0.25 mm
Shape and size of work piece	Circular piece of diameter 50 mm and thickness 3 mm
Dielectric fluid	Deionized water

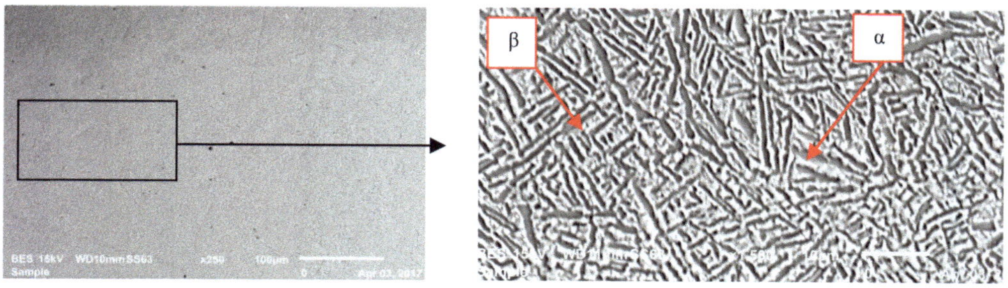

Fig. 14.1 Microstructure of Ti17 after WEDM machining (S (mm/min) =36, P (bar) =80)

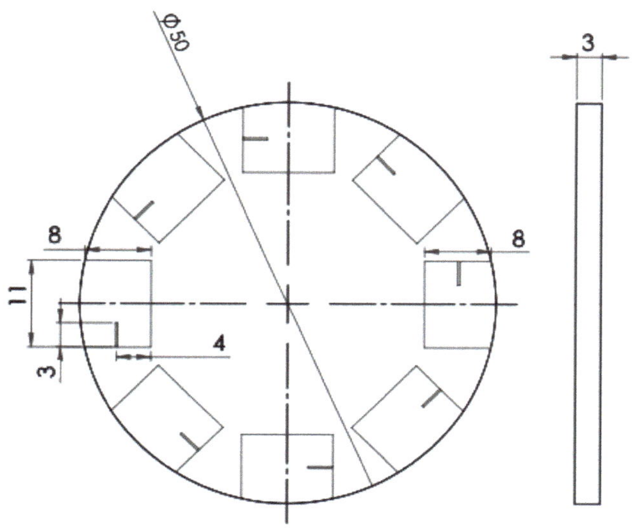

Fig. 14.2 Experimental configuration for work-piece

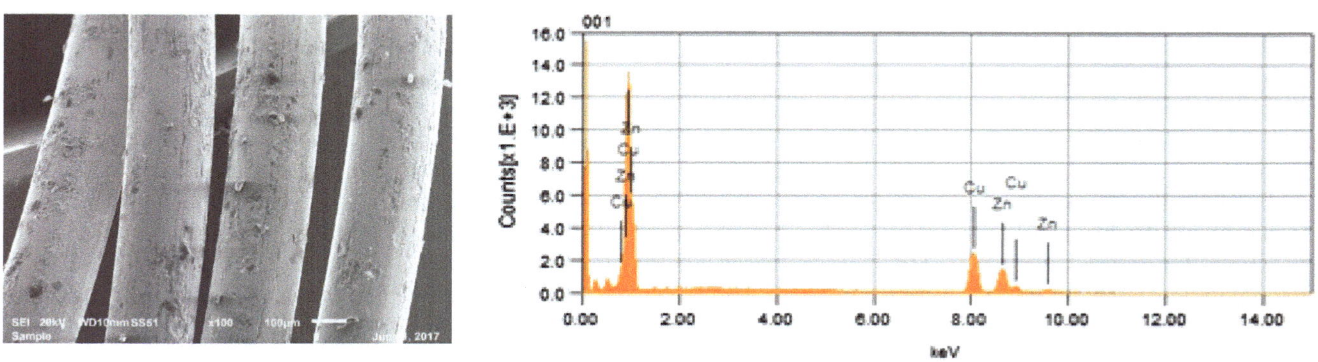

Fig. 14.3 SEM picture of wire tool and composition of wire tool from SEM analysis

composite (an alloy Ti17 which are known to be the most difficult materials to machine, so tests of cutting by the machine of the wire EDM were carried out on the material. In order to optimize this method, an integrated method was used which brings together the Taguchi method and the RSM response surface methodology (Figs. 14.2 and 14.3).

Table 14.4 The basic Taguchi L9 (3^4) orthogonal array

Run	Control factors and levels				Coding				Results	
	U(V)	Ton(μs)	S(mm/min)	P(bar)	X1	X2	X3	X4	Kerf(μm)	MRR (μm)
1	80	0.8	29	60	−1	−1	−1	−1	341.83	39.65
2	80	0.9	36	80	−1	0	0	0	341.36	49.116
3	80	1	43	100	−1	1	1	1	346.83	59.654
4	100	0.8	36	100	0	−1	0	1	345.07	49.690
5	100	0.9	43	60	0	0	1	−1	341.06	58.662
6	100	1	29	80	0	1	−1	0	331.86	38.495
7	120	0.8	43	80	1	−1	1	0	309.88	53.299
8	120	0.9	29	100	1	0	−1	1	317.78	36.862
9	120	1	36	60	1	1	0	−1	317.03	45.652

Table 14.5 Attribution of the levels to the factors

Levels	Pulse on time Ton(μs)	Servo voltage U(V)	Speed of advance S (mm/min)	Flushing pressure P(bar)
	T	U	S	P
1	0.8	80	29	60
2	0.9	100	36	80
3	1	120	43	100

In this work we used The Taguchi method and the Response Surfaces Methodology (RSM) to efficiently analyze the experimental data and seek satisfactory solutions. First, the Taguchi method is used to collect experimental data and to study in a preliminary way the link between the objectives and the parameters. In this case, the RSM models are designed to approximate the target functions, so that a multi-objective model is easy to acquire.

The standard experimental model based on Taguchi an orthogonal matrix L 9 (3^4) was used in this study and presented in Table 14.4. This basic design uses up to four control factors, with three levels each. A total of nine experimental runs must be performed, using the combination of levels for each control factor as shown in Table 14.5. Four control factors including pulse (Ton), startup voltage (U), feed rate or speed advance (S) and flushing pressure (P) for kerf width and material removal rate MRR have been selected.

Taguchi method is statistical method, or sometimes called robust design method, developed by Genichi Taguchi to improve the quality of manufactured goods, to evaluate losses due to poor quality and to seek to minimize noise sensitivity for any product and / or process, and more recently also applied to engineering, biotechnology and marketing. The application of this technique has become widespread in many American and European industries after the 1980s. The advantage of Taguchi design is the study of multiple factors can be considered at once. Therefore, not only the controlled factors can be considered, but also noise factors for example. Although similar to experimental design (DOE), the Taguchi design or method only leads to balanced (orthogonal) experimental combinations, making the Taguchi design even more efficient than a fractional factor design. By using Taguchi techniques, industries are able to dramatically reduce the development cycle time of the product for design and production, improve manufacturing efficiency and performance reliability, thereby reducing costs and increasing profits. In addition, the Taguchi design allows us to examine the variability caused by noise factors, which are generally ignored in the traditional DOE approach. The Taguchi method requires only a smaller number of orthogonal experimental combinations that dramatically improve design efficiency. There are several researchers who have used the Taguchi method (Farnaz Nourbakhsh et al. (2013) [16] is a study where they used a Taguchi L18 plane to optimize input parameters (current, pulse time, Tension and wire tension) to improve surface integrity, Boujelbene et al. [21] have done an optimization of MRR material removal rate, TWR wire wear rate and surface roughness using the L25. Two basic tools in the Taguchi technique are the orthogonal network (OA) and the signal-to-noise ratio (S/N). OA is used in the Taguchi method to save time and cost experiments. The S/N ratio is used to measure the deviation of the quality characteristics from the desired values, including the highest Higher-The-Better (HTB), Nominal-The-Better (NTB) and Lower-The-Better LTB). In [17] they studied the variation of kerf and rate of removal of MRD WEDM materials by using ANOVA and SN ration. In this study we aim to optimize the kerf and the material removal rate MRR so that the most objective type of objective function is High-The-Better) And Lower-The-Better (STB) were used. The exact relationship between the S/N ratio and the signal is given by the following Eqs. (14.1) and (14.2):

HTB: Eq. (14.1); LTB: Eq. (14.2)

$$\frac{S}{N} = -10 * \log\left(\sum_{i=0}^{n} Yi^2/n\right) \quad (14.1)$$

$$\frac{S}{N} = -10 * \log\left(\sum_{i=0}^{n} \left(1/Yi^2\right)/n\right) \quad (14.2)$$

Where n is the number of experiments, and y_i is the value of the kerf or MRR.

The Response Surface Method (RSM) proposed by Box and Wilson (1951) in the early 1950s received considerable attention because of its good empirical performance in modeling. It is a set of mathematical and statistical techniques that provide adapted models between input parameters and responses to develop improve and optimize a process. Several researches have used the RSM method to improve and optimize the performance of machining processes such as surface integrity (roughness, heat affected zone, residual stress), MRR material removal rate, thread wear TWR, kerf [19–20]. The most extensive application of RSM is found in the industrial world, especially in the case where several input variables potentially influence the performance measures or the product or process quality characteristics [18]. This study seeks an appropriate approximation method for analyzing the surface roughness relationship Ra, the thermally affected zone HAZ with respect to independent input parameters. A mathematical equation of the second order polynomial response surface is used as indicated in the equation; the coefficients of the function can be obtained by the least squares method

$$Y = \beta_0 + \sum_{i}^{n} \beta_i X_i + \sum_{i<j}^{n} \sum \beta_{ij} X_i X_j + \sum \beta_{ij} X_i^2 + \varepsilon \quad (14.3)$$

Where y denotes kerf or MRR, x represents the parameters of WEDM (T_{on}: pulse time, U: start up voltage or servo voltage, S: feed rate or speed advance and P: flushing pressure or dielectric injection pressure), B_{ij} is the coefficients of each term, ε is a residual error. Table 14.3 shows the basic Taguchi L9 orthogonal array.

14.2.3 Influence of Machining Parameters on Performance of WEDM Process

14.2.3.1 Influence of Machining Parameters on Kerf Width

The Kerf is defined as the width of the piece that is removed by a cutting process. When you cut parts on a CNC machine, you would make precision cuts with finished dimensions as close as possible to the programmed shape. This drives us to study the kerf and to look for how we can act on this performance from the machining conditions. Thus gap is determined by the intensity of the spark energy. Figure 14.4 shows the different slots effected by wire of electric discharge machine whie machining of Ti17 alloy. Their slots are machined by differents machining conditions.

Figure 14.5 shows the effect of flushing pressure on the kerf width in three values of servo voltage U = 80 V, U = 100 V, and U = 120 V, in three cases kerf width has the same trend, it decrease with increase of flushing pressure from 60 bar to 80 bar after that it increase.

Figure 14.6 shows the effect of advance speed on the kerf width in tow values of servo voltage U = 100 V and U = 120 V, from this figure kerf widh knew a little increasing winth increase of advanced speed with servo voltage U = 100 V.

Kerf width presented a constant profile us function a pulse on time (Ton) with tow different values of servo voltage U = 80 V and U = 120 V, it is mean that pulse on time don't have a great influence on the kerf width see Fig. 14.7. In the other hand, Fig. 14.8 showed the great influence of servo velotage (U) then, the kerf width decreases with the increase of servo voltage.

14.2.3.2 Influence of Machining Parameters on Material Removal Rate MRR

Material removal machining is a process for manufacturing mechanical parts.

Remove material to give the raw piece the desired shape and size, using a machine tool. By this technique, we obtains a parts with great precision.

This processing method makes it possible to manufacture parts with very complex geometry and small gaps or small internal rays. We obtain surfaces of extreme finesse and precision.

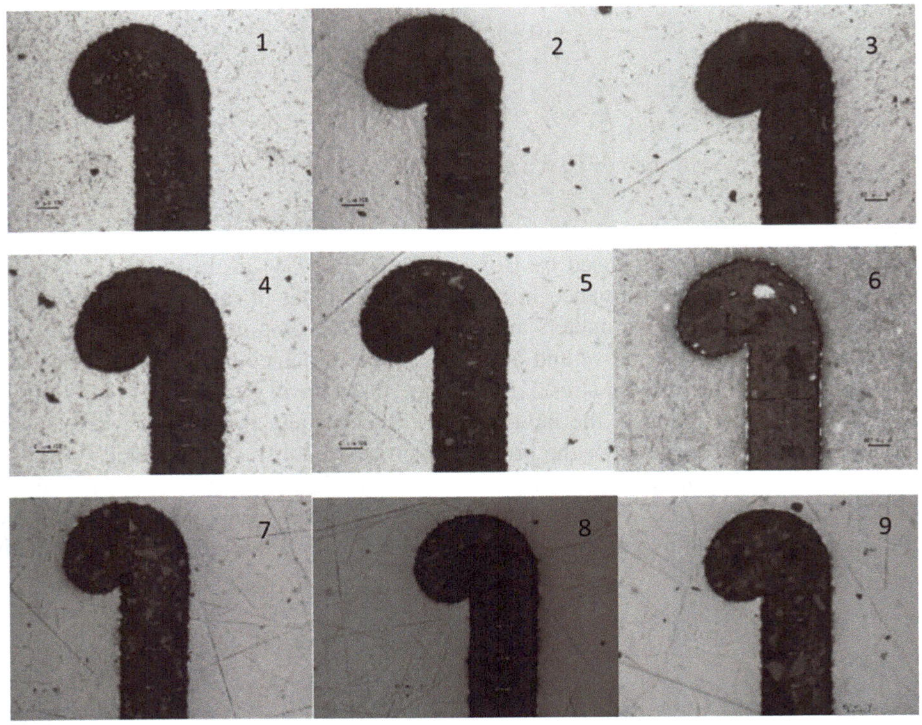

Fig. 14.4 Kerf width for 9 different samples

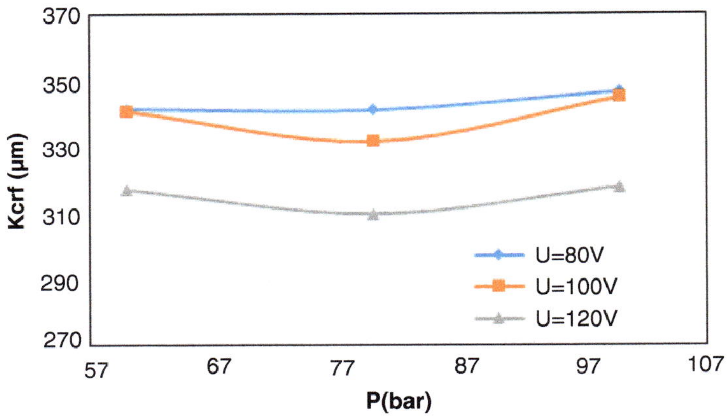

Fig. 14.5 Effect of flushing pressure P on kerf width

To measure the removal rate of material MRR from wire EDM machining of the aluminum titanium based MRR intermetallic composite; a slit at the edge of the sample was made as shown in Fig. 14.9 to calculate the MRR of each sample with Eq. (14.4).

$$MRR = S^{*} e^{*} p \tag{14.4}$$

S = advance speed mm/min

e = thickness of the workpiece in mm

p: Depth of the slot made by the wire in mm

Figures 14.10, 14.11, 14.12, and 14.13 show the evolution of material removal rate according to the WEDM machining parameters during the machining of Ti17 alloy. From curves in Fig. 14.12 we notice that MRR increases with a big pent with the increase of advanced speed or feed rate (S) with the two differents values of flushing pressure. In other hand MRR

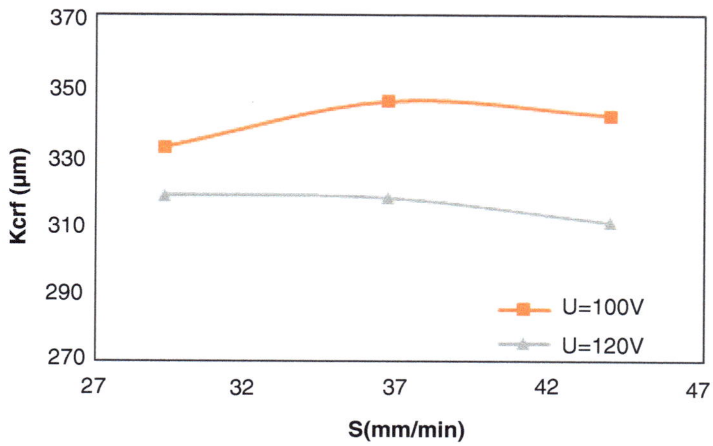

Fig. 14.6 Effect of advance speed on kerf width

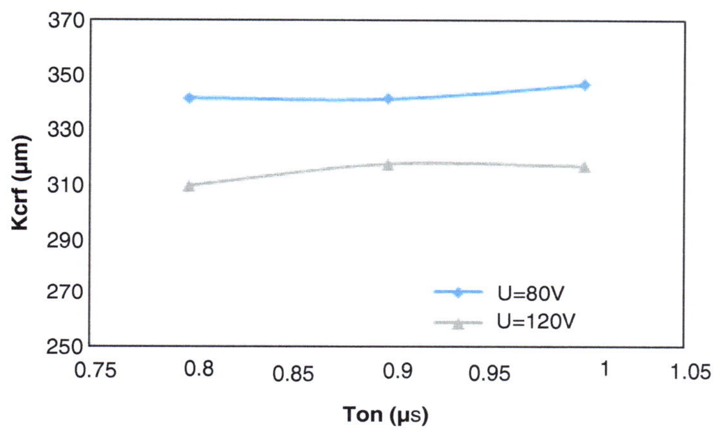

Fig. 14.7 Effect of pulse on time Ton on kerf width

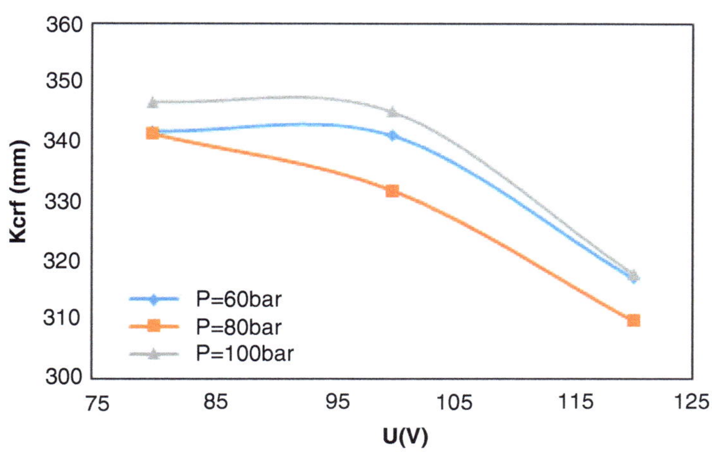

Fig. 14.8 Effect of servo voltage U on kerf width

decreases a very little with increase of servo voltage (Fig. 14.10) and increases a very little with increase of flushing pressure (Fig. 14.13).

The surface topographical images of microchannels fabricated by WEDM in recycled Titanium based composite (an alloy Ti17) material are shown in Fig. 14.14. The effects of machining parameters of cutting on channel width are shown

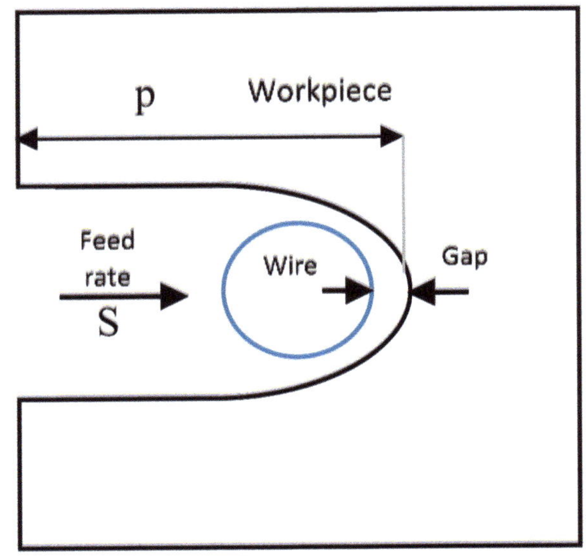

Fig. 14.9 Kerf width

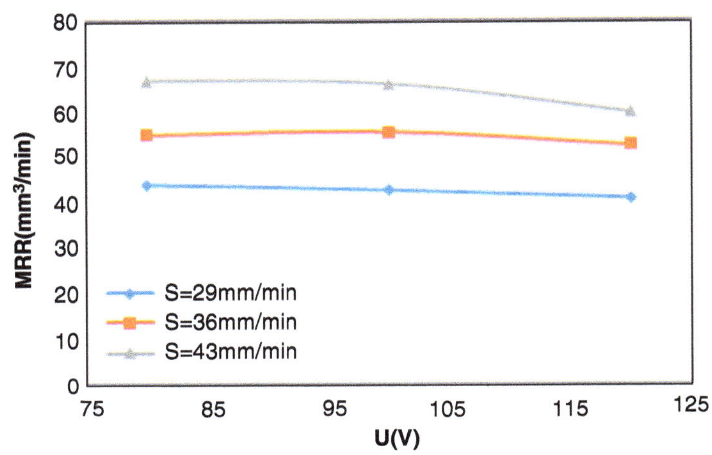

Fig. 14.10 Effect of servo voltage on MRR

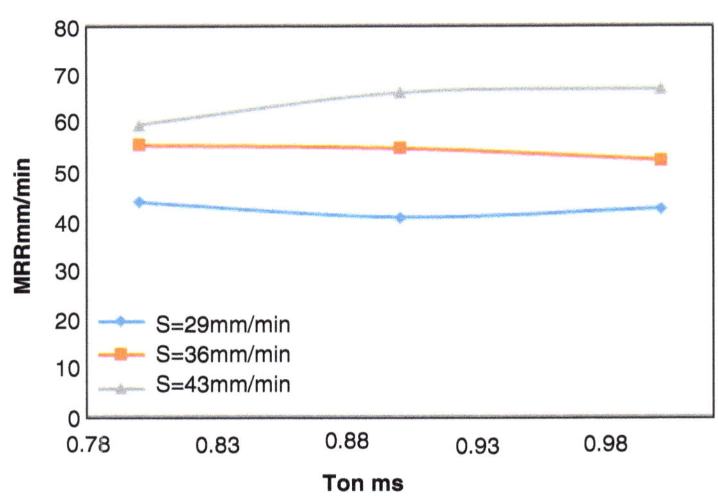

Fig. 14.11 Effect of pulse on time on MRR

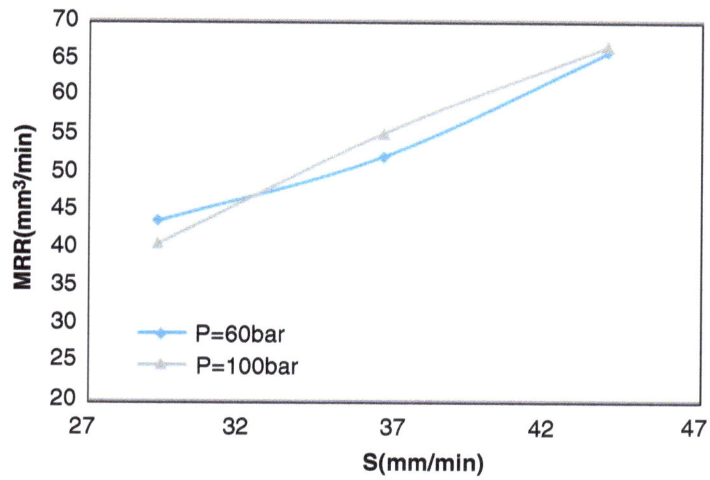

Fig. 14.12 Effect of advance speed on MRR

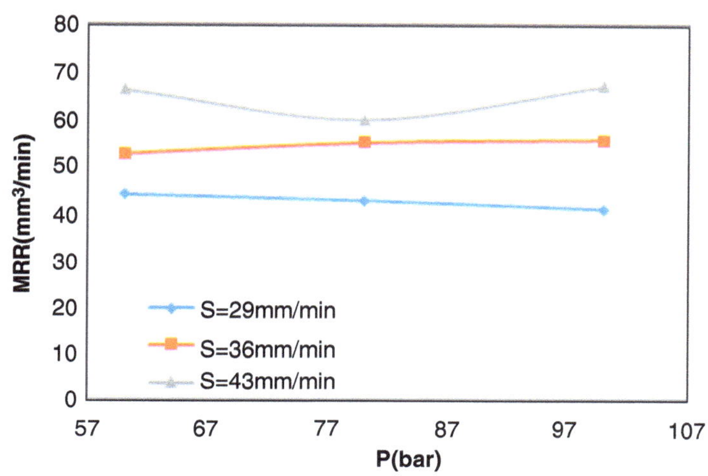

Fig. 14.13 Effect of flushing pressure on MRR

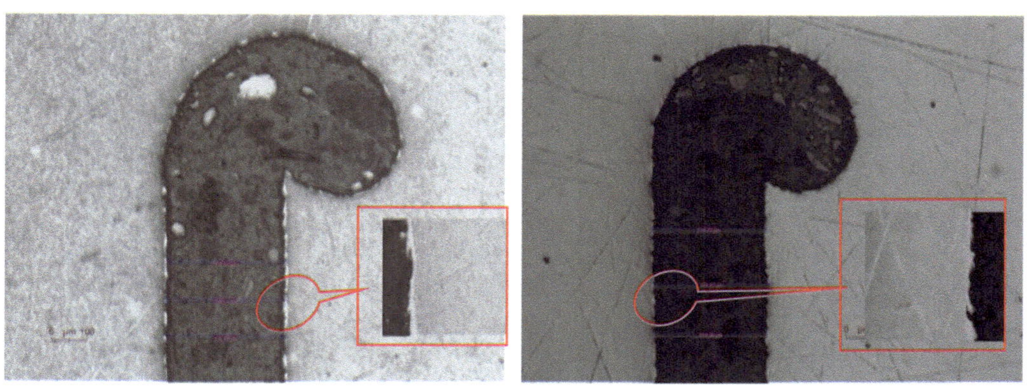

Fig. 14.14 optic picture in the internal surface of the Ti17 alloy Kerfs machined by WEDM with: (**a**) U = 100 V, Ton = 1 μs, S = 29 mm/min, P = 80bars; (**b**) U = 120 V, Ton = 0.9 μs, S = 29 mm/min, P = 100bars

in Fig. 14.14a, b, we observe a big variation on the surface and subsurface and the presence of the "white ayer", because WEDM process uses heat from electrical sparks to cut the material, the spark creat a heat affected zone that contains a thin layer of recast called'White layer" like in the Fig. 14.14 and the depth of the heat affected zone and recast depends of energy and power.

Table 14.6 Response table for signal to noise ratios smaller is better and for means of kerf

	Level	Means				S/N ratio			
Process parameters		U(V)	Ton(μs)	S(mm/min)	P(bar)	U(V)	Ton(μs)	S(mm/min)	P(bar)
Average value	L1	343.3	332.3	330.5	333.3	−50.71	−50.42	−50.38	−50.45
	L2	339.3	333.4	334.5	327.7	−50.61	−50.45	−50.48	−50.30
	L3	314.9	331.9	332.6	336.6	−49.96	−50.41	−50.43	−50.53
	Delta	28.4	1.5	4.0	8.9	0.75	0.04	0.10	0.23
	Rank	1	4	3	2	1	4	3	2

Table 14.7 Response table for signal to noise ratios larger is better and for means of MRR

		Means				S/N ratio			
Process parameters	Level	U(V)	Ton(μs)	S(mm/min)	P(bar)	U(V)	Ton(μs)	S(mm/min)	P(bar)
Average value	L1	49.48	47.55	38.34	47.99	33.77	33.48	31.67	33.51
	L2	48.95	48.21	48.15	46.97	33.67	33.51	33.65	33.36
	L3	45.27	47.93	57.21	48.74	33.02	33.47	35.14	33.59
	Delta	4.20	0.67	18.87	1.76	0.75	0.04	3.47	0.23
	Rank	2	4	1	3	2	4	1	3

Table 14.8 S/N ratio means and predicted S/N ratio and predicted means of Kerf

	Control factors and levels				Results					
Run	U(V)	Ton(μs)	S(mm/min)	P(bar)	kerf	Pred kerf	SNRA	MEAN	PMEAN	PSNRA
1	80	0.8	29	60	341.83	341.83	*50.6763*	*341.833*	*341.833*	*−50.6763*
2	80	0.9	36	80	341.36	341.362	*−50.6645*	*341.367*	*341.367*	*−50.6645*
3	80	1	43	100	346.83	346.83	*−50.8024*	*346.830*	*346.830*	*−50.8024*
4	100	0.8	36	100	345.07	345.07	*−50.7582*	*345.070*	*345.070*	*−50.7582*
5	100	0.9	43	60	341.06	341.06	*−50.6566*	*341.060*	*341.060*	*−50.6566*
6	100	1	29	80	331.86	331.86	*−50.4191*	*331.860*	*331.860*	*−50.4191*
7	120	0.8	43	80	309.88	309.88	*−49.8239*	*309.880*	*309.880*	*−49.8239*
8	120	0.9	29	100	317.78	317.78	*−50.0425*	*317.780*	*317.780*	*−50.0425*
9	120	1	36	60	317.03	317.03	*−50.0220*	*317.030*	*317.030*	*−50.0220*

14.3 Statistical Analysis

Tables 14.6 and 14.7 show response table for means and signal to noise ratio for kerf and MRR of Recycled Titanium based composite (an alloy Ti17) material successive. This response table represents the effects of various input factors on kerf and MRR. Higher the slope in the main effects plot corresponding values of delta is higher in the response table. The rank represents directly the level of effect of input based on the values of delta. Here according to ranks, the effects of various input factors on kerf in sequence of its effect are servo voltage and flushing pressure and the effects of various input factors on MRR in sequence of its effect are advanced speed and servo voltage. That means servo voltage affects the kerf at highest level and at lowest level. And advanced speed affects MRR at highest level and lowest level.

Tables 14.8 and 14.9 shows the values of signal to noise ratio (SNRA) and Predicted signal to noise ratio (PSNRA) for kerf and MRR of recycled Titanium based composite (an alloy Ti17) material successive. The values of predicted signal to noise is very much close to the calculated signal to noise values hence the analysis of Taguchi for signal to noise ratio is correct, it's perfect. The representation of effects of various parameters on kerf and MRR and optimize condition is very much nearby.

Figures 14.14 and 14.15 show the influence of each cutting parameter: the pulse on time, the servo voltage, the feed rate or advanced speed and the injection pressure or flushing pressure. However, according to the curves of the effects (Fig. 14.14), the servo voltage has the greatest influence on the kerf values and the value U = 120 V corresponds to the smallest kerf value compared with U1 and U2. In addition, the speed S1 corresponds to the smallest kerf value with compared with S3 and S2, the pulse time $T_{on}3$ corresponds to the smallest kerf value compared with $T_{on}1$ and $T_{on}2$ and the flushing prssure P2 corresponds to the smallest kerf value compared with P1 and P3. In other hand, according to the curves of the effects (Fig. 14.15), the advanced speed has the greatest influence on the MRR values and the value S 3 = 43 mm/min corresponds to the

Table 14.9 S/N ratio means and predicted S/N ratio and predicted means of MRR

	Control factors and levels				Results					
Run	U(V)	Ton(μs)	S(mm/min)	P(bar)	Ra	Pred Ra	SNRA	MEAN	PMEAN	PSNRA
1	80	0.8	29	60	2.92	2.9288	31.9653	39.6527	39.6527	31.9653
2	80	0.9	36	80	1.93	1.9475	33.8247	49.1184	49.1184	33.8247
3	80	1	43	100	1.74	1.7400	35.5128	59.6548	59.6548	35.5128
4	100	0.8	36	100	1.96	1.9425	33.9253	49.6901	49.6901	33.9253
5	100	0.9	43	60	2.48	2.4713	35.3671	58.6623	58.6623	35.3671
6	100	1	29	80	2.01	1.9838	31.7081	38.4958	38.4958	31.7081
7	120	0.8	43	80	2.48	1.8487	34.5344	53.2994	53.2994	34.5344
8	120	0.9	29	100	1.8	1.8175	31.3316	36.8625	36.8625	31.3316
9	120	1	36	60	2.84	2.8400	33.1892	45.6523	45.6523	33.1892

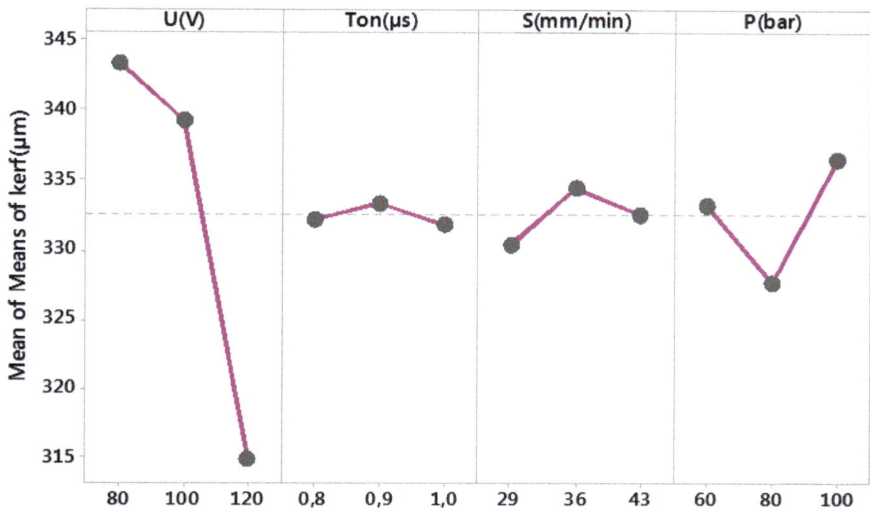

Fig. 14.15 The effect of machining parameters on kerf (μm)

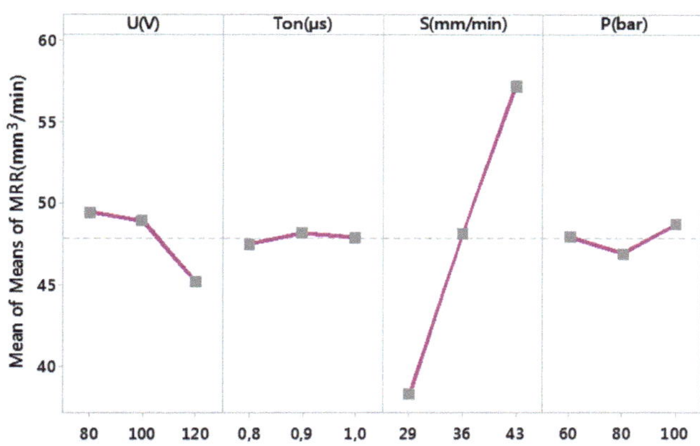

Fig. 14.16 The effect of machining parameters on MRR

bigest MRR value compared with S1 and S2 the servo voltage U1 corresponds to the greast MRR value wcompared with U3 and U2, the flushing pressure P2 corresponds to the biggest MRR value compared with P1 and P2.

The S/N ratio was used to determine the optimum parameters for a smaller value of kerf in WEDM machined surface of Ti17 alloy and according to Fig. 14.16 and Table 14.6 the optimal level of the machining parameters is the level with the greatest S/N ratio.

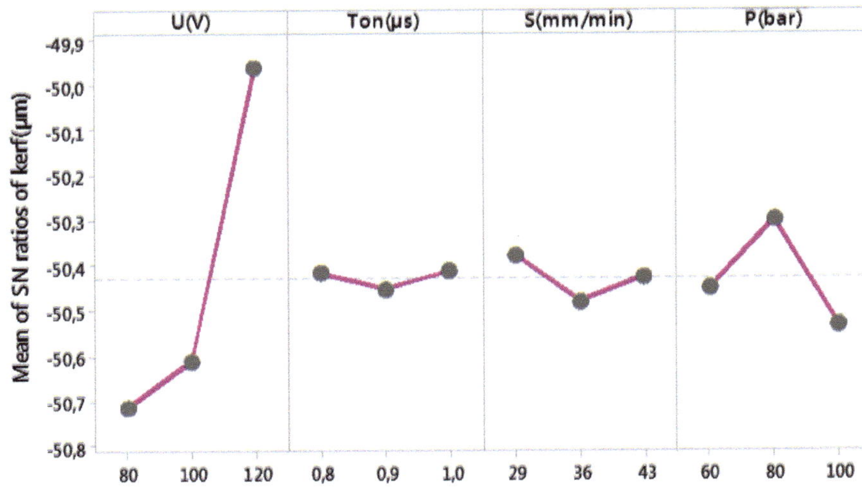

Fig. 14.17 Main effects plot for SN ratios for kerf

Table 14.10 ANOVA result for Kerf

Source	DF	Seq SS	Cont%	Adj SS	Adj MS	F-Value	P-Value	Remarks
Model	8	4711.27	99.60%	4711.27	588.909	555.19	0.000	Significant
U(V)	1	3641.46	76.98%	461.80	461.799	435.36	0.000	Significant
T_{on}(µs)	1	0.57	0.01%	10.26	10.262	9.67	0.006	–
S (mm/min)	1	19.82	0.42%	55.65	55.653	52.47	0.000	Significant
P (bar)	1	47.60	1.01%	294.85	294.855	277.97	0.000	Significant
U^2 (V)	1	625.46	13.22%	625.46	625.465	589.65	0.006	–
Ton^2(µs)	1	10.43	0.22%	10.43	10.428	9.83	0.000	Significant
S^2	1	52.16	1.10%	52.16	52.156	49.17	0.000	Significant
P^2(bar)	1	313.78	6.63%	313.78	313.782	295.81	0.000	Significant
Error	18	19.09	0.40%	19.09	1061			
Total	26	4730.37	100.00%					
S			R-sq		R-sq(adj)		R-sq(pred)	
1.02992			99.60%		99.42%		99.09%	

In Fig. 14.17 The S/N ratio was used to determine the optimum parameters for a great value of MRR in WEDM machined surface of Ti17 alloy and according to Fig. 14.17 and Table 14.7 the optimal level of the machining parameters is the level with the greatest S/N ratio.

Analysis of variance (ANOVA) is used to evaluate the model developed, as shown in Tables 14.10 and 14.11, analysis of the ANOVA variance is a result which makes it possible to see if the variables Explanations give significant information to the model or not.

Larger F-Value indicates that the variation of the process parameter makes a big change on the performance characteristics. F-Values of the machining parameters are compared with the appropriate confidence table.

According to F_{test} analysis the significant parameters on the kerf *are* flushing pressure P, servo voltage U, feed rate S and P^2, T^2 and S^2, in addition on the MRR all parameters are significants. The percent contributions on the machining parameters on the kerf and MRR are shown in Tables 14.7 and 14.8.

Equations 14.5 and 14.6 presented the regression Equation of ker and MRR models:

$$Kerf\ (\mu m) = 73,9 + 4,394 \times U + 235,5T_{on} + 4,482 \times S - 2,811 \times P - 0,02553 \times U^2 - 131,8 \times {T_{on}}^2 - 0,06017 \times S^2 + 0,01808 \times P^2 \quad (14.5)$$

Table 14.11 ANOVA result for MRR

Source	DF	Seq SS	Cont%	Adj SS	Adj MS	F-Value	P-Value	Remarks
Model	8	1713.54	99.97%	1713.54	214.193	8899.64	0.000	Significant
U(V)	1	79.53	4.64%	11.16	11.156	463.54	0.000	Significant
T_{on}(μs)	1	0.67	0.04%	1.41	1.406	58.42	0.000	Significant
S (mm/min)	1	1602.09	93.47%	10.10	10.102	419.73	0.000	Significant
P (bar)	1	2.51	0.15%	10.79	10.791	448.38	0.000	Significant
U^2 (V)	1	14.90	0.87%	14.90	14.90	619.24	0.000	Significant
Ton^2(μs)	1	1.35	0.08%	1.35	1.35	55.91	0.000	Significant
S^2	1	0.88	0.05%	0.88	0.88	36.45	0.000	Significant
P^2(bar)	1	11.61	0.68%	11.61	11.61	482.54	0.00	Significant
Error	18	0.43	0.03%	0.43	0.024		0.00	Significant
Total	26	1713.98	100%					
S			R-sq		R-sq(adj)		R-sq(pred)	
0.155137			99.97%		99.96%		99.94%	

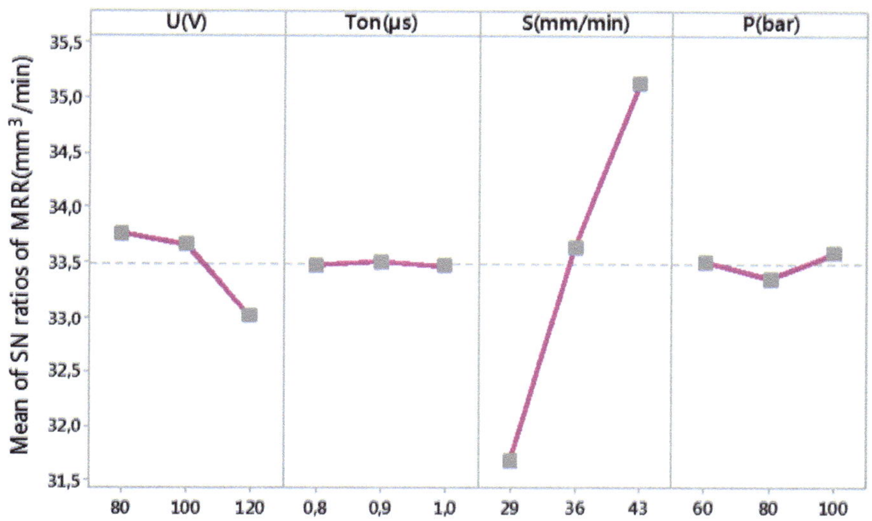

Fig. 14.18 Main effects plot for SN ratios for MRR

$$\text{MRR}\ (\mu m) = -58,27 + 0,6829 \times U + 87,2 \times T_{on} + 1,9096 \times S - 0,5378 \times P - 0,003940 \times U^2 - 47,36 \times {T_{on}}^2 - 0,00780 \times S^2 + 0,003478 \times P^2 \tag{14.6}$$

Figures 14.18 and 14.19 show that the residuals are distributed approximately in a straight line, showing a good relationship between the experimental and predicted values for all *Kerf* performances, and MRR, and the variable follows the normal distribution. Therefore, the developed models (Eqs. 14.5 and 14.6) are considered to be fairly adapted to the observed values. Likewise, these figures show that the residues found are randomly dispersed but are independent.

Figures 14.20 and 14.21 shows the effect of servo voltage and flushing pressure P on kerf, and the effects of advanced speed and servo voltage on MRR. As is clear, smaller values of kerf can be obtained by selecting greater servo voltage. In other hand, greater values of MRR can be obtained by slecting great advanced speed. s

The RSM models are carried out with 9 experimental tests and on evry sample we affected three test of kerf value and MRR value. For each combination of the input factors, the prediction value of the response $Y_{j,pred}$ is compared with the experimental value of the response $Y_{j,exp}$. Figures 14.21 and 14.22 illustrate the comparison between the experimental and predictive response values, comparing predictive and experimental value of kerf and comparing predictive and experimental value of MRR, to test the interpolation of the prediction of the models developed by RSM (Eqs. 14.5 and 14.6). We conclude that the predictive results with RSM are very close to the experimental results (Figs. 14.23 and 14.24).

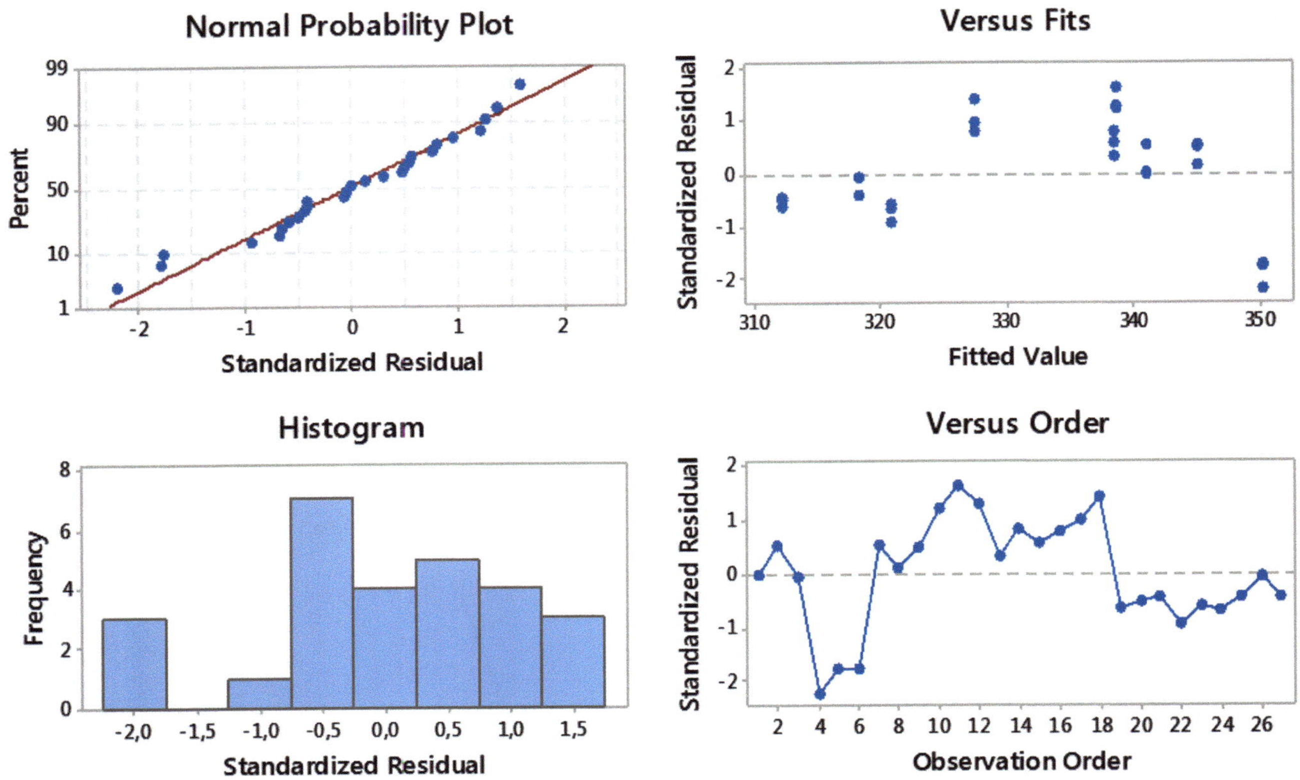

Fig. 14.19 Residual plots for Kerf

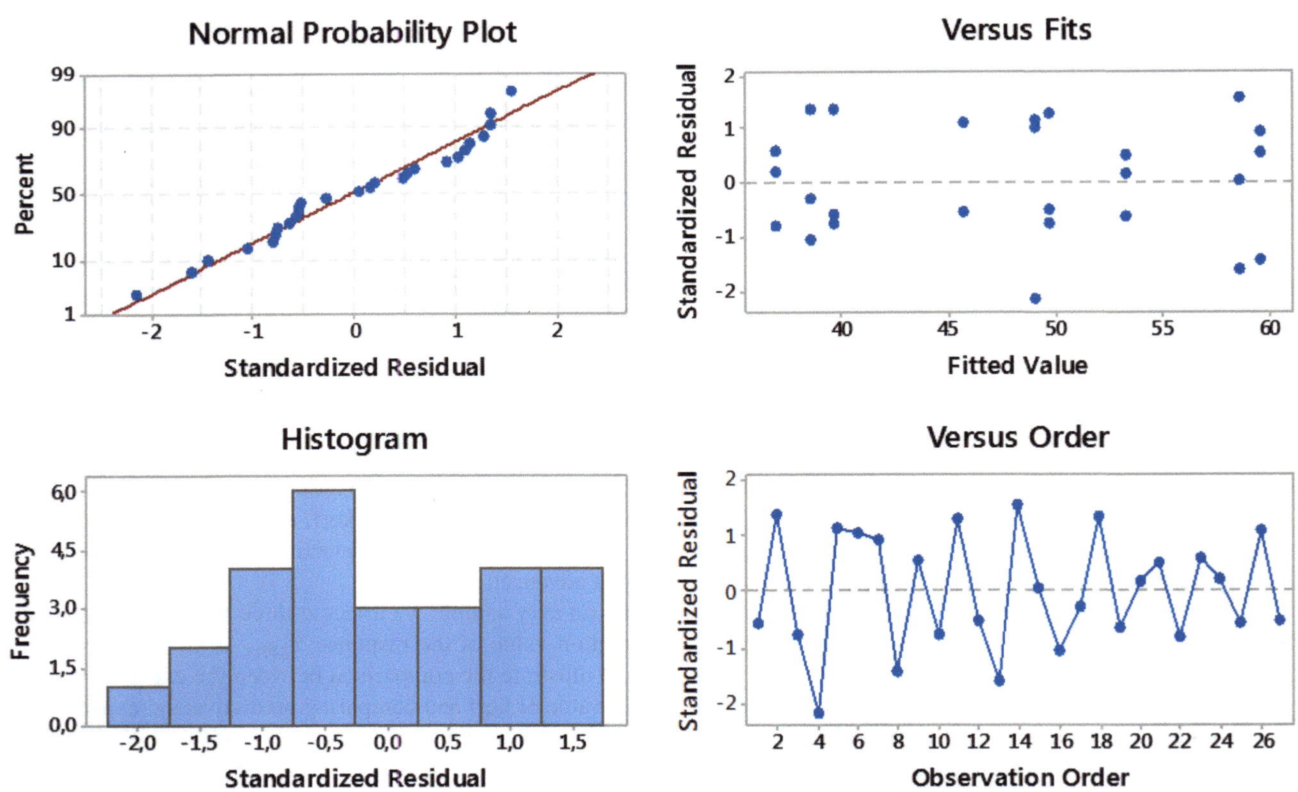

Fig. 14.20 Residual plots for MRR

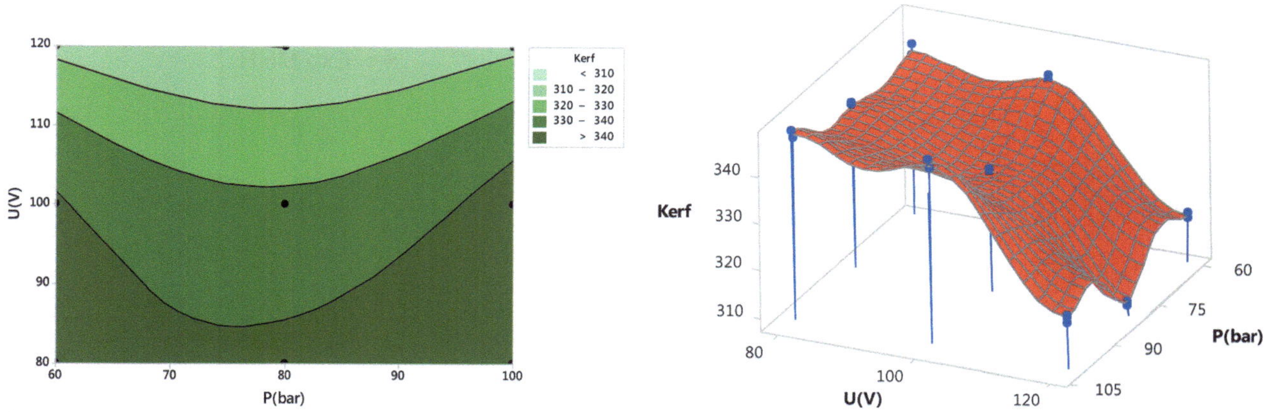

Fig. 14.21 Contour and 3D plots for kerf as function servo voltage and flushing pressure

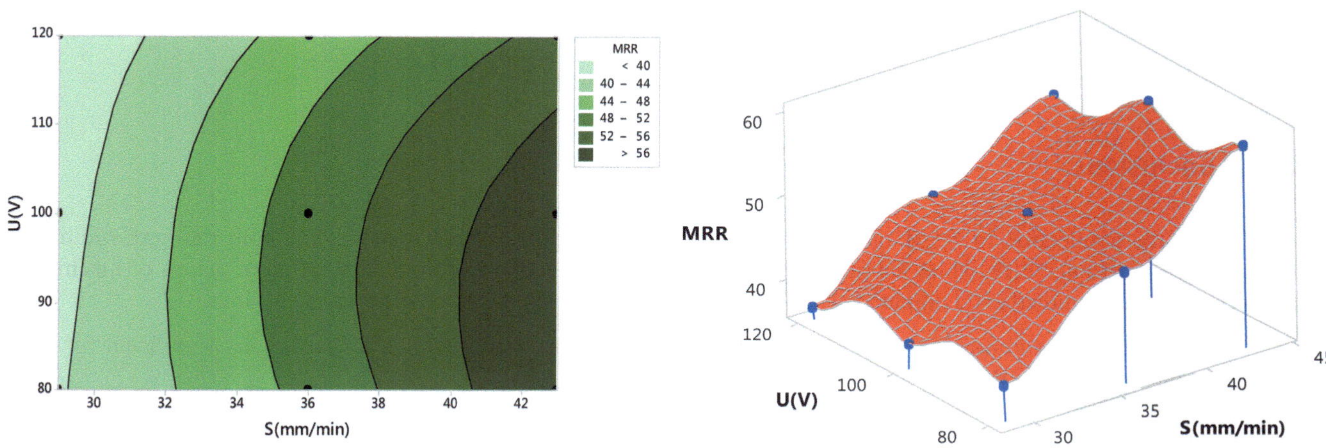

Fig. 14.22 Contour and 3D plots for MRR as function advance speed S and servo voltage

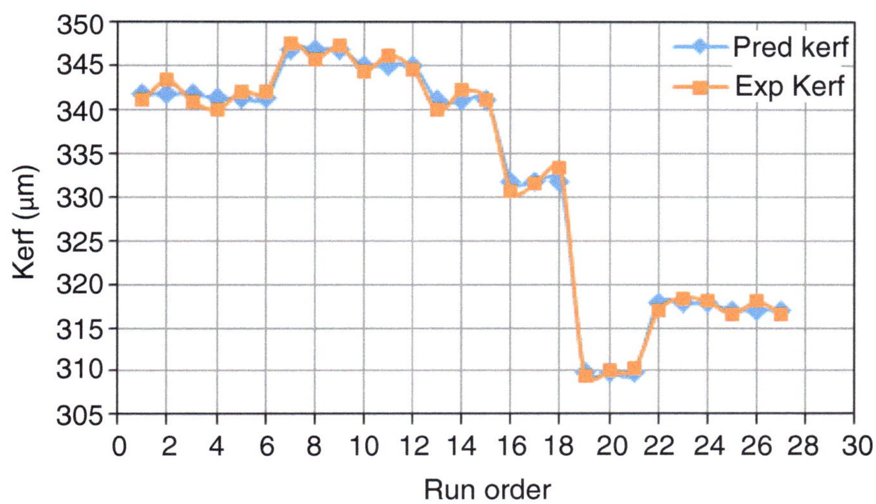

Fig. 14.23 Comparison between measured and predicted values for surface roughness Ra

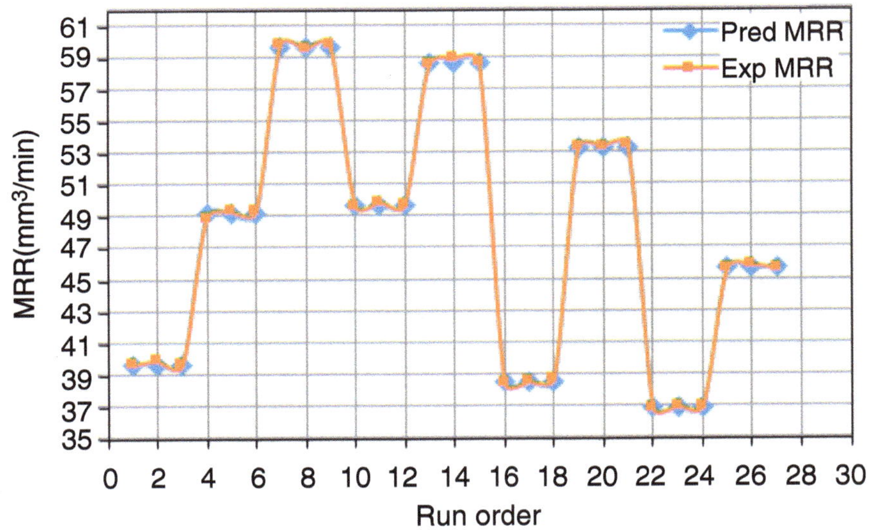

Fig. 14.24 Comparison between measured and predicted values for surface roughness Ra

14.4 Conclusions

In this paper, the application of RSM on the WEDM of v to obtain mathematical model for both the kerf width and the material removal rate MRR while investigating the influences of machining parameters optimum values of machining parameters have been studied and computed. The foremost conclusions which can be draw are as follows:

- All the input parameters considered for the experimental investigation were found to be statistically significant for their effects on the kerf and MRR.
- Smaller values of kerf can be obtained by selecting greater servo voltage. In other hand, greater values of MRR can be obtained by selecting great advanced speed.
- the servo voltage has the greatest influence on the kerf values and the value U = 120 V corresponds to the smallest kerf value compared with U1 and U2
- the advanced speed has the greatest influence on the MRR values and the value S 3 = 43 mm/min corresponds to the bigest MRR value compared with S1 and S2
- The values of predicted signal to noise is very much close to the calculated signal to noise values hence the analysis of Taguchi for signal to noise ratio is correct, it's perfect. The representation of effects of various parameters on kerf and MRR and optimize condition is very much nearby.

References

1. Mouralova, K., Kovar, J., Klakurkova, L., Prokes, T., Horynova, M.: Comparison of morphology and topography of surface of WEDM machined structural materials. Measeurement. **104**(12–20 (2017)
2. Knight, W.A., Boothroyd, G.: Fundamentals of Metal Machining and Machine Tools, 3rd edn. CRC Press (2005). Roca Baton, Florida, USA, ISBN 1574446592
3. Jain, V.K.: Advanced Machining Process. Allied Publishers, Ballard Estate, Mumbai, India (2009)
4. E.C. Jameson. Electrical Discharge Machining, 2001., ISBN08-726-3521-X
5. Tai, T.Y., Lu, S.J.: Improving the fatigue life of electro-discharge. Machined SDK1111 tool steel via the suppression of surface cracks. Int. J. Fatigue. **31**(3), 433–438 (2009)
6. Lee, S.H., Li, X.: Study of the surface integrity of the machined workpiece in the EDM of tungsten carbide. J. Mater. Process. Technol. **139**(1), 315–321 (2003)
7. Zeid, O.A.: On the effect of electro discharge machining parameters on the fatigue life on AISID6 tool steel. J. Mater. Process. Technol. **68**(1), 27–32 (1997)
8. Chalisgaonkar, R., Kumar, J.: Multi-response optimization and modeling of trim cut WEDM operation of commercially pure titanium (CPTi) considering multiple user's performances. Eng. Sci. Technol. Int. J. **18**, 125–134 (2015)

9. Medfar, A., Boujelebene, M., Bayraktar, E.: A mathematical model to choose effective cutting parametrs in electroerosion, EDM. J. Achiev. Mater. Manifact. Eng. **47**(1), (2011)
10. Tebni, W., Boujelebene, M., Bayraktar, E., Ben Salem, S.: Parametric approach model for determing electrical discharge machining (EDM) conditions, effect of cutting parameters on the surface integrity. Arabian J. Sci. Eng. **34/1c**, 101–114 (2009)
11. Ozkul, I., Berat Baris, B., Adnan, A.: Machinability of sleipner cold work stell with wire electric discharge machining. Int. J. Elect. Mech. Mechatronic Eng. **2**(3), 252–260
12. A. A. Khan, S.E. Saifuddin, Wear characteristics of copper and aluminium electrodes during EDM of stainless steel and carbide, Proceeding of the international conference on Mechanical Engineering Dhaka, Bangladesh (2005)
13. Prabhu, S., Vinayagam, B.K.: Analysis of surface characteristics of AISI D2 tool steel material using electric discharge machining process with single wall carbon nano tubes. IACSIT Int J. Eng. Technol. **2**(1), (2010)
14. Guu, Y.H.: AFM surface imaging of AISI D2 steel machined by the EDM process. Apple. Surface Sci. **242**(2004.) (ICME05-AM-09]), 245–250 (2005)
15. AY, M., Aydogdu, D.: Tel erozyomda Kesma paranehinin parçain bayut olçuusume et kilerinin denyesel ineelennesi. Makine teknolojileri Electronik Deigisililt. **7**(13), 31–44 (2010)
16. Farnaz Nourbakhsh, K.P., Rajurkar, A.P.M., Cao, J.: Wire electro-discharge machining of titanium alloy. Procedia CIRP. **5**, 13–18 (2013)
17. Tosun, N., Cogun, C., Tosun, G.: A study on kerf and material removal rate in wire electrical discharge machining based on taguchi method. J. Mater. Process. Technol. **152**, 316–322 (2004)
18. Rajesh, R., Dev Anand, M.: The optimization of electro-discharge machining process using response surface methodology and genetic algorithms. Procedia Eng. **38**, 3941–3950 (2012)
19. Chalisgaonkar, R., Kumar, J.: Multi-response optimization and modeling of trim cut WEDM operation of commercially pure titanium (CPTi) considering multiple user's perferences. Eng. Sci. Technol. Int. J. **18**, 125–134 (2015)
20. GS Paix Méthodes Taguchi, une approche pratique. Addision-Wesley, MA (1992)
21. Boujelbene, M., Bayraktar, E., Tebni, W., Ben Salem, S.: Influence of machining parameters on the surface integrity in electrical discharge machining. Arch. Mater. Sci. Eng. **37**(2), 110–116 (2009)

Chapter 15
Alternative Composite Design from Recycled Aluminum Chips for Mechanical Pin-Joint (Knuckle) Applications

D. Katundi, A. B. Irez, E. Bayraktar, and I. Miskioglu

Abstract In this work, an alternative aluminum matrix composite (AMCs) was designed from the recycled chips of the aluminium series of AA7075 (90 wt %) and AA1050 (10 wt %) given by French aeronautic company through combined method of powder metallurgy followed by Sintering + Forging. We aimed for the application for the connection link in a mechanism to transfer motion, for example; between the two railways wagons etc. and also some connecting link in aeronautical pieces as an alternative replacement for conventional alloys used in this area. First of all, A typical Al-Zn-Mg-Si-Ni matrix was developed and reinforced basically with B_2O_3 (5, 10%). Chip milling was performed using high energy milling in a planetary ball mill with an inert argon atmosphere to prevent oxidation of the powders. Two compositions were prepared with two different percentages of B_2O_3 and also one composition was kept without reinforcement for comparison with the reinforced ones. Mechanical properties, compression and dynamic drop weight tests of these composites designed here can be improved with the doping process and doping volume fractions. Micro-hardness results were compared according to the optimization conditions of the doping and the reinforcement. Static compression and impact-drop weight tests were carried out. The microstructure analyses have been carried out by Scanning Electron Microscope (SEM).

Keywords Pin-joint composites · Sinter+ forging · Damage analyses · Static-dynamic tests · SEM analyses

15.1 Introduction

The design policy of the new composite materials should improve the mechanical and physical properties of the pieces used in new technical applications, considerably better than the current ones. Therefore, in the his stage of the research project that is going on, we proposed a new design aluminium matrix composites for the mechanical joints such as the pin – joint very often known as knuckle components, as rod-couplings, recently a wind shield wipers for automotive industries as transmitting pieces. Here, the basic idea is to create a heterogeneous structure actually for the pin joint. In fact some of the industrial partners use extensively knuckle component as pin joint, etc. As well known, heterogeneous and anisotropic materials are useful structures for decreasing damage zone by creating dynamic instabilities such as local plastic deformation, buckling, etc., it means that the behaviour of the materials can be changed by the heterogeneous structures, [1–8].

Again, the Powder Metallurgy (PM) route is known as most commonly used method for the preparation of reinforced MMCs. This method is generally used as low – medium cost to produce small objects (especially round), tough, the high strength and resistant materials. Since no melting is involved, there is no reaction zone developed, showing high strength properties. For this reason, in the present work, a simple idea was developed on the production of pin joint pieces from the high resistance aluminium based composites [9–13].

The composites developed in the frame of the research that is going on were based on aluminum matrix originally from the recycled chips of the aluminium series of AA7075 (90 wt %) and AA1050 (10 wt %) given by French aeronautic company. We aimed for the application for the connection link in a mechanism to transfer motion, for example; between the two railways wagons etc. and also some connecting link in aeronautical pieces as an alternative replacement for conventional

D. Katundi · E. Bayraktar (✉)
Supmeca-Paris, School of Mechanical and Manufacturing Engineering, Paris, France
e-mail: bayraktar@supmeca.fr

A. B. Irez
CentraleSupélec, Université Paris-Saclay, Gif-sur-Yvette, France

I. Miskioglu
Michigan Technology University, Engineering Mechanics Department, Houghton, MI, USA

P. R. Thakre et al. (eds.), *Mechanics of Composite, Hybrid and Multifunctional Materials, Volume 5*, Conference Proceedings of the Society for Experimental Mechanics Series, https://doi.org/10.1007/978-3-319-95510-0_15

Table 15.1 Composition of the three groups developed

Aluminium matrix composites	JB	J1	J2	J3
Reinforcement B_203 (wt %)	0	5	10	15

alloys used in this area. As an alternative composite was produces through combined method of powder metallurgy followed by Sintering + Forging [1, 4, 6–8, 10, 13–18]. This composite contains basically zinc, silicon, magnesium and further reinforcements and auxiliary elements, where the Zn and Mg and Mn contents are 8, 0, 15 and 3 wt % respectively. For the sake of the mutual diffusion of certain reinforcements in the matrix, a few percentages of nickel was also added in the structure. This structure can be accepted as hypoeutectic structure [1, 4, 17].

15.2 Experimental Conditions

15.2.1 Materials Processing

In the present work, an alternative aluminum matrix composite (AMCs) was designed from the recycled chips of the aluminium series of AA7075 (90 wt %) and AA1050 (10 wt %) in the atomized form given by French aeronautic company through combined method of powder metallurgy (Sintering) followed by Forging. First of all, A typical Al-Zn-Mg-Si-Ni matrix was developed and reinforced basically with B_2O_3 (5, 10%). Chip milling was performed using high energy milling in a planetary ball mill with an inert argon atmosphere to prevent oxidation of the powders. Two compositions were prepared with two different percentages of B_2O_3, here called after J1 (5%) and J2 (10%) also one composition was kept without reinforcement (JB) for comparison with the reinforced ones. For easy wettability of the reinforcement of B_2O_3, certain amount of pure nano aluminium was added in the mixture. The composites produced with this novel combined method have certain advantages regarding to conventional manufacturing processes such as low cost, capability of manufacturing of the pieces with complex shapes, and processing simplicity, etc. The compositions of the three groups developed in the present work were given in Table 15.1.

This composite contains basically zinc, silicon, magnesium and further reinforcements and auxiliary elements, where the Zn and Mg and Mn contents are 8, 0, 15 and 3 wt % respectively. For the sake of the mutual diffusion of certain reinforcements in the matrix, a few percentages of nickel was also added in the structure. This structure can be accepted as hypoeutectic structure.

Microstructural characterization and Mapping analyses was done by means of scanning electron microscope (SEM). The dispersion of reinforcement particles in the matrix and interface at matrix/reinforcements was evaluated. Micro hardness tests ($HV_{0.1}$) have been carried out on the polished and etched specimens.

All the density measurements of the specimens were carried out by using *Archimedes* method. These values change between 2.72, 2.88 and 2.90 $\pm$ 3 for the JB, J1 and J2 respectively. Three cylindrical specimens (H/D $\geq$ 1.5) for each composition were tested under quasi-static compression conditions according the DIN 50106 norm by using a constant test speed as 0.0067 1/s.

Dynamic test known as drop weight tests have been carried out on a universal drop weight test device (Dynatup Model 8200 machine) with a total weight of 10.5 kg, punch height of 600 mm and with an impact velocity of 3 m/s.

Wear resistance was measured by scratch wear tests at a frequency of 15 Hz. All of the compositions were tested in two different numbers of cycles, $50{*}10^3$, $100{*}10^3$ cycles. After scratch test, damaged zone was investigated by 3D optical roughness meter. Surface and volume loss/time and maximum depth were evaluated for damage characterization.

15.3 Results and Discussions

15.3.1 Microstructure and Mapping Analyses of the Three Compositions Produced by "Sintering" and "Sinter+ Forging Process"

General microstructures of the three compositions (JB, J1 and J2) were presented in the Fig. 15.1. It is noted that the main structure of JB without reinforcement is quasi hypoeutectic structure with high Zn (8%) and Mg (3%) contents. Distribution

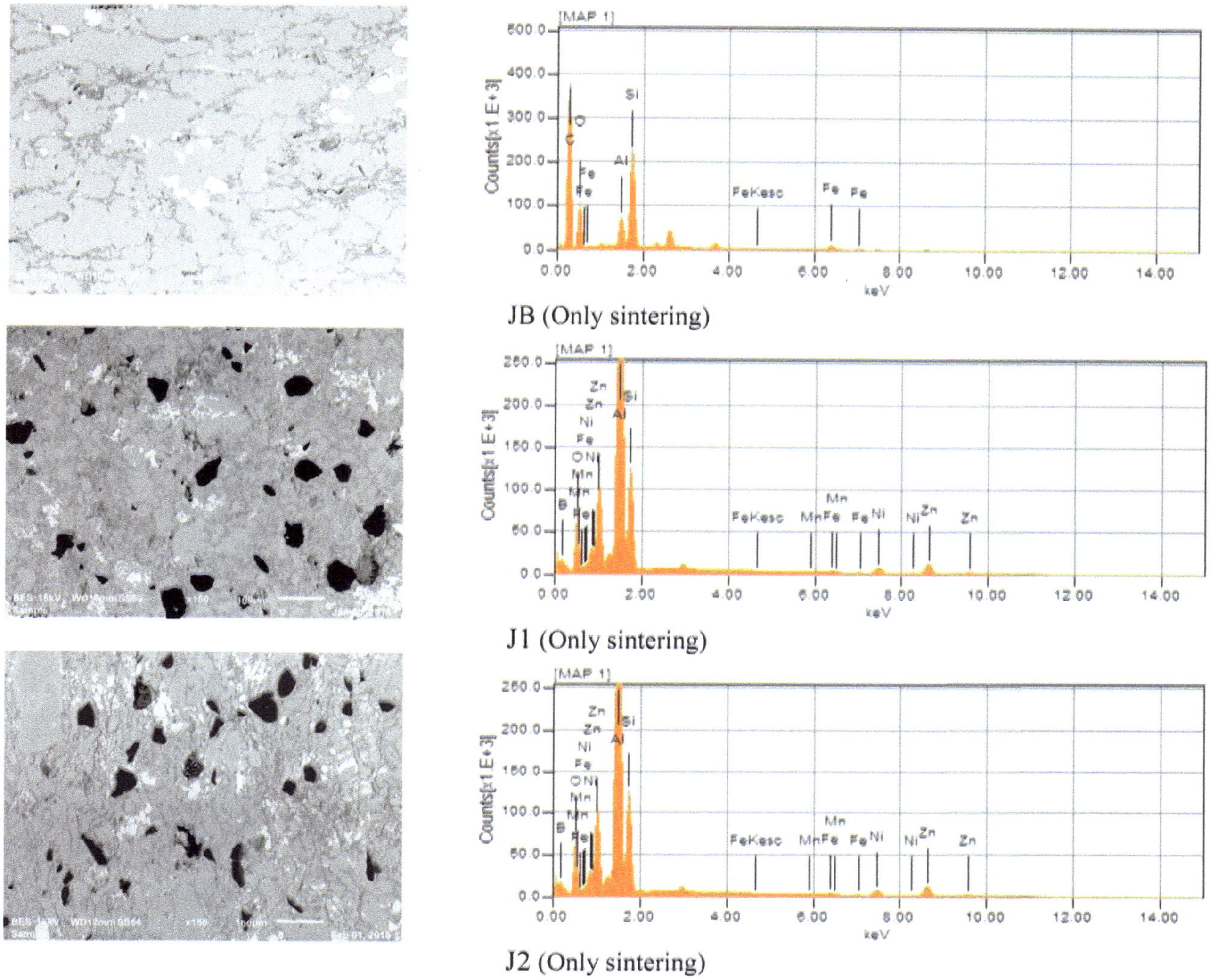

Fig. 15.1 General microstructure of the compositions produced by only sintering process, JB, JI and J2 respectively

of auxiliary elements is more and less homogenous. It seems that some of the certain elements are precipitated around the grain boundaries. "EDS" analyses for the three compositions were also given in the Fig. 15.1.

However, local agglomeration is also observed. This case is directly related of mixture conditions. All of the specimens were prepared under our laboratory conditions, therefore, it should be improved operational conditions. As for the reinforcements, boron oxide and boron and also nickel were appeared on the microstructure. Major reinforcement elements such as B_2O_3, was well seen that its distribution is homogenous. However, the particles size of the boron oxide (black color) is stayed big may be found agglomeration locally like a small island. This is related to the sintering process. It is not sufficient to prevent this type of agglomeration, etc.

In the Fig. 15.2, the microstructure obtained thorough the combine method sinter + Forging. All of the reinforcement particles and microstructure have been improved well. Distribution and the size of the reinforcements have been changed successfully. Small size and homogenous distribution of the reinforcements and also the grain size of the matrix have been carried out by application of the combined methods; sintering + forging. At the second stage of the process, it means that hot forging influences considerably. For this reason, this method has been accepted as more advantageous than the conventional methods. In comparison of this method as cost effect point of view, the combined method is seen much more quality, simple and economic manufacturing method.

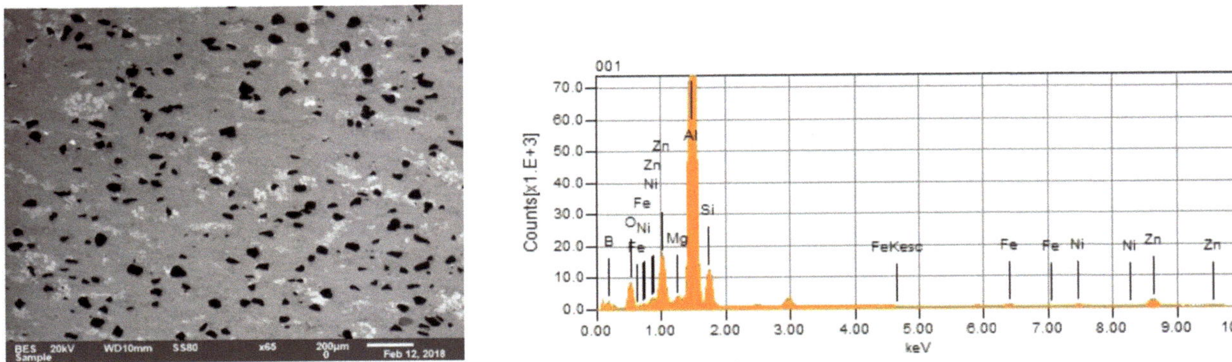

J1 (sintering+ Forging)

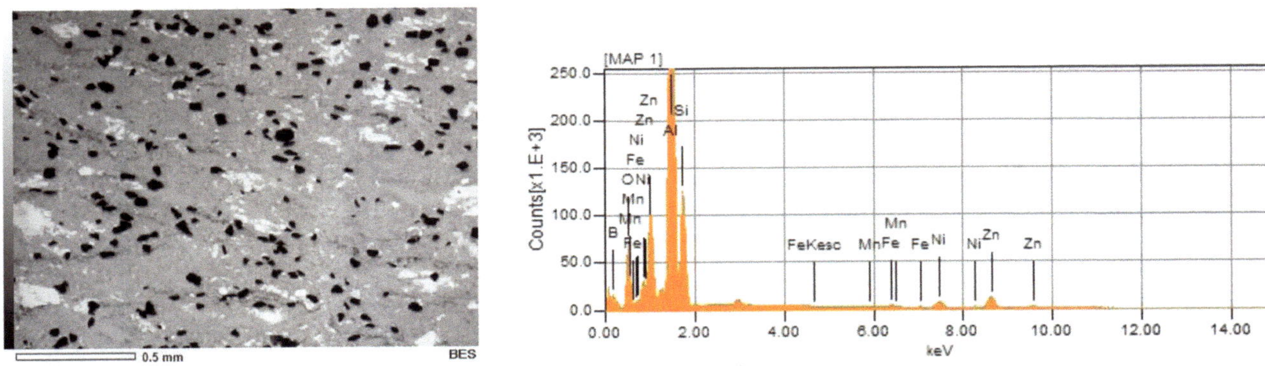

J2 (sintering+ Forging)

Fig. 15.2 General microstructure of the compositions produced by double process; sintering + Forging), JI and J2 respectively

15.3.2 Mapping Analyses of the Three Compositions Produced by Sinter+ Forging Process

As obtain vey homogenous structure with the combined method, the microstructure and mapping quantitative analyses have been carried out by SEM to see the distribution of the reinforcement and further auxiliary elements. Figures 15.3 and 15.4 present the general distribution of the micro and nano size particles in the microstructures of J1 and J2 compositions, respectively. It is noted that the distribution of certain auxiliary elements, for example, Fe, Mn, and Ni show small agglomeration in the matrix but the main particles are very smoothly distributed. This is main advantage of the combined method; sinter+forging. In any case, these structures can be improved by means of operating parameters it means that experimental parameters such as milling conditions, sintering and forging temperature, etc. should be controlled for obtaining ideal microstructures.

Micro hardness evaluation has been carried out on the mounted and polished specimens obtained only by the combined method; Sinter + Forging. It is noted that the hardness values have shown an increase but these are not so considerable increase as shown in the Table 15.2. This increase was observed directly with the percentage of the reinforcement elements. In fact, the reinforcements used here, such as boron oxide B_2O_3, Mn and Mg have a considerable effect on the mechanical properties, among them, vibrational fatigue properties can be improved by using these composites.

15.3.3 Static Compression Test Results

The idea for novel composite design can be applied very well on the industrial parts in an economic way. Actually, sintered forging process is a novel process mainly called as near-net shape process for the manufacturing of the pieces [13–16]. Mainly, this process is used for bulk materials in industrial applications. In fact, low-cost sinter-forging approach to processing of particle-reinforced metal matrix composites gives always high performance applications of the industrial pieces (fatigue-creep, static and impact compression, etc.).

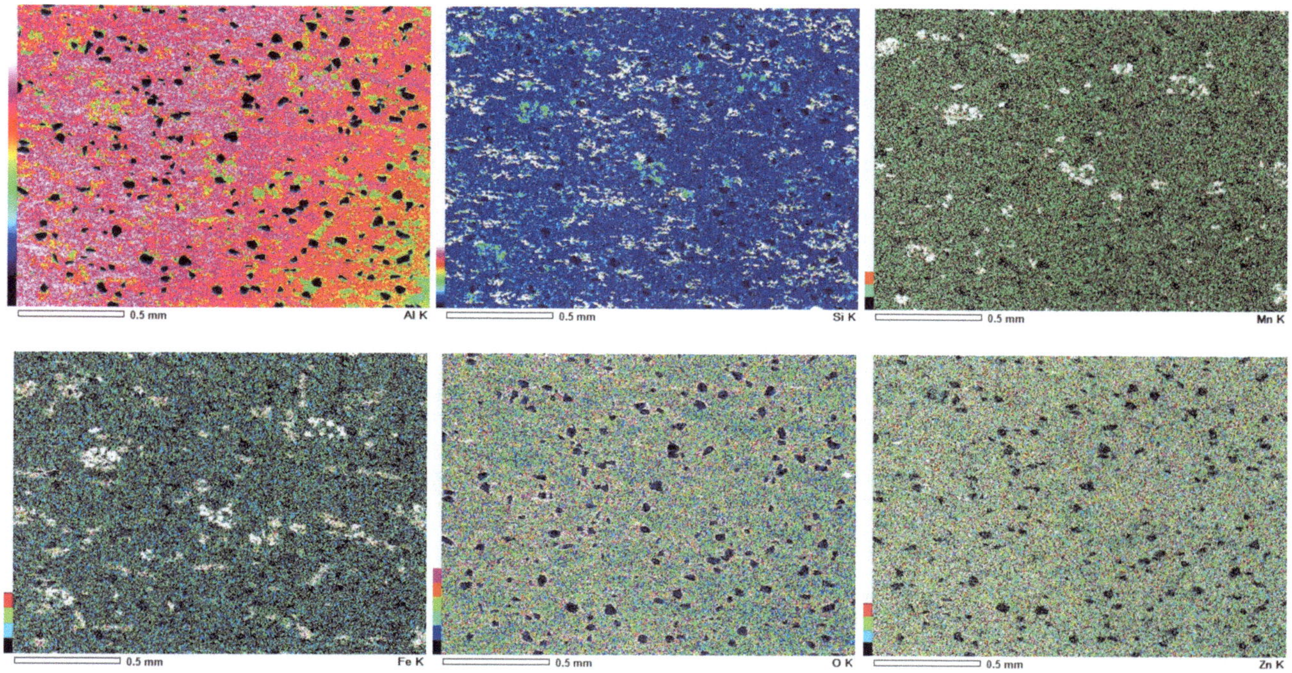

Fig. 15.3 Mapping analyses of the composite for J1 produced by combined process; Sinter + Forging

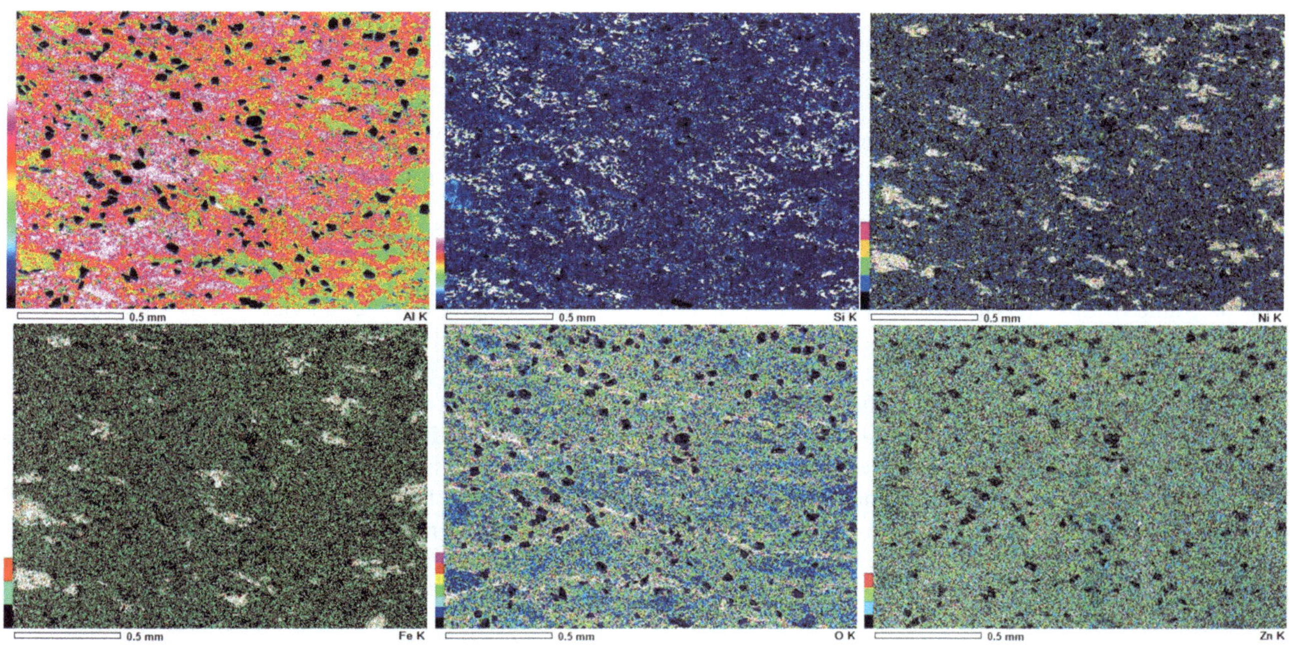

Fig. 15.4 Mapping analyses of the composite for J2 produced by combined process; Sinter + Forging

Table 15.2 Micro hardness (HVN) measurements for three compositions as mean values

Composite name	Composition		
	JB	J1	J2
Micro hardness, mean values ($HV_{0.1}$)	97	134	168

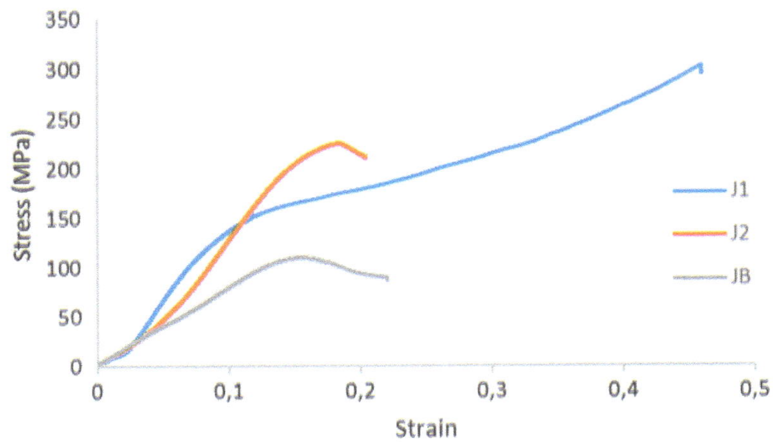

Fig. 15.5 Static compression test for the test specimens containing three compositions, J1, J2 and J3

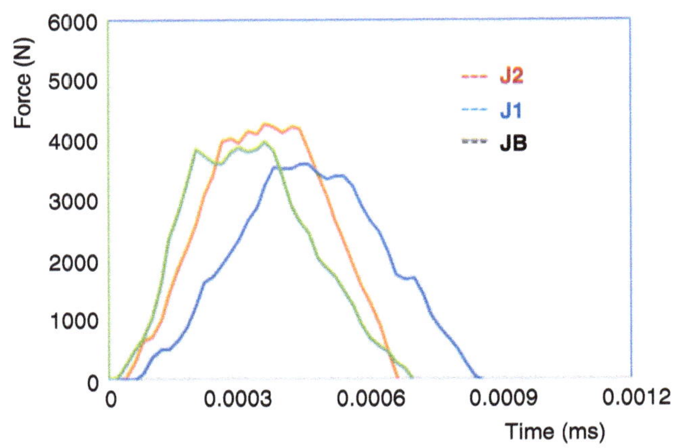

Fig. 15.6 Impact behaviour for three different specimens obtained by dynamic compression test, JB, J1 and J2 respectively

In the frame of this present work, a simple static compression test was given in the laboratory scales for mechanical behaviour of this novel composite produced with sinter+forging processes. This idea should be developed very well on the many other composites in the industrial scale. It means that very tough and strong pieces can be obtained with this combined process but cheaper than the other manufacturing processes.

Figure 15.5 gives static compression test results obtained on the specimens produced by sinter + forging processes. These values are mean values obtained from three tests for each composition. Evidently standard deviation is variable around ±20–25 MPa. One may observe that the higher reinforcement element gives always higher resistance regarding to the simple composition without reinforcement (as compared with JB, J1 and J2). Ultimate tensile strength values in static compression tests for the specimens of J1 and J2 have been found around 220–250 MPa whereas the specimens of JB stayed at the level of 100 MPa.

15.3.4 Low Velocity or Dynamic Compression (Drop Weight) Test Results

Low velocity or dynamic compression tests results were given in the Fig. 15.6. Maximum force was evaluated there by the values from both support data points. Here, a series of impact tests were performed at room temperature at the center of cylindrical specimens using the instrumented drop weight test device as explained in detail in 2nd the section (experimental conditions). Three specimens were tested for each composition (Fig. 15.5).

First of all, the effect of combined effect of sinter+forging processing on the impact resistance of the specimens (J1, and J2) can be detected very clearly from these graphs. The specimens of JB (without reinforcements) have shown relatively low values regarding to the compositions with the reinforcements.

It seems from these graphs, impact resistance is related to absorbed energy. In fact, all of the specimens have shown that the most part of the impact force is used to maintain the balance with the inertia force, and only a small portion of the impact force is actually used to deform and fracture of the specimen. Absorbed energy should be related with the process used here that this energy increases considerably in the structure obtained with sinter+forging. These results are only obtained in laboratory scales and should be improved with detail analyses.

15.3.5 Wear (Scratch) Test Results

Reduction of friction rests the key task for wear-resistant composites. Evaluation of the wear resistance of this composite developed in this work have been carried out in two different numbers of cycles, $50*10^3$, and $100*10^3$ cycles. Influence of reinforcement elements and essentially influence of the manufacturing processes, sinter + Forging are observed for three different specimens in the Fig. 15.7 and also wear (scratch) test results were given in Table 15.3 for the specimens of JB, J1 and J2 respectively.

The surface damage (mm^2), volume lost (μm^3) and the depth (μm) values are presented in the same figures for each test condition. Effects of reinforcements and obviously, the effects of the sinter + Forging process are observed as an advantage over others.

As shown in the results of scratch tests, the composites processed with sinter+forging has shown always higher wear resistance regarding to the simple structures. In this particular case, the size of the reinforcements and their dispersion on the matrix should have contributed to improve the wear resistance under experimental conditions carried out in the present work.

Total evaluations of the microstructure and mechanical properties (static compression, dynamic- drop weight and also wear-scratch tests) give a clear idea about the combined process (sintered + forging). This process applied for these types of composites is caused by bonding diffusion at the interface between matrix and reinforcement and some of the particles were forced into the grains during the forging (second) stage of this process. For this reason, very tough, solid and homogeneous structure could be obtained. Porosity and other structural – micro defects were quasi eliminated.

15.4 Conclusions

A novel composite was designed design aluminium matrix composites for the mechanical joints such as the pin – joint very often known as knuckle components as alternative composite against the conventional pieces used for example as rod-couplings, for automotive industries as transmitting pieces. In the frame of this common project, novel composites have been developed from aluminium AA7075 powder obtained with atomization of fresh scrap-chips as the initial form reinforced with B2O3, Mn, Ni, Mg etc. particles as main reinforcements in an economic way. Low cost manufacturing of these composites have been successfully managed through the combined method of sinter + forging.

Microstructural analysis has shown that a good bonding at interface of matrix-reinforcement essentially in the specimens manufactured with combined process sinter + forging; a tough and complete microstructure was obtained without porosity. Wear resistance and ductility of these structures should be improved with doping process and good pretreatment conditions; ball milling in longer time is needed for helping the fine and homogeneous distribution of the particles in the matrix. Mechanical behaviour of these specimens produced with sinter + forging process are better than the specimens produced without reinforcement. This process seems very confident values for future work of the production of alternative pieces used in joint parts and also for other tribological applications. Optimizations of the certain parameters such as processing parameters, reinforcement content, etc. need much more experimental work to create real parts in the industrial scales. Here, only limited measurements at room temperature were presented as they are indicative parameters for better understanding the effect of the reinforcements on the optimization of the mechanical, (static and dynamic) and wear properties of the composites produced in the present work.

Acknowledgments This work has been carried out on the frame of collaboration between SUPMECA/PARIS and MICHIGAN TECHNICAL UNIVERSITY/ HOUGHTON-MI-USA. Authors acknowledge and appreciate so much Dr. G. ZAMBELIS from Airbus-Helicopter/Paris for valuable technical helps and discussions.

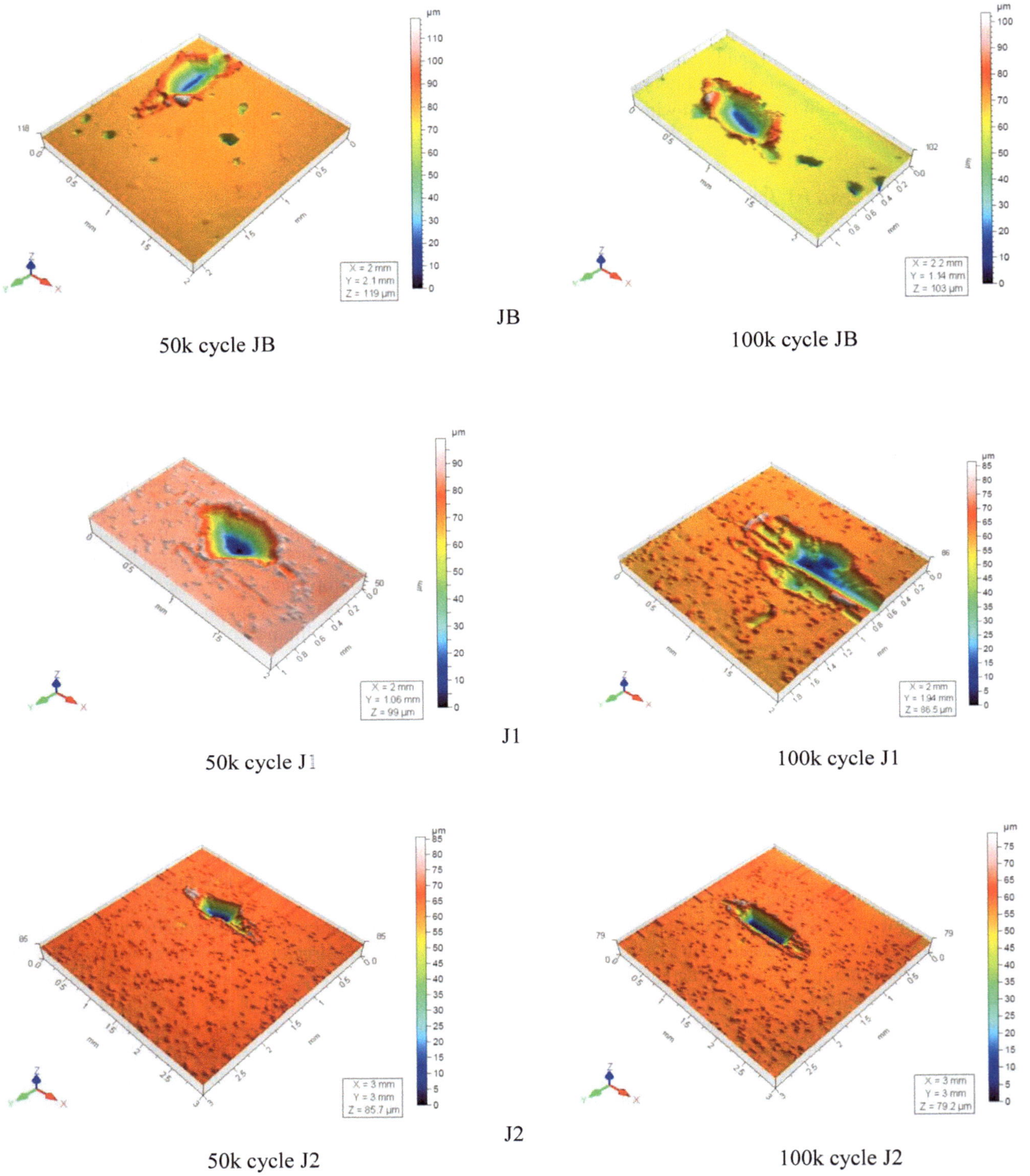

Fig. 15.7 Images of surface damage zones produced by wear – scratch test results for the three different composition in the cycle of $50*10^3$ and 100.10^3 cycles for the comparison of damage zone defined with surface and volume lost

Table 15.3 Wear (macro scratch) test results for the specimens of JB, J1 and J2 respectively

Wear test	50.10^3 cycles		100.10^3 cycles	
Composites	Worn surface (mm^2)	Worn volume (μm^3)	Worn surface (mm^2)	Worn volume (μm^3)
JB	0.301	6,071,537	0.478	9,398,134
J1	0.246	5,539,514	0.302	9,113,252
J2	0.187	4,815,100	0.283	7,067,373

References

1. Kurşun, A., Ferreira, L.F.P., Bayraktar, E., Miskioglu, I.: Design of hybrid composites from scrap aluminum reinforced with (SiC+TiO2+Gr+Ti+B), Springer link. Mech. Composite Multi Funct. Mater. **7**, 225–232 (2016). https://doi.org/10.1007/978-3-319-41766-0, ISBN 978-3-319-41766-0
2. Shashanka, R., Chaira, D.: Development of nano-structured duplex and ferritic stainless steels by pulerisette planetary milling followed by pressureless sintering. Mater. Charact. **99**, 220–229 (2015)
3. Zhao, H., Barber, G.C., Liu, J.: Friction and wear in high speed sliding with and without electrical current. Wear. **249**, 409–914 (2001)
4. Ferreira, L.F.P., Bayraktar, E., Robert, M.H., Miskioglu, I.: Particles reinforced scrap aluminum based composites by combined processing; sintering + thixoforming, Springer link. Mech. Composite Multi Funct. Mater. **7**, 145–152 (2016). https://doi.org/10.1007/978-3-319-41766-0, ISBN 978-3-319-41766-0
5. Hymn, C.L.: Tribological properties of unidirectional poly – Phenylene sulfide-carbon fiber laminated composites. Wear. **117**, 147–159 (1987)
6. Kurşun, A., Bayraktar, E., Enginsoy, H.M.: Experimental and numerical study of alumina reinforced aluminum matrix composites: Processing, microstructural aspects and properties. Compos. Part B. **90**, 302–314 (2016)
7. Vicen, M., Fabian, P., Tillova, E.: Arch. Foundry Eng. **17**(3), 139–142 (2017). ISSN (1897-3310)
8. Katundi, D., Ferreira, L.-M.P., Bayraktar, E., Robert, M.-H.: Design and microstructural evolution, mechanical and physical properties of fine particles reinforced aluminium matrix composites. JAMPT Adv. Mater. Process. Technol. **2**(4), 566–577 (2016). Taylor & Francis-USA
9. Groth, K., Heidenfelder, F., Holinski, R.: Advancements of tribological performance of carbon brushes in electrical motors. Ind. Lubrication Tribology. **53**(1), 5–9 (2001)
10. Ferreira, L.-M.-P., Bayraktar, E., Robert, M.-H.: Magnetic and electrical properties of aluminum matrix composite reinforced with magnetic nano iron oxide. JAMPT J Adv. Mater. Process. Technol. **2**(1), 165–173 (2016). Taylor and Francis-USA
11. Lim, S.C., Gupta, M., Ren, L., Kok, J.K.M.: Tribological properties of Al-Cu/SiCp metal matrix composites fabricated using the rheo-casting technique. J. Mater. Process. Technol. **89-90**, 591–596 (1999)
12. Torralba, J.M., Dacost, C.E., Velasco, F.: P/M Alu- minum matrix composites: An overview. J. Mater. Process. Technol. **133**(1–2), 203–206 (2003)
13. Viana, M.-F., Robert, M.-H., Bayraktar, E.: Structure analysis of graded AA2011–Al_2O_3/$NaAlSi_3O_8$ composites produced by thixoforming. JAMPT Adv. Mater. Process. Technol. **3**(2), 231–239 (2017). Taylor & Francis-USA
14. Ferreira, L.-M.-P., Bayraktar, E., Miskioglu, I., Robert, M.H.: Design and physical properties of multifunctional structural composites reinforced with nanoparticles for aeronautical applications. JAMPT Adv. Mater. Process. Technologies. **3**(1), 33–44 (2017). Taylor & Francis-USA
15. Dasgupta, R.: Aluminum alloy-based metal matrix composites: a potential material for wear resistant applications, International Scholarly Research Network., ISRN Metallurgy. **2012**, Article ID 594573, 14 pages (2012). https://doi.org/10.5402/2012/594573
16. Gatamorta, F., Bayraktar, E., Miskioglu, I., Katundi, D., Robert, M.H.: Design of low composites from recycled copper + aluminium chips for tribological applications. In: Mechanics of composite and multi-functional materials, vol. **6**, pp. 101–110. Springer (2017)
17. Ferreira, L.F.P., Bayraktar, E., Robert, M.H., Miskioglu, I.: Recycling of scrap aluminium (AA7075) chips for low cost composites, Springer link. Mech. Composite Multi Funct. Mater. **7**, 19–26 (2016). https://doi.org/10.1007/978-3-319-41766-0, ISBN 978-3-319-41766-0
18. Ferreira, L.P.: Production of aluminum metal matrix composites by thixoforming of recycled chips, Thesis for Master of Science, University of Campinas, UNICAMP, Mechanical and Manufacturing engineering, Campinas –SP, Brazil (2013)

Chapter 16
Manufacturing of Copper Based Composites Reinforced with Ceramics and Hard Intermetallics for Applications of Electric Motor Repair Parts

G. Zambelis, E. Bayraktar, D. Katundi, and I. Miskioglu

Abstract In the present work, a recycled copper based composites reinforced with ceramic as an alternative replacement for the application of electric motor repair parts with the use of novel processing techniques.

A practical solution was proposed as cost effective economic manufacturing of the composites for this type of applications. Copper based composite design (Cu-Al-Nb_2Al) was based on the ceramic reinforcements such as titanium carbide (TiC) in different percentages and niobium aluminate intermetallics (Nb_2Al). Because TiC and Nb_2Al make a good combination of thermal and electrical conductivities, microstructural stability and strength retention at elevated temperatures, etc. These reinforcements increase considerably wear resistance of the composites for electrical contact applications. Otherwise, certain percentage of fresh scrap aluminium powder, the mixture of AA1050 (80 wt% + AA7075 (20 wt %) chips were used to create an exothermic combustion reaction in the process for helping diffusion bonding process of the ceramics to the copper matrix. At the first stage of the present work a preliminary study has been carried out for developing a cost effect and high wear resistant electrical brushes for aeronautical applications. Microstructural and wear analyses have been carried out to optimize the process conditions for a practical tool that will be used for final industrial applications. Three basic compositions were prepared depending on the percentage of TiC. The microstructure and damage analyses have been carried out by Scanning Electron Microscope (SEM).

Keywords Copper matrix composites · Titanium carbide · Ceramic reinforcements · SEM-microstructure

16.1 Introduction

Safety design of the high resistance and high toughness composites is a vital requirement for operational materials that will be manufactured for critical and safety applications, in there early fracture cannot be accepted. The development of high resistance but high toughness composites has usually been a bridge between hardness vs. ductility. In the present work, a recycled copper based composites (Cu-Al-Nb_2Al) reinforced with ceramic as an alternative replacement for the application of electric motor repair parts with the use of novel processing techniques. Our motivation is on the interpretation of the microstructure of this composite developed in this work and explain the interface relationship between the matrix and reinforced elements and micro-mechanisms at the interface. This idea can help the significance of the relation between strength and toughness in the composites. As well known, copper is very important materials for this application especially for the electrical contact such as for the manufacturing of the brush used in the electrical motors as an important parts and/or other electronic devices. However, the copper based composites have lower wear resistance.

In the literature, different kind of compositions have been proposed for finding a solution to this problem. Generally, these solutions proposed as high strength and high wear resistant composites for electronics applications in the commercial markets are expensive and takes a long production time. For this reason, we have tried to propose a solution as cost effective economic manufacturing of the composites for this type of applications. Copper based composite design was based on

G. Zambelis
Airbus-Helicopter, Research and Development Center, Paris, France

E. Bayraktar (✉) · D. Katundi
Supmeca-Paris, School of Mechanical and Manufacturing Engineering, Paris, France
e-mail: bayraktar@supmeca.fr

I. Miskioglu
Michigan Technology University, Engineering Mechanics Department, Houghton, MI, USA

P. R. Thakre et al. (eds.), *Mechanics of Composite, Hybrid and Multifunctional Materials, Volume 5*, Conference Proceedings of the Society for Experimental Mechanics Series, https://doi.org/10.1007/978-3-319-95510-0_16

the ceramic reinforcements such as titanium carbide (TiC) in different percentages and niobium aluminate intermetallics (Nb_2Al). Because TiC and Nb_2Al make a good combination of thermal and electrical conductivities, microstructural stability and strength retention at elevated temperatures, etc. These reinforcements increase considerably wear resistance of the composites for electrical contact applications. Otherwise, certain percentage of fresh scrap aluminium powder, the mixture of AA1050 (80 wt% + AA7075 (20 wt %) chips were used to create an exothermic reaction in the process for helping diffusion bonding process of the ceramics to the copper matrix.

The first stage of the present work a preliminary study has been carried out for developing a cost effect and high wear resistant electrical brushes for aeronautical applications. Microstructural and wear analyses have been carried out to optimize the process conditions for a practical tool that will be used for final industrial applications.

16.2 Experimental Conditions

In the present work, an alternative copper based composite was designed. As the basic ceramic reinforcements such as titanium carbide (TiC) in different percentages and niobium aluminate intermetallics (Nb_2Al). These ceramic reinforcements make a good combination of thermal and electrical conductivities, microstructural stability and strength retention at high temperatures, etc. Additionally they increase considerably wear resistance of the composites. Copper and aluminium fresh scrap recycled chips in the atomized form given by French aeronautic company. The composites was fabricates through the combined method of powder metallurgy (Sintering) followed by Forging at high temperature.

A typical Cu-Al-Nb_2Al matrix was developed and reinforced basically with TiC (10, 20 and 30 wt %) and Nb_2Al.

Mixture has been carried out by using high energy milling in a planetary ball mill for 4 h.

Three composites were prepared with three different percentages of TiC, here called after S1, S2 and S3 also one composition was kept without TiC reinforcement (SB) for comparison with the reinforced ones. Otherwise, for easy wettability of the reinforcement of TiC and Nb_2Al, percentage of fine graphite +5 wt % GnPs (graphene nano particles and fine aluminium powder was added in the mixture. In fact, the mixture of AA1050 (90 wt % + AA7075 (10 wt %) chips were used to create an exothermic reaction in the process for helping diffusion bonding process of the ceramic reinforcement and Nb_2Al intermetallics to the copper matrix. Hot compaction was made at 300 MPa and sinter + forging has been carried out at 850 °C under inert atmosphere. The composites produced with this novel combined method have certain advantages regarding to conventional manufacturing processes such as low cost, capability of manufacturing of the pieces with complex shapes, and processing simplicity, etc. The compositions of the three groups developed in the present work were given in Table 16.1.

The first stage of the present work a preliminary study has been carried out for developing a cost effect and high wear resistant electrical brushes for aeronautical applications. Microstructural and wear analyses have been carried out by Scanning Electron Microscope to optimize the process conditions for a practical tool that will be used for final industrial applications.

Microstructural characterization and Mapping analyses was done by means of scanning electron microscope (SEM). The dispersion of reinforcement particles in the matrix and interface at matrix/reinforcements was evaluated. Micro hardness tests ($HV_{0.1}$) have been carried out on the polished and etched specimens.

All the density measurements of the specimens were carried out by using *Archimedes* method. These values change between 7.82, 6.88, 7.65 and 7.90 ± 3 for the SB, S1, S2 and S3 respectively.

Wear resistance was measured by scratch wear tests at a frequency of 15 Hz. All of the compositions were tested in two different numbers of cycles, $50*10,^3$ $100*10^3$ cycles. After scratch test, damaged zone was investigated by 3D optical roughness meter. Surface and volume loss/time and maximum depth were evaluated for damage characterization.

Table 16.1 Copper based composites and distribution of the reinforcements in the matrix

Cu based composites Cu-Al-Nb_2Al	SB	S1	S2	S3
Major reinforcementTiC (wt %)	–	10	20	30

16.3 Results and Discussions

16.3.1 *Microstructure and Mapping Analyses of the Compositions Produced by "Sinter + Forging Process"*

General microstructures of the four compositions (SB, S1, S2 and S3) were presented in the Fig. 16.1. It is noted that the main structure of SB without reinforcement is eutectic structure with aluminium, TiC and Nb_2Al contents. Distribution of elements in the matrix quasi homogenous with mutual diffusion can be observed very clearly the reinforcements with matrix. It seems that certain elements are smoothly precipitated around the grain boundaries.

However, local agglomeration of the graphite with graphene is also observed in certain zones. This case is directly related of mixture conditions. All of the specimens were prepared under our laboratory conditions, therefore, it should be improved operational conditions. As for the reinforcements, Titanium carbide was appeared on the microstructure.

The microstructure of the composites with reinforcements obtained thorough the combine method sinter + Forging are seems very homogeneous and improved with a good diffusion bonding due to the addition of the aluminium in the matrix that creates an exothermic reaction. The eutectic structure was observed in the structure and mainly the reinforcement particles made a good diffusion chemical bonding with matrix by means of the exothermic reaction of the aluminium in the matrix. At the second stage of the process, it means that hot forging influences the microstructure considerably. For this reason, this novel techniques should be accepted as more advantageous than the conventional methods.

As obtain vey homogenous structure with the combined method, the microstructure and mapping quantitative analyses have been carried out by SEM to see the distribution of the reinforcement and further auxiliary elements in the composites. Mapping Figs. 16.2, 16.3, 16.4 and 16.5 present the general distribution of the micro and nano size particles in the microstructures of the four compositions, respectively. It is noted that the distribution of certain auxiliary elements, for example, Nb_2Al, and Graphite and GnPs show small agglomeration in the matrix but the main particles are very smoothly distributed. This is main advantage of the combined method; sinter+forging. In any case, these structures can be improved by means of operating parameters it means that experimental parameters such as milling conditions, sintering and forging temperature, etc. should be controlled for obtaining ideal microstructures. For each composition, "EDS" analyses were given in the same mapping figures.

16.3.2 *Wear (Scratch) Test Results*

Reduction of friction is always a significant task for wear-resistant composites. The actual composite (Cu-Al-Nb_2Al) was developed for the future electronic devices, especially for the electrical brushes, etc. where the wear resistance is needed. For this reason, all of the specimens taken for each composition were measured their wear resistance experimentally. Evaluation of the wear resistance of these composites have been carried out in two different numbers of cycles, $50*10^3$, and $100*10^3$ cycles. Influence of reinforcement elements and essentially influence of the manufacturing processes, sinter + Forging are observed for four different specimens in the Fig. 16.6 and also wear (scratch) test results were given in Table 16.2 for the specimens SB, S1, S2 and S3 respectively.

The surface damage (mm^2), volume lost (μm^3) and the depth (μm) values are presented in the same figures for each test condition. Effects of reinforcements and obviously, the effects of the sinter + Forging process are observed as an advantage over others.

As shown in the results of scratch tests, the composites reinforced with ceramics, TiC and hard intermetallics Nb_2Al has shown always higher wear resistance regarding to the simple structures. In this particular case, the size of the reinforcements and their dispersion on the matrix should have contributed to improve the wear resistance under experimental conditions carried out in the present work.

Micro hardness evaluation has been carried out on the mounted and polished specimens obtained only by the combined method; Sinter + Forging. It is noted that the hardness values have shown an increase but these are not so considerable increase as shown in the Table 16.3. This increase was observed directly with the percentage of the reinforcement elements. In fact, the reinforcements used here, such as boron oxide TiC and also auxiliary Nb_2Al have a considerable effect on the mechanical properties that can be optimized by using these composites.

Total evaluations of the microstructure and also wear-scratch tests give a clear idea about the combined process (sintered + forging). This process applied for these types of composites is caused by bonding diffusion at the interface between matrix and reinforcement and some of the particles were forced into the grains during the forging (second) stage of this process. For

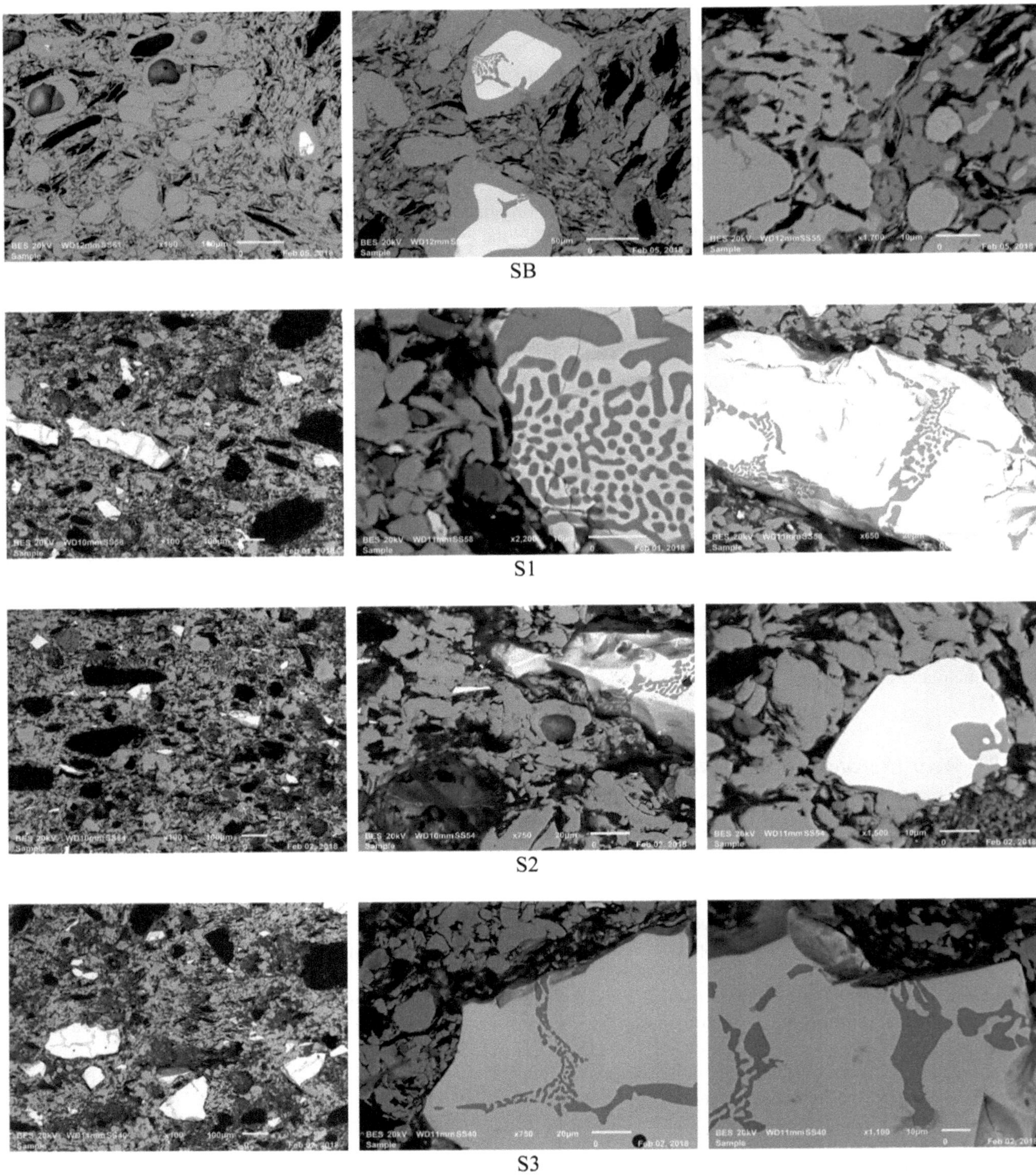

Fig. 16.1 General microstructure of the compositions produced by double process; sintering + Forging), SB, S1, S2vand S3, respectively

this reason, very tough, solid and homogeneous composite structure could be obtained. Porosity and other structural – micro defects were quasi eliminated.

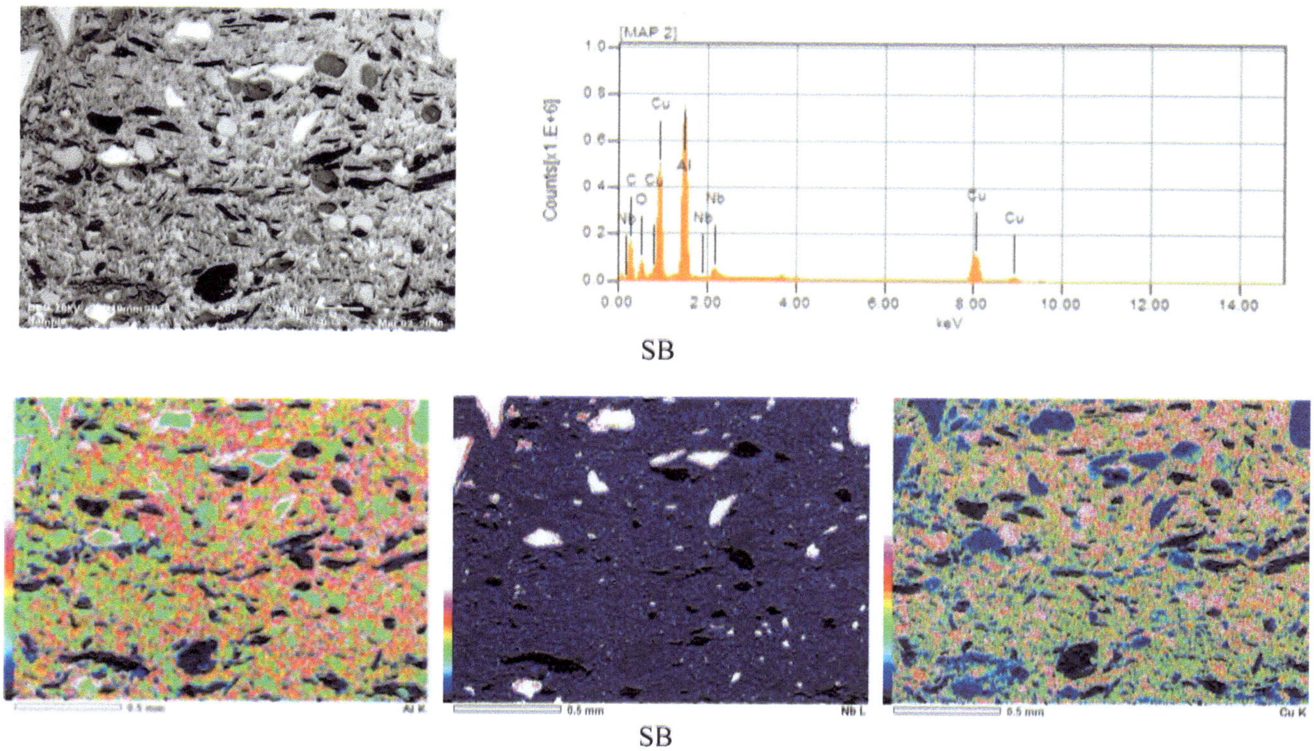

Fig. 16.2 Mapping analyses of the composite for SB produced by combined process; Sinter + Forging

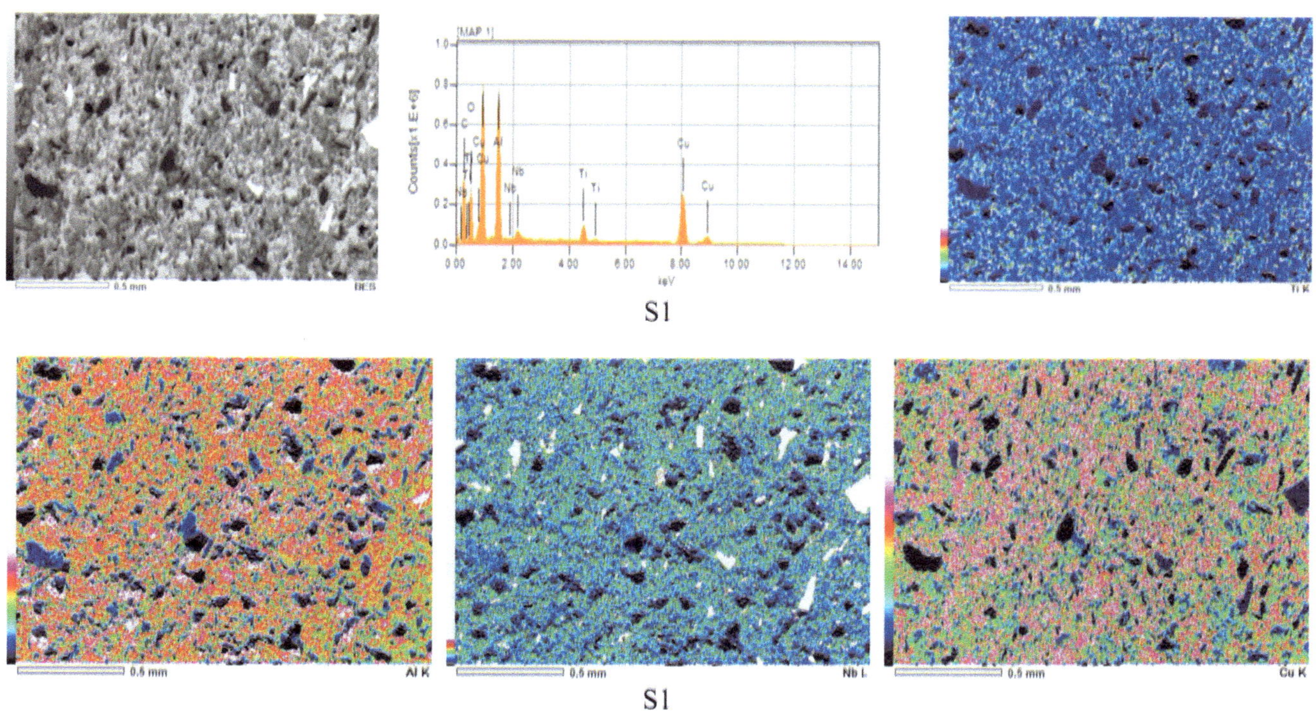

Fig. 16.3 Mapping analyses of the composite for S1 produced by combined process; Sinter + Forging

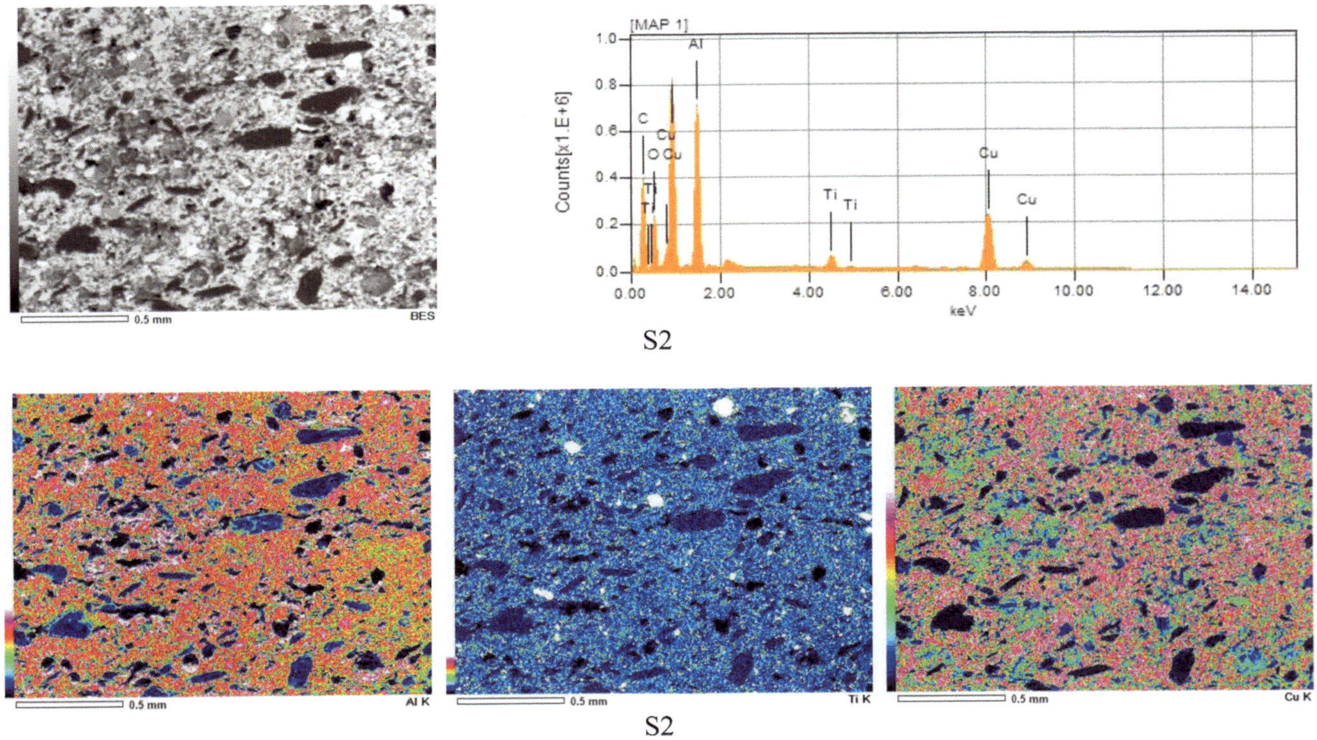

Fig. 16.4 Mapping analyses of the composite for S2 produced by combined process; Sinter + Forging

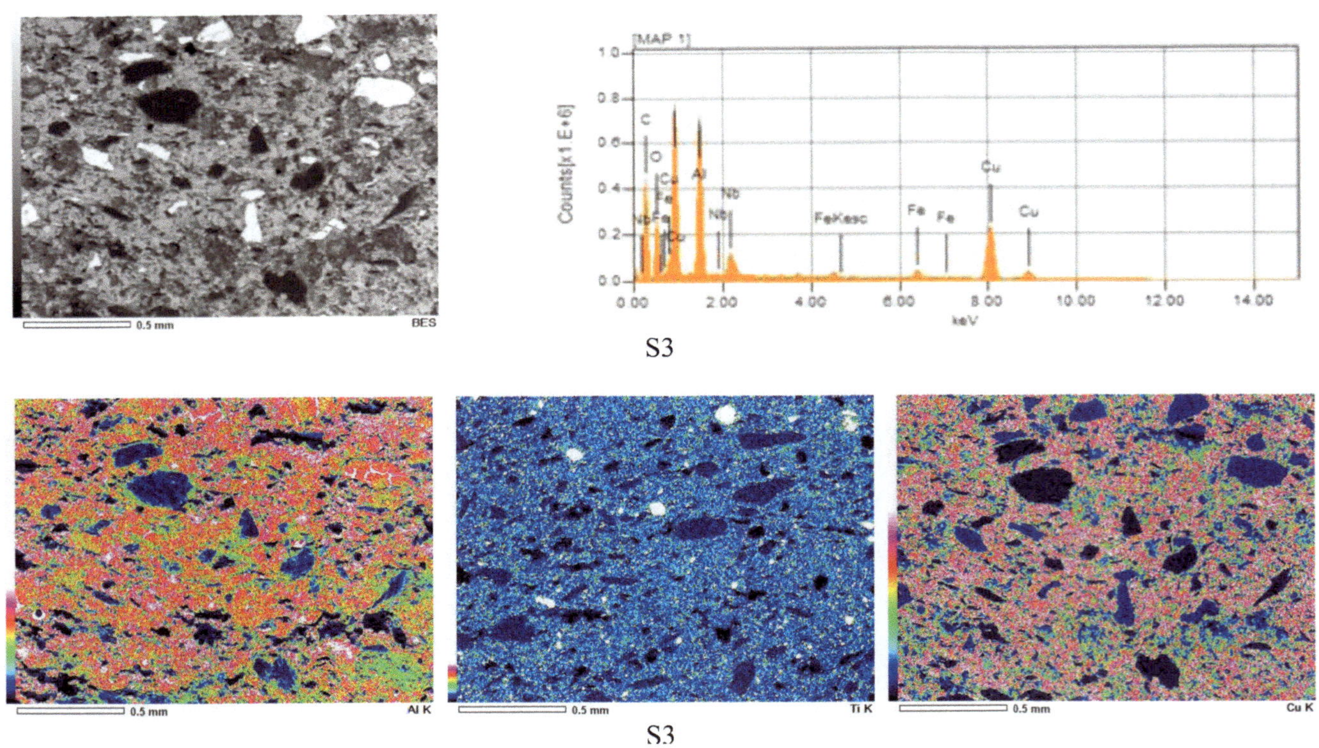

Fig. 16.5 Mapping analyses of the composite for S3 produced by combined process; Sinter + Forging

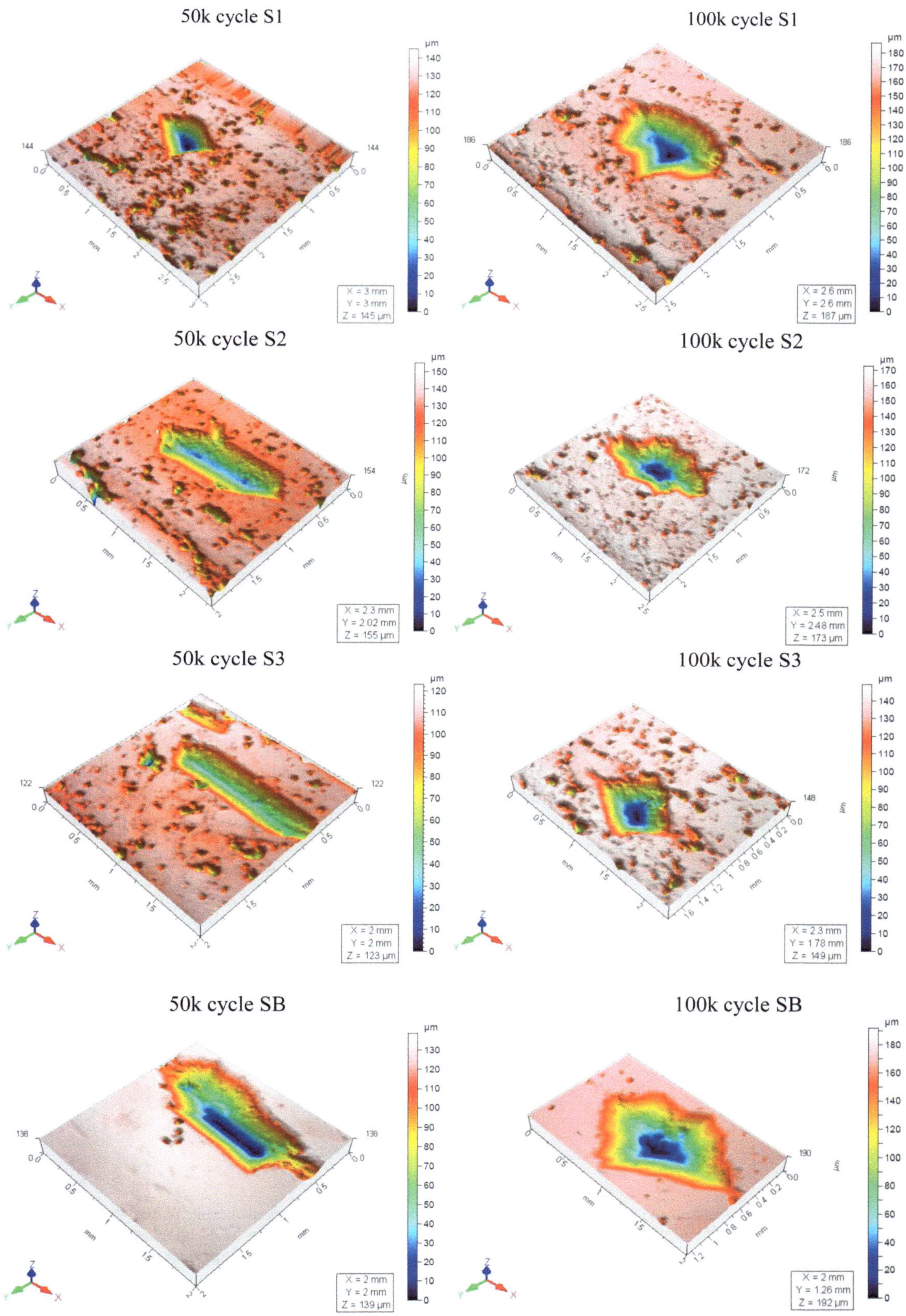

Fig. 16.6 Images of surface damage zones produced by wear – scratch test results for four different compositions in the cycle of $50*10^3$ and 100.10^3 cycles for the comparison of damage zone defined with surface and volume lost

Table 16.2 Wear (macro scratch) test results for tested specimens in this work

Wear test	50.10^3	Cycles	100.10^3	Cycles
Composition	Worn surface mm^2	Worn volume μm^3	Worn surface mm^2	Worn volume μm^3
SB	1.18	52,213,634	1.09	62,269,062
S1	0.859	37,358,445	1.06	58,333,931
S2	0.655	17,914,695	0.531	26,512,388
S3	0.549	23,261,026	1.12	62,225,704

Table 16.3 Micro hardness (HVN) measurements for the compositions as mean values

Cu based composites	Composition			
	SB	S1	S2	S3
Micro hardnessMean values (HVN)	95 ± 15	141 ± 12	174 ± 11	182 ± 24

16.4 Conclusion

In the present work, a recycled copper based composites reinforced with ceramic as an alternative replacement for the application of electric motor repair parts with the use of novel processing techniques.

A practical solution was proposed as cost effective economic manufacturing of the composites for this type of industrial applications. Copper based composite design (Cu-Al-Nb_2Al) was based on the ceramic reinforcements such as titanium carbide (TiC) in different percentages and niobium aluminate intermetallics (Nb_2Al). Because TiC and Nb_2Al make a good combination of thermal and electrical conductivities, microstructural stability and strength retention at elevated temperatures, etc. These reinforcements increase considerably wear resistance of the composites for electrical contact applications. Otherwise, certain percentage of fresh scrap aluminium powder, the mixture of AA1050 (80 wt% + AA7075 (20 wt %) chips were used to create an exothermic reaction in the process for helping diffusion bonding process of the ceramics to the copper matrix. At the first stage of the present work a preliminary study has been carried out for developing a cost effect and high wear resistant electrical brushes for aeronautical applications. By this way, low cost manufacturing of these composites have been successfully managed through the combined method of sinter + forging.

The microstructure of the composites with reinforcements obtained thorough the combine method sinter + Forging are seems very homogeneous and improved with a good diffusion bonding due to the addition of the aluminium in the matrix that creates an exothermic reaction. The eutectic structure was observed in the structure and mainly the reinforcement particles made a good diffusion chemical bonding with matrix by means of the exothermic reaction of the aluminium in the matrix. At the second stage of the process, it means that hot forging influences the microstructure considerably. For this reason, this novel techniques should be accepted as more advantageous than the conventional methods.

Optimizations of the certain parameters such as processing parameters, reinforcement content, etc. need much more experimental work to create real parts in the industrial scales. Here, only limited measurements at room temperature were presented as they are indicative parameters for better understanding the effect of the reinforcements on the microstructure and wear properties of the composites aimed for electrical field area.

Acknowledgments This work has been carried out on the frame of collaboration between Supmeca/Paris and Michigan technical university Houghton-MI-USA. Authors acknowledge and appreciate so much Dr. H-A. Alhas from Airbus-Space/London for valuable technical helps and discussions.

References

1. Zhao, H., Barber, G.C., Liu, J.: Friction and wear in high speed sliding with and without electrical current. Wear. **249**, 409–914 (2001)
2. Ferreira, L.F.P., Bayraktar, E., Robert, M.H., Miskioglu, I.: Particles reinforced scrap aluminum based composites by combined processing; sintering + thixoforming, Springer link. Mech. Composite Multi.-func Mater. **7**, 145–152 (2016). https://doi.org/10.1007/978-3-319-41766-0, ISBN 978-3-319-41766-0
3. Kurşun, A., Bayraktar, E., Enginsoy, H.M.: Experimental and numerical study of alumina reinforced aluminum matrix composites: processing, microstructural aspects and properties. Compos. Part B. **90**, 302–314 (2016)
4. Y-J Kwon, M., Kobashi, N.K.: Fabrication of copper matrix composites and simultaneous bonding with Aluminium alloy using combustion reaction of cu-Ti-B system. Mater. Trans. **45**(11), 3163–3171 (2004)

5. Varol, T., Çanakçi, A.: The effect of type and ratio of reinforcement on the synthesis and characterization cu-based nanocomposites by flake powder metallurgy. J. Alloys Compd. **649**, 1066–1074 (2015)
6. Ferreira, L.-M.-P., Bayraktar, E., Miskioglu, I., Robert, M.H.: Design and physical properties of multifunctional structural composites reinforced with nanoparticles for aeronautical applications. JAMPT Adv. Mater. Process. Technologies. **3**(1), 33–44 (2017). Taylor & Francis-USA
7. Katundi, D., Ferreira, L.-M.-P., Bayraktar, E., Miskioglu, I., Robert, M.-H.: Design of magnetic aluminium (A356) based composites through combined 2 method of sinter + forging, Springer link. Mech. Composite Multi Funct. Mater. **6**, 89–100 (2017). https://doi.org/10.1007/978-3-319-63408-1, ISBN 978-3-319-63408-1
8. Gatamorta, F., Bayraktar, E., Miskioglu, I., Katundi, D., Robert, M.H.: Design of low composites from recycled copper + aluminium chips for tribological applications, Springer link. Mech. Composite Multi Funct. Mater. **6**, 101–110 (2017). https://doi.org/10.1007/978-3-319-63408-1, ISBN 978-3-319-63408-1

Chapter 17
Damping and Toughening Effect of the Reinforcements on the Epoxy Modified Recycled + Devulcanized Rubber Based Composites

A. B. Irez, E. Bayraktar, and I. Miskioglu

Abstract This paper discusses damping effect of the reinforcements on the epoxy modified (10, 20 and 30 wt. %) devulcanized recycled rubber based composites. Within this study, the influence of Boron, γ-alumina and hollow glass microspheres (HGM) added in the matrix as the reinforcement elements were examined. As well known, devulcanization process is an efficient method of micronized rubber that can be used as useful products in a highly valued form. In this study, a combined process (chemical treatment + microwave) was used in order to modify the structure of the scrap recycled rubber. To characterize toughening and impact behaviour, quasi static 3P bending tests and choc tests were used for obtaining the principal values K_{Ic} and G_{Ic} and also damping capacity due to the reinforcement particles. Additionally, Shore-D hardness tests were measured. Scanning Electronic microscopy (SEM) was used to check dispersion quality and fracture surfaces, in scanning mode.

Keywords Epoxy · Recycled rubber · Wear resistance · SEM · Fracture toughness · Hollow glass microspheres

17.1 Introduction

In the modern technology, several conventional materials are being replaced by composites due to their several properties. Major of these replacements are essentially based on the composites with recycled fresh and clean materials for economic and low cost engineering applications. These are finding a very wide range of application in many different areas, from domestic, automotive to aircraft engineering, etc. [1–8].

In these applications, energy dissipation by mechanical damping is of considerable importance for reducing vibrations with absorbing energy. For this reason, damping in the composite structures can be achieved by reflexive damping that is a characteristic/intrinsic property of a composite structure with a higher consistency [6–11].

The enhancement in reflexive damping of the composite structures can also be achieved by mixture of recycled and devulcanized rubber as a matrix with epoxy as a binding- sticking under a special treatment conditions for damping improvement.

As known well, epoxy is relatively brittle, having poor resistance to crack initiation and growth. However rubber as formulated, is ductile with high absorbing property. For this reason, recycled and devulcanized rubber can be toughened by the addition of epoxy [2–13], for binding and sticking of rubber matrix in order to obtain favorite physical and mechanical properties. By addition of the certain reinforcements such as extra pure and fine gamma-alumina and boron and also micro size hollow glass bubbles, damping and toughening capacities could be increased.

A. B. Irez
CentraleSupélec, Université Paris-Saclay, Gif-sur-Yvette, France

E. Bayraktar (✉)
Supmeca-Paris, School of Mechanical and Manufacturing Engineering, Paris, France
e-mail: bayraktar@supmeca.fr

I. Miskioglu
Michigan Technology University, Engineering Mechanics Department, Houghton, MI, USA

P. R. Thakre et al. (eds.), *Mechanics of Composite, Hybrid and Multifunctional Materials, Volume 5*, Conference Proceedings of the Society for Experimental Mechanics Series, https://doi.org/10.1007/978-3-319-95510-0_17

17.2 Experimental Conditions

17.2.1 Materials Processing

Today, rubbers are widely used in countless areas thanks to their advantageous properties. However, the virgin form of rubber usually limits its area of applications. For this reason, a procedure named as vulcanization is applied to the rubbers to improve their materials properties. Vulcanization is described as an irreversible rubber crosslinking process using sulphur or sulphur containing compound [6–14]. As a result, transverse bonds attach the free chains of rubber to create cross-linked rubber structure. Therefore, vulcanized rubbers are elastic, insoluble and infusible materials that cannot be reprocessed as is the case with thermoplastics. On the other hand, this favourable process poses an obstacle when the subject is composite manufacturing by using formerly vulcanized rubbers or waste rubbers. Since, vulcanized rubbers do not carry any free chains to form new bonding with the composite matrix and lack of bonding between rubbers and matrix affects negatively the final properties. Thus, in order to improve the interface quality between matrix and rubber, scrap rubber should be devulcanized.

In fact, devulcanization is known to be an efficient practice for manufacture of recycled rubbers to increase their flowing capacity and interaction with other substances. During this process, sulphur links are tried to be broken and also generated new other links, it means that the structure of the material is modified entirely as renewable process [7, 9–15].

In the frame of the present study, a new combined method was used to devulcanize fresh scrap rubber (SBR) with a particle size varying from 10 to 130 μm [6–8]. This combined method contains a pre-chemical treatment followed exposing to microwave heating in a short time that is applied in two stages in practical point of view. By this way, a good cohesion is supplied at the interface between epoxy resin and rubber powders to improve the properties of the recycled rubber coming from the sportive affaires. It means that they are fresh clean and completely different from ground tire.

The basis of the chemical treatment consists in a short silanization (vinyltriethoxysilane (2%)) process followed by acrylic acid and a small amount of toluene solution to activate the surface of the rubber particles [16]. After drying of the chemical treated rubber powders, they are exposed to short microwave heating in two stages during 3 min in order to avoid degradation of the main chains.

At the final step, devulcanized rubber particles were mixed with bisphenol A type solid epoxy resin powders (Huntsman™ GT 6097) as well as its solid hardener. This mixture used as a matrix after that the new designed composites are manufactured by using classical powder metallurgy methods. After the mixture of the reinforcements (dry boron and alumina powders (micro-scale particles)) in the matrix, milling process is carried out during 4 h. Then, HGM (also called glass bubbles, manufactured by 3M with a density of 0.227 g/cm^3, specified as S38HSS & K1) were added and mixed to obtain homogenous distribution of the constituents (ultrasound) for an additional 2 h [6–8, 12, 13, 15–17].

After having a homogeneous powder compound, the composite specimens were manufactured by using double uniaxial action press at a temperature of 180 °C under a pressure of 70 MPa during the heating of 30 min. All of the specimens (30/50 mm in diameter) were cooled slowly in the press. All of the specimens were post-cured isothermally at 80 °C for 24 h.

The compositions of boron + γ alumina and HGM reinforced epoxy-recycled rubber based composites were given in Table 17.1.

Table 17.1 Composition of the epoxy-rubber based composites

Recycled rubber based composites		Matrix (wt. %)		Reinforcements (%wt.)		
Group	Name of the composition	Epoxy	Rubber	Alumina	Boron	HGM
A	A I	10	90	5	5	–
	A II	20	80			
	A III	30	70			
B	B I	10	90	5	5	5
	B II	20	80			
	B III	30	70			

Table 17.2 Shore D Hardness values

Hardness measurement			
Specimen	Shore D	Specimen	Shore D
A I	80.5	B I	79.6
A II	81.6	B II	80
A III	81.8	B III	81.3

Table 17.3 Density values of the specimens (Shore D)

Hardness measurement			
Specimen	Density (g/cm^3)	Specimen	Density (g/cm^3)
A I	1491	B I	1482
A II	1493	B II	1485
A III	1495	B III	1498

17.2.2 Microstructure: Fracture Surface Analyses and Shore-D Hardness Measurements

Fracture surface damage analyses and microstructural observation have been realized by means of optical microscope (OM) and scanning electron microscopy (SEM). SEM observation was realized on fracture surface of the tested specimens with Scope/JSM-6010LA Jeol® electron microscope.

Surface hardness measurements of the specimens were performed after post curing. Shore D hardness test measurements on the polished flat surfaces of the specimens were carried out according to ASTM D 2240 using Shore D hardness tester, (type HBD-100-0). Hardness results were given in Table 17.2.

Three-point bending tests (3PB) have been carried out according to the ASTM D790 standards. Tests were realized with the machine Zwick Proline Z050TN and during the tests crosshead speed was selected as 1 mm/min. Flexural stress and strain were obtained from the test results. At least three specimens for each composition were used and standard deviation and average values were given in results chapter with standard deviation values.

In addition, fractural properties such as plain strain fracture toughness (K_{Ic}) and critical strain energy release rate (G_{Ic}) were investigated with SENB specimens and the tests were realized according to ASTM D5045 standard. Notches were introduced by tapping a fresh razor blade. Charpy impact tests were realized by means of Zwick 5102 pendulum impact tester with a 1 J pendulum configuration. After preparation of specimens, densities of the compositions were measured and they were given in Table 17.3.

17.2.3 Damage Analysis by Means of Scratch Test and 3D Optical Roughness Meter

In the current research basic idea on the tribological behaviour of the epoxy and recycled rubber based composites was evaluated performing scratch tests. A 3D optical surface scanner was utilized to assess the damaged zone after the scratch test in terms of scratch depth and average scratch roughness.

The contact between the sliding diamond indenter and the surface of the composite material during scratch test is analyzed. The normal and tangential forces on indenter are recorded. However, main focus is given to the damage area and volume. In the frame of the current research, the resistance to scratch deformation is evaluated in terms of scratch depth, surface and worn volume subsequent to scratching only under dry conditions and 50,000 and 100,000 number of cycles of wear.

17.3 Results and Discussions

17.3.1 Specimens and Microstructure Analyses of the Composites

In Fig. 17.1, Initial stage of rubber-epoxy powder were presented in order to prepare hot compacted test specimens from composites studied in this research. Two different specimens were shown in macro scale in different diameters for characterization of the composites that were manufactured by means of hot compaction technics. These specimens were tailored for different characterization of the composites such as 3P-Bending test and dynamic choc test such as Charpy resilience test in order to evaluate basically toughening mechanism in detail. Other specimens were used for microstructural

Fig. 17.1 Initial stage of rubber-epoxy powder to prepare hot compacted test specimens from composites studied in this research

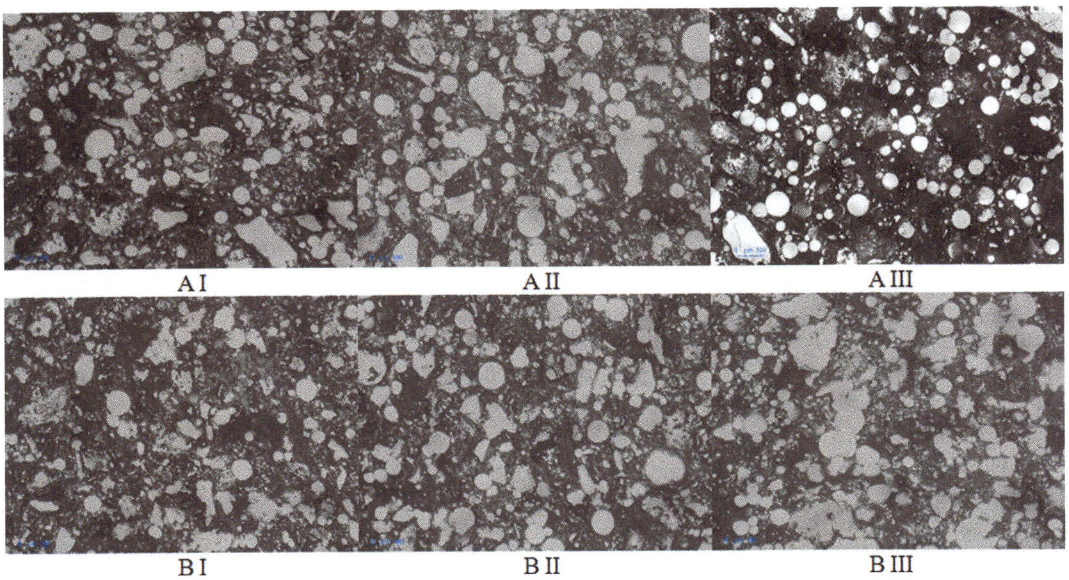

Fig. 17.2 Microstructure of composites prepared for two groups A and B respectively

analyses and also for evaluation of the physical and other mechanical properties. For each test, 4 and 5 specimens were used to optimize test results.

Microstructural observation was realized with OM and they were shown in Fig. 17.2. Small size cylindrical specimens were sectioned transversally then polishing were realized after they were mounted. It seems that all of the microstructures show a extensively homogenous distribution of the reinforcements in the structure.

17.3.2 Three Point Bending (3PB) Test Results and Fracture Surface Observation

Three-Point Bending (3PB) tests have been carried out for each different type of composites.

17.3.2.1 Flexural Testing and Fracture Toughness Determination

Flexural stress is calculated during three-point bending test according to the Eq. 17.1:

$$\sigma = \frac{3 \times P \times l}{2 \times b \times h^2} \tag{17.1}$$

In this formula, l is the span length, P is the maximal bending load, b and h are the sample width and thickness, respectively.

Flexural strain, ε_f, was determined according to the Eq. 17.2:

$$\varepsilon_f = \frac{6Dh}{l^2} \tag{17.2}$$

Where D is the maximum deflection at the center of the specimen.

E_B is the modulus of elasticity in bending and it is expressed with the Eq. 17.3 as follows:

$$E_B = \frac{l^3 m}{4bh^3} \tag{17.3}$$

where m is the tangent of the initial straight portion of the stress-strain curve.

The mode I fracture toughness, K_{Ic}, was determined by testing of the SENB specimens and K_{Ic} was calculated according to the Eq. 17.2:

$$K_{Ic} = \frac{F}{B\ w^{1/2}} f(x); \quad x = \frac{a}{W}\ , 0 < \frac{a}{W} < 1 \tag{17.4}$$

where F is the maximum force from the load-elongation trace; B is the thickness of the specimen; W is the width and "a" is the total notch length.

f(x) is the geometry correction factor and is expressed with the Eq. 17.3 as follows:

$$f(x) = 6(x)^{0.5} \left\{ \frac{\left[1.99 - x(1-x)\left(2.15 - 3.93x + 2.7x^2\right]\right.}{(1+2x)(1-x)^{1.5}} \right\} \tag{17.5}$$

Critical strain energy release rate (fracture energy) G_{Ic} was calculated using the expression Eq. 17.4:

$$G_{Ic} = \frac{{K_{Ic}}^2}{E} \tag{17.6}$$

where E is the elasticity modulus for plane stress approach examined for thin specimens.

Table 17.4 indicates the mechanical properties in bending mode. Fracture toughness values were also presented in the table.

Table 17.4 Comparison of mechanical properties of A-B group specimens

	Flexural stress (MPa)	Flexural modulus (MPa)	Strain in break (ε %)	K_{Ic} (MPa m1/2)	G_{Ic} (kJ/m^2)
A I	13.09 ± 1.49	1071 ± 119	0.42 ± 0.03	0.615 ± 0.12	0.360 ± 0.14
A II	11.53 ± 1.52	906 ± 180	0.41 ± 0.05	0.554 ± 0.012	0.339 ± 0.01
A III	10.70 ± 2.20	985 ± 78	0.36 ± 0.001	0.587 ± 0.009	0.349 ± 0.01
B I	11.92 ± 2.11	955.68 ± 114	0.43 ± 0.02	0.535 ± 0.007	0.300 ± 0.001
B II	8.87 ± 1.83	779 ± 224	0.41 ± 0.06	0.654 ± 0.169	0.567 ± 0.28
B III	5.99 ± 1.65	601 ± 98	0.28 ± 0.07	0.581 ± 0.034	0.562 ± 0.07

In summary, three-point Bending (3PB) tests have been carried out for each different type of composites in order to fracture toughness determination. Fracture toughness (K_{Ic}) and critical strain energy release rate (G_{Ic}) are compared. It is clear that the first group (AI, AII and AIII) have shown higher toughness behaviour which do not contain hollow glass bubbles. These values were only obtained under the laboratory conditions that can be improved during the project that is going on.

Fracture surfaces analyses have been carried out on the broken specimens with 3PB tests. Two groups of the composites with three compositions were analyzed in scanning electron microscopy (SEM). All of the fracture surfaces have been shown in Fig. 17.3 with different magnification. One may be observe the devulcanized rubber matrix and fine and homogeneous distribution of the reinforcement elements.

It seems that chemical treatment followed by devulcanization process result in very well interface between the rubber matrix and reinforcements. Epoxy has played an important role on the adhesion of these element with matrix. Even, devulcanized rubber and epoxy has shown a good cohesion with a strong chemical bonding thanks to the applied chemical and devulcanization treatments. These pictures indicate typical fracture surfaces from 3PB testing. Large hyperbolic markings open in the direction of crack propagation. Brittle fracture is observed in some smooth areas.

17.3.3 Charpy Impact Testing

These composites structures are sensitive to impact damage. Charpy resilience test results have given more precise detail on the broken specimen under choc-dynamic response. Each composition group was tested with at least four specimens. Absorbed energies during impact testing of the composites were presented in Fig. 17.4 for each composition separately. It seems that there is a considerable dispersion in the results with high standard deviation varying between 10% and 20%) due to heterogeneous structure. These results are only carried out under laboratory test conditions. The real test conditions will be compared in next stage of this research project that is going on.

17.3.4 Wear Resistance by Scratch Test and Damage Analyses by Means of 3D Optical Roughness Meter

Decrease of friction contact reposes on the main task for wear-resistant composites. Evaluation of the wear resistance of this composite developed in this work have been carried out in two different numbers of cycles, $50*10^3$, $100*10^3$ cycles. It seems that the influence of reinforcement elements and essentially influence of two stages of the manufacturing processes (chemical and devulcanization+ hot compaction) has played an important role on the damage behaviour of the compositions indicated in Fig. 17.5 for AI, AII, AIII and BI, BII and BIII respectively.

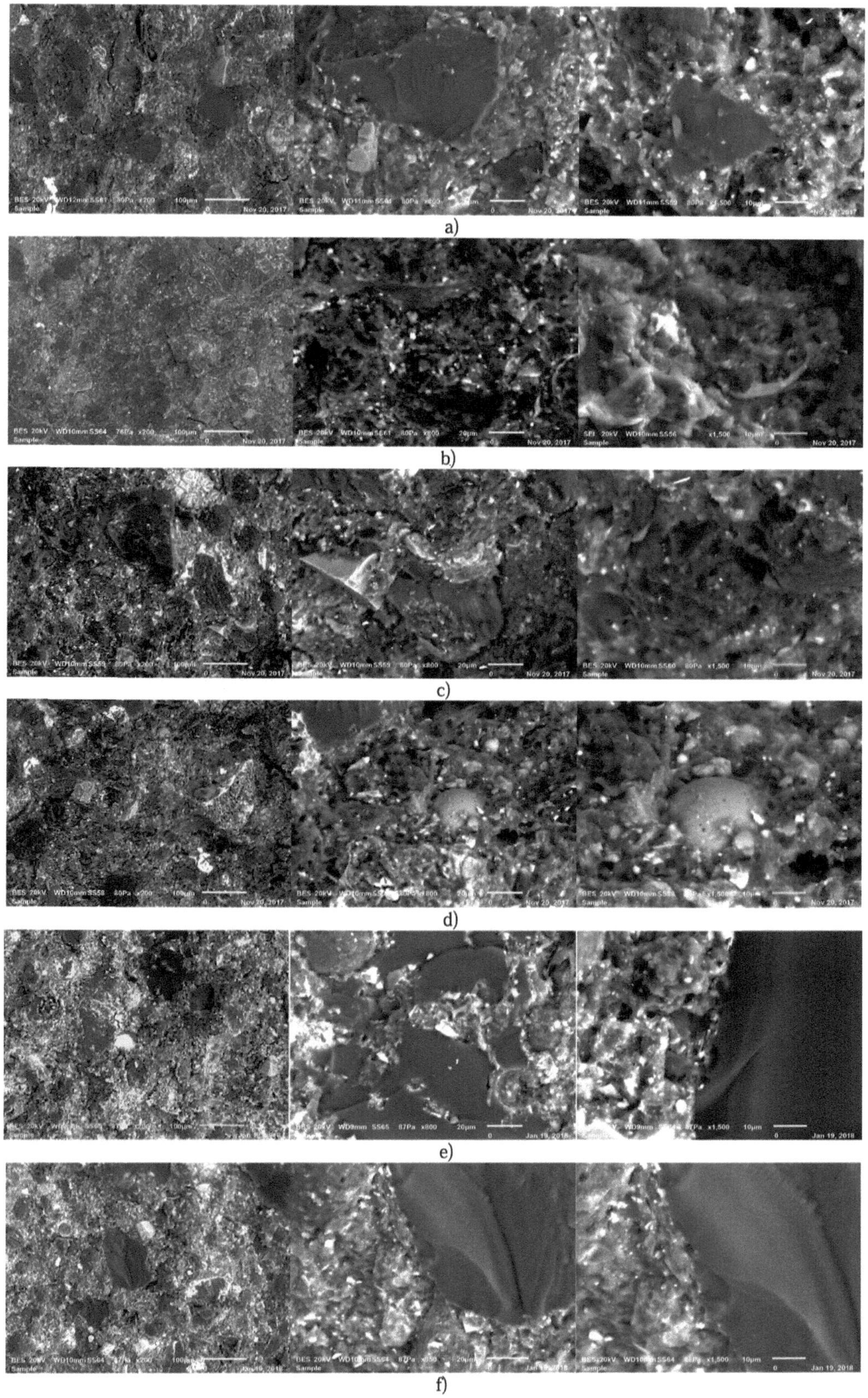

Fig. 17.3 Fracture surfaces of the specimens broken through 3P-Bending testing. (**a**) A I; (**b**) A II; (**c**) A III; (**d**) B I; (**e**) B II; (**f**) B III

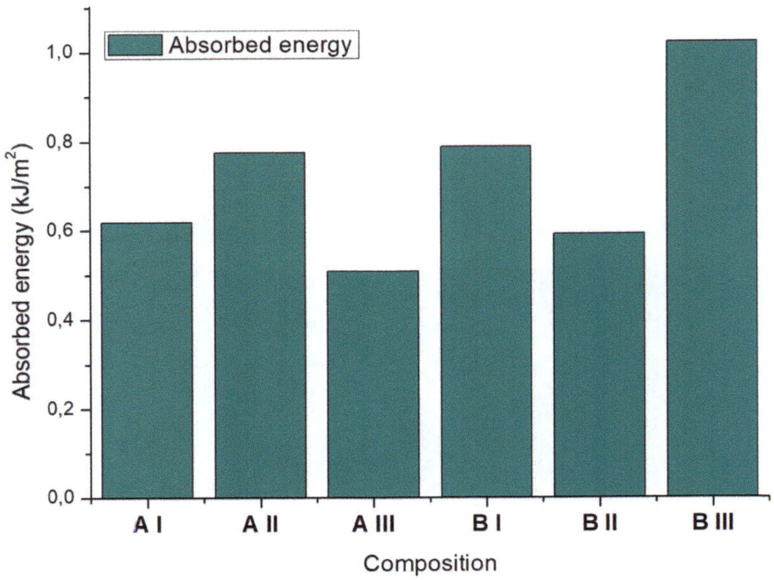

Fig. 17.4 Absorbed energy obtained with Charpy Impact Tests for A and B group compositions

The surface, volume and the depth values are presented in the same figures for each test condition. Effects of reinforcements mainly gamma alumina (d < 1 μm) and boron (d < 1–5 μm), and evidently their effect on the process have shown obvious advantages on the specimens because their dispersion on the matrix should have contributed to improve the wear resistance under experimental conditions carried out in the present work. However, all of these compositions have also shown a high standard deviation due to heterogeneous structure. For this reason, these scratch test results have been summarized in Fig. 17.6a, b for two compositions series (AI, AII, AIII and also BI, BII, BIII). In these figures, the volume lost and traces of the surface damages are shown as column for each composition separately.

17.4 Conclusion

The following conclusions can be drawn from the first stage of this research project:

Microstructural observations of the six compositions show very successfully adhesion to reinforcements with rubber matrix undergone chemical and devulcanization treatments. A good mixture of devulcanized rubber with Epoxy added to the matrix has improved toughening behaviour. The addition of glass bubbles tend to promote decrease in density and to increase plastic deformation of the material. Fracture surfaces of the specimens taken from six compositions indicate regular fracture deviations due to the effect of added reinforcements even if some local agglomerations observed on in the structure; play strong role as barriers against the propagation of the fracture surface. In general, increasing of the rubber percentage (as a matrix) in the structure the density values decrease that are lighter.

Three-Point Bending (3PB) tests have been carried out for each different type of composites in order to fracture toughness determination. Fracture toughness (K_{Ic}) and critical strain energy release rate (G_{Ic}) are compared. It is clear that the first group (AI, AII and AIII) have shown higher toughness behaviour which do not contain hollow glass bubbles.

These values were only obtained under the laboratory conditions that can be improved during the project that is going on.

From the wear scratch tests have been carried out that the wear lost on the compositions increases with increase of the scratch cycles that is mean with sliding distance and velocity. These composites can be considered as suitable damping materials under similar conditions. In the frame of this research, only dry scratch behaviour has been studied. But, wet scratch behaviour of these composites should be studied to deeper wear analyses. These tests may be conducted in the further stage of this research project.

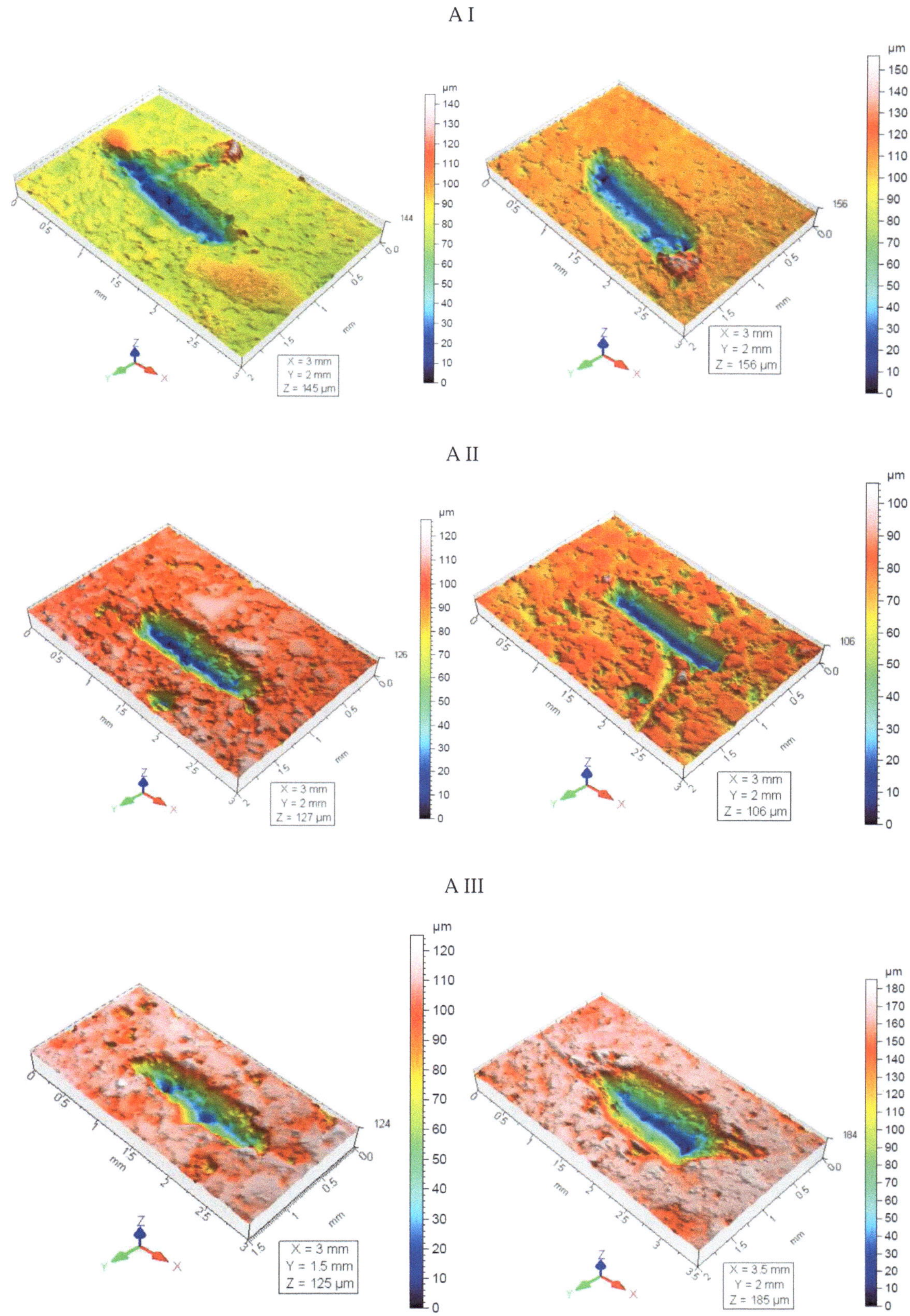

Fig. 17.5 Three dimensional damage traces obtained in the direction of width and length for A I-II-III and B I-II-III for 50 k cycles shown on the left column and for 100 k cycles on the right column respectively

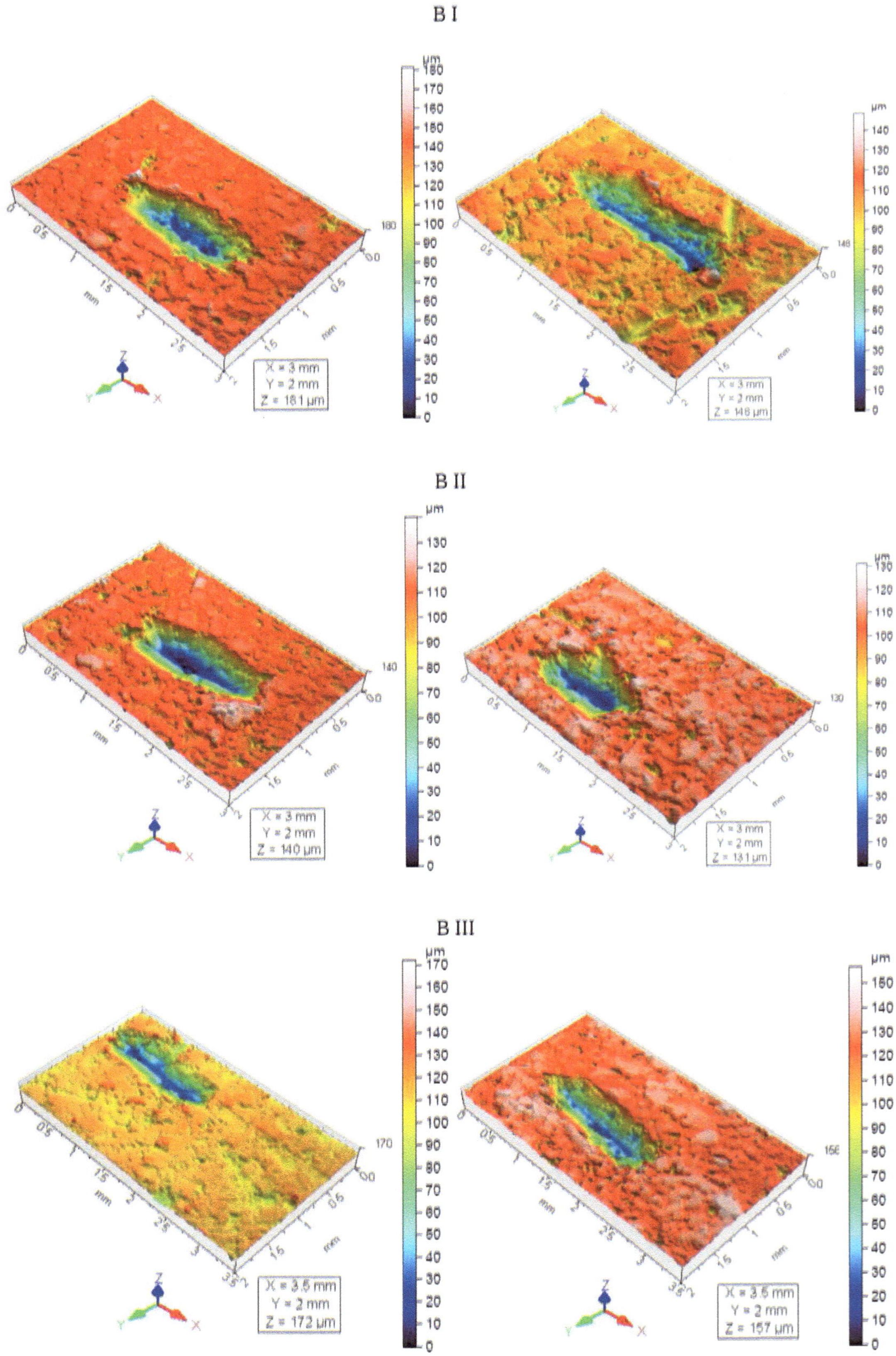

Fig. 17.5 (continued)

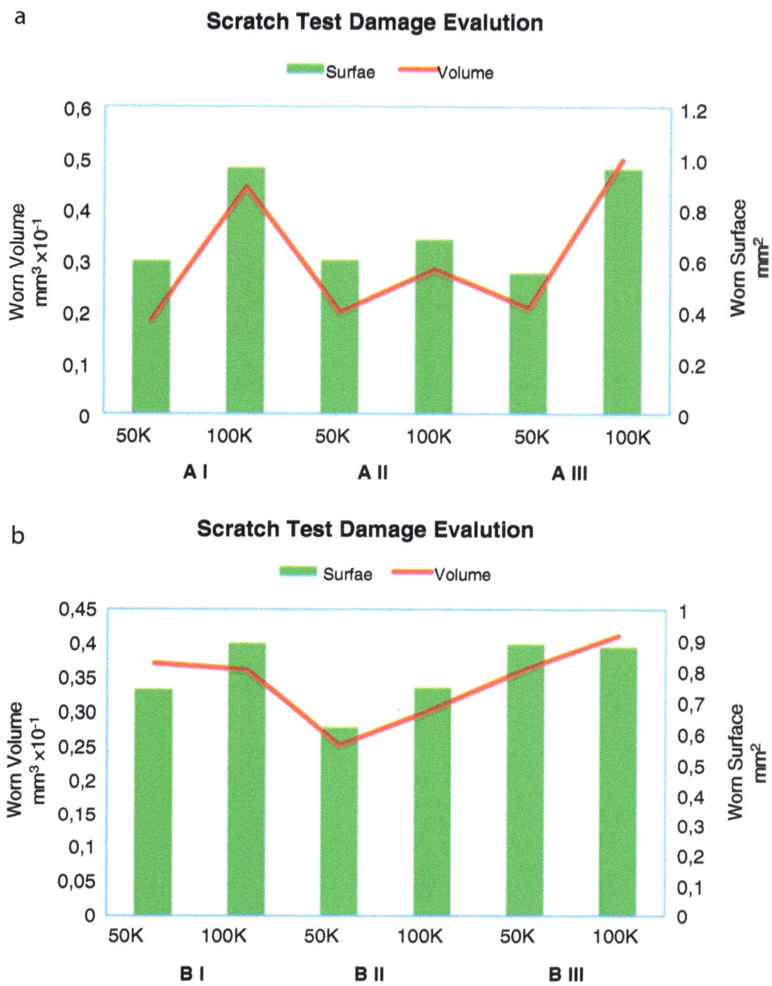

Fig. 17.6 Damage traces obtained in the two directions width and length of the specimens (**a**) A I-II-III and (**b**) B I-II-III for 50 k and 100 k cycles respectively

Acknowledgement The authors would like to thank to Dr. H-A. ALHAS from Airbus Space Centre, London UK for his kind helps for the materials and supporting this research project.

References

1. Balakrishnan, S., Start, P.R., Raghavan, D., Hudson, S.D.: The influence of clay and elastomer concentration on the morphology and fracture energy of preformed acrylic rubber dispersed clay filled epoxy nanocomposites. Polymer. **46**, 11255–11262 (2005)
2. Hunston, D.L., He, J., Raghavan, D., Hoffman, D.K.:. Load history dependence of fracture in rubber-toughened epoxies. Proceedings of Adhesion Society Meeting. Blacksburg, Virginia, USA, pp. 10–15 (2000)
3. Hunston, D.L., He, J., Raghavan, D., Hoffman, D.K.: Limits on toughening of epoxies. Proceedings of Adhesion Society Meeting. p. 200 (1998)
4. Riew, C.-K.: Rubber-Toughened Plastics Advances in Chemistry Series, vol. 222. ACS, Washington, DC (1989)
5. Zaimova, D., Bayraktar, E., Tan, M.-J., Miskioglu I., Katundi, DD.: Design of multifunctional energetic structural composites: A preliminary study on an epoxy-rubber matrix with exothermic mixture reinforcements. SEM-2013, Annual Conference – Experimental and Applied Mechanics, The Westin Lombard Yorktown Center Lombard, IL, USA, June 3–5, 2013, vol. 1, no. 1, pp. 1–8, ed. SEM/Springer-Link (2013)
6. Akiba, M., Hashim, A.S.: Vulcanization and crosslinking in elastomers. Prog. Polym. Sci. **22**, 475–521 (1997)
7. Zaimova, D., Ferreira, L.-M.P., Bayraktar, E., Miskioglu, I.: Scrap rubber based composites reinforced with ceramic oxides and silica, Springer link. Mech. Compos. Multi Funct. Mater. **7**, 27–36 (2016). https://doi.org/10.1007/978-3-319-41766-0, ISBN 978-3-319-41766-0
8. Ferreira, L.M.P., Miskioglu, I., Bayraktar, E., Katundi, D.: Mechanical and Tribological Properties of Scrap Rubber Reinforced with Al2O3 Fiber, Aluminium and TiO2, Springer link. Mech. Compos. Multi Funct. Mater. **7**, 37–44 (2016). https://doi.org/10.1007/978-3-319-41766-0, ISBN 978-3-319-41766-0

9. Wetzel, B., Haupert, F., Zhang, M.Q.: Epoxy nanocomposites with high mechanical and tribological performance. Compos. Sci. Technol. **63**, 2055–2067 (2003)
10. Ferabolia, P., Masini, A.: Development of carbon/epoxy structural components for a high performance vehicle. Compos. Part B. **35**, 323–330 (2004)
11. ASTM D 790, Plastics (I). Standard test methods for flexural properties of unreinforced and reinforced plastics and electrical insulating materials, Annual book of ASTM standards. American Society for Testing and Materials; (1999)
12. Ferreira, L.M.P., Miskioglu, I., Bayraktar, E., Katundi, D.: Mechanical and Tribological Properties of Scrap Rubber Reinforced with Al2O3 Fiber, Aluminium and TiO2, Springer link. Mechanics of Composite and Multi-functional Materials. **7**, 37–44 (2016). https://doi.org/10.1007/978-3-319-41766-0, ISBN 978-3-319-41766-0
13. Irez, A.B., Hay, J., Miskioglu, I., Bayraktar, E.: Scrap-rubber based composites reinforced with boron and alumina, Springer link. Mech. Compos. Multi Funct. Mater. **6**, 1–10 (2017). https://doi.org/10.1007/978-3-319-63408-1, ISBN 978-3-319-63408-1
14. Marouf, B.T., Pearson, R.A., Bagheri, R.: Anomalous fracture behaviour in an epoxy based hybrid composite. Mater. Sci. Eng. A. **515**, 49–58 (2009). [6mposites Reinforced with Nanoparticles, Springer link. Mech. Compos. Multi Funct. Mater. **6**, 33–44 (2017). https://doi.org/10.1007/978-3-319-63408-1, ISBN 978-3-319-63408-1
15. Irez, A.B., Miskioglu, I., Bayraktar, E.: Mechanical characterization of epoxy – scrap rubber based composites reinforced with nano graphene, Springer link. Mech. Compos. Multi Funct. Mater. **6**, 45–58 (2017). https://doi.org/10.1007/978-3-319-63408-1, ISBN 978-3-319-63408-1
16. Shokoohi, S., Arefazar, A., Khosrokhavar, R.: Silane coupling agents in polymer-based reinforced composites: A review. J. Reinf. Plast. Compos. **27**(5), 473–485 (2008)
17. Irez, A.B., Bayraktar, E., Miskioglu, I.: Mechanical characterization of epoxy – scrap rubber based composites reinforced with alumina fibers, Springer link. Mech. Compos. Multi Funct. Mater. **6**, 59–70 (2017)

Chapter 18
Impact and Post-impact Behavior of Composite Laminates Reinforced by Z-Pins

L. Francesconi and F. Aymerich

Abstract Delamination, i.e. the separation between layers that occurs by failure of the resin-rich interlaminar interface, is a direct consequence of the lack of out-of-plane reinforcements and, undoubtedly, the most common damage mode in laminated composites.

Delaminations induced by low-velocity impacts are of primary concern in structural applications, since impact damage, often miss-detected, may propagate, impairing the load bearing capacity of the component, particularly under compressive loads. Preventing delamination, as well as delaying and limiting its propagation, are thus key issues in the design of composite structures. Over the last decades, many strategies have been proposed to address this problem. Among them, the introduction of through-thickness reinforcements (stitching, pinning and stapling) has proven to be effective in improving the interlaminar properties of composite materials. In particular, recent studies have shown that Z-pinning, which consists in inserting high stiffness pins through the thickness of uncured laminates, may significantly enhance the delamination resistance of laminated components.

In this study, conventional and Z-pin reinforced $[0_2/90_2]_s$ graphite/epoxy laminates were subjected to low-velocity impact and compression after impact (CAI) tests in order to examine the effect of the reinforcements on the impact response and the residual post-impact properties of the laminate. The results show that Z-pinning significantly reduces the extent of delamination induced by impact, while, on the contrary, it appears to only marginally improve the post-impact compressive strength of the laminates.

Keywords Z-pin · Composite · Laminated structures · Through-thickness reinforcements · Delamination

18.1 Introduction

The unique combination of desirable features, such as lightness and corrosion resistance, and higher specific mechanical properties with respect to metals, have given to Carbon Fiber Reinforced Polymers (CFRPs) a key role in several design scenarios (such as aerospace, aircraft and automobile construction) in the last decades.

The CFRPs exceptional features derive from apposite planar disposition of the reinforcing fibres that allows these structures to bear loads along particular directions. However, in contrast with their high in-plane performances, due to the lack of reinforcement elements [1] CFRPs provide limited load carrying capability on the direction perpendicular to the plane and, therefore, they are particularly vulnerable to impacts and prone to delamination.

The deficiency of strength in the through thickness direction represents the primary weakness and also the first limitation for a wider spreading of this class of structures.

Furthermore, their low resistance to delaminations becomes even more critical when low velocity impact damage is taken into account: barely visible (or not detectable with the unaided eye) damages and delaminations, may propagate underneath/undetected, degrading the laminate's mechanical strength and leading to the ultimate failure of components.

Several approaches [2, 3] have been proposed in the last decades to increase the out-of-plane strength of composite laminates to prevent and to limit the propagation of delaminations [4–9].

Among them, starting with the early work on stitching [10–15], the insertion of translaminar reinforcements has proven to be a successful strategy to enhance the interlaminar properties of composite laminates [9] by increasing the impact and

L. Francesconi (✉)
Department of Mechanical Engineering, Santa Clara University, Santa Clara, CA, USA
e-mail: lucafrancesconi@icloud.com

F. Aymerich
Department of Mechanical, Chemical and Materials Engineering, University of Cagliari, Cagliari, Italy

P. R. Thakre et al. (eds.), *Mechanics of Composite, Hybrid and Multifunctional Materials, Volume 5*, Conference Proceedings of the Society for Experimental Mechanics Series, https://doi.org/10.1007/978-3-319-95510-0_18

the post-impact resistance of laminated composites [16–18] and providing an effective strategy to restraining the generation and the propagation of the delamination front [16, 19, 20, 21].

More recently, pinning (also known with the name of "Z-Pin" or "Z-Fibers®") a new technology of through-thickness discontinuous reinforcement, [2, 21–23], has attracted considerable attention from researches and industries.

The Z–Pins are usually manufactured using high stiffness/strength materials (titanium alloy, steel or carbon) by pulling and pultruding a continuous fibre tow which is then cut to length and inserted into a foam carrier [22].

Commonly the pins have a nominal diameter of 0.2–1.0 mm, and a density ranging from 0.5 to 5.0 vol% and they are embedded in a foam carrier that provides the support, the relative position and the right orientation during the insertion process, performed prior to the consolidation stage, usually by applying pressure or by vibrating the pins using dedicated ultrasonic guns [22].

Experimental investigations [11, 22, 24] concluded that among several manufacturing approaches the ultrasonically assisted insertion (often called UAZ) significantly facilitates the pins' insertion limiting the distortion of the base panel. Several authors [22, 25] highlighted that the introduction of through-thickness high stiffness pins reduces the delamination induced by impact, as a result of the improved interlaminar fracture and delamination properties of the material under static and impact loadings.

Other researchers [23–28] demonstrated that the damage performance improvements for the Z-Pinning reinforced components are attributable to the closure tractions applied at the delamination interfaces by the discontinuous inclusions that limit the driving force available for cracks propagation (bridging effect) and to the modification of the typical failures modes of this class of structures.

Despite many conveniences, and similarly to what has already been documented for other TRL methods, such as stitching, the pins' insertion process severely alters the base structure of the laminate, introducing various defects such as fibre waviness, fibre crimp and breakage, resin-rich zones, cure stresses, and micro-cracking during the consolidation phase. It was found [11, 23, 24, 29, 30] that these defects -depending upon the stacking sequence of the laminate and the areal density of the pinning- act as potential crack initiation sites or trigger additional damage modes, altering the structural response of the laminates impairing basic elastic properties such as the tensile strength and the Young's modulus.

Different studies [24, 29, 30] investigated the performance of pinned carbon/epoxy specimens with nominal thicknesses of 2, 4, and 6 mm under low-velocity impact and compression-after-impact (CAI) loading situations.

The pinned samples exhibit impact damage areas reduced by up to 60% (depending on the specimen thickness and impact energy) and an increase of more than the 40% was recorded in the compression-after-impact (CAI) strength.

Furthermore, higher failure strains have been found for all the pinned specimens with respect to the unpinned ones, suggesting that the use of Z-pins could substantially enlarge the design limits in many critical applications.

A common conclusion observed in all these studies is that the pinning is a successful yet a promising strategy to increase the delamination resistance of composite structures but more reliable experimental data are required to better understand their role in improving the impact and the post-impact laminates' performances.

This paper investigates the damage response of pinned and unpinned composite laminates subjected to low-velocity impacts to examine in deep the modification of the damage and the failure modes of the considered structures. Impact tests have been performed with several energies (2–35 J) and the projected delaminated areas were measured using X-ray inspections. Post-impact compression tests were then performed to evaluate the residual mechanical strength and the influence of the Z-Pins in the after impact performances of the considered components.

18.2 Materials and Testing Methods

Cross-ply composite $[0_2/90_2]_s$ panels of 350 by 400 by 2.45 mm were laminated using unidirectional Seal Texipreg® HS300/ET223 carbon/epoxy prepreg tapes. Before the consolidation phase, the panels were subdivided in smaller coupons of 65 by 87.5 mm; some of them were reinforced using carbon Z-Pins (Fig. 18.1), manufactured by Albany Engineered Composites (AEC).

The pins were inserted in a 2 by 6 cm central portion of the sub-samples with an overall pin density of 2% and a diameter of 0.28 mm. The Z-Pins came out from the manufacturer pre-embedded in double density foams that support and orient them during the insertion procedures, performed by using the Branson LPX20 ultrasonic gun, operating in a manual hand-held mode using high frequency sinusoidal compressive waves (1000 J/cycle and 20 kHz).

The pins (Fig. 18.1a, b) were vibrated and driven into the laminate by the rectangular sonotrode of the gun (Fig. 18.1c) while the foam collapses immediately under the effect of the ultrasonic stress waves. After the insertion processes, to match the exact thickness of the laminates, the Z-Pins were manually shaved using a sharp blade.

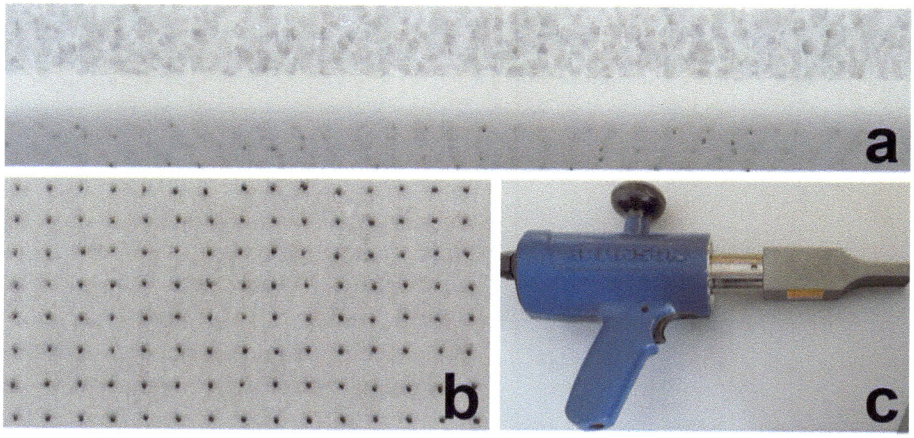

Fig. 18.1 (**a**) Side view of the 2% Z-pins embedded in the double density foam carrier. (**b**) Frontal view of a Z-pin mat (2% density, 12 mil diameter, 15 mm height). (**c**) Branson ultrasonic gun model LPX20

To conclude the fabrication stage, the panels were vacuum bagged and consolidated in autoclave following the curing cycle suggested by the manufacturer, (without any modification due to the presence of the pins) consisting in an initial 3.0 °C/min ramp up, 60 min at 125 °C, 4 °C/min cooling down ramp to room temperature.

The cured panels were cut using a diamond wheel saw to finally obtain the subsamples of 65 × 87.5 × 2 mm in size; no significant differences in thickness were found between pinned and unpinned coupons.

18.3 Impact and CAI Tests

The pinned and the unpinned laminates were impact tested at different velocity/energy using an instrumented drop-weight testing machine provided with an hemispherical-ended tup of 12.5 mm in diameter and a total mass of 2.34 kg.

During the experimental tests the samples were simply supported on a steel plate having a 45 mm × 67.5 mm rectangular opening.

The impact behaviour of reinforced and unreinforced samples were determined through drop weight impact tests conducted at several energy levels (2–30 J), to investigate the damage response of the two types of materials and to elucidate the role of the z-pins in the modification of the impact resistance of composite laminates.

The impact force between the impactor and the sample was measured by a semiconductor strain-gauge bridge bonded to the ended rod of the impactor, while the velocity of the impactor, immediately before the collision and at the rebound, was extracted by an infra-red sensor.

The histories of the velocity and of the displacement of the impactor were finally calculated by double integration of the contact force signal and the damage was examined externally by visual inspection of the samples and internally by penetrant enhanced X radiography.

All the samples, after being impacted, have been subjected to CAI (Compression after Impact) tests to assess the residual strength of the coupons and to evaluate the influence of the Z-Pin in the residual life and in the residual mechanical performances.

To avoid buckling and damage localization at the leading edge the impacted samples were tagged and an anti-buckling testing system (illustrated in Fig. 18.2) was used during the tests to support the lateral edge of the coupons during the application of the compressive load.

The samples were placed in the anti-buckling system of Fig. 18.2 with the 0° fibers parallel to the loading axis of an universal testing machine equipped with a load cell of 250 kN manufactured by Instron, and the compression tests were performed with a stroke rate of 1 mm/min.

To avoid damage localization at the loading interfaces of the samples, small tags were applied at the top and at the bottom loading edges.

After the Compression-After-Impact analysis the internal and the external damage of the samples were revaluated by mean of visual inspection and penetrant enhanced X radiography.

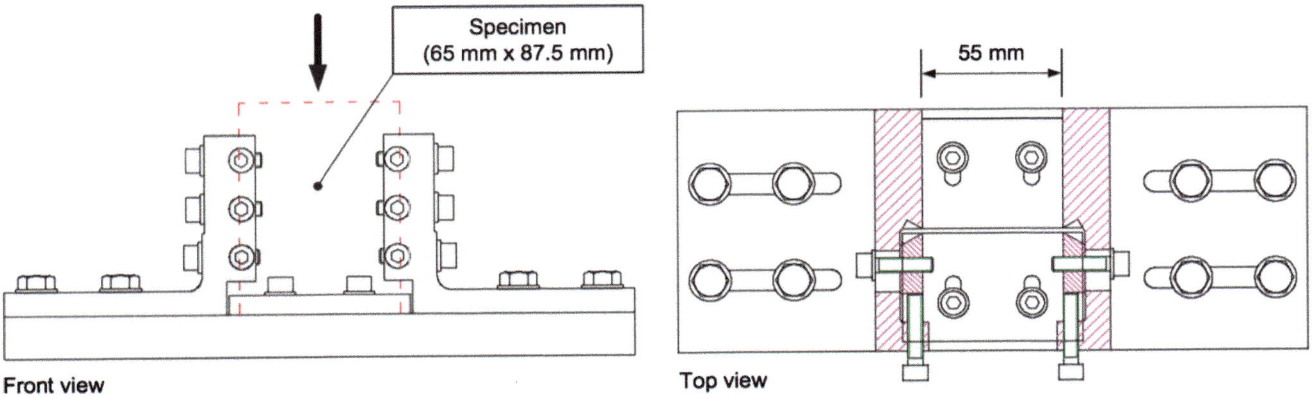

Fig. 18.2 Front and Top view of the anti-buckling feature used during the CAI tests

18.4 Results

Figure 18.3 reports the force-time and the force displacement traces for the reinforced "CPS" (plotted in red) and the unreinforced "CP" (plotted in black) $[0_2/90_2]_s$ samples impacted at several energies between 8 and 24 J.

The curves present a regular and smooth behavior for impact energies up to 8 J in both the diagrams of Fig. 18.3; after this threshold we observed a sudden force drop combined with high frequency oscillations indicating a major loss of strength and a sudden propagation of the damage.

Force-time and force–deflection curves show a smooth increasing in the peak forces for impact energies in the range of 2–16.5 J, with the pinned laminates having a maximum values of about 6 kN, while the unpinned counterparts, for the same impact energies, reached about 5 kN.

With increasing the impact energy to 16.5–30 J, a plateau of the maximum force was reported for the two families of laminates. Despite different absolute values, the plots of Fig. 18.3 have comparable trends for the entire range of impact energies tested, indicating that the pins' insertion doesn't twist the essence of the mechanical response of the samples.

This fact is also supported by the analysis of selected post-impact radiographies representative of the evolution of the internal damage of the tested coupons: Fig. 18.4 presents the X-rays for the unreinforced (left column) and the reinforced (right column) laminates impacted respectively at 4 and 30 J.

The X-rays (Fig. 18.4a, b) performed after a 4 J impact on the samples shown an evident two-lobes delamination profile at the lower 90°/0° interface, while no evident signs of delaminations can be observed at the upper 0°/90° interface.

The damage starts with tensile matrix cracks in the bottom 0° layers and with shear matrix cracks in the 90° layers for the pinned and the unpinned samples: for energy as low as 4 J the two classes of composites have a comparable damage behaviour (and comparable projected delaminated areas), thus suggesting that the effect on the Z-pins is marginal.

Increasing the impact energy, reinforced and unreinforced samples still present a two-lobed delamination that mainly propagates along the direction of the fibres and, with a minor contribution, in the 90° layers.

While on the mid 90° layers diffused tensile matrix cracks can yet be observed, the delamination propagates mainly on the 0° directions. In particular, the unreinforced samples present a significant propagation of the delamination at the lower interface on the direction of the fibres while, in contrast, in the reinforced samples the pins modify and limit the damage propagation and the projected delamination appears significantly smaller than that of the unreinforced samples.

Further increasing the impact energy to about 30 J (Fig. 18.4c, d) a complete perforation has been reported for the unreinforced samples (Fig. 18.4c) while the reinforced laminates (Fig. 18.4d), even if characterized by a severe damage around the impact area, still present a residual structural strength.

In the unreinforced samples visual inspections and X-rays of (Fig. 18.4d) reveal large strips of the bottom 0° layers entirely detached by the rest of the laminates, while in the reinforced ones, the damage is gathered around the impact site and limited within the pinned area.

No evident sign of perforation can yet be observed for the reinforced samples but several broken fibres, that appear as short fragmented irregular black lines that start from the pins' spots/tips of the X-rays of Fig. 18.4c, d are present around the impact site and around the pins' locations.

Two additional diagrams, describing the evolution of some parameters commonly used to characterize the damage response of impacted laminates are presented in Fig. 18.5.

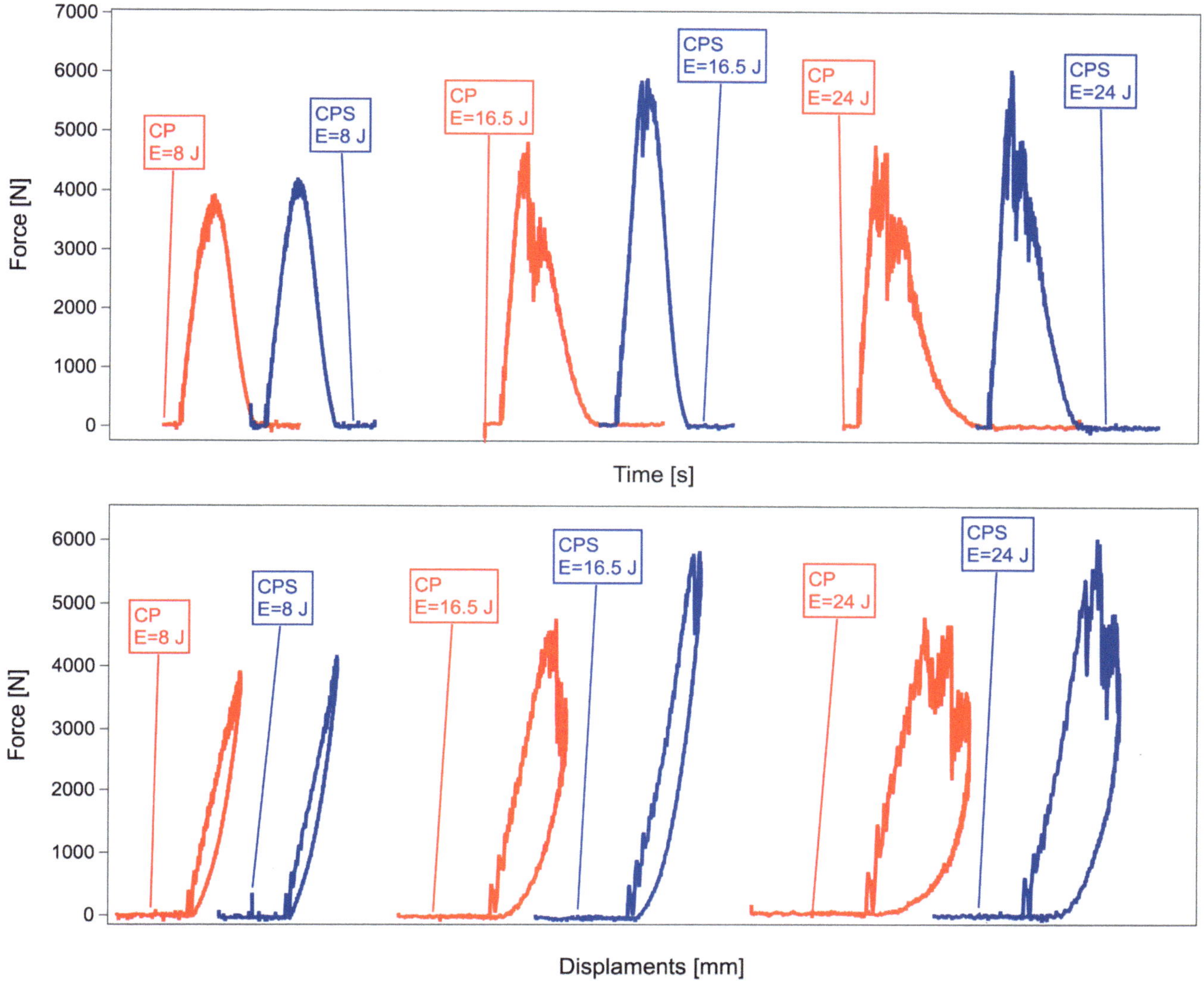

Fig. 18.3 Force-Time (top) and Force-Displacement (bottom) diagrams for the reiforced ("CPS", plotted in red) and the unreiforced (CP, plotted in black) samples impacted at different energies

The peak force vs. impact energy plot (Fig. 18.5, right side) shows comparable tendency in the interval 2–8 J, while higher peak forces (about 20% higher) were reported for the pinned laminates once exceeded the threshold energy level of 16.5 J. Both the curves possess a "plateau": after reaching the maximum force value (corresponding at an impact energy of about 15 J) the peak force no longer increases even when increasing the impact energy for both the pinned and the unpinned laminates.

These considerations were also validated by the graph reported on the left side of Fig. 18.5 where the projected delaminated areas are plotted as a function of the impact energies. The trend of the curves in the diagram of Fig. 18.5 (left side) indicates that the two types of samples possess a similar behavior for the lower impact energies, since the small size of the delaminations is not large enough to activate the effect of the reinforcements. In contrast, starting from impacts of about 8 J up to the complete structural collapse of the samples, the bridging effect of the pins leads to about 40–45% smaller projected delaminated areas.

When the impact energy increases in the interval 16.5–30 J the delaminations of the two classes of laminates appear different in size, with the reinforced samples having about a 60–70% smaller delaminated areas than that of the unreinforced ones indicating a positive action of the Z-pins in limiting the separation of the interface's layers.

To evaluate the residual strength of the coupons CAI tests were performed on all the impacted samples and the traces of the maximum normalized force as a function of the impact energy have been reported in the chart presented in Fig. 18.6.

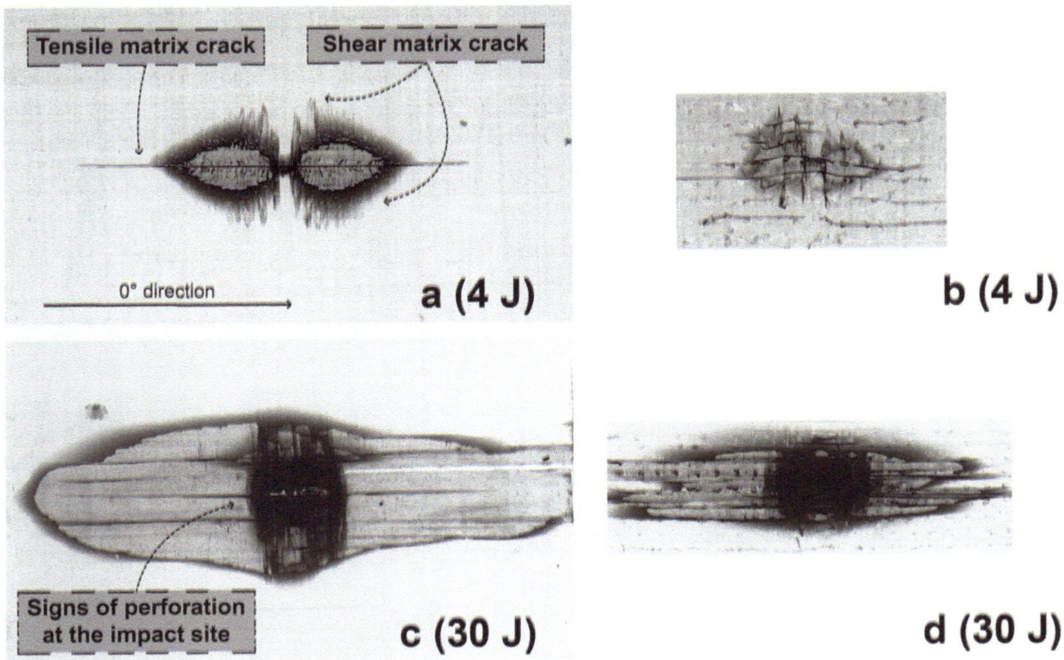

Fig. 18.4 X-rays representative of different damage state of the pinned (right side) and unpinned (left side) impacted at 4 J (**a**, **b**) and 30 J (**c**, **d**)

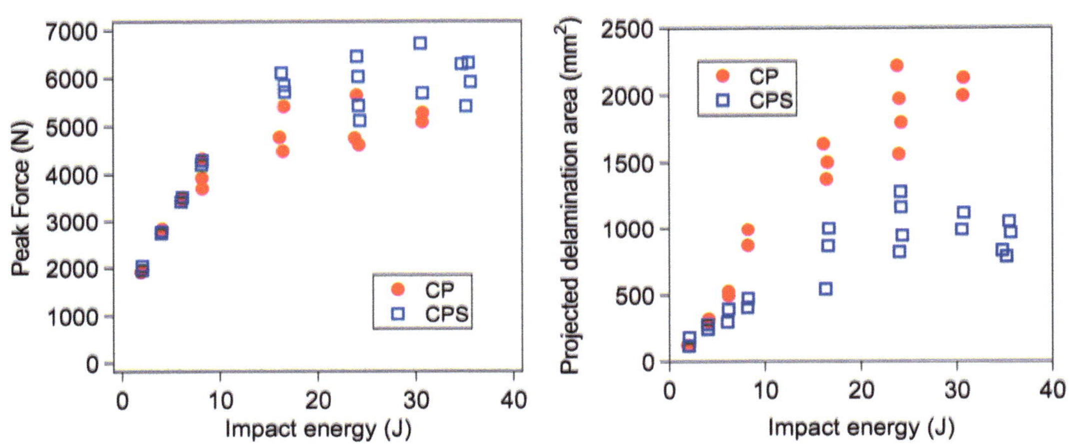

Fig. 18.5 Projected delaminated area (left chart) and peak forces (right chart) plotted as a function of the impact energy

The graph of Fig. 18.6 shows that for the entire range of impact energies considered the reinforced cross ply samples possess better post impact load bearing capacity even if this effect is not remarkable as the one obtained in the reduction of the delaminated areas. In fact, the macroscopic CAI failure modes of the two types of tested laminates are also similar in nature and despite the different delamination areas pinned and unpinned samples are characterized by comparable values of CAI strengths.

For comparable energy levels the effect of the pins, increasing the delamination resistance, leads to smaller delamination area and a more localized damage profile.

For the layup considered, where the damage is characterized by big size delaminations at the two 0/90 and 90/0 interfaces, this is probably the main reason of a slightly better post impact response under a compressive load (without essential modifications in the failure modes). This was experimentally observed during the CAI test by a visual inspection of the samples and validated by the very close behaviour of the curves of Fig. 18.6 and by additional radiographs that were taken after the CAI tests.

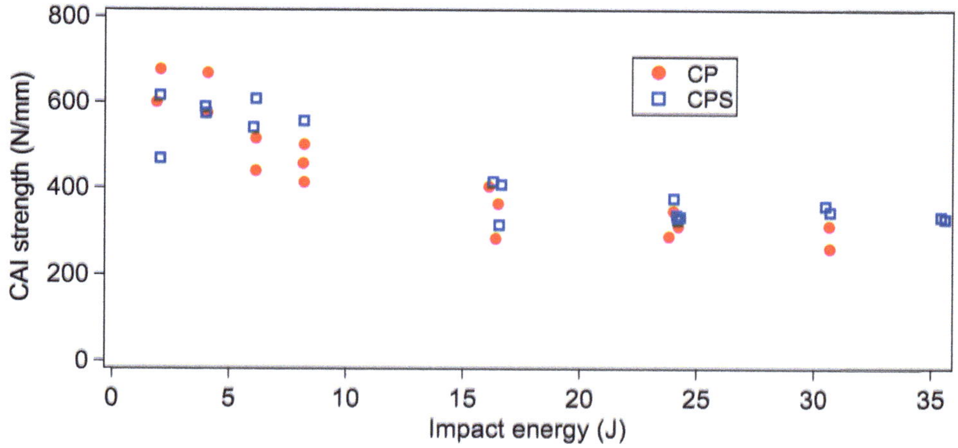

Fig. 18.6 CAI tests: Maximum force for the different impact energies for the CP (unreinforced) and CPS (pinned) samples

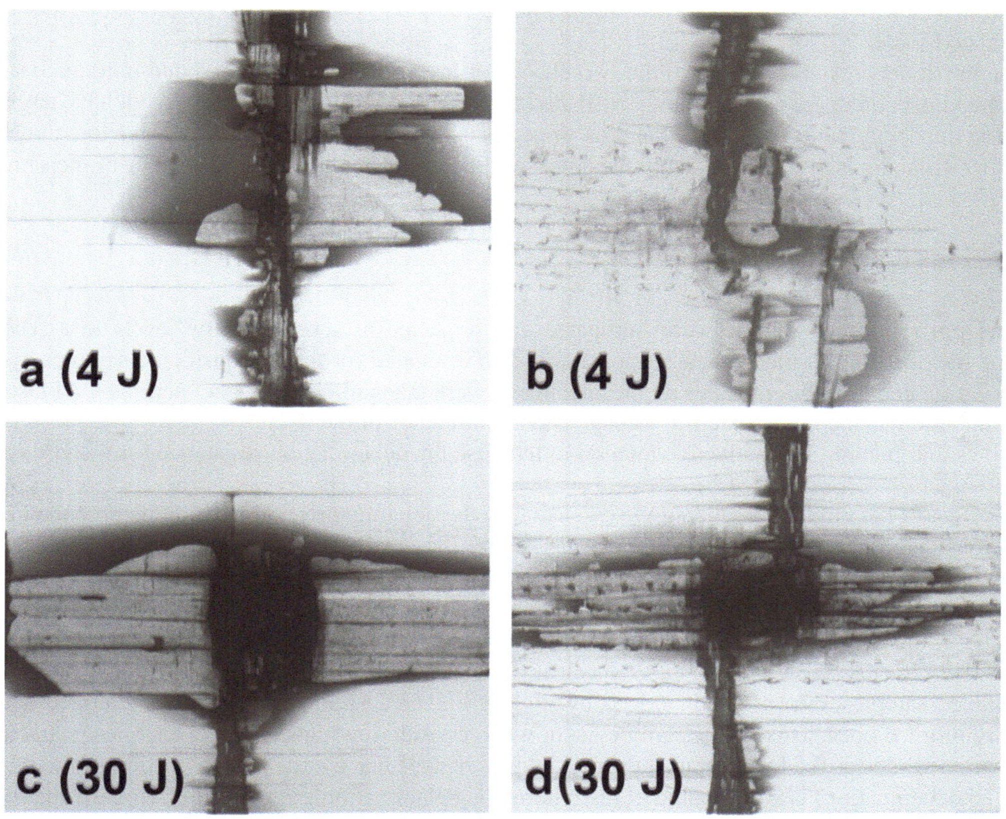

Fig. 18.7 Radiographs taken for both the pinned (right side) and unpinned (left side) samples after the CAI test to investigate the internal damage and failure for impact energy of 4 J (**a**, **b**) and 30 J (**c**, **d**)

Immediately after the CAI tests, the samples were inspected by means of X-rays to monitor the internal post-impact damage and failure modes of the CAI tested coupons. A selected set of images for the same energy levels considered previously in this study is presented in Fig. 18.7.

The x-rays of Fig. 18.7 taken respectively at an energy level of 4 J (Fig. 18.7a, b) and 30 J (Fig. 18.7c, d) for both the reinforced (right side) and the unreinforced (left side) samples clearly show a similar failure mode for all the samples without meaningful change due to the extension of the damage caused by the original impact.

Both reinforced and unreinforced specimens exhibit a sudden collapse (probably triggered and led by the unstable propagation of fiber fractures) once that the maximum normalized force is reached, with a crack orthogonal to the loading axis that propagates from the delaminated region to the lateral undamaged ligaments.

Increasing the load it is still possible to appreciate the effect of the Z-Pins in restrain the growth of the delamination and, furthermore, the radiographs taken after the CAI tests show that the localized damages and resin-rich areas induced by the pins' insertion affect the post-impact cracks' propagation, acting as stress risers and potential fracture initiation sites.

All the tested CP and CPS have a similar post impact behavior: they very well sustain a compressive load up to their maximum force (which is connected to the original impact energy and to the extension of the damage/delamination) and once that this threshold level is reached, they suddenly fail by a quick elastic propagation of major fibre fractures (black solid zig-zag lines) and matrix breakage (blurred lines parallel to the long side of the coupon) directed perpendicularly to the loading directions.

The minor differences that can be found between the reinforced and the unreinforced samples is again that, especially when the original impact damage is small and contained inside the pinned area, the pins drive the propagation of the fractures since the crack can use the pin's location as a concentration point to advance and bring the sample to the final collapse (Fig. 18.7b, d).

This effect, that affect in a negative way the after impact life of the sample, is counterbalanced by the original smaller delamination and that is why the overall behavior of the pinned sample is characterized by an higher (even if marginally) maximum normalized force.

Increasing the impact energy (Fig. 18.7c, d) the two classes of investigated samples respond in a much similar way: the larger extent of the impact damage, which is now distributed on the entire reinforced region, do not have any preferable path to propagate (since the delamination covers all the pins' spots) causing a very similar fracture behavior, characterized by the presence of one or two major cracks starting from the edges of the impact point to the lateral sides of the samples.

18.5 Conclusions

Unreinforced and pin reinforced sample were fabricated and impacted at several energy levels to assess their damage performances and the post-impact mechanical response was also evaluated for the all considered samples by means of CAI tests. Visual inspection and X-radiographs of all the specimens were taken after the impact and the CAI tests to investigate the extension of the internal and of the external damage and the influence of the Z-Pins.

The study shows that pinning is an effective method to increase the delamination resistance for the considered $[0_2/90_2]s$ laminates.

The bridging effect of the pins acts reducing the extension of the delamination and limiting its propagation for the entire range of energy considered in this study (2–35 J).

The reduction in the projected delaminated area is marginal at very low impact energies (2–4 J) due to the small extension of the delamination (not capable of triggering the bridging and the closure effect exerted by the pins) but it gets more incisive, (50%–80% smaller of the unreinforced samples) increasing the impact energy from 8 J to 35 J.

While the positive effect on the impact response is clear it is not possible to make the same conclusions considering the CAI tests: the pinned samples, even if characterized by slightly higher CAI strengths, are characterized by similar failure modes of the unreinforced ones. This is especially evident when considering higher impact energies characterized by large internal delamination while for the lower impact energy the presence of inactivated pins (a portion of the reinforced area is not affected by the delamination) is still capable of restraining the delamination growth and driving the macro-cracks.

From the data presented in this study it is possible to conclude that while it is clear that the Z-pins improve the delamination resistance of cross-ply carbon fiber composites, its effect on the post-impact life of the samples still need to be investigated.

More post-impact analysis for different energy levels and, in particular, different stacking sequences need to be considered in order to understand the influence of the layups and the position of the interfaces/delaminations in the determination of the impact and post impact performances of pinned composite laminates.

References

1. Abrate, S.: Impact on Composite Structures. Cambridge University Press (UK), Cambridge (1998)
2. Tong, L., Mouritz, A.P., Bannister, M.K.: 3D Fibre Reinforced Polymer Composites. Elsevier, Oxford (2002)
3. Kim, J.K., Mai, Y.W.: Engineered Interfaces in Fibre-Reinforced Composites. Elsevier, Oxford (1998)

4. Koricho, E.G., Khomenko, A., Haq, M., Drzal, L.T., Belingardi, G., Martorana, B.: Effect of hybrid (micro- and nano-) fillers on impact response of GFRP composite. Compos. Struct. **134**, 789–798 (2015)
5. Walker, L., Sohn, M.S., Hu, X.Z.: Improving impact resistance of carbon-fibre composites through interlaminar reinforcement. Compos. A: Appl. Sci. Manuf. **33**(6), 893–902 (2002)
6. Falzon, B.G., Hawkins, S.C., Huynh, C.P., Radjef, R., Brown, C.: An investigation of mode I and mode II fracture toughness enhancement using aligned carbon nanotubes forests at the crack interface. Compos. Struct. **106**, 65–73 (2013)
7. Zeng, Y., Liu, H.Y., Mai, Y.W., Du, X.S.: Improving interlaminar fracture toughness of carbon fibre/epoxy laminates by incorporation of nano-particles. Compos. Part B. **43**(1), 90–94 (2012)
8. Saghafi, H., Zucchelli, A., Minak, G., Palazzetti, R.: The effect of interleaved composite nanofibrous mats on delamination behavior of polymeric composite materials. Compos. Struct. **109**(1), 41–47 (2014)
9. Dickinson, L., Farley, G., Hinders, M.: Translaminar reinforced composites: A review. J. Compos. Technol. Res. **21**(1), 3–15 (1999)
10. Dransfield, K., Baillie, C., Mai, Y.W.: Improving the delamination resistance of CFRP by stitching—a review. Compos. Sci. Technol. **50**(3), 305–317 (1994)
11. Mouritz, A.P., Leong, K.H., Herszberg, I.: A review of the effect of stitching on the in-plane mechanical properties of fibre-reinforced polymer composites. Compos. A: Appl. Sci. Manuf. **28**(12), 979–991 (1997)
12. Cheng, X., Ali, A.M., Li, Z.: Residual strength of stitched laminates after low velocity impact. J. Reinf. Plast. Compos. **28**(14), 1679–1688 (2009)
13. Tan, K.T., Watanabe, N., Iwahori, Y.: Impact damage resistance, response, and mechanisms of laminated composites reinforced by through-thickness stitching. Int. J. Damage Mech. **21**, 51–80 (2012)
14. Aymerich, F., Pani, C., Priolo, P.: Effect of stitching on the low-velocity impact response of [03/903]s graphite/epoxy laminates. Compos. A: Appl. Sci. Manuf. **38**, 1174–1182 (2007)
15. Francesconi, L., Aymerich, F.: Impact damage resistance of thin stitched carbon/epoxy laminates. J. Phys. Conf. Ser. **628**(1), 012099 (2015)
16. Liu, D.: Delamination resistance in stitched and unstitched composite plates subjected to impact loading. J. Reinf. Plast. Compos. **9**(1), 59–69 (1990)
17. Jain, L.K., Mai, Y.W.: On the effect of stitching on mode I delamination toughness of laminated composites. Compos. Sci. Technol. **51**(3), 331–345 (1994)
18. Jain, L.K., Mai, Y.W.: Analysis of stitched laminated ENF specimens for interlaminar mode II fracture toughness. Int. J. Fract. **68**(3), 219–244 (1994)
19. Tan, K.T., Watanabe, N., Iwahori, Y.: X-ray radiography and micro-computed tomography examination of damage characteristics in stitched composites subjected to impact loading. Compos. Part B. **42**(4), 874–884 (2011)
20. Aymerich, F., Pani, C., Priolo, P.: Damage response of stitched cross-ply laminates under impact loadings. Eng. Fract. Mech. **74**(4), 500–514 (2007)
21. Francesconi, L., Aymerich, F.: Numerical simulation of the effect of stitching on the delamination resistance of laminated composites subjected to low-velocity impact. Compos. Struct. **159**(1), 110–120 (2017)
22. Mouritz, A.P.: Review of z-pinned composite laminates. Compos. A: Appl. Sci. Manuf. **38**(12), 2383–2397 (2007)
23. Chang, P., Mouritz, A.P., Cox, B.N.: Properties and failure mechanisms of z-pinned laminates in monotonic and cyclic tension. Compos. A: Appl. Sci. Manuf. **37**(10), 1501–1513 (2006)
24. Partridge, I.K., Cartié, D.D.R.: Delamination resistant laminates by Z-Fiber pinning: Part I manufacture and fracture performance. Compos. A: Appl. Sci. Manuf. **36**(1), 55–64 (2005)
25. Mouritz, A.P.: Delamination properties of z-pinned composites in hot–wet environment. Compos. A: Appl. Sci. Manuf. **52**, 134–142 (2013)
26. Pegorin, F., Pingkarawat, K., Mouritz, A.P.: Comparative study of the mode I and mode II delamination fatigue properties of z-pinned aircraft composites. Mater. Des. **65**, 139–146 (2015)
27. Pegorin, F., Pingkarawat, K., Daynes, S., Mouritz, A.P.: Influence of z-pin length on the delamination fracture toughness and fatigue resistance of pinned composites. Compos. Part B. **78**, 298–307 (2015)
28. Cartié, D.D.R., Troulis, M., Partridge, I.K.: Delamination of Z-pinned carbon fibre reinforced laminates. Compos. Sci. Technol. **66**(6), 855–861 (2006)
29. Yoshimura, A., Nakao, T., Yashiro, S., Takeda, N.: Improvement on out-of-plane impact resistance of CFRP laminates due to through-the-thickness stitching. Compos. A: Appl. Sci. Manuf. **39**(9), 1370–1379 (2008)
30. Chang, P., Mouritz, A.P., Cox, B.N.: Flexural properties of z-pinned laminates. Compos. A: Appl. Sci. Manuf. **38**(2), 244–251 (2007)

Chapter 19
Layered Jamming Multifunctional Actuators

Hugh A. Bruck, Ruben Acevedo, Jasmin Rohwerder, Lena Johnson, and Satyandra K. Gupta

Abstract Recent advances in additive manufacturing have enabled the realization of concepts in layered materials that were not previously viable. We have created a layered manufacturing process using robots to fabricate actuators using pre-fabricated materials, e.g. solar cells and batteries. This has allowed us to design and fabricate multifunctional structures that can encompass new capabilities for applications, such as pneumatic actuators for robotics. One way to enhance the functionality of these actuators is to use layers that are not fused together to create "variable stiffness" components when a vacuum is drawn. Studying the mechanics of these components is important in understanding how to configure the layers of the jamming multifunctional actuators. We have prototyped and characterized the mechanics of a layered jamming multifunctional actuator to model the effects of layering on the multifunctional performance of these new actuation materials. The proposed model is for a soft actuator with opposing jamming layers, and the predictions of actuation performance were found to be consistent with measurements from the prototype structures. Three different types of composite jamming materials were investigated, consisting of polyurethane rubber, silicone, or paper with paper. It was found that the order of magnitude difference in the stiffness of the polyurethane rubber and silicone with paper reduced the predicted curvature of the more compliant material under transverse load by approximately 50%, which increases the shear strain between the stiffer materials due to the reduced shear stiffness of the more compliant material. Two prototype actuation structures were fabricated: (1) a programmable array of layered jamming actuators to control the 3D shape of a flexible solar cell for energy harvesting, and (2) a multi-mode actuator capable of bending and extensional deformations. We have also demonstrated the viability of these actuators on a robotic platform known as "ArmadilloBot" that is capable of walking, and then morphing into a rolled-up structure, just like a real Armadillo.

Keywords Multifunctional materials · Soft actuators · Layered jamming · Robotics

19.1 Introduction

Recent advances in materials and manufacturing are expected to have profound changes on the way we design and manufacture products. Here are representative examples:

- Advancements in materials synthesis have enabled development of materials with remarkable properties. These innovations are expected to revolutionize energy harvesting, energy storage, and enable realization of self-powered materials.
- Advances in additive manufacturing are enabling fabrication of multi-material composite structures with unprecedented geometrical complexity.
- Multifunctional materials are enabling new design possibilities by integrating multiple functional capabilities into a single structure to reduce the parasitic mass and enhance performance.

We have previously created multifunctional wings with integrated flexible solar cells and batteries using an additive manufacturing process to enhance the endurance of flapping wing air vehicles (FWAVs) (see Fig. 19.1) [1–9].

Additive manufacturing (AM) has also emerged as a popular process in a wide variety of applications [10–24]. It can be used to create structures that would have been impossible to create using other processes while possessing a

H. A. Bruck (✉) · R. Acevedo · J. Rohwerder · L. Johnson
Department of Mechanical Engineering, University of Maryland, College Park, MD, USA
e-mail: bruck@eng.umd.edu

S. K. Gupta
Department of Mechanical and Aerospace Engineering, University of Southern California, Los Angeles, CA, USA

P. R. Thakre et al. (eds.), *Mechanics of Composite, Hybrid and Multifunctional Materials, Volume 5*, Conference Proceedings of the Society for Experimental Mechanics Series, https://doi.org/10.1007/978-3-319-95510-0_19

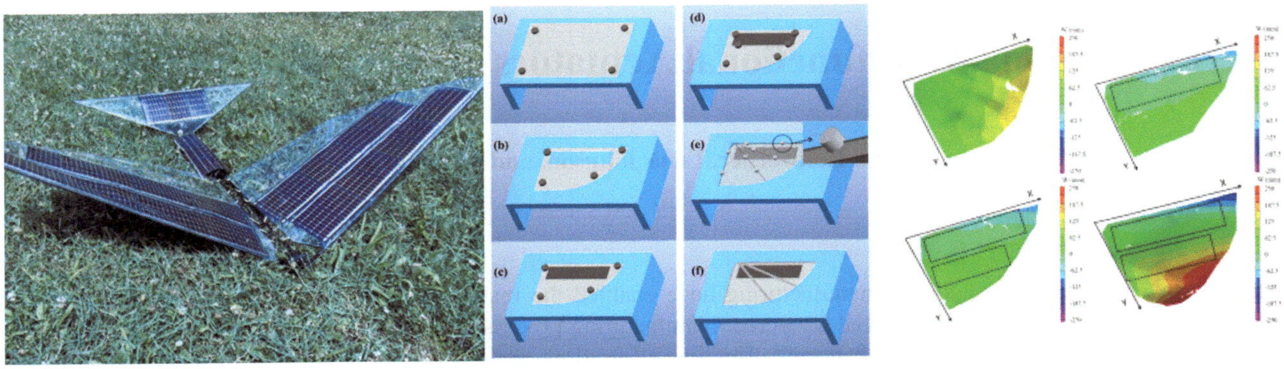

Fig. 19.1 (Left) Robo Raven III: a flapping wing air vehicle with solar cells integrated in wings using (middle) new additive manufactruing process composed of steps (**a**)-(**f**) and with (right) 3D DIC characterization of deformation changes due to outlined solar cell configurations [5]

variety of novel materials [25–33] Fabrication processes for multifunctional power generating materials have focused on embedding prefabricated components into different polymers and polymer composites with two distinct types of structural functionality: (1) heat dissipation, and (2) electrical conductivity [34–36]. Several processes are being developed for fabricating multifunctional materials [37–41]. Representative examples include shape deposition manufacturing, multi-material selective laser sintering, and 3D Printing. Rapid prototyping techniques have been used to create a worm robot module, where the embedded actuators served not only as the actuation source, but also as the structure frame. In order to develop high-performance power that are based on novel, sustainable/renewable and affordable materials, researchers developed all-solid-state flexible supercapacitors using compressed CNF/RGO/CNT aerogel film as the electrode with rapid charge/discharge times and lower energy densities [42]. Low-cost screen printing technology has been used to manufacture stretchable electrochemical sensors, biofuel cell arrays, and tactile strain sensors which are used in wearable devices, robotics, or bionics [43–46]. A number of researchers have also explored using similar fabrication approaches to integrate solar cell technology into flexible structures to make them more multifunctional by enabling "energy harvesting" [47–54].

In this paper, we present a layered design and manufacturing process that we have developed for realizing layered jamming multifunctional actuators for a variety of applications, including robotics. Details of the structural analysis for the layered jamming structures are first developed. Then, the effects of different jamming materials without multifunctional capability are investigated. A prototypical structure with an array of jamming layers is presented to demonstrate the capability for programming the stiffness in 2D via NURBS-like structures for an energy harvesting structure consisting of a flexible solar cell. The layered jamming layers are then used to create a soft actuator with multi-mode deformation capability consisting of bending and extension. Finally, we integrate the actuators into a new robotic platform known as "ArmadilloBot", which enables it to walk and then morph into a rolled-up shape, just like a real Armadillo.

19.2 Designing and Manufacturing Layered Multifunctional Materials

Layered multifunctional materials can be created from the following building block materials: (1) thin film solar cells, (2) thin film batteries, (3) flexible circuits, (4) polymer composite structural materials, We synthesize the layered materials by optimally placing the building block components within the material. The following list presents preliminary criteria that are relevant during the design: (1) deformations under mechanical loading, (2) mass per unit area, (3) power generated by per unit area, (4) energy stored by per unit area, and (5) load bearing capacity. For solar cells and batteries, we can also evaluate energy harvesting and storage efficiency, in addition to Dielectric constants and thermal conductivity. The appropriate criteria to impose as constraints or to impose as an objective function depend upon the specific problem being considered. For example, in an energy harvesting tile, the objective could be to maximize average power generated. Size, mass, energy generated, deformation, and load bearing requirements are imposed as constraints.

We utilize a new robot assisted layered manufacturing process we have developed for realizing flexible, thin multifunctional structures [55]. The process has to be able to work with heterogeneous materials and realize geometrically complex interfaces. Each layer is generated in the right shape and attached to the previous layer. The building block for our process includes the following steps: (1) layer shaping, (2) layer coating, (3) adhesive dispensing, (4) layer assembly, and (5) electrical connections. Figure 19.2 shows the schematic of the process used for making the material, and fabrication of a layered

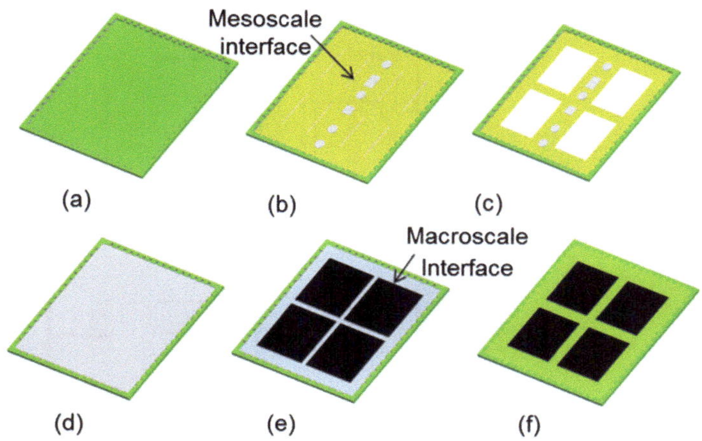

Fig. 19.2 Schematic of design and fabrication process for layered construction of multifunctional materials. (**a**) Base material, (**b**) flexible circuit, (**c**) flexible batteries, (**d**) insulation layer, (**e**) flexible solar cells, (**f**) completed tile

multifunctional wing structure can be seen at https://www.youtube.com/watch?v=RznNFBMZloU. A model for identifying process parameters is used for joining the layers to previously constructed material. This model can identify the temperature and pressure conditions needed to provide the adequate joining. Interfacial features are selected to ensure the applied pressure and temperature does not cause damage to the interface. The process parameters along with the underlying building block of the material are therefore used to characterize the constraint on the interface shape and size.

19.3 Modeling Structural Response of Soft Actuator with Jamming Layers

To model the structural response, a standard cantilever beam structure is analyzed with a rectangular cross-section for the soft actuator, and opposing jamming layers. A schematic of the cross-sectional area of the multi-mode actuator with its design variables labeled can be seen in Fig. 19.3. The top and bottom blue sections are the jamming layers that can have programmable Young's Modules E_1 and E_3. The gray section is the linear extending actuator that has a constant Young's Modules E_2.

The force, F, due to the input Pressure, P, is given by

$$F = PA = Pa^2 \tag{19.1}$$

where A is the cross-sectional area of the air channel. The strain, $\varepsilon(y)$, in the actuator varies with y and is defined by

$$\varepsilon(y) = ky + \varepsilon_o \tag{19.2}$$

where k is the curvature of the actuator and ε_o is the strain at the location $y = 0$. Following Bernoulli-Euler beam theory, the flowing equations govern the actuation deformation:

$$\Sigma F_x = 0: \quad Pa^2 = \int_0^{A_1} E_1 \varepsilon(y)\, dA + \int_0^{A_2} E_2 \varepsilon(y)\, dA + \int_0^{A_3} E_3 \varepsilon(y)\, dA \tag{19.3}$$

$$\Sigma M = 0: \quad Pa^2 c = \int_0^{A_1} E_1 \varepsilon(y) y\, dA + \int_0^{A_2} E_2 \varepsilon(y) y\, dA + \int_0^{A_3} E_3 \varepsilon(y) y\, dA \tag{19.4}$$

where,

$$c = \frac{h_2}{2} + h_1 \tag{19.5}$$

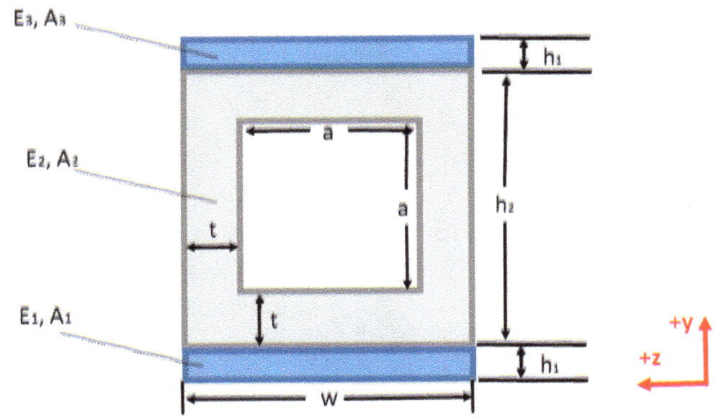

Fig. 19.3 A schematic of the cross-sectional area of the multi-mode soft actuator with its design variables labeled

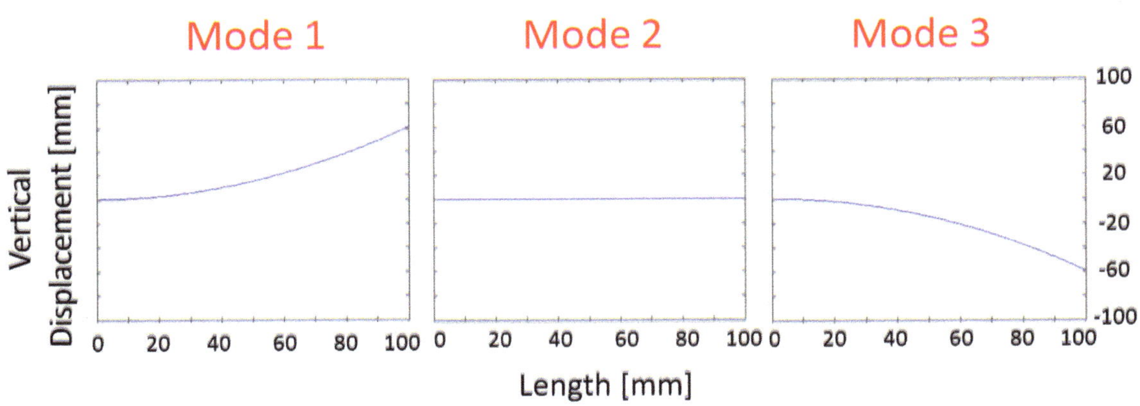

Fig. 19.4 Vertical Displacement prediction graphs of the three modes of actuation

and y = 0 is chosen so the bending moment is taken about the bottom of the beam. By plugging (19.2) into (19.3) you can solve for $\varepsilon_o = \varepsilon_o(k)$. Then by plugging in $\varepsilon_o(k)$ and (19.5) into Eq. (19.4) you can solve for k.

For a clamped end condition, the vertical displacement ofthe actuator, v, will be given by:

$$v = -\frac{k}{2}x^2 \tag{19.6}$$

where x is the distance along the length of the actuator.

By keeping E_2 constant and varying E_1 and E_3 in the model you can predict the vertical displacement of the actuator. Figure 19.4 shows the model's vertical displacement predictions for the three modes of actuation. For all three modes, $P = 30$ kPa and $E_2 = 30$ MPa. For mode 1, the top layer is jammed and assumed to have a Young's Modules of $E_3 = 300$ MPa, with the bottom jammed layer assumed to have the same Young's Modulus as the soft actuator $E_1 = 30$ MPa. For mode 2, no layer is jammed and all layers are assumed to have the same Young's Modulus as the actuator, $E_1 = E_2 = E_3 = 30$ MPa. Finally for mode 3, the bottom layer is jammed with $E_1 = 300$ MPa and $E_3 = 30$ MPa.

19.4 Layered Jamming Multifunctional Actuators

19.4.1 Layered Jamming Structure

Thin layers of flexible materials, such as the layers of a LiPo battery, can be placed in a stack and sealed. Then, under a vacuum, the stiffness of the resulting composite structure will increase due to the frictional force between layers generated by the atmospheric pressure acting on the sealed package. We can then arrange and attach these composites in a number of

Type of Actuator	Vertical Displacement [cm]
Polyurethane Sheets (10 layers)	1.94
Silicone Sheets (10 layers)	1.42
Paper Sheets (10 layers)	7.81

Fig. 19.5 (Left) A table of the vertical displacements of three different types of variable stiffening jamming composite cantilever beams stiffened by an 80 kPa vacuum under a 100 g load applied to at the end of the cantilever. (Right) A figure of the experimental set up for the 100 layer Polyurethane variable stiffening composite

Table 19.1 The predicted vertical displacements for each jamming layer material, and the associated reduction needed in the predicted curvature to account for the loss of strain transfer across layers due to the orders of magnitude difference in stiffness between the paper and the polyurethane rubber or silicone

Layer material	Predicted vertical displacement (cm)	% Difference from measured	% Reduction needed in predicted curvature of more compliant layer
Polyurethane rubber/paper	0.75	61.3	60.3
Silicone/paper	0.80	43.7	46.3
Paper	7.95	1.8	−1.8

orientations and surfaces to improve substrate performance, such as attaching them to soft actuators where they can be used not only to control stiffness, the resulting actuation force, and/or displacement, but also to provide power for actuation.

Initially, we studied the variable stiffness phenomena for layered jamming without the use of energy storage structures as the jamming material. Not only does the number of layers in the composite affect stiffness under a vacuum, but also the inherent stiffness of the jamming material used, as indicated in Fig. 19.5. The relation between the stiffness before and after the vacuum is applied is N^2, where N is the number of jamming layers. Therefore, for 10 layers, the change in stiffness will be 100×.

We created three different types of variable stiffening composites made of: (1) polyurethane sheets, (2) silicone sheets, and (3) paper, all sealed in a *polyethylene pouch with a tube inlet for negative pressure.* All three composites were composed of 5 layers of bond paper that were 0.1 mm thick per layer, along with another 5 layers of paper or 5 layers of polyurethane rubber of Shore Hardness 90A or silicone of Shore Hardness 40A, both of thickness 0.8 mm per layer. The jamming materials were tested in a cantilever beam configuration that was 13 cm long and 2.4 cm wide. The composites are stiffened by an 80 kPa vacuum, and a 100 g load is applied at the tip. We could determine the difference in stiffness by the vertical displacements each composite experiences under the same load, and compare it with the predicted values. The properties of the layers and the corresponding vertical displacements that were predicted can be seen below in Table 19.1. The correlation between the predicted and measured displacements indicated that the effects of the jamming material were not accurately captured assuming perfect strain transfer across each layer. Instead, there is an increase in the shear strain between the stiffer paper layers due to the order of magnitude differences in the stiffness of the polyurethane rubber and silicone compared to paper that significantly reduces the shear stiffness between the paper layers. The resulting reduction needed in the predicted curvature of the softer layer to account for these differences is also reported in Table 19.1, and is approximately 50%.

19.4.2 Prototype for a Layered Jamming Multifunctional Actuator

In exploring the benefits that variable stiffening elements can have on actuators, we created a prototype for a layered jamming multifunctional actuator consisting of: (a) a of a flexible solar cell for harvesting solar energy, and (b) a programmable array of layered jamming multifunctional actuators to allow for more complex 3D shape control. By selectively actuating different elements in the array, it is possible to program the sheet stiffness at different locations in the array with NURBS-like control. This allows for us to program in a deformed shape for the structure under an applied loading, such as wind. This is particularly useful in creating deformable wings for FWAVs like Robo Raven, where the wing deformation can be controlled at various stages during the flapping cycle, or even at a fixed position, to maximize net lift and thrust forces. The programmable sheet array consists of six variable stiffening composites: three in a vertical orientation and three in a horizontal orientation. All of the multifunctional actuators had five paper layers as the jamming material. The multifunctional actuator was affixed to our 6 axis force sensing platform we have previously used to characterize lift and thrust forces for FWAVS. We placed the

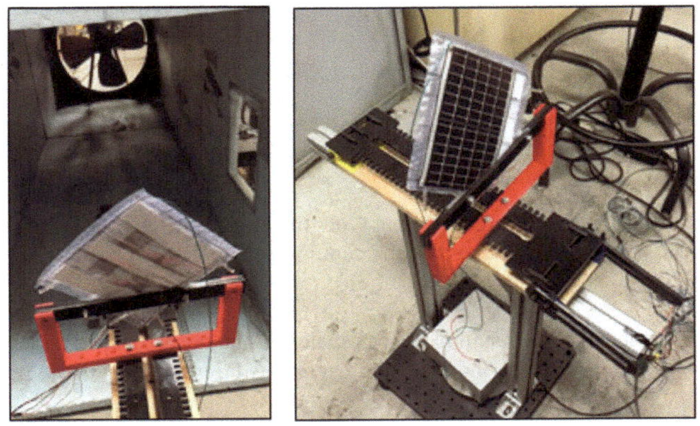

Fig. 19.6 Experimental set up for the variable stiffness sheet array

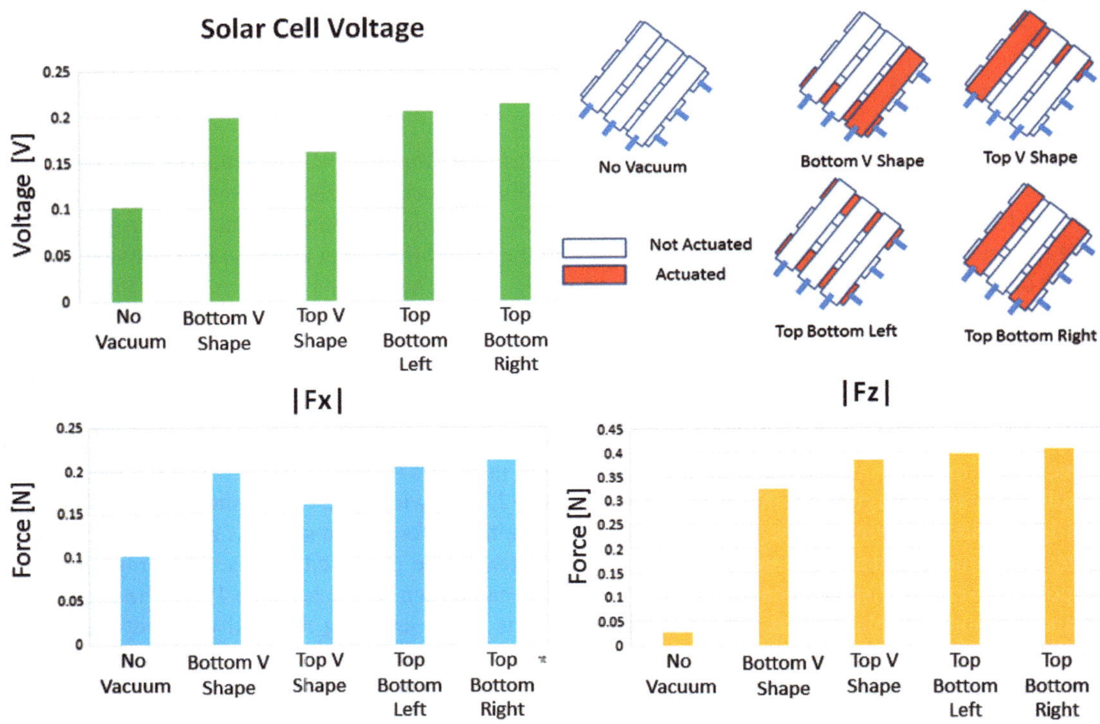

Fig. 19.7 Variable stiffness array outputting different forces at different programmable deformation scenarios

actuator in our wind tunnel to measure the change in the aerodynamic drag forces due to the associated shapes we created by program in the stiffness (Fig. 19.6).

Five different programmable stiffness scenarios were tested: (i) the array with no elements actuated, (ii) the array with its bottom v elements actuated, (iii) the array with its top v elements actuated, (iv) the array with the top and bottom elements pointing left actuated, and (v) the array with the top and bottom elements pointing right actuated. The voltage output from the solar panel can be seen in Fig. 19.7, indicating that different shapes were achieved at different actuation scenarios and that each scenario outputted different Fx and Fz forces to the force sensor. These results indicate that it is possible to utilize layered jamming multifunctional actuators in programmable arrays as FWAV wings to minimize the drag on the upstroke, and to maximize the lift force on the down stroke.

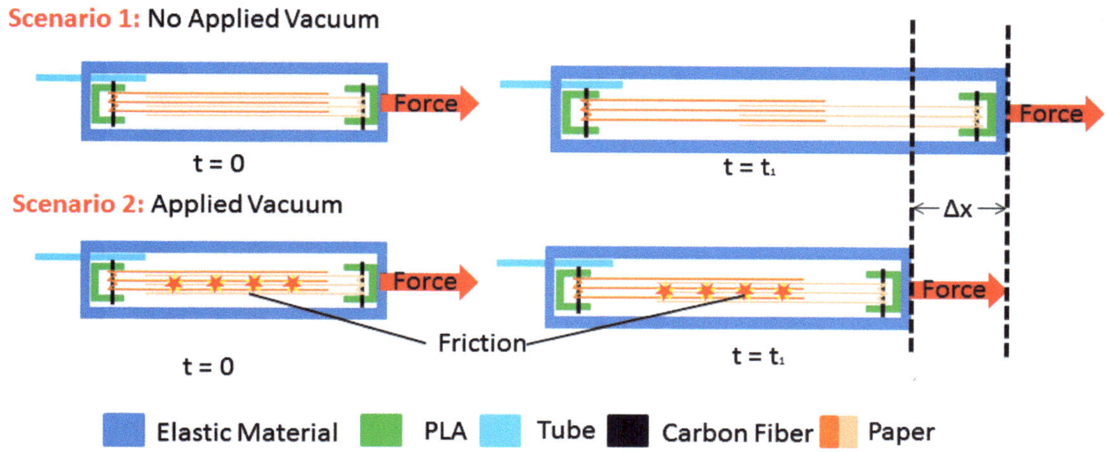

Fig. 19.8 A schematic of the layer jamming concept for variable elongation stiffness

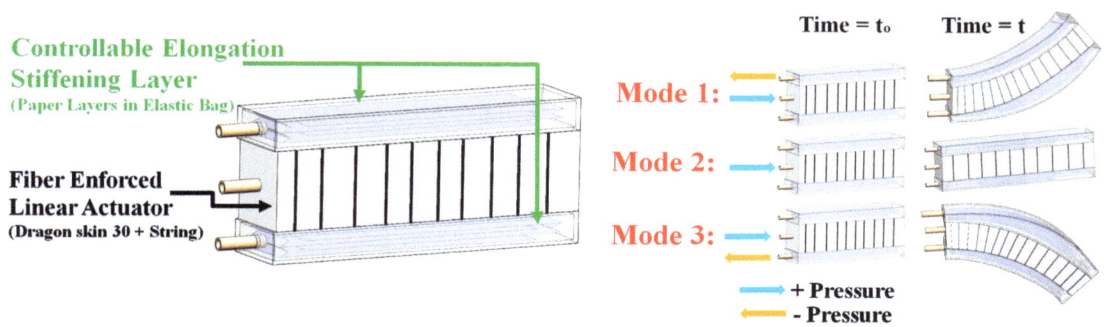

Fig. 19.9 A schematic of a multi-mode soft actuator that can be programmed to bend up, linearly extend, or bend down

19.4.3 Layered Jamming for Multi-mode Control of Extension and Bending for Soft Actuators

Previous work has shown that layered jamming is a promising solution for variable stiffness to normal loads [56]. We expanded this research to use layer jamming for variable elongation stiffness. Figure 19.8 is a schematic of the layer jamming concept for variable elongation stiffness. By interweaving the jamming layers and sealing them in an elastic pouch, when a stretching force is applied to the composite, the paper layers are free to slide against each other to allow for extension. However, when a vacuum is applied to the composite, the friction between the layers resists the layers from sliding against each other to extend thus increasing its elongation stiffness.

With this variable stiffening elongation composite, we can combined it with a soft linear extending actuator to create a multi-mode soft actuator that can bend up, linearly extend, or bend down all while variably controlling the actuators stiffness at the same time. This gives layer jamming a multifunctional purpose of not only variably stiffening a soft actuator but also making is deformation more controllable. Figure 19.9 shows a schematic of the multi-mode soft actuator.

Pressure applied to the silicone actuator controls the extension, while the vacuum applied to the top and/or bottom layer controls the bending direction. Due to this structure, there are three modes of actuation. First, applying a vacuum to the top layer increases its elongation stiffness, which results in an upward deflection under an applied pressure (Fig. 19.9 – *mode 1*). Second, in the absence of vacuum, an applied pressure will result in the whole actuator extending linearly (Fig. 19.9 – *mode 2*). Lastly, applying a vacuum to the bottom layer produces a downward deflection (Fig. 19.9 – *mode 3*). In combination, this approach allows for not only the direction of bending to be controlled, but also the degree of bending (*i.e.*, by regulating the magnitude of the vacuum).

Figure 19.10 shows a prototype proof of concept where all there modes were achieved. A $\sim$30 kPa pressure was applied to the linear actuator and ~ -80 kPa was applied to the jamming layers. The total length of the actuator was 100 mm with the silicone linear actuator having a height and width of 20 mm. The jamming layer were composed of 20 paper layers each with a total thickness of 2 mm. For mode 1, there was a vertical displacement of about $\sim$50 mm. For mode 2, the actuator extended by 50%. For mode 3, there was also vertical displacement of about $\sim$50 mm.

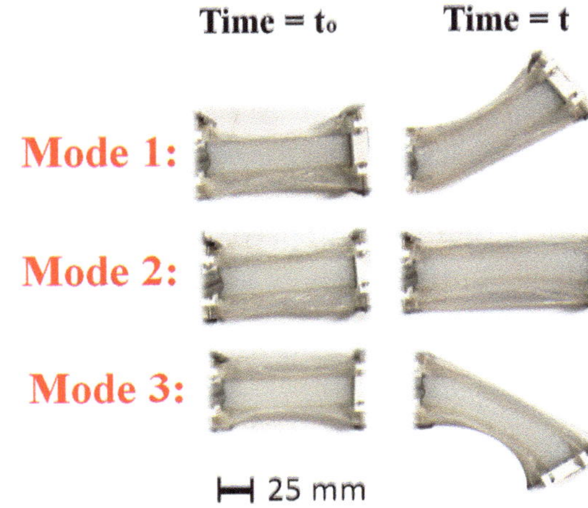

Fig. 19.10 Prototype for a multi-mode soft actuator that can be programmed to bend up, linearly extend, or bend down

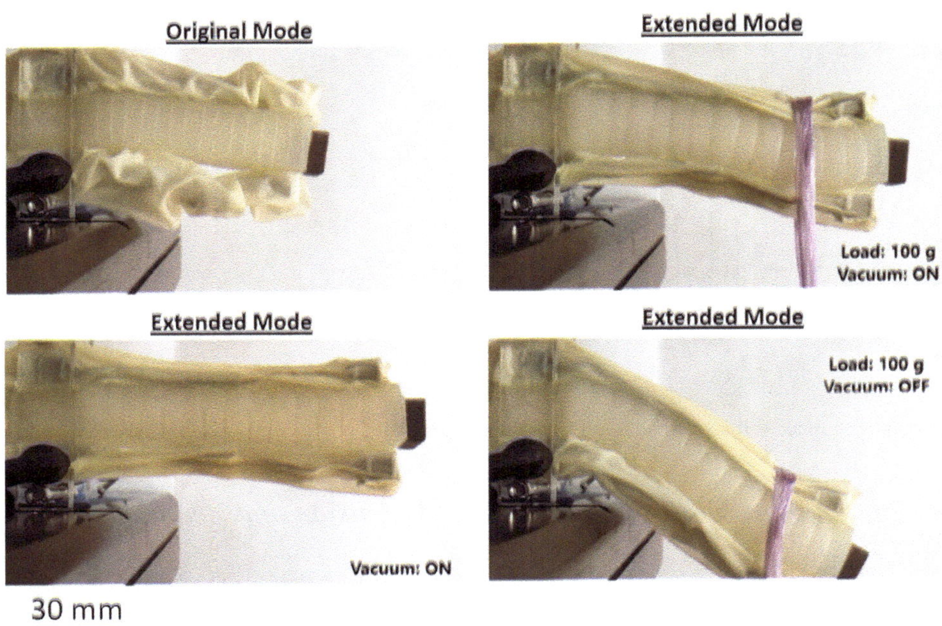

Fig. 19.11 A sequence of photos showcasing a soft actuator that can be variably stiffened in an extended state

The experimental vertical displacement results after actuation correlated closely with our predictions in Eq. (19.6). For mode 1, we expected to see a maximum upward vertical displacement of 60 mm. and obtained a maximum upward vertical displacement of 50 mm. For mode 2, we expected to see no vertical displacement and an extension of 45 mm, while the measured extension was 40 mm. Finally, for mode 3, because of the symmetry of the jamming layers, a maximum downward vertical displacement of −60 mm was predicted, while the measured value was −50 mm, the same difference as observed with the upward vertical displacement. Our proposed model can predict the 3 modes of actuation with reasonable accuracy, indicating the ability to use the model for designing structures using the soft actuators with opposing jamming layers for multi-mode control using different material properties, E_1, E_2, and E_3, as well as varying the geometry of the actuators.

An added benefit of this jamming actuator is that not only can we control the elongation stiffness of various layers, but we can still variably control the stiffness of the actuator in the extended state. A sequence of photos showcasing a soft actuator that can be variably stiffened in an extended state is shown in Fig. 19.11. The top left picture is the actuator at its original state. 40 kPa pressure is inputted to the linear actuator causing it to extend by 50% as shown in the bottom left picture. A −80 kPa vacuum is applied to both the top and bottom jamming layers to stiffen the actuator in its elongated state and a 100 g load is applied to the tip of the actuator as shown in the top right picture. The actuator is able to hold its position until the vacuum is removed from the jamming layers, thus reducing the stiffness of the actuator causing it to bend down as seen in the bottom right picture.

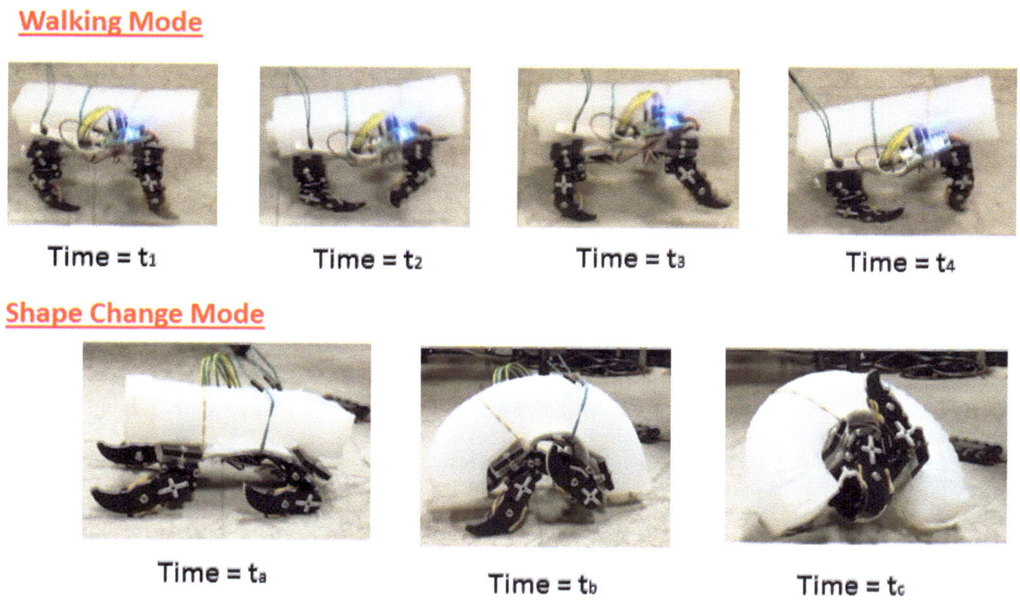

Fig. 19.12 A sequence of photos showcasing ArmadilloBot in its two modes: (i) walking mode and (ii) shape changing mode.

19.4.4 Integration of Layered Jamming Actuators in a Robot: ArmadilloBot

Finally, we explored the feasibility of the jamming principle for robotic applications in ArmadilloBot, a new shape changing legged robot platform that we have developed capable of walking, and then rolling up into a ball, just like a real Armadillo. ArmadilloBot is composed of three main components. (i) the front and back legs made out of Delrin actuated by servos and rubber band passive joints, (ii) a variable stiffening layer that acts like the robots spine connecting the front and back legs of the robot, and (iii) the soft bending actuator shell that shape changes the robot into a ball. As seen in Fig. 19.12, the robot has two modes, a walking mode and a shape change mode.

For the walking mode the robots spine must be stiffen to allow for the forward motion produce by the legs to transfer throughout the body. If the spine was too flexible the legs would just move and the robots main body would never move forward or be able to balance itself. In the shape changing mode, the spine needs to be flexible to allow for the soft actuator shell to bend itself into a ball. If the robot's spine was always stiff, the robot would not be able to change shape, and if the robot's spine was always flexible, the robot would not be able to move forward. Therefore, variable stiffness enables both functions to be realized.

19.5 Conclusions

We have presented a new design and manufacturing process to realize layered jamming multifunctional actuators. A wide range of "building blocks" for these actuators can be utilized in a new AM process we have developed for fabricating the actuators using robots. A structural modeling approach has been developed as well, which is capable of predicting the actuation performance for multi-mode control using a soft actuator with opposing jamming layers. Prototype layer composite jamming materials were investigated, including: (1) polyurethane, (2) silicone, and (3) paper, all with paper. It was found that the order of magnitude difference in the stiffness of the polyurethane rubber and silicone with paper reduced the predicted curvature of the more compliant material under transverse load by approximately 50%, which increases the shear strain between stiffer materials due to the lower shear stiffness of the more compliant material. Prototypical structures were then fabricated and characterized, including: (1) a programmable array of layered jamming actuators for energy harvesting using flexible solar cells, and (2) a multi-mode actuator capable of bending and/or extensional deformation. Measurements of the actuation performance were consistent with the predictions from the structural model, which enables design of these actuators for different building block materials and geometric configurations. Layered jamming actuators were also integrated onto a new robotic platform known as "ArmadilloBot", which is capable of a walking mode and a shape changing mode where it can "roll up" like a real Armadillo.

Acknowledgements This work was sponsored by AFOSR grant FA9550-15-1-0350 with Dr. Byung-Lip "Les" Lee program manager. Opinions expressed are those of the authors and do not necessarily reflect opinions of the sponsors.

References

1. Perez-Rosado, A., Philipps, A., Barnett, E., Roberts, L., Gerdes, J., Gupta, S.K., Bruck, H.A.: Compliant multifunctional wing structures for harvesting solar energy", Proceedings of 19th International Conference on Composite Materials, Montreal, Canada, 1–9 (2013)
2. Perez-Rosado, A., Philipps, A., Barnett, E., Roberts, L., Gerdes, J., Gupta, S.K., Bruck, H.A.: Compliant multifunctional wing structures for flapping wing MAVs. Exp. Mech. Compos. Hybrid Multifunct. Mater. **6**, 77–83 (2014)
3. Perez-Rosado, A., Greisinger, A.G.J., Bruck, H.A., Gupta, S.K.: Performance characterization of multifunctional wings with integrated solar cells for unmanned Ariel vehicles, Proceedings of the ASME 2014 International Design and Engineering Technical Conferences & Computers and Information in Engineering Conference, 1–11 (2014)
4. Perez-Rosado, A, Bruck, H.A., Gupta, S.K.: Enhancing the design of solar-powered flapping wing air vehicles using multifunctional structural components. ASME Mechanism and Robotics Conference, Boston, MA, (August 2015)
5. Perez-Rosado, A., Gelhar, R.D., Nolen, S., Gupta, S.K., Bruck, H.A.: Design, fabrication, and characterization of multifunctional wings to harvest solar energy in flapping wing air vehicles. Smart Mater. Struct. **24**, 065042 (2015)
6. Perez-Rosado, A., Bruck, H.A., Gupta, S.K.: Integrating solar cells into flapping wing air vehicles for enhanced flight endurance. J. Mech. Robot. Trans. ASME. (2015). https://doi.org/10.1115/1.4032411
7. Holness, A., Steins, E., Bruck, H.A., Peckerar, M., Gupta, S.K.: Performance characterization of multifunctional wings with integrated flexible batteries for flapping wing unmanned aerial vehicles. ASME 2016 International Design Engineering Technical and Computers and Information in Engineering Conference, Charlotte, NC, 1–10 (2016)
8. Holness, A.E., Perez-Rosado, A., Bruck, H.A., Peckerar, M., Gupta, S.K.: Multifunctional wings with flexible batteries and solar cells for robotic birds. *Proceedings of the 2016 SEM Annual Conference and Exposition,* Orlando, FL, 1–8 (2016)
9. Holness, A.E., Bruck, H.A., Gupta, S.K.: Chapter 7. Flexible energy harvesting/storage structures for flapping wing air vehicles. In: Fracture, Fatigue, Failure and Damage Evolution Proceedings of the Society for Experimental Mechanics Series, pp. 35–45 (2017). https://doi.org/10.1007/978-3-319-62831-8_6
10. Dapino, M.J.: Smart structure integration through ultrasonic additive manufacturing. Conference: 7th Annual ASME Conference on Smart Materials, Adaptive Structures and Intelligent Systems (SMASIS), Newport, RI, ASME, New York, NY (2014)
11. Duro-Royo, J., Mogas-Soldevila, L., Oxman, N.: Flow-based fabrication: An integrated computational workflow for design and digital additive manufacturing of multifunctional heterogeneously structured objects. Comput. Aided Des. **69**, 143–154 (2015)
12. Kaipa, K.N., Shriyam, S., Kumbla, N.B., Gupta, S.K.: Resolving occlusions through simple extraction motions in robotic bin picking. ASME Manufacturing Science and Engineering Conference, Blacksburg, USA, ASME, New York, NY (2016). doi:https://doi.org/10.1115/MSEC2016-8661
13. Keum, H., Kim, S.: Micro-masonry for 3D additive micromanufacturing. J. Visualized Exp. (90), Article Number: e51974 (2014)
14. Khoda, A. K. M. B., Ozbolat, I.T., Koc, B.: Spatially multi-functional porous tissue scaffold. 3rd ECCOMAS Thematic Conference on Tissue Engineering (ICTE), Leiria, Portugal (2013)
15. Kimionis, J., Isakov, M., Koh, B.S., et al.: 3D-printed origami packaging with inkjet-printed antennas for RF harvesting sensors. IEEE Trans. Microw. Theory Tech. **63**(12), 4521–4532 (2015)
16. Kirleis, M.A., Simonson, D., Charipar, N.A., et al.: Laser embedding electronics on 3D printed objects. Conference on Laser 3D Manufacturing, San Francisco, CA (2014). doi: 10.1117/12.2044222
17. Li, J., Monaghan, T., Masurtschak, S., et al.: Exploring the mechanical strength of additively manufactured metal structures with embedded electrical materials. Mater. Sci. Eng. A Struct. Mater. Properties Microstructure Process. **639**, 474–481 (2015)
18. Mathew, J., Havermann, D., Polyzos, D., et al.: SS316 structure fabricated by selective laser melting and integrated with strain isolated optical fiber high temperature sensor. 24th International Conference on Optical Fibre Sensors (OFS) Location: Curitiba, Brazil (2015)
19. Meisel, N.A., Elliott, A.M., Williams, C.B.: A procedure for creating actuated joints via embedding shape memory alloys in PolyJet 3D printing. J. Intell. Mater. Syst. Struct. **26**(12), 1498–1512 (2015)
20. Pa, P., Larimore, Z., Parsons, P., et al.: Multi-material additive manufacturing of embedded low-profile antennas. Electron. Lett. **51**(20), 1561–1562 (2015)
21. Pal, D., Patil, N., Zeng, K., et al.: An efficient multi-scale simulation architecture for the prediction of performance metrics of parts fabricated using additive manufacturing. Metallurgical Mater. Trans. A Phys. Metallurgy Mater. Sci. **46A**(9), 3852–3863 (2015)
22. Panesar, A., Brackett, D., Ashcroft, I., et al.: Design framework for multifunctional additive manufacturing: Placement and routing of three-dimensional printed circuit volumes. J. Mech. Des. **137**(11), Article Number: 111414 (2015)
23. Vatani, M., Lu, Y., Engeberg, E.D., et al.: Combined 3D printing technologies and material for fabrication of tactile sensors. Int. J. Precision Eng. Manufacturing. **16**(7), 1375–1383 (2015)
24. Velasquez-Garcia, L.F.: SLA 3-D printed arrays of miniaturized, internally fed, polymer electrospray emitters. J. Microelectromechanical Syst. **24**(6), 2117–2127 (2015)
25. Calignano, F., Tommasi, T., Manfredi, D., et al.: Additive manufacturing of a microbial fuel cell-a detailed study. Sci. Rep. **5**, Article Number: 17373 (2015)
26. Cappelleri, D., Efthymiou, D., Goswami, A., et al.: Towards mobile microrobot swarms for additive micromanufacturing. Int. J. Adv. Robot. Syst. **11**, 150 (2014)
27. Carrico, J.D., Traeden, N.W., Aureli, M., et al.: Fused filament 3D printing of ionic polymer-metal composites (IPMCs). Smart Mater. Struct. **24**(12), 125021 (2015)

28. Hengsbach, S., Lantada, A.D.: Rapid prototyping of multi-scale biomedical microdevices by combining additive manufacturing technologies. Biomed. Microdevices. **16**(4), 617–627 (2014)
29. Hoerber, J., Glasschroeder, J., Pfeffer, M., et al.: Approaches for additive manufacturing of 3D electronic applications. Conference: 47th CIRP Conference on Manufacturing Systems, Windsor, Canada (2014)
30. Hou, S., Tyrer, J.: 3D printing of structures with embedded circuit boards using novel holographic optics. IEEE 63rd Electronic Components and Technology Conference (ECTC), Las Vegas, NV (2013)
31. Laumer, T., Stichel, T., Amend, P., et al.: Simultaneous laser beam melting of multimaterial polymer parts. J. Laser Appl. **27**(SI), Article Number: S29204 (2015)
32. Strantza, M., Aggelis, D.G., de Baere, D., et al.: Evaluation of SHM system produced by additive manufacturing via acoustic emission and other NDT methods. Sensors. **15**(10), 26709–26725 (2015)
33. Street, R.A., Ng, T.N., Schwartz, D.E., et al.: From printed transistors to printed smart systems. Proc. IEEE. **103**(4), 607–618 (2015)
34. Bejgerowski, W., Gupta, S.K., Bruck, H.A.: A systematic approach to designing multi-functional thermally conducting polymer structures with embedded actuators. J. Mech. Des. **131**, 111009 (2009). https://doi.org/10.1115/1.4000239
35. Bejgerowski, W., Gerdes, J., Gupta, S.K., Bruck, H.A.: S. Wilkerson, "design and fabrication of a multi-material compliant flapping wing drive mechanism for miniature air vehicles." *2010 ASME Mechanisms and Robotics Conference*, Montreal. Canada. 1–12 (2010)
36. Gershon, A.L., Gyger Jr., L.S., Bruck, H.A., Gupta, S.K.: In situ characterization of residual strains near electronic components embedded in thermoplastic polymers during processing and operation. In: Banks-Sills, L., Gilat, R. (eds.) Advances in Mathematical Modeling and Experimental Methods for Materials and Structures the Jacob Aboudi Volume, pp. 145–160. Springer Dordrecht Heidelberg, London, New York (2010)
37. Ma, R.R., Belter, J.T., Dollar, A.M.: Hybrid deposition manufacturing: design strategies for multimaterial mechanisms via three-dimensional printing and material deposition. J. Mech. Robot. **7**(2), Article Number: 021002 (2015)
38. Maier, R.R.J., Macpherson, W.N., Barton, J.S., et al.: Embedded Fiber optic sensors within additive layer manufactured components. IEEE Sens. **13**(3), 969–979 (2013)
39. Ma, S., Ribeiro, F., Powell, K., et al.: Fabrication of novel transparent touch sensing device via drop-on-demand inkjet printing technique. ACS Appl. Mater. Interfaces. **7**(39), 21628–21633 (2015)
40. MacCurdy, R., Mcnicoll, A., Lipson, H.: Bitblox: Printable digital materials for electromechanical machines. Int. J. Robot. Res. **33**(10), 1342–1360 (2014)
41. Maier, R.R.J., Havermann, D., Schneller, O., et al.: Optical fibre sensing in metals by embedment in 3D printed metallic structures. 23rd Int. Conf. Opt. Fibre Sens. Book Ser. Proc. SPIE. **9157**, 915707 (2014)
42. Zheng, Q., Cai, Z., Ma, Z., et al.: Cellulose Nanofibril/reduced graphene oxide/carbon nanotube hybrid aerogels for highly flexible and all-solid-state supercapacitors. ACS Appl. Mater. Interfaces. **7**(5), 3263–3271 (2015)
43. Bandodkar, A.J., Jeerapan, I., You, J.M., Nuñez-Flores, R., Wang, J.: Highly stretchable fully-printed CNT-based electrochemical sensors and biofuel cells: Combining intrinsic and design-induced stretchability. Nano Lett. **16**, 721–727 (2016)
44. Wissman, J., Perez-Rosado, A., Edgerton, A., Levi, B.M., Karakas, Z.N., Kujawski, M., Philipps, A., Papavizas, N., Fallon, D., Bruck, H.A., Smela, E.: New compliant strain gauges for self-sensing dynamic deformation of flapping wings on miniature air vehicles. Smart Mater. Struct. **22**, 085031 (2013)
45. Chen, Y., Yu, M., Bruck, H.A., Smela, E.: Stretchable touch-sensing skin over padding for co-robots. Smart Mater. Struct. **25**, 055006 (2016)
46. Barnett, E.M., Lofton, J.J., Yu, M., Bruck, H.A., Smela, E.: Targeted feature recognition using mechanical spatial filtering with a low-cost compliant strain sensor. Nat. Sci. Rep. **7**, 5118 (2017). https://doi.org/10.1038/s41598-017-05341-w
47. Asp, L.E., Greenhalgh, E.S.: Structural power composites. Compos. Sci. Technol. **101**, 41–61 (2014)
48. Chou, T.P., Zhang, Q.F., Fryxell, G.E., Cao, G.Z.: Dye-sensitized hierarchically structured ZnO solar cells for highly efficient energy harvesting. Adv. Mater. **19**, 2588–2592 (2007)
49. Graetzel, M.: The advent of mesoscopic injection solar cells. Prog. Photovolt. **14**(5), 429–442 (2006)
50. Krebs, F.C.: Fabrication and processing of polymer solar cells: A review of printing and coating techniques. Sol. Energy Mater. Sol. Cells. **93**(4), 394–412 (2009)
51. Krebs, F.C.: Polymer solar cell modules prepared using roll-to-roll methods: Knife-over-edge coating, slot-die coating and screen printing. Sol. Energy Mater. Sol. Cells. **93**(4), 465–475 (2009)
52. Krebs, F.C., Jorgensen, M., Norrman, K., Hagemann, O., Alstrup, J., Nielsen, T.D., Kristensen, J.: A complete process for production of flexible large area polymer solar cells entirely using screen printing-first public demonstration. Sol. Energy Mater. Sol. Cells. **93**(4), 422–441 (2009)
53. Huang, J., Li, G., Yang, Y.: A semi-transparent plastic solar cell fabricated by a lamination process. Adv. Mater. **20**, 415–419 (2008)
54. van den Donker, M.N., Gordijn, A., Stiebig, H., Finger, F., Rech, B., Stannowski, B., Bartl, R., Hamers, E.A.G., Schlatmann, R., Jongerden, G.J.: Flexible amorphous and microcrystalline silicon tandem solar modules in the temporary superstrate concept. Sol. Energy Mater. Sol. Cells. **91**(7), 572–580 (2007)
55. Bhatt, P.M., Peralta, M., Bruck, H.A., Gupta, S.K.: Robot assisted additive manufacturing of thin multifunctional structures, submitted to Proceedings of ASME 13th Manufacturing Science and Engineering Conference, College Station, Texas, MSEC 2018-6620 (2018)
56. Lin, K.-Y., Gupta, S.K.: Soft fingers with controllable compliance to enable realization of low cost grippers. In: Conference on Biomimetic and Biohybrid Systems. Springer, Cham (2017)

Chapter 20
2D Microscale Observations of Interlaminar Transverse Tensile Fracture in Carbon/Epoxy Composites

Austin J. Smith, Caitlin M. Arndt, Danielle Benson, and Michael W. Czabaj

Abstract This study presents a new experiment to image and analyze the evolution of transverse tensile fracture in tape-laminate carbon/epoxy composites at the microscale. To this end, a miniature double cantilever beam specimen is developed to produce stable transverse tensile fracture and various amounts of crack turning through a stack of 90° plies. The specimen is wedge-loaded with a custom micromechanical tester, while the crack growth is optically monitored at both the micro- and macro-scale. Post-test, microscale images are analyzed to determine the transverse crack path relative to the applied loading, and to directly measure local strains using 2D digital image correlation. The experimental data sets obtained in this study enable direct validation of finite element simulations of transverse tensile cracking in fibrous composites at the microscale and provide impetus for their further model development.

Keywords Transverse fracture · Polymer matrix composites · Microscale testing · Double cantilever beam test · Digital image correlation

20.1 Introduction

The mechanisms that govern initiation and evolution of transverse tensile fracture in tape-laminate carbon/epoxy composites are not well understood. This is particularly true at the microscale where fracture is governed by the (1) stochastic distribution of the reinforcing fibers, (2) existence of flaws, and (3) local variability in material properties. To date, several studies have used finite element (FE) analysis to gain further insight into transverse fracture at the microscale [1–6]; however, the necessary experimental data to validate such models is still lacking.

20.2 Experimental Methodology

In this study, a new experiment is developed to analyze and image the evolution of transverse tensile fracture in tape-laminate carbon/epoxy composites at the microscale. To this end, a miniature double cantilever beam (DCB) specimen is developed to produce stable transverse fracture through a stack of 90° tape-laminate plies. The miniature DCB specimen (Fig. 20.1) utilizes a symmetric $[90_5/0_N/90_5 \,||\, 90_5/0_N/90_5]$ stacking sequence, where the "||" symbol designates the location of a pre-implanted $12.5\mu m$ thick Teflon film which serves to create an artificial starter crack. The "0_N" defines a stack of 0° reinforcing plies, with N = 5, 10, and 15, which are used to vary the flexural rigidity of the arms thereby controlling the macroscopic crack path. Specimens with relatively compliant arms (N = 5) exhibit almost immediate mixed-mode crack turning, while specimens with stiff arms (N = 15) exhibit mode-I dominated horizontal crack extension (Fig. 20.1).

Prior to testing, the X-Z planes of each specimen (see Fig. 20.1) are polished to a mirror finish allowing for *in situ* surface observations of crack growth in the central 90° ply stack during loading. To achieve a microscopically-sharp crack tip, all specimens are fatigue precracked beyond the pocket of resin that typically forms ahead of the Teflon insert during cure [7]. Prior to testing, high magnification reference images of the microstructure surrounding the crack tip are captured and precise locations and radii of fiber cross-sections are determined using an in-house template matching algorithm [8]. This extracted microstructure can be used to construct realistic microscale FE models. During testing the microstructure surrounding the crack tip is monitored with a traveling optical microscope using a Mitutoyo 20× lens, while the opposite side is monitored

A. J. Smith · C. M. Arndt · D. Benson · M. W. Czabaj (✉)
Department of Mechanical Engineering, University of Utah, Salt Lake City, UT, USA
e-mail: m.czabaj@utah.edu

P. R. Thakre et al. (eds.), *Mechanics of Composite, Hybrid and Multifunctional Materials, Volume 5*, Conference Proceedings of the Society for Experimental Mechanics Series, https://doi.org/10.1007/978-3-319-95510-0_20

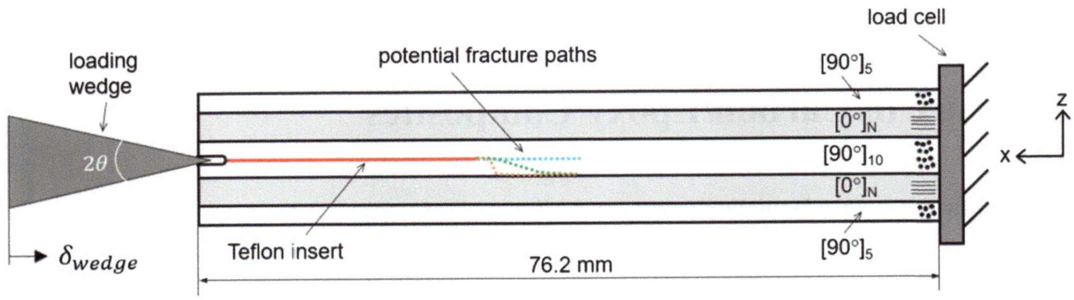

Fig. 20.1 Geometry of the miniature DCB specimen

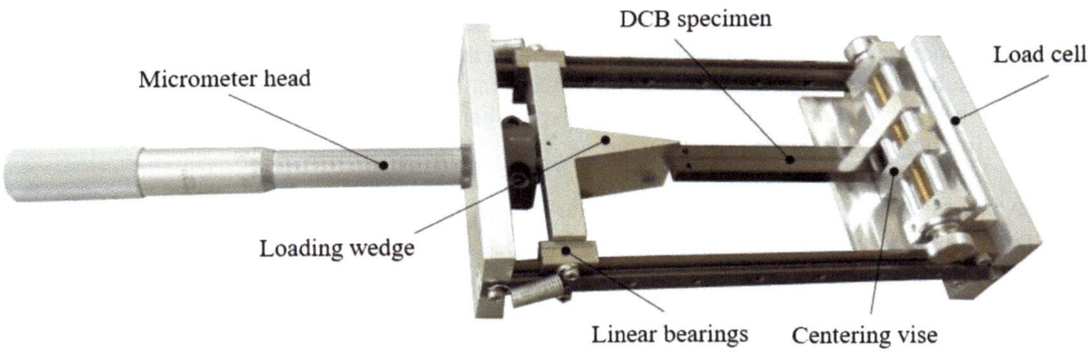

Fig. 20.2 Micromechanical test fixture with sample mounted. The micrometer head drives the wedge which applies mode-I (opening) load to the sample. The horizontal force is sensed by a load cell located behind the centering vise

with a macroscopic 17 mm lens to capture the global specimen response. The specimen is manually wedge loaded with a custom micromechanical tester shown in Fig. 20.2.

The applied displacement and resulting force of the wedge are recorded along with corresponding optical images of the evolving crack path. The load and displacement of the wedge (x-direction in Fig. 20.1) are converted to the equivalent load and displacement of the arms (z-direction in Fig. 20.1). Displacement of the arms is obtained directly from macroscale optical images. The corresponding wedge load is converted to the z-direction load using the following expression:

$$P_{arm} = 2P_{wedge}\left(\frac{\cos\theta - \mu\sin\theta}{\sin\theta + \mu\cos\theta}\right), \tag{20.1}$$

where θ is the half-angle of the wedge (see Fig. 20.1) and μ is the static coefficient of friction between the wedge and specimen material. The right-hand side of the expression is multiplied by two to account for both arms of the specimen.

Post-test, the optical images are used to determine the change in crack length relative to the applied loading. They also provide direct measure of the local displacement and strains in the crack tip region using 2D digital image correlation (DIC), with the fiber cross-sections acting as a random speckle pattern (Fig. 20.3).

20.3 Results and Discussion

Exploratory tests on practice specimens have provided encouraging data that validates the proposed approach. Steady crack growth has been observed in quasi-static loading, and the varying flexural rigidity in the specimen arms appears to influence the angle of transverse fracture and subsequent crack turning as hypothesized. Interestingly, specimens with stiffer arms did exhibit straighter crack growth, but also, more stable crack growth than specimens with more compliant arms.

The data thus far suggests that at the microscale the transverse crack path is governed by the local distribution of fibers and favors growth along fiber/matrix interfaces as seen in Fig. 20.3a. At the macroscopic (i.e. ply) level, the transverse fracture appears to follow a path that is perpendicular to the direction of maximum principal stresses as determined from ply-level stress analysis.

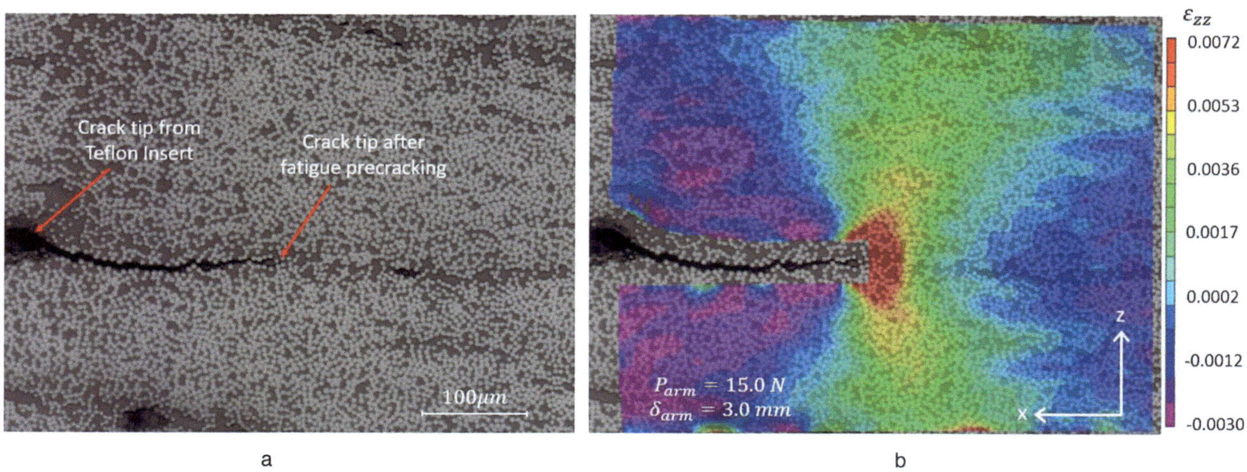

Fig. 20.3 (**a**) Image of the microstructure surrounding the crack tip where light-grey circles correspond to fiber cross-sections, dark grey regions are the matrix material, and black areas are voids/cracks. (**b**) Local strain field (ε_{zz}) surrounding the crack tip obtained using fiber cross-sections as natural random DIC speckle pattern

Currently, tests are being conducted on specimens with N = 5, 10, and 15 (3 per geometry), and the following data is being extracted:

1. Initial microstructure (fiber radii and location) for construction of realistic FE models
2. Far-field loading data including wedge and specimen load-displacement responses
3. Local DIC-derived strain and displacement fields surrounding the crack tip at various crack lengths
4. Local crack path and global crack turning angle
5. Crack tip location as a function of the loading obtained by combining microscopic optical images with corresponding load-displacement data

20.4 Conclusion

The results obtained in this study will provide a comprehensive set of experimental data necessary to (1) construct realistic FE microscale models of fiber reinforced composites, and (2) validate simulations of transverse tensile fracture in those models.

References

1. Gonzalez, C., Llorca, J.: Mechanical behavior of unidirectional fiber-reinforced polymers under transverse compression: Microscopic mechanisms and modeling. Compos. Sci. Technol. **67**, 2795–2806 (2007)
2. Yang, L., Yan, Y., Liu, Y., Ran, Z.: Microscopic failure of fiber reinforced polymer composites under transverse tension and compression. Compos. Sci. Technol. **72**, 1818–1825 (2012)
3. Okabe, T., Imamura, H., Sato, Y., Higuchi, R., Koyanagi, J., Talreja, R.: Experimental and numerical studies of initial cracking in CFRP cross-ply laminates. Compos. Part A. **68**, 81–89 (2015)
4. Naya, F., Gonzalez, C., Lopes, C.S., Van der Veen, S., Pons, F.: Computational micromechanics of the transverse and shear behavior of unidirectional fiber reinforced polymers including environmental effects. Compos. Part A. **92**, 146–157 (2017)
5. Herraez, M., Mora, D., Naya, F., Lopes, C.S., Gonzalez, L.J.: Transverse cracking of cross-ply laminates: A computational micromechanics perspective. Compos. Sci. Technol. **110**, 196–204 (2015)
6. Herraez, M., Gonzalez, C.: Lopes CS. A numerical framework to analyze fracture in composite materials: From R-curves to homogenized softening laws. Int. J. Solids Struct. **000**, 1–13 (2017)
7. Czabaj, M.W., Ratcliffe, J.G.: Comparison of intralaminar and interlaminar mode I fracture toughnesses of a unidirectional IM7/8552 carbon/epoxy composite. Compos. Sci. Technol. **89**, 15–23 (2013)
8. Czabaj, M.W., Riccio, M.L., Whitacre, W.W.: Numerical reconstruction of graphite/epoxy composite microstructure based on sub-micron resolution X-ray computed tomography. Compos. Sci. Technol. **105**, 174–182 (2014)

Chapter 21
Electro-Mechanical Response of Polymer Bonded Surrogate Energetic Materials with Carbon Nanotube Sensing Networks for Structural Health Monitoring Applications

Samantha N. Rocker, Nishant Shirodkar, Tanner A. McCoy, and Gary D. Seidel

Abstract The addition of carbon nanotubes throughout the binder phase of energetic particulate composites is investigated in an effort to develop real-time embedded sensing networks for detection of small-scale damage in polymer bonded explosives undergoing mechanical load. The experimental effort herein focuses on the exploration of multi-walled carbon nanotube (MWCNT) concentration in energetic composite material; the fabrication of such specimen include the substitution of sugar as a mock energetic for the crystal particulate Ammonium Perchlorate (AP), an oxidizer most often used in solid rocket propellants. Further explored was the particulate embedded into a binder of PDMS, a polymeric silicone, as well as the addition of aluminum powder, a common combustive fuel in solid propellants, in the particulate. Electrical and mechanical properties of neat (no MWCNTs) energetics and MWCNT hybrid energetics were quantitatively evaluated under monotonic compression, and tension to failure. Noteworthy electro-mechanical response was obtained for these MWCNT AP hybrid energetics, justifying further study of CNT binder network formation for real-time electro-mechanical sensing in an effort for structural health monitoring (SHM) in energetics.

Keywords Energetic materials · Carbon nanotubes · Nanocomposite · Piezoresistivity · Structural health monitoring · Strain sensing · Electromechanical coupling · Damage detection

Nomenclature

PBX	=	Polymer-bonded explosive
AP	=	Ammonium perchlorate
$PDMS$	=	Polydimethylsiloxane
CNT	=	Carbon nanotube
$MWCNT$	=	Multi-walled carbon nanotube
$wt\%$	=	Weight percent
SHM	=	Structural health monitoring
LCR	=	Inductance (L), Capacitance(C), Reactance(R)
L	=	Distance between electrodes
R	=	Resistance
X	=	Reactance
Z	=	Impedance
σ	=	Stress
ϵ	=	Strain
λ	=	Slip angle
ϕ	=	Slip angle

S. N. Rocker · N. Shirodkar (✉) · T. A. McCoy · G. D. Seidel
Aerospace Structures and Materials Lab, Kevin T Crofton Aerospace and Ocean Engineering Department, Virginia Tech, Blacksburg, VA, USA
e-mail: srocker@vt.edu; nish26@vt.edu; tmccoy01@vt.edu; gary.seidel@vt.ed

P. R. Thakre et al. (eds.), *Mechanics of Composite, Hybrid and Multifunctional Materials, Volume 5*, Conference Proceedings of the Society for Experimental Mechanics Series, https://doi.org/10.1007/978-3-319-95510-0_21

21.1 Introduction

Polymer bonded explosives, or PBXs, are composite materials made up of energetic particulates combined in a polymeric binder matrix. Typically used for solid explosives unable to be manufactured using standard casting procedures, PBXs have found application in solid rocket propellants, where materials are expected to maintain precise engineering shapes while also safely withstanding and absorbing shock to avoid accidental detonation. In order to achieve such specific structural properties, the particulate and binder components of these composites are often directed to materials which can be pressed and machined at room temperature.

With safety as a primary consideration in the manufacturing and handling of PBXs, there are concerns of accidental impact and its resulting internal damage, in the form of micro-fractures. Mechanical loading of this nature has the potential to compromise the stability of the material during ignition by allowing hot spot initiation, the development of localized high-temperature regions within the material.

As such, it has become a critical venture of study to determine methods of internal damage sensing in energetic materials. To ensure the safety of such experimentation, surrogate particulates have been identified by the research community to replicate expected mechanical characteristics of typical energetics. Sivior found that ammonium perchlorate (AP), a solid rocket propellant oxidizer, held certain monoclinic particulate geometry and mechanical properties which allowed sugar to serve as an appropriate surrogate [1].

Researchers have performed extensive experimentation on the mechanical characterization of energetic and mock energetic material systems such as Balzer et al. [2], who used X-ray tomography and scanning electron microscopy to observe fracture patterns of both AP and sugar specimen following mechanical load testing. These tests, however, only explored characteristics of damage post-loading.

Instead, incorporating the piezoresistive properties of multi-walled carbon nanotubes into a particulate composite without significant change to the chemical and mechanical properties of such materials was identified as a potential solution to sensing small scale internal damage in real time [3–11]. Multi-functional nanocomposites have been studied for their piezoresistive damage sensing potential in greater detail in recent years. As these materials incur internal damage on the micro-scale, the conductive pathways formed by CNTs through an otherwise non-conductive material are disrupted, causing inherent increases in resistance through the specimen. Experiments performed by Sengezer and Seidel [4] have concluded that in-situ structural health monitoring (SHM) can be achieved for PBXs, specifically ammonium percholate and epoxy, undergoing low-velocity impact by using weight concentrations of MWCNTs just below composite percolation and a specific frequency range for measurement of resistance characteristics.

While low velocity impact [4] and thermal loads [6] have been tested on energetic nanocomposites and successfully demonstrated active piezoresistivity in more brittle material systems such as that which incorporates an epoxy binder phase, studies have also been done by researchers to test vinyl ester nano-composites, finding that peak piezoresistive performance occurred during compression testing when the material system acted most ductile [12]. A ductile material system in all directions of loading, such as that formed by a PDMS binder phase [13], could reveal microscale damage throughout the elastic region of strain during monotonic loading [14].

These studies lay a promising foundation for the experimental effort herein, which explores the effects of monotonic compression and tension-to-failure on a nanocomposite bonded material system composed of sugar as the energetic surrogate and aluminum powder as a fuel simulant in the particulate mixture, PDMS as the ductile binder phase, and the addition of CNTs with a weight concentration just below percolation. Understanding the piezoresistive response through this effort is a first step towards applying nanocomposite binders in electrical impedance tomography as a structural health monitoring approach.

21.2 Experimental Setup

21.2.1 Materials Selection

The surrogate energetic used for the particulate phase of this study is monoclinic crystalline sugar sold as a product by Domino, and certain specimen included fine Aluminum Powder (Atomized 3 µm 1 lb) to simulate metalized energetics.The multi-walled carbon nanotubes (PD15L1-5) dispersed in the binder phase of this composite were grown by NanoLab using proprietary chemical vapor deposition techniques, to obtain 95% purity, an approximate diameter of 15 nm, and lengths from 1 to 5 µm. The polymeric matrix in which these MWCNTs are distributed is polydimethylsiloxane, or PDMS, a silicon-

Table 21.1 Weight percentage of material system compositions

Specimen material	% CNT	% Particulate	% PDMS	Ratio of Sugar/Al	Ratio of resin/Curing agent
Neat Sugar PDMS	0	60	40	100:0	10:1
Neat Sugar-Al PDMS	0	60	40	80:20	10:1
CNT Sugar PDMS	0.5	60	40	100:0	10:1
CNT Sugar-Al PDMS	0.5	60	40	80:20	10:1

based organic polymer which is a visco-elastic substance similar to rubber for this study's application. It was obtained from Dow Corning Corporation and is sold under the registered name of "Sylguard 184" and has two parts: Part A, which is the resin/base and Part B, which is the curing agent.

21.2.2 *Fabrication*

During this stage of the study, sugar was used as a mock energetic in all the test specimens. In total, four different materials systems were fabricated. Their compositions are outlined in Table 21.1.

Following the guidelines outlined on the data sheet for Slyguard 184 PDMS by Dow Corning Corporation, the PDMS in all the specimens was cured for 60 min at 135 °C and then an additional 24 h room temperature curing was done. The particulate mixture, in its appropriate quantity, was first mixed with the resin or Part A of the PDMS. This was done manually using tongue depressors and visually verified for attaining a homogenous mixture. As is observed in all particulate composites with more than 50% particulate composition by weight, there is lots of void formation in the specimens. For this purpose, the mixture was degassed in two stages: once after mixing particulate and Part A, and second time after the curing agent or Part B is mixed. Both the times the degassing was performed for 25 min at 60 MPa vacuum pressure. The degassed mixture was then pressed into ASTM standard tension and compression molds with tongue depressors and compressed with a counter mold with 1 kg weight on it. The counter mold ensured minimal void formation in the test specimen during the curing stage.

However, this procedure takes a slight deviation while using MWCNTs in the specimens. The dispersion of CNTs must be uniform. This is often difficult as they have a tendency to agglomerate due to due to van der Waals interactions. Hence, there arises a need to first disperse the CNTs using sonication and then mix them with the binder. In this study, the CNTs were dispersed as per the procedure outlined in Sengezer and Seidel [4]. First, the CNTs were weighed and dispersed in acetone. The low viscosity and non-reactive nature of acetone ensures minimal effect on the properties of the developed test specimen. The mixture of acetone and MWCNTs was sonicated at 20 kHz and 25% amplitude using a "QsonicaSonicator 4000" for an hour. Then Part A or binder was added and mixed using a magnetic stirrer before it was sonicated for an hour again with the same settings. After this, a uniform CNT dispersion is achieved and the acetone in this mixture is removed by distillation using a IKA RV-10 rotary evaporator. Now a uniformly dispersed blend of PDMS binder and MWCNTs was obtained; to which the particulate was added. The rest of the procedure follows the same steps as in the neat material systems i.e. the specimens without MWCNTs. Both compression and tension ASTM standard specimen are pictured in Figs. 21.1 and 21.2, respectively. Note that in the images, electrical tape secured the wires while the electrodes cured.

21.2.3 *Test Apparatus and Procedure*

Compressive and tensile loading on the specimen were performed using an ADMET MTESTQuattro Universal Testing Machine using prescribed procedure as outlined in ASTM D638 for tension and ASTM D695 for compression. As additional assurance from the results of previous studies on composites with a PMDS binder [15], a prescribed strain rate of 5 millimeters per minute was selected for this experiment in order to develop data with substantial regions of strain, micro-fracture, and macro-fracture. The apparatus' software gathered load and extension data for compression and tension tests. This data was processed with the dimensions of each specimen to obtain stress and strain data for every specimen which was then averaged for every material system.

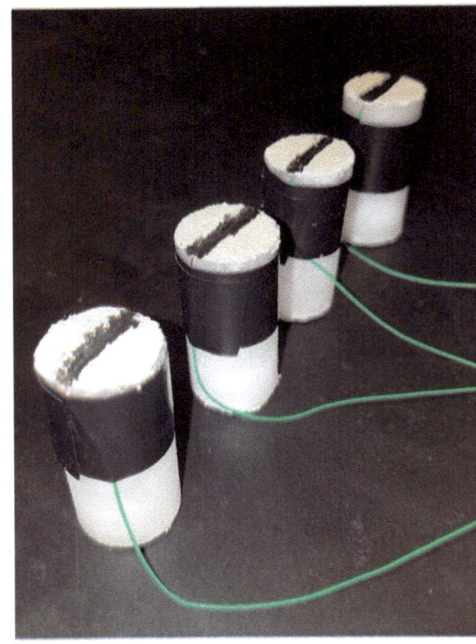

Fig. 21.1 Image of compression Neat Sugar PDMS specimen with top/bottom electrode placement

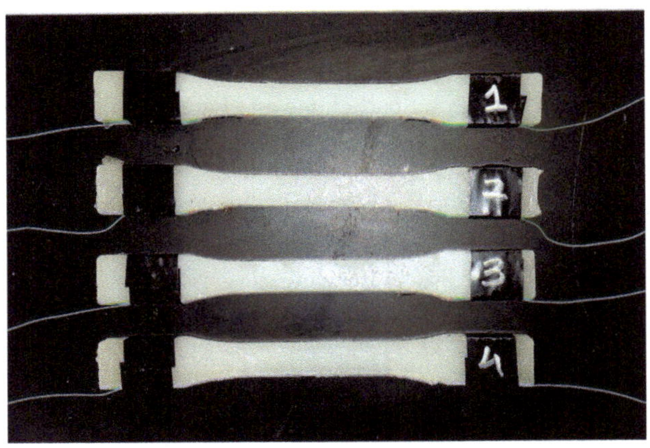

Fig. 21.2 Image of tension Neat Sugar PDMS specimen with side electrode placement

For electrical characterization of the test specimens, two electrodes were attached to every specimen and then the static resistance and admittance data was collected using an LCR meter (Agilent Technologies A4980A Precision Impedance, Capacitance, and Resistance meter). A layer of silver epoxy (MG Silver Conductive Epoxy – 14 g) was coated over the specimens at the tentative location of electrode. This was done as the silver epoxy, due to high conductivity and paste form, minimizes the contact resistance and micro air voids in the application of the coating. After the layer cures, a conductive paint (Bare Conductive Paint Pen – 10 mL) was used to glue copper wires to the silver epoxy layer. The LCR meter collected resistance and admittance data for every specimen for 1 min to enable identification of the difference between static resistance and resistance change due to strain sensing. An important caution necessary to get accurate strain sensing is to eliminate or minimize the frequency dependent resistance change of a test specimen. In this study, the frequency dependence arises due to the very low conductivity of the sugar crystals and PDMS. Following the observations by Sengezer and Seidel [4], a frequency of 10 kHz was found to be the most ideal frequency for LCR meter data acquisition of our test specimens; there was significant noise in the data acquired at low and very high frequencies for neat systems and so the range of 10 kHz gives the most stable results. The frequency dependence of CNT based specimens is not affected to that extent as their conductivity is higher than the neat specimens.

21.3 Results and Discussion

21.3.1 Visual Fracture Analysis

Compressive samples are shown in Fig. 21.3 after testing was completed. Following an ideal failure mode, all samples were shown to fracture along a 45° shear plane. The plane shows where maximum slippage has occurred between the crystals and the material, and is shown in Eq. 21.1 (Fig. 21.3):

$$\tau_R = \sigma \cos(\lambda) \cos(\phi) \tag{21.1}$$

where the resolved shear stress is a maximum at $\lambda = \phi = 45°$. The testing demonstrated that the samples fractured at or around maximum shear stress and electro-mechanical properties could be tested adequately. This also implies that the specimens demonstrated sensing abilities at their maximum stress bearing capacity. The stress-strain curve for the different material systems under mechanical compression is shown in Fig. 21.5.

Tensile specimens followed the elastomer stress-strain pattern i.e. they have large extensions but do not yield noticeably, and rather stay in the elastic region till fracture shown in Fig. 21.6. As necking begins to occur during testing, the coalescence of micro voids initiated the formation a crack. The crack is propagated through shear deformation until ultimate failure.

Fig. 21.3 Image of compression specimen for each material system – post failure. From left to right: Sugar PDMS, Sugar Aluminum PDMS, Sugar CNT PDMS, Sugar Aluminum CNT PDMS

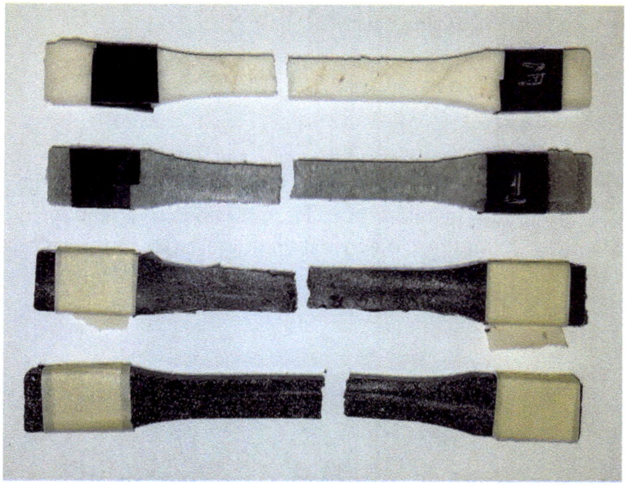

Fig. 21.4 Image of tension specimen for each material system – post failure. From top to bottom: Sugar PDMS, Sugar Aluminum PDMS, Sugar CNT PDMS, Sugar Aluminum CNT PDMS

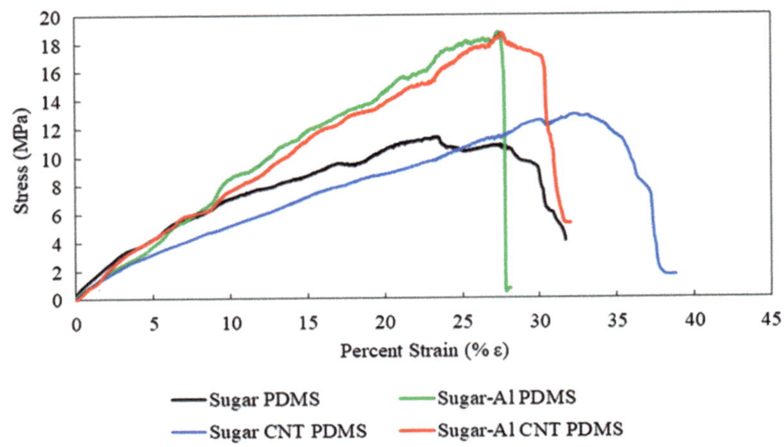

Fig. 21.5 Average stress-strain behavior for all material systems: compression

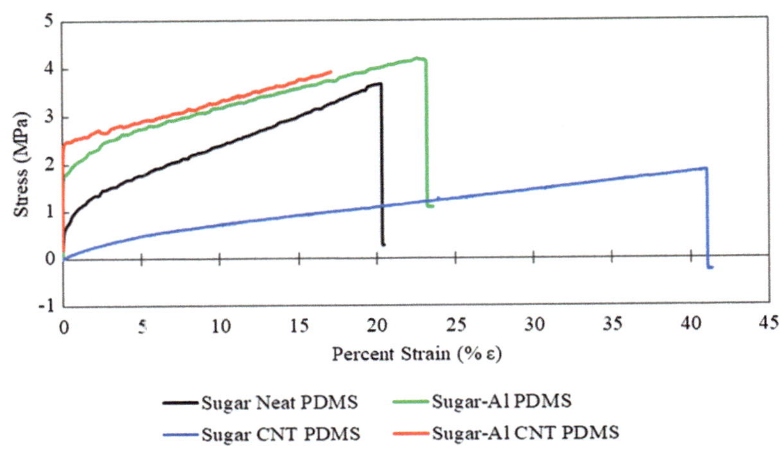

Fig. 21.6 Average stress-strain behavior for all material systems: tension

21.3.2 Piezoresistive Sensing

Coupled with the visible demonstration of strain and irreversible failure in all four material systems, analysis was performed on the real-time electrical characterization data of resistance (R) and capacitive reactance (X), gathered by the LCR meter. The normalized impedance was calculated from these electrical characteristics in the following manner:

$$Z^2 = R^2 + X^2 \tag{21.2}$$

$$\frac{\Delta Z}{Z_i} = \frac{Z - Z_i}{Z_i} \tag{21.3}$$

Normalized impedance change is a property which captures the counteracting changes of resistance and reactance, and removes any dependency on static electrical properties, bringing the results to a rational state of comparison between all material systems. In this experimental effort, the use of impedance for empirical comparison still delivers dominance to the characteristic with larger magnitude, capacitive reactance. Capacitive reactance has a negative and inverse relationship with capacitance, of the form:

$$X_C = -(\omega C)^{-1} = -(2\pi f C)^{-1} \tag{21.4}$$

where X_C is capacitive resistance, f is frequency, and C is capacitance. Therefore, a normalized impedance driven dominantly by capacitive reactance would behave oppositely and inversely to the capacitive state of the material system. Piezoresistive sensing comes from a material system with conductive components experiencing damage such that it tends to a less positive capacitive state, which translates to an increasingly negative normalized impedance change. These material systems act similarly to a parallel electrode capacitor: as the plates get closer, the capacitance increases, but once the dielectric medium has changed (due to damage), it becomes more dielectric, causing the capacitance to increase.

The range of normalized impedance change was averaged between the four specimens tested for each one of the four material systems, and is related to its corresponding percent strain, as shown in Figs. 21.7 and 21.8.

Neat Sugar PDMS showed negligible sensing of impedance change in tension and minimal sensing in compression, excluding that which would be experienced from the reduction of distance between the electrodes during compression. This insulative material system functions as a baseline, for which there are minimal free electrons available to allow for any substantial change in capacitive state.

The addition of conductive aluminum powder into the composite particulate increased the magnitude of sensing as compared to its entirely sugar counterpart; just before specimen failure, Sugar Al PDMS had a noticeably linear curve of impedance change. Minimal changes in the linearity of the impedance change for either tensile or compressive loading is indicative that, while the specimens were sensing the macro-scale damage and separation of the composite, it was not able to sense the stages of micro-scale damage, which would be observed as impedance regimes of different slopes.

Following suit, MWCNTs were added at 0.5 wt % to the Sugar PDMS composite alone, and sensing was further notable in both tension and compression. In Fig. 21.8, the tension results indicate that the capacitive state of the material system experienced minimal change until reaching a 3% strain, at which point the impedance change decreases at a faster rate,

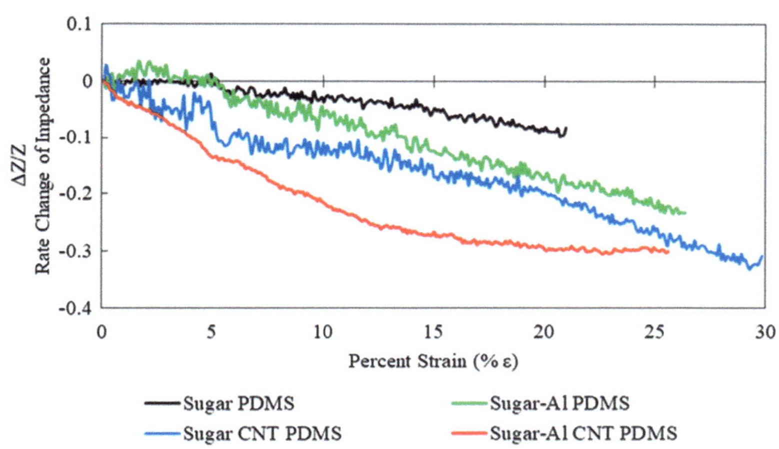

Fig. 21.7 Average change in normalized impedance with percent applied strain for different material systems: compression

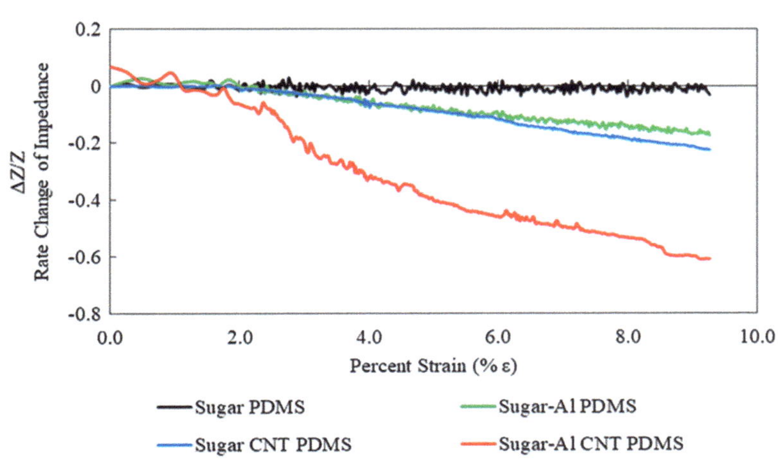

Fig. 21.8 Average change in normalized impedance with percent applied strain for different material systems: tension

before settling to a smaller slope starting at 9% strain until specimen failure. For compression, the results indicate a quick change in impedance as soon as strain is applied which is indicative of strain sensing abilities being imparted by the CNTs to the specimen. Thereafter, a linear decline of impedance can be seen; note that on comparison with the neat sugar specimen results, it was observed that the slope is much higher with CNT addition to the binder. Thus, this linear decline cannot be attributed to the decrease in distance of the electrodes but rather to the piezoresistive sensing of CNTs.

21.4 Conclusion

The primary purpose of this experimental study was to observe the strain sensing effects of MWCNTs embedded in Polymer Bonded Explosives (PBXs). In an effort to understand the impact of MWCNTs on piezoresistive properties for such composites, four different Sugar-PDMS based material systems were tested: two with MWCNTs and two without. Tension and compression test specimens of each material system were tested under pure mechanical loading at standard room temperature and pressure. For both tension and compression, the capacitive reactance dominates over the resistance and thus the impedance negatively decreases as strain increases. A very common trend is observed in the test results: is that the material system with Sugar-Al and MWCNTs shows the highest strain sensing with respect to the other material systems. But in general, the strain sensing response of material systems for both tension and compression follows as (from highest to lowest): Sugar-Al CNT > Sugar CNT > Sugar-Al > Sugar (where all the material systems have PDMS as the binder). The material systems without MWCNTs have minimal change in the slope of impedance variation, which indicates their lack of strain sensitivity to microscale damage. However, the MWCNT based systems have a non-linear response to increasing strain, showing sharp slope changes which indicate the strain sensitivity of the MWCNT network dispersed in the composite. This experimentation, thus demonstrates the electro-mechanical characterization of mock PBXs in order to provide proof of concept for strain sensing abilities of PBXs embedded with CNTs under mechanical loading to facilitate real time structural health monitoring.

Acknowledgements The authors would like to acknowledge the support of Air Force Office of Scientific Research (grant FA9550-16-1-0198) in the Dynamic Materials and Interactions Program. The authors would like to acknowledge Dr. Michael Philen for providing an access to use the ADMET machine in the Aerospace Structures and Materials Lab at Virginia Tech.

References

1. Siviour, C., Laity, P., Proud, W., Field, J., Porter, D., Church, P., Gould, P., Huntingdon-Thresher, W.: High strain rate properties of a polymer-bonded sugar: their dependence on applied and internal constraints. In: Proceedings of the Royal Society of London A: Mathematical, Physical and Engineering Sciences, No. 464, pp. 1229–1255 (2008)
2. Balzer, J.E., Siviour, C.R., Walley, S.M., Proud, W.G., Field, J.E.: Behaviour of ammonium perchlorate–based propellants and a polymer–bonded explosive under impact loading. Proc. Math. Phys. Eng. Sci. **460**(2043), 781–806 (2004). https://doi.org/10.1098/rspa.2003.1188
3. Prakash, N., Seidel, G.D.: Effects of microscale damage evolution on piezoresistive sensing in nanocomposite bonded explosives under dynamic loading via electromechanical peridynamics. Model. Simul. Mater. Sci. Eng. **26**(1), 015003 (2018). http://stacks.iop.org/0965-0393/26/i=1/a=015003
4. Sengezer, E., Seidel, G.: Structural health monitoring of nanocomposite bonded energetic materials through piezoresistive response. AIAA J. **56**, 1–14 (2017)
5. Prakash, N., Seidel, G.D.: Computational electromechanical peridynamics modeling of strain and damage sensing in nanocomposite bonded explosive materials (NCBX). Eng. Fract. Mech. **177**, 180–202 (2017). https://doi.org/10.1016/j.engfracmech.2017.04.003, http://www.sciencedirect.com/science/article/pii/S0013794416307007
6. Rocker, S., Wade Pearrell, T., Sengezer, E., Seidel, G.: Thermo-electromechanical response of polymer-bonded explosives for structural health monitoring of energetic materials, p. V001T01A009 (2017)
7. Chaurasia, A.K., Sengezer, E.C., Talamadupula, K.K., Povolny, S., Seidel, G.D.: Experimental characterization and computational modeling of deformation and damage sensing through the piezoresistive response of nanocomposite bonded surrogate energetic materials. J. Multifunct. Compos. **2**, 227–253 (2014)
8. Talamadupula, K.K., Chaurasia, A.K., Seidel, G.D.: Multiscale Modeling of Effective Piezoresistivity and Damage Response in Nanocomposite Bonded Explosives (2017)
9. Prakash, N., Seidel, G.: Coupled electromechanical peridynamic modeling of strain and damage sensing in granular energetic materials. In: Conference: 58th AIAA/ASCE/AHS/ASC Structures, Structural Dynamics, and Materials Conference (2017). https://doi.org/10.2514/6.2017-0126
10. Sengezer, E.C., Seidel, G.D.: Real time in-situ sensing of damage evolution in nanocomposite bonded surrogate energetic materials. Proc. SPIE **9800**, 9800–9810 (2016). https://doi.org/10.1117/12.2218856
11. Talamadupula, K.K., Chaurasia, A.K., Seidel, G.D.: 2-Scale hierarchical multiscale modeling of piezoresistive and damage response in polymer nanocomposite bonded explosive. In: ASME 2016 Conference on Smart Materials, Adaptive Structures and Intelligent Systems (2016). https://doi.org/10.1115/SMASIS2016-9234

12. Ku-Herrera, J., Avilés, F.: Cyclic tension and compression piezoresistivity of carbon nanotube/vinyl ester composites in the elastic and plastic regimes. Carbon **50**(7), 2592–2598 (2012) https://doi.org/10.1016/j.carbon.2012.02.018, http://www.sciencedirect.com/science/article/pii/S000862231200139X
13. Ando, M., Noda, N.-A., Kuroshima, Y., Ishikawa, Y., Takeda, H.: Impact properties of polydimethylsiloxane copolymerized polycarbonate and application of the time-temperature superposition principle. Trans. JSME **80**, SMM0149 (2014)
14. Vadlamani, V.K., Chalivendra, V.B., Shukla, A., Yang, S.: Sensing of damage in carbon nanotubes and carbon blackembedded epoxy under tensile loading. Polym. Compos. **33**(10), 1809–1815 (2012). https://doi.org/10.1002/pc.22326
15. Khanafer, K., Duprey, A., Schlicht, M., Berguer, R.: Effects of strain rate, mixing ratio, and stress-strain definition on the mechanical behavior of the polydimethylsiloxane (PDMS) material as related to its biological applications. Biomed. Microdevices **11**, 503–508 (2009)

Chapter 22
Strength and Energy Absorption Capability of Porous Magnesium Composites Reinforced by Carbon Nanofibers

Huiru Xu and Qizhen Li

Abstract Porous magnesium-carbon nanofiber (CNF) composites were manufactured to investigate the variation of compressive mechanical properties with the change of porosity and CNF concentration. When the CNF concentration changed from 0% to 2%, the average yield strength significantly increased for the porous composites with the porosity of either 24%, 34%, or 50%. The yield strength and the ultimate compressive strength decreased at an increasing rate with the increase of porosity. For each studied porosity, the addition of CNF to porous magnesium composite samples increased energy absorption capability when the samples underwent any given strain level. Four theoretical strengthening models were utilized to estimate yield strength of the studied porous composites and the results indicated that the shear lag model and the rule of mixture model provided better yield strength estimations than the Strengthening factor model and the Zhang & Chen model.

Keywords Porous magnesium · Composites · Strength · Energy absorption capability · Mechanical properties

22.1 Introduction

Although the density of magnesium is low (i.e. 1.738 g/cm^3), the addition of pores into magnesium forms porous magnesium and can further reduce the density. On one hand, the strength of porous magnesium were usually deteriorated from that of dense magnesium and some reinforcing phases such as alumina, CNF, and carbon nanotube (CNT) were employed to improve the strength of porous magnesium composites [1–9]. On the other hand, porous magnesium is expected to possess high energy absorption capability. Therefore, it is interesting to investigate the effect of the reinforcing phase on energy absorption capability and strength of porous magnesium composites with various porosities. Although the reference [2] reported the enhancement of mechanical properties of porous magnesium composites with a variety of CNF concentrations, this reference focused on one porosity and a wide range of CNF concentrations. It is critical to explore the variation of strength and energy absorption capability with the change of porosity. This study manufactured porous magnesium composites with a range of porosities and CNF concentrations, and investigated their energy absorption capability and strength. This study also explored the various theoretical models for yield strength prediction of composites, i.e. Shear Lag model [10, 11], Rule of mixture model [6, 12], Strengthening factor model [6, 13, 14], and Zhang & Chen model [15].

22.2 Experimental Methods

Porous magnesium composites were manufactured through powder metallurgy. The raw materials are 99.8% pure magnesium powders ($\sim$ 40 μm) serving as the matrix, 95% pure carbon nanofibers with a diameter of 200 $\sim$ 600 nm and a length of 5 $\sim$ 50 μm serving as the reinforcing phase, and 99.3% pure urea serving as the space holder. The manufacturing process is the same as that utilized in the references [1–5, 7], i.e. the mixing-compacting-sintering three-step process. Three concentrations (i.e. 0%, 0.2%, and 2%) of carbon nanofiber were obtained. Three different porosities (i.e. 24%, 34%, and 50%) were

H. Xu · Q. Li (✉)
School of Mechanical and Materials Engineering, Washington State University, Pullman, WA, USA
e-mail: qizhen.li@wsu.edu

P. R. Thakre et al. (eds.), *Mechanics of Composite, Hybrid and Multifunctional Materials, Volume 5*, Conference Proceedings of the Society for Experimental Mechanics Series, https://doi.org/10.1007/978-3-319-95510-0_22

achieved for each carbon nanofiber concentration through varying the weight fraction of urea in the magnesium-CNF-urea mixtures. The porous magnesium composites were compressively tested at a strain rate of $\sim 10^{-3}$/s and at room temperature. For each type of porous magnesium composites, three repeats of compressive tests were conducted. The stress and strain data were obtained and analyzed to investigate the energy absorption capability and strength of the studied porous magnesium composites.

22.3 Results and Discussion

Table 22.1 reports yield strength (σ_y), ultimate compressive strength (σ_{ucs}), strain at ultimate compressive strength (ε_{ucs}), and energy absorption (E_{abs}) of the porous magnesium composites with various CNF concentrations (i.e. 0%, 0.2%, and 2%) and different porosities (i.e. P = 24%, 34%, and 50%) respectively. Energy absorption capability can be calculated through the following integration,

$$E_{abs} = \int_0^{\varepsilon} \sigma d\varepsilon \tag{22.1}$$

Where σ and ε are the compressive stress and the compressive strain respectively that were experienced by a sample. Figure 22.1 reports the calculated energy absorption history with the increase of experienced deformation/strain for each studied sample, showing that energy absorption increased with the increase of deformation/strain. By setting the integration upper limit in Eq. (22.1) as ε_{ucs}, the corresponding energy absorption was obtained and reported in Table 22.1 and Fig. 22.2. The overall trend is that the increase of porosity reduced the energy absorption capability of the samples. Among the studied CNF concentrations, the samples with 0.2% CNF showed higher energy absorption capability than the samples with the other two CNF concentrations (i.e. 0% and 2%) for each of the studied porosities.

The integration limit for Eq. (22.1) was also set to be 0.05, 0.10, and 0.15 to obtain the energy absorption at different strain states for the samples and the data were in Table 22.2 and Fig. 22.3. Similarly to the energy absorption at ε_{ucs}, the samples with 0.2% CNF showed higher energy absorption capability than the samples with either 0% or 2% CNF for each of the studied porosities (i.e. 24%, 34%, and 50%) and each strain level (i.e. 0.05, 0.10, and 0.15). Table 22.2 also listed the energy absorption for the Zn-22Al alloy foam with 60% porosity, and the data were 0.9, 1.7, and 3.0 MJ/m^3 for the strain level of 0.05, 0.10, and 0.15 respectively [16]. These data are close to the data (0.82, 1.70, and 2.61 MJ/m^3 for the strain level of 0.05, 0.10, and 0.15 respectively) for our porous magnesium with 0.2% CNF and 50% porosity. The density of Zn-22Al alloy foam with 60% porosity should be about 2.47 g/cm^3, while our porous magnesium with 0.2% CNF and 50% porosity has a density of about 0.87 g/cm^3. Therefore, the porous magnesium composite processed in this study has higher specific energy absorption capability than the Zn-22Al alloy foam with 60% porosity.

Table 22.1 Yield strength (σ_y), ultimate compressive strength (σ_{ucs}), strain at ultimate compressive strength (ε_{ucs}), and energy absorption (E_{abs}) at ε_{ucs} for the porous magnesium composites with various CNF concentrations (i.e. 0%, 0.2%, and 2%) and different porosities (i.e. P = 24%, 34%, and 50%) respectively. (Note: the CNF concentrations are in terms of weight percentage. P refers to the porosity of the studied composites)

CNF (%)	P (%)	σ_y (MPa)	σ_{ucs} (MPa)	ε_{ucs}	E_{abs} (MJ/m^3)
0	24	62 ± 4	73 ± 8	0.059 ± 0.021	2.99
	34	48 ± 7	63 ± 4	0.049 ± 0.009	2.19
	50	13 ± 3	15 ± 2	0.047 ± 0.022	0.33
0.2	24	85 ± 6	108 ± 5	0.076 ± 0.006	5.29
	34	62 ± 7	72 ± 2	0.051 ± 0.005	2.22
	50	20 ± 2	21 ± 2	0.044 ± 0.024	0.70
2	24	85 ± 5	104 ± 5	0.070 ± 0.013	4.62
	34	37 ± 9	51 ± 7	0.054 ± 0.010	1.88
	50	16 ± 2	20 ± 4	0.050 ± 0.017	0.58

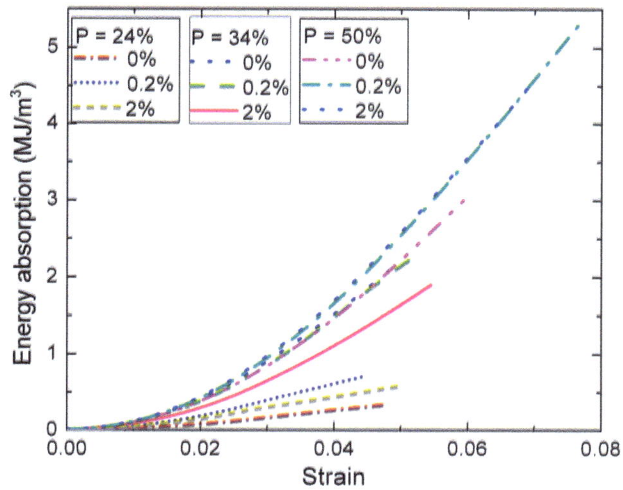

Fig. 22.1 Energy absorption of the porous magnesium composites with various CNF concentrations (i.e. 0%, 0.2%, and 2%) and different porosities (i.e. P = 24%, 34%, and 50%) respectively. (Note: the CNF concentrations are in terms of weight percentage)

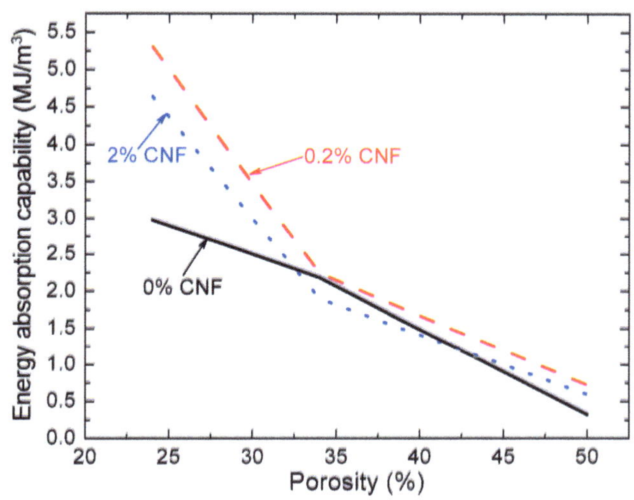

Fig. 22.2 Energy absorption at ε_{ucs} of the porous magnesium composites with various CNF concentrations (i.e. 0%, 0.2%, and 2%) and different porosities (i.e. P = 24%, 34%, and 50%) respectively. (Note: the CNF concentrations are in terms of weight percentage)

Four different theoretical models (i.e. Rule of mixture model σ_y^{FOM}, Shear Lag model σ_y^{S-L}, Strengthening factor model σ_y^{STR}, and Zhang & Chen model σ_y^{Z-C}) were utilized to predict yield strength of these studied porous magnesium composites in this work. The details about the mathematical expressions and the choices of the theoretical computation parameters for these four models are in the reference [2]. The predicted yield strengths were presented in Table 22.3 and Figs. 22.4, 22.5, and 22.6. When there is 0% CNF, the predictions from all four theoretical models overlapped with the experimental data. Zhang-Chen model and strengthening factor model consistently gave a higher strength value than the experimental strength for each type of the studied porous magnesium composites. For the composites with 0.2% CNF, the predicted strength using the shear lag model matched well with the experimental value for all the studied porosities. For the composites with 2% CNF, the shear lag model provided the best strength predictions when the porosity is higher than ~35%, while the rule of mixture model provided the best strength predictions when the porosity is lower than ~35%.

Table 22.2 Energy absorption (E_{abs}) at $\varepsilon = 0.05$, 0.10, and 0.15 for the porous magnesium composites with various CNF concentrations (i.e. 0%, 0.2%, and 2%) and different porosities (i.e. P = 24%, 34%, and 50%) respectively. (Note: the CNF concentrations are in terms of weight percentage. P refers to the porosity of the studied composites)

CNF (%)	P (%)	E_{abs} at $\varepsilon = 0.05$ (MJ/m^3)	E_{abs} at $\varepsilon = 0.10$ (MJ/m^3)	E_{abs} at $\varepsilon = 0.15$ (MJ/m^3)
0	24	2.21	5.13	7.67
	34	2.20	4.56	6.90
	50	0.35	0.89	1.29
0.2	24	2.53	7.96	11.38
	34	2.14	4.83	7.77
	50	0.82	1.70	2.61
2	24	2.58	6.46	9.51
	34	1.63	4.19	6.14
	50	0.57	1.26	1.86
ZA22 foam [16]	60	0.9	1.7	3.0

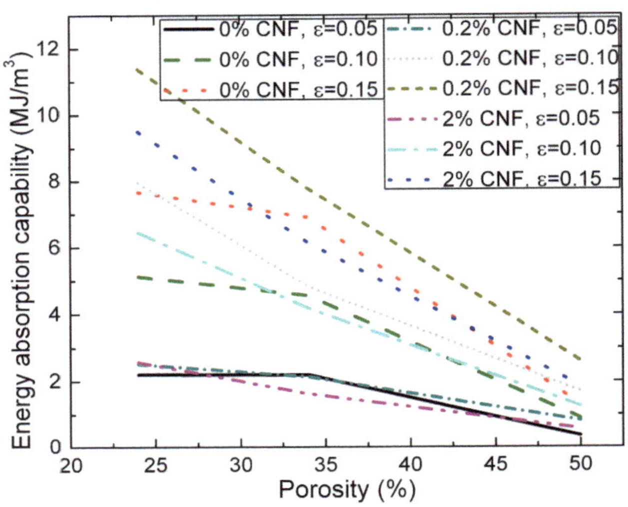

Fig. 22.3 Energy absorption at $\varepsilon = 0.05$, 0.10, and 0.15 of the porous magnesium composites with various CNF concentrations (i.e. 0%, 0.2%, and 2%) and different porosities (i.e. P = 24%, 34%, and 50%) respectively. (Note: the CNF concentrations are in terms of weight percentage)

Table 22.3 Experimental yield strength (σ_y) and theoretical yield strength (σ_y^{ROM} from Rule of mixture model, σ_y^{S-L} from Shear Lag model, σ_y^{STR} from Strengthening factor model, and σ_y^{Z-C} from Zhang & Chen model) of the porous magnesium composites with various CNF concentrations (i.e. 0%, 0.2%, and 2%) and different porosities (i.e. P = 24%, 34%, and 50%) respectively

CNF (%)	P (%)	σ_y (MPa)	σ_y^{ROM}(MPa)	σ_y^{S-L}(MPa)	σ_y^{STR}(MPa)	σ_y^{Z-C}(MPa)
0	24	62 ± 4	62	62	62	62
	34	48 ± 7	48	48	48	48
	50	13 ± 3	13	13	13	13
0.2	24	85 ± 6	67	63	77	70
	34	62 ± 7	53	49	63	62
	50	20 ± 2	18	13	27	27
2	24	85 ± 5	69	73	89	87
	34	37 ± 9	55	57	75	74
	50	16 ± 2	21	15	40	40

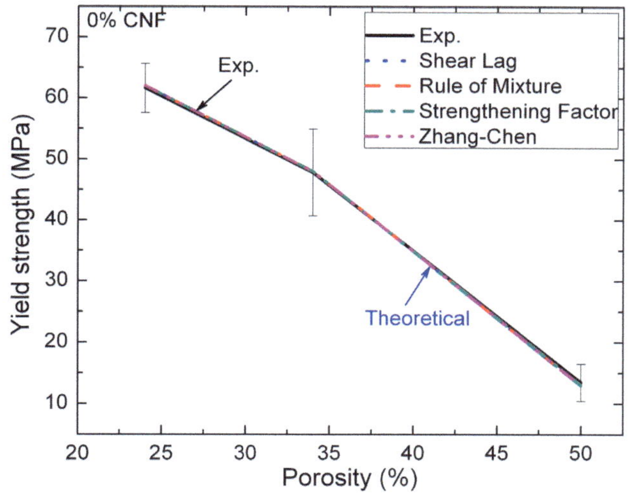

Fig. 22.4 Experimental and theoretical yield strength (from Rule of mixture model, Shear Lag model, Strengthening factor model, and Zhang & Chen model) of the porous magnesium composites with the CNF concentration of 0%

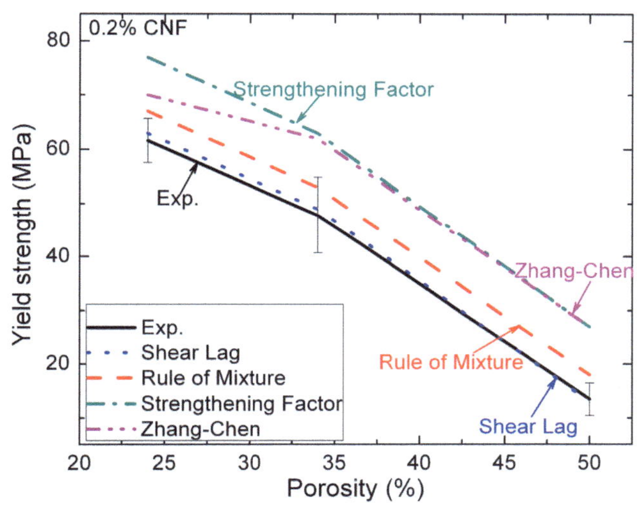

Fig. 22.5 Experimental and theoretical yield strength (from Rule of mixture model, Shear Lag model, Strengthening factor model, and Zhang & Chen model) of the porous magnesium composites with the CNF concentration of 0.2%

22.4 Conclusion

Porous CNF reinforced magnesium composites were manufactured through a powder metallurgical approach. The manufactured samples had the CNF concentrations of 0%, 0.2%, and 2%, while the porosity of these samples was either 24%, 34% or 50%. The energy absorption data showed that the porous composites with 0.2% CNF had the highest absorption capability for each studied porosity. The increase of porosity resulted in the decrease of the energy absorption capability for each studied CNF concentration. The investigation of the strengthening models showed that the shear lag model provided the best yield strength predictions for the porous composites with 0.2% CNF, while both the rule of mixture model and the shear lag model are good models for estimating the yield strength for the porous composites with 2% CNF.

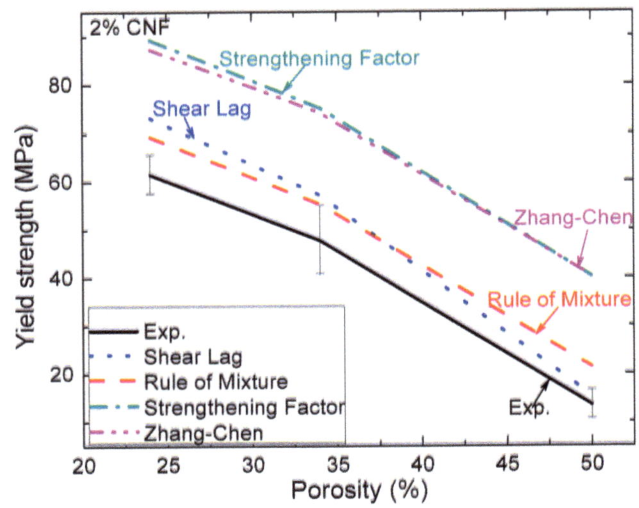

Fig. 22.6 Experimental and theoretical yield strength (from Rule of mixture model, Shear Lag model, Strengthening factor model, and Zhang & Chen model) of the porous magnesium composites with the CNF concentration of 2%

Acknowledgement The support for the research from the National Science Foundation under Award No. 1449607 is greatly appreciated.

References

1. Xu, H., Zou, N., Li, Q.: Effect of ball milling time on microstructure and hardness of porous magnesium/carbon nanofiber composites. JOM. **69**(7), 1236–1243 (2017)
2. Xu, H., Li, Q.: Effect of carbon nanofiber concentration on mechanical properties of porous magnesium composites: experimental and theoretical analysis. Mater. Sci. Eng. A. **706**(Supplement C), 249–255 (2017)
3. Li, Q.: Effect of porosity and carbon composition on pore microstructure of magnesium/carbon nanotube composite foams. Mater. Des. **89**, 978–987 (2016)
4. Zou, N., Li, Q.: Compressive mechanical property of porous magnesium composites reinforced by carbon nanotubes. J. Mater. Sci. **51**(11), 5232–5239 (2016)
5. Cay, H., Xu, H., Li, Q.: Mechanical behavior of porous magnesium/alumina composites with high strength and low density. Mater. Sci. Eng. A. **574**(Supplement C), 137–142 (2013)
6. Li, Q., Tian, B.: Compression behavior of magnesium/carbon nanotube composites. J. Mater. Res. **28**(14), 1877–1884 (2013)
7. Li, Q.: Carbon nanotube reinforced porous magnesium composite: 3D nondestructive microstructure characterization using x-ray micro-computed tomography. Mater. Lett. **133**, 83–86 (2014)
8. Zou, N., Li, Q.: Mechanical properties of lightweight porous magnesium processed through powder metallurgy. JOM. **70**, 650–655 (2018)
9. Xu, H., Li, Q.: Deformation mechanisms and mechanical properties of porous magnesium/carbon nanofiber composites with different porosities. J. Mater. Sci. **53**(20), 14375–14385 (2018)
10. Ryu, H.J., Cha, S.I., Hong, S.H.: Generalized shear-lag model for load transfer in SiC/Al metal-matrix composites. J. Mater. Res. **18**(12), 2851–2858 (2011)
11. Fougere, G.E., Riester, L., Ferber, M., Weertman, J.R., Siegel, R.W.: Young's modulus of nanocrystalline Fe measured by nanoindentation. Mater. Sci. Eng. A. **204**(1), 1–6 (1995)
12. Hertzberg, R.W.: Deformation and Fracture Mechanics of Engineering Materials. John Wiley and Sons, New York (1989)
13. Dieter, G.E., Bacon, D.J.: Mechanical Metallurgy. McGraw-Hill, New York (1986)
14. Arsenault, R.J., Shi, N.: Dislocation generation due to differences between the coefficients of thermal expansion. Mater. Sci. Eng. **81**, 175–187 (1986)
15. Zhang, Z., Chen, D.L.: Consideration of Orowan strengthening effect in particulate-reinforced metal matrix nanocomposites: a model for predicting their yield strength. Scr. Mater. **54**(7), 1321–1326 (2006)
16. Yu, S., Liu, J., Wei, M., Luo, Y., Zhu, X., Liu, Y.: Compressive property and energy absorption characteristic of open-cell ZA22 foams. Mater. Des. **30**(1), 87–90 (2009)

Chapter 23
Mechanical Characterization of Open Cell Aluminum Foams Using X-ray Computed Tomography

Kristoffer E. Matheson and Michael W. Czabaj

Abstract Open-cell aluminum foams show excellent potential for use in a variety of applications. In order to accelerate the use of aluminum foams in engineering industry, it is important to accurately understand the relationship between the manufacturing processes and the resulting foam properties. Two tests are developed to support development of finite element (FE) models that will be used to design foam geometries optimized to meet specific design criteria. A bulk crush test is created for small foam blocks which uses in situ X-ray computed tomography (CT) imaging. A tensile experiment is developed to test individual foam ligaments, which measures the aluminum's mechanical behavior and will be used to improve the simulation's accuracy beyond what is possible using published bulk material properties of aluminum.

Keywords Aluminum foam · Computed tomography · Cellular materials · Mechanical characterization · Open cell

23.1 Introduction

To date, much work has been done to characterize the compressive response of metallic foams [1–6]. However, the stochastic structure of these foams causes behavior to vary between nominally identical foams [7]. This makes it difficult to predict, for example, a foam's precise stress/strain behavior under compressive loading.

A project is underway to develop accurate finite element (FE) models that will be used to design open cell foam configurations tailored to specific customer applications. Accurate material properties are required for such a model to produce correct results, and published mechanical properties of bulk aluminum have been found to not be representative of the aluminum in the foam [8]. Work has been done to accurately characterize the aluminum within the foam [2, 4, 6, 8]. Zhou has used a tension test to measure the strength of individual ligaments [8]. A similar tensile test is created here that will measure the nonlinear/plastic response of actual aluminum foam ligaments to provide material input data for the simulation.

A crush test is created to characterize the foam structure in bulk to validate the FE modeling approach. In this bulk crush test, the compressive response of a block of open-cell foam is captured by measuring the force vs. displacement performance as well as collecting sequential 3D images taken in situ using X-ray computed tomography (CT) at progressive compression increments. In conjunction with this compression test framework, an FE model is produced by deriving a mesh from an X-ray CT image of the pristine uncrushed foam. Boundary conditions that simulate the compression test are applied and material properties obtained from the ligament tension test are used. The FE model can be validated by comparing both the force-displacement curve from the experiment and the order and manner of ligament failure within the foam using the 3D X-ray CT images.

23.2 Experimental Methods

Specimens for compression testing are extracted from larger blocks of foam using electrical discharge machining (EDM). All specimens have a diameter of 10 mm and height ranging between 10 and 20 mm. In this particular work, a 6061-T6 alloy foam with 10.5% relative density, 2 pores-per-cm cell density, and an initial height of 18.3 mm is used. The compression tests are performed using a custom in-house-built load frame. A linear stepper motor with a resolution of 0.0079 mm/step is used for actuation, a built-in encoder provides displacement data, and a 1.3 kN load cell provides force measurements.

K. E. Matheson · M. W. Czabaj (✉)
Department of Mechanical Engineering, University of Utah, Salt Lake City, UT, USA
e-mail: m.czabaj@utah.edu

P. R. Thakre et al. (eds.), *Mechanics of Composite, Hybrid and Multifunctional Materials, Volume 5*, Conference Proceedings of the Society for Experimental Mechanics Series, https://doi.org/10.1007/978-3-319-95510-0_23

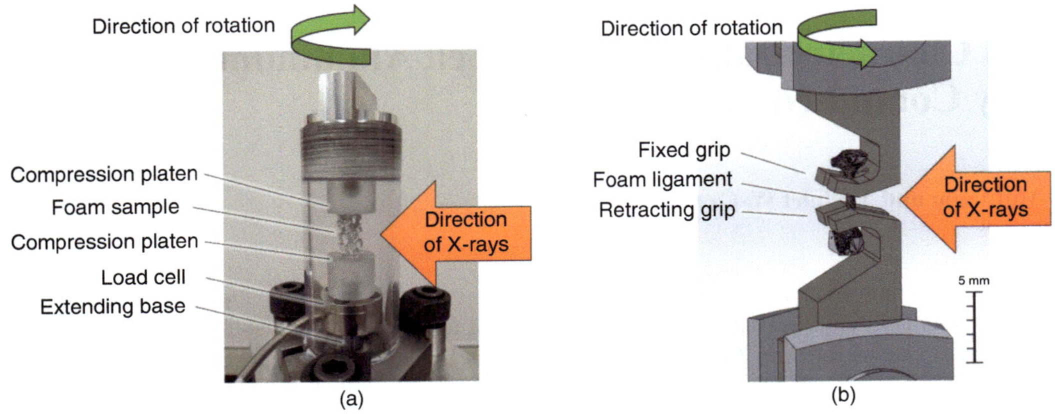

Fig. 23.1 (**a**) Compression test setup, showing transparency of load frame and how foam sample is held in place (**b**) Close view of ligament tension test setup

In addition to being compact, the load frame is transparent on all sides, allowing it to rotate a full 360° without obstructing the specimen when placed inside an X-ray CT system. A picture of the load frame prior to a compression test is shown in Fig. 23.1a. Prior to testing, each pristine specimen is imaged at 13 microns/voxel to obtain a baseline X-ray CT image. The sample is then quasi-statically compressed at a strain rate of 3.5×10^{-3} s^{-1}. During compression, force and displacement data are continuously collected at a rate of 20 samples/second. After achieving a global strain between 10–12%, the test is paused, and without unloading the specimen, another X-ray CT scan is performed. This process is repeated in approximately 10%–12% strain increments until the sample is fully densified or a maximum force of 600 N is reached.

The ligament tension test is performed in a manner similar to the compression test outlined above. Ligaments are extracted from foam blocks using wire EDM. The compression platens are replaced with custom stainless steel grips shown in Fig. 23.1b. For all tests the ligament displacement is measured using the stepper motor's built-in encoder, and tensile force is measured using a 111 N load cell. Once an X-ray CT scan of the pristine ligament is collected (10 microns/voxel), small increments of tension are quasi-statically applied at a constant strain rate of 10^{-2} s^{-1}. Force and displacement data are continuously collected at 30 samples/second. X-ray CT scans are performed at regular strain intervals of 25%, until the sample has fully ruptured.

23.3 Results and Discussion

An example set of reconstructed X-ray CT images at various stages of compression are shown in Fig. 23.2a. In this figure, ligaments that underwent excessive deformation are identified and are highlighted using different colors. At each increment, ligaments that experienced plastic collapse are colored blue, while ligaments that experienced brittle fracture are colored red. Ligaments that remained mostly unchanged are left silver. The force vs. displacement curve for the compression test is shown in Fig. 23.3a. Stress and strain for the homogenized bulk material are calculated based on the initial dimensions of the pristine foam. The set of CT images and the force-displacement data are compared to the results of the FE simulation described above to validate the simulation results.

An example set of X-ray CT images obtained during tensile testing of a single ligament is shown in Fig. 23.2b. Due to density of the stainless steel grips, it was not possible to use X-rays to image the portions of the ligament that were obscured by the grips (see Fig. 23.1b). Note that necking can be observed in the ligament prior to rupture. The force versus displacement curve for a ligament tension test is shown in Fig. 23.3b. A cross-sectional area of the ligament at the point of failure can be measured in the pristine X-ray CT image to calculate engineering stress, or the cross-sectional area in each successive X-ray CT image can be measured to calculate true stress. A first-order estimate of strain can be generated based on the relative displacement of the ends of the gauge region. Using this approach, stress and strain are plotted alongside force and displacement in Fig. 23.3b. The accuracy of these strain measurements is limited, however, and requires refinement before correct performance metrics can be measured from the stress-strain curve. One indication that the strain measurements require refinement is that the elastic modulus shown in the Fig. 23.3b is significantly lower than aluminum's bulk elastic modulus. Currently, a new method is being developed to apply tracking features to the surface of the ligament to obtain elongation directly from the X-ray CT imagery.

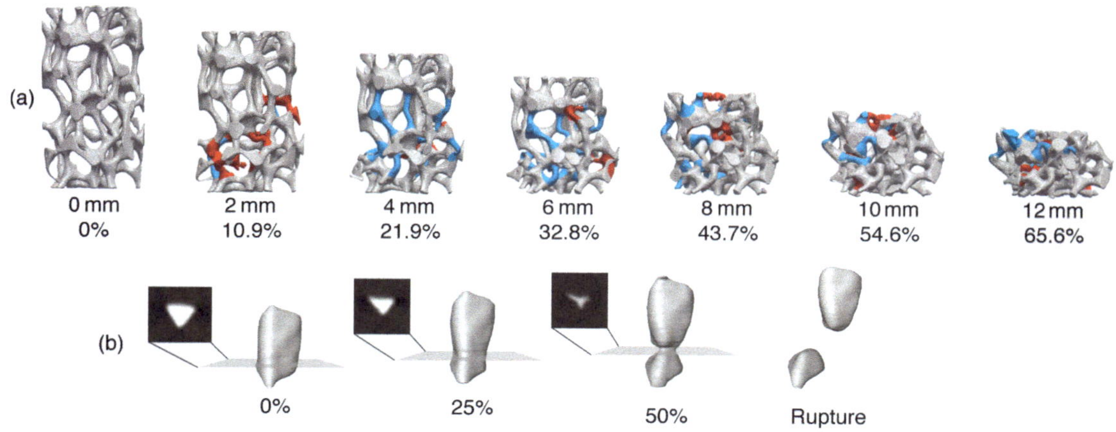

Fig. 23.2 (**a**) Reconstructed X-ray CT images of foam at progressing stages of compression, with colored ligaments showing where significant deformation in the current compression increment is observed (**b**) Reconstructed X-ray CT images of ligament gauge region, with X-ray CT image slices showing cross-section of ligament at center of gauge region

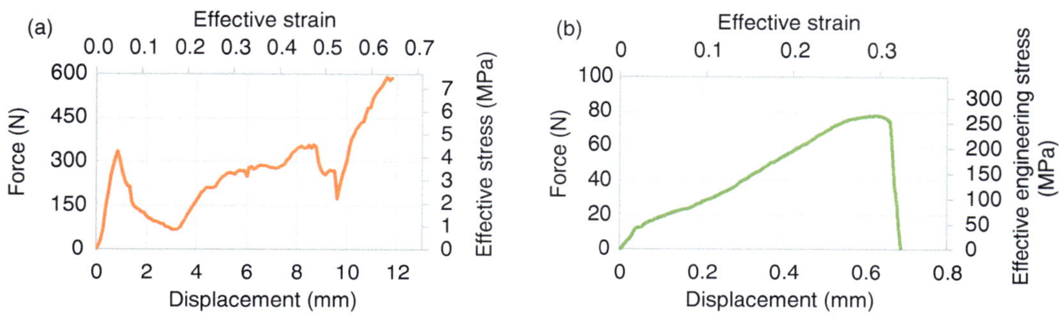

Fig. 23.3 Force vs. displacement curves from (**a**) compression test, and (**b**) ligament tension test

23.4 Conclusions

The experiment methodology described herein provides a way to: (a) measure the global compressive force/displacement response of bulk aluminum open-cell foams, (b) visualize yielding, collapse, and fracture of individual ligaments during compression, (c) collect 3D image data that can be used to generate realistic finite element models of as-manufactured foams, and (d) directly obtain the stress-strain behavior of individual ligaments. The method of measuring strain in the single ligament tension tests is currently undergoing further refinement.

References

1. Ashby, M.F., Evans, A.G., Fleck, N.A., Gibson, L.J., Hutchinson, J.W., Wadley, H.N.G.: Metal Foams—A Design Guide. Butterworths Heinemann, London (2000)
2. Simone, A.E., Gibson, L.J.: Aluminum foams produced by liquid-state processes. Acta Mater. **46**(9), 3109–3123 (1998)
3. Nieh, T.G., Higashi, K., Wadsworth, J.: Effect of cell morphology on the compressive properties of open-cell aluminum foams. Mater. Sci. Eng. A. **283**(1-2), 105–110 (2000)
4. Markaki, A.E., Clyne, T.W.: The effect of cell wall microstructure on the deformation and fracture of aluminium-based foams. Acta Mater. **49**(9), 1677–1686 (2001)
5. Schüler, P., Fischer, S.F., Bührig-Polaczek, A., Fleck, C.: Deformation and failure behaviour of open cell Al foams under quasistatic and impact loading. Mater. Sci. Eng. A. **587**, 250–261 (2013)
6. Zhou, J., Mercer, C., Soboyejo, W.O.: An investigation of the microstructure and strength of open-cell 6101 aluminum foams. Metall. Mater. Trans. A. **33**(5), 1413–1427 (2002)
7. Matheson, K.E., Cross, K.K., Nowell, M.M., Spear, A.D.: A multiscale comparison of stochastic open-cell aluminum foam produced via conventional and additive-manufacturing routes. Mater. Sci. Eng. A. **707**, 181–192 (2017)
8. Zhou, J., Allameh, S., Soboyejo, W.O.: Microscale testing of the strut in open cell aluminum foams. J. Mater. Sci. **40**, 429–439 (2005)

Chapter 24
Damage Detection and Visco-Elastic Property Characterization of Composite Aerospace Panels Using Ultrasonic Guided Waves

M. Capriotti, R. Cui, and F. Lanza di Scalea

Abstract Composite materials are widely used in commercial and military aircraft. A major concern is the detection and quantification of damage in these panels caused, for example, by foreign object impacts. A related issue is the characterization of the composite visco-elastic properties, whether the ply-by-ply properties or the laminate's engineering properties. Both of the aforementioned tasks can be accomplished by careful use of ultrasonic guided waves that are multimode and dispersive waves propagating in the composite waveguide. The test piece for these studies is a carbon-reinforced plastic (CFRP) panel with co-cured stiffeners representative of modern commercial aircraft construction (e.g. B787). The wave dispersive properties of this panel were first determined from broadband ultrasonic tests and 2D Fourier Transform techniques. A Semi-Analytical Finite Element (SAFE) analysis was then performed to calculate dispersion curves and cross-sectional mode shapes of relevant guided modes propagating in the panel. The SAFE analysis allowed to iteratively change the elastic properties of each layer in the panel so as to identify the layer-by-layer properties from the experimental extraction of the dispersive guided waves propagating in the test panel. A scanning inspection system using air-coupled ultrasonic transducers operating at specific frequencies was also developed to detect and quantify impact-induced damage in the skin or the stringer of the panel.

Keywords Aerospace · Ultrasonic guided waves · SHM · Composites · Outlier analysis

24.1 Introduction

One of the major problems related to aerospace structures is the detection of defects and their characterization. Aircrafts are often subjected to a wide variety of environmental and operational conditions that cause damage in the structure and can severely impair its strength and life-span estimation. Wide area impact damage due to Ground Service Equipment is one of the most dangerous and frequent scenario: defects of different type, location and size are formed and their detection and characterization is crucial to the aerospace industry.

24.2 Problem Statement and Approach

The inspection of aerospace structures represents a challenging problem. The increased use of composite materials in commercial aircrafts, due to their higher stiffness to weight ratio and properties that can be tailored, requires material characterization and novel investigation techniques. Moreover, the geometry of the structure itself and the assembly of multiple structural components necessitates additional modeling and inspection efforts.

The above mentioned complexity together with the waveguide-geometry of the structure and the large size to be inspected suggests the use of Ultrasonic Guided Waves (UGW). Their low attenuation over long distances allows the inspection of wide areas and the penetration of multiple interfaces. The multi-mode and dispersive behavior, also, can be exploited in the post-processing phase, both for the laminate and defects characterization.

A through-transmission technique is used to exploit the physics of the guided wave propagation and its interaction with the interfaces and defects.

M. Capriotti · R. Cui · F. Lanza di Scalea (✉)
NDE/SHM Laboratory, Department of Structural Engineering, University of California San Diego, La Jolla, CA, USA
e-mail: flanza@ucsd.edu

P. R. Thakre et al. (eds.), *Mechanics of Composite, Hybrid and Multifunctional Materials, Volume 5*, Conference Proceedings of the Society for Experimental Mechanics Series, https://doi.org/10.1007/978-3-319-95510-0_24

24.3 Characterization

Carbon-Fiber Reinforced Plastic (CFRP) panels of different sizes, lay-up configurations and assembly components were the test pieces for this study. They represented modern commercial aircrafts and real impact damage scenarios.

A Finite Element Method modeling requires large computational effort to properly characterize the full structure, its properties and the damage formation and its effects. Hence, an experimental approach of UGW testing was attempted at first, using the 2D Fourier Transform technique [1]. This technique allows to retrieve the guided-wave dispersion behavior of a waveguide of arbitrary material and geometry in terms of phase velocity, establishing the waveguide dispersive relation between the temporal and spatial frequency domains (angular frequency ω and wavenumber k). Figure 24.1 shows the results in a 2D surface map (left) and the corresponding dispersion curve (right) for a CFRP stringer panel.

On the modeling side, a Semi-Analytical Finite Element (SAFE) analysis [2] was implemented to compute phase and group velocity dispersion curves and cross-sectional mode shapes of different composite panel lay-ups. Parametric studies and inversion with respect to the experimental extraction can be performed to solve for laminate thickness, ply orientation and elastic constants, iteratively inputting different layer by layer properties. Figure 24.2 shows the phase and group velocity dispersion curves (left) for the skin of an aerospace structure, represented by a composite plate, 16 plies T800/3900 Graphite/Epoxy Prepreg Tape, symmetric. Also shown in this figure are the cross sectional mode shapes for the A0 and S0 modes at 170 kHz (right) in terms of in plane (blue) and out of plane (red) displacement.

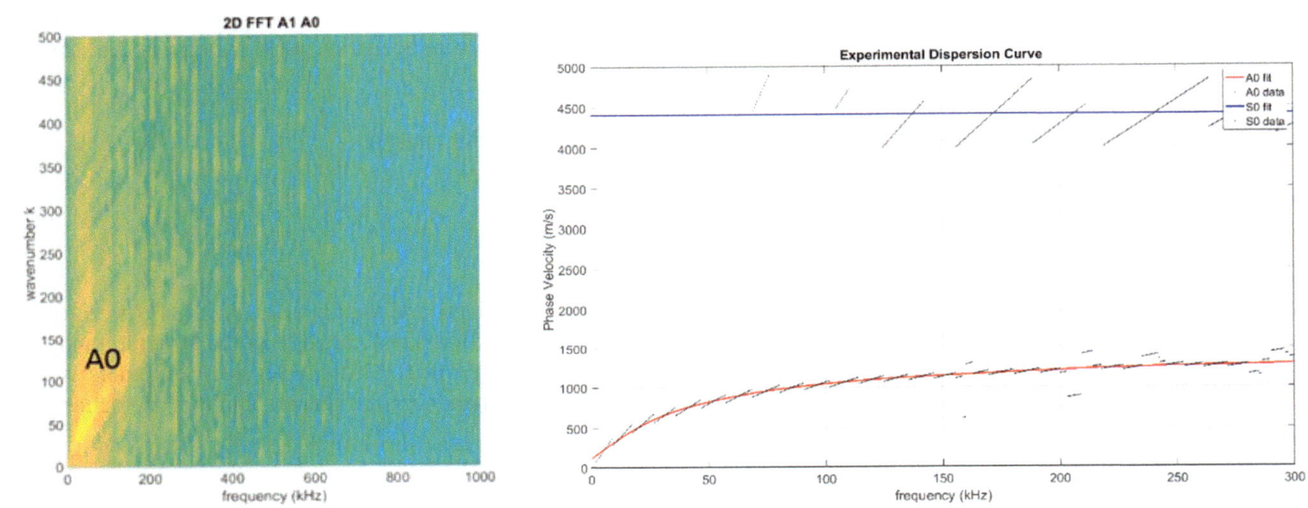

Fig. 24.1 (Left) 2DFFT map; (right) phase velocity dispersion curves of SH0 and A0 modes

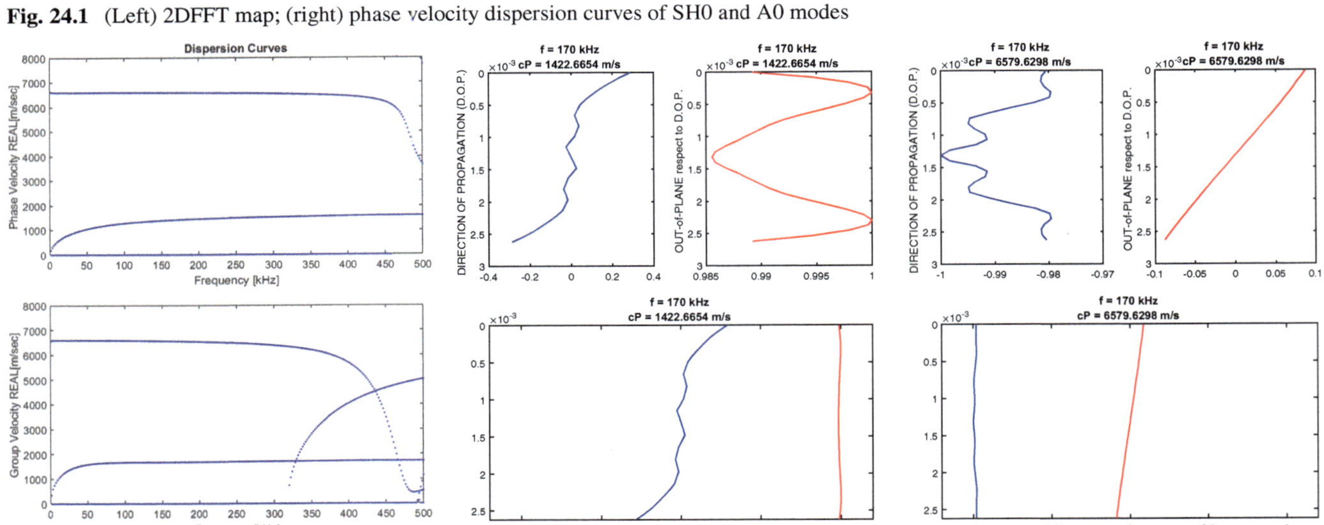

Fig. 24.2 Skin (left) dispersion curves; (right) cross-sectional mode shapes of A0 and S0 modes at 170 kHz

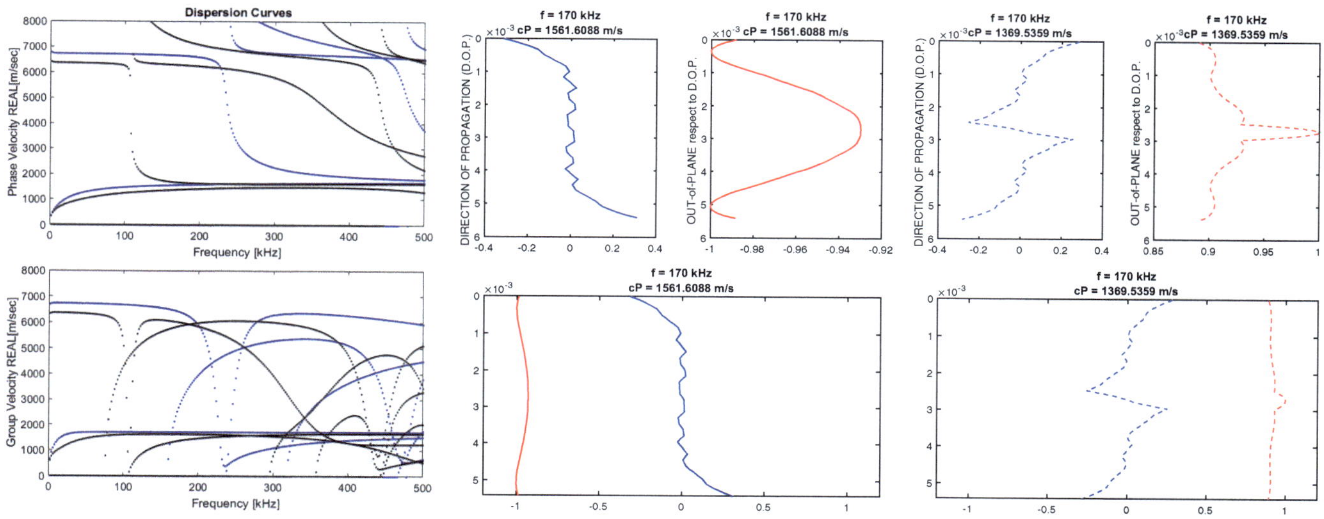

Fig. 24.3 Skin-Stringer (left) dispersion curves; (right) cross-sectional mode shapes of A0 mode at 170 kHz for pristine (solid) and damaged (dashed) laminate

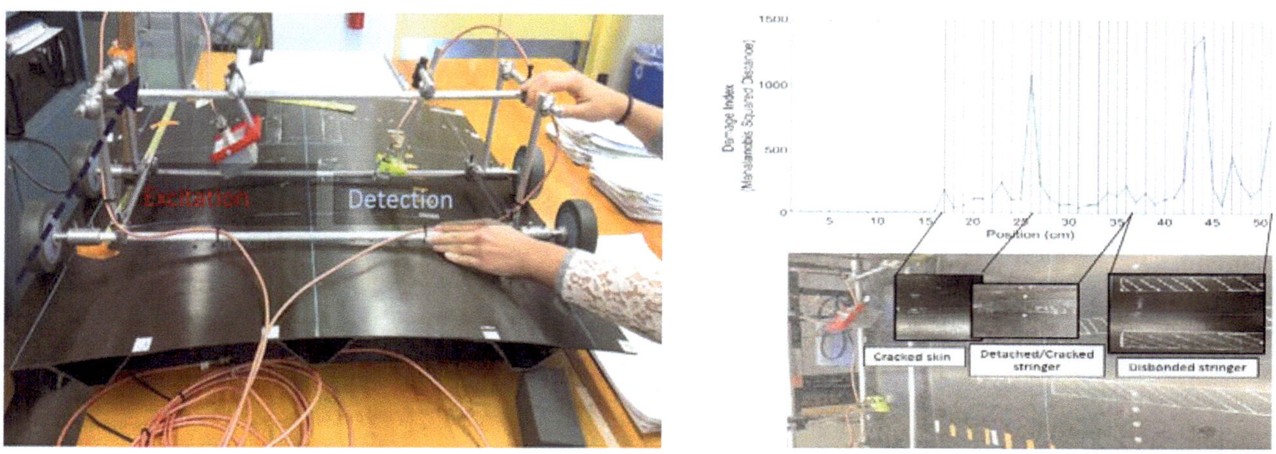

Fig. 24.4 Aerospace panel inspection: (left) NDE technique experimental set-up; (right) DI vs position results

Figure 24.3 shows the results for the skin-stringer interface in a pristine condition (blue) and a damaged condition (black), represented by a 50% reduction of the elastic properties in three plies at the interface.

24.4 Inspection

A non-contact NDE system has been developed relying on ultrasonic air-coupled transducers mounted on a moving cart that allows the scanning of the aerospace structure in an efficient and relatively rapid manner [3]. The recorded waveforms are post-processed using a statistical multivariate outlier analysis, that enables defect detection and localization through a Damage Index (DI). Figure 24.4 shows the scanning system on an impacted panel (left) and the DI plot, where skin crack, stringer crack and skin to stringer disbonds have been detected (right).

A second generation prototype has also been developed where a dual output has been implemented to extract the skin-stringer Green's function and then relate this to the presence of damage.

24.5 Conclusions

UGWs can be successfully used to detect defects and characterize complex composite aerospace structures. Modeling and experiments were carried out in this work as they apply to the detection of damage in realistic composite panels representative of modern aircraft construction (e.g. Boeing 787 fuselage).

References

1. Alleyne, D., Cawley, P.: A two-dimensional Fourier transform method for the measurement of propagating multimode signals. J. Acoust. Soc. Am. **89**, 1159–1168 (1991)
2. Bartoli, I., Marzani, A., Lanza di Scalea, F., Viola, E.: Modeling wave propagation in damped waveguides of arbitrary cross-section. J. Sound Vib. **295**(3–5), 685–707 (2006)
3. Capriotti, M., Kim, H.E., Lanza Di Scalea, F., Kim, H.: Nondestructive inspection of impact damage in composite aircraft panels by ultrasonic guided waves and statistical processing. Mat. J. Spec. Issue Struct. Health Monit. Aerosp. Appl. **10**(616), 1–12 (2017)

Chapter 25
Microscale Investigation of Transverse Tensile Failure of Fiber-Reinforced Polymer Composites

Caitlin M. Arndt, Paige DaBell, and Michael W. Czabaj

Abstract In this study, two experimental approaches are used to understand fracture mechanisms that govern transverse tensile failure of fiber-reinforced polymer composites at the microscale. These observations are used to directly measure, and indirectly estimate, the magnitude and scatter of the transverse tensile strength, Y_T, and the associated effective flaw size, a_0. To this end, static three-point bend tests are performed on pristine and notched 90° unidirectional IM7/8552 carbon-epoxy samples. In the pristine specimen study, tensile microcracks are observed to initiate well before the ultimate failure load used to compute Y_T is reached. In the notched specimen study, a comparison is made between experimentally measured strengths with known notch lengths and a linear elastic fracture mechanics solution. The fracture mechanics solution significantly over-predicts the apparent Y_T for notch lengths less than a ply thickness, suggesting that this approach may not be appropriate for estimation of transverse tensile strength at the microscale. The observations made in this study suggest that Y_T may not be a true material property, but rather, a structural property dependent on specimen geometry and microstructural variability.

Keywords Transverse tensile strength · Linear elastic fracture mechanics · Microstructure · Optical microscopy · Matrix cracking

25.1 Introduction

Prediction of transverse tensile failure of tape-laminate fiber-reinforced polymer (FRP) composites is often based on the experimentally measured transverse tensile strength (Y_T). Despite the accepted use of Y_T for such predictions, past studies have demonstrated that this apparent material property cannot be reliably measured using existing test methods. This is particularly evident from the large scatter in Y_T and its dependence on specimen geometry [1–4]. Possible sources of variability in Y_T include the perceived sensitivity of matrix fracture to microstructural features such as the random fiber distribution, resin-rich regions, and pre-existing flaws. In addition, some tests (e.g. transverse uniaxial tension) exhibit inherent stress concentrations near the specimen grips, which can introduce additional variability into measured values of Y_T.

In this study, two experiments are used to understand, at the microscale, the fracture mechanisms that govern transverse tensile failure of FRPs. A set of static three-point bend (3 PB) flexure tests are performed on pristine FRP samples to determine the actual microscale mechanisms that lead to transverse tensile failure in tape laminate composites. Additionally, 3 PB flexure tests are performed on specimens with artificially inserted microscale starter cracks to investigate whether it is possible to estimate Y_T based on linear elastic fracture mechanics.

25.2 Specimen Preparation and Experimental Methods

The specimens used in this study were fabricated from Hexcel's IM7/8552 26-ply unidirectional laminate with 35% resin content and 190 gsm areal weight. Prior to testing, all specimens were polished on the 1–2 and 2–3 material planes to a mirror finish using silicon carbide sandpaper followed by a lapping cloth coated with a 0.05 micron alumina powder suspension. After polishing, all specimens were nominally 5.22 mm wide and 4.64 mm thick, with standard deviation of 3.48% and 3.29%, respectively. The specimen geometry and corresponding 1–2/2–3 material planes are shown in Fig. 25.1. For a selected number of samples, microscopically-sharp starter cracks were milled into the top 1–2 surface using a FEI

C. M. Arndt · P. DaBell · M. W. Czabaj (✉)
Department of Mechanical Engineering, University of Utah, Salt Lake City, UT, USA
e-mail: m.czabaj@utah.edu

P. R. Thakre et al. (eds.), *Mechanics of Composite, Hybrid and Multifunctional Materials, Volume 5*, Conference Proceedings of the Society for Experimental Mechanics Series, https://doi.org/10.1007/978-3-319-95510-0_25

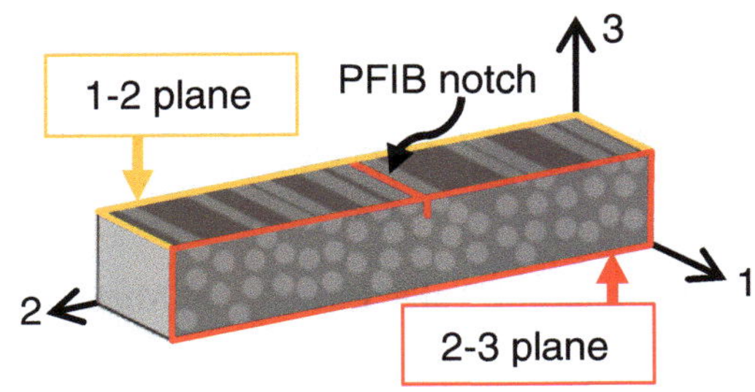

Fig. 25.1 Graphic of specimen reference planes and PFIB notch location. Fiber cross-section areas not to scale

Helios Plasma Focused Ion Beam (PFIB). The notch depth ranged from 10 to 100 μm, and spanned the entire width along the 1-dirction, as shown in Fig. 25.1. The size and geometry of the PFIB notches were intended to mimic naturally occurring tensile cracks observed during 3 PB testing of pristine specimens.

The experimental setup consisted of an inverted 3 PB test fixture with a support span of 25.4 and 4 mm diameter support and loading pins. The test fixture was inverted in order to keep the tensile side of the specimen stationary during testing, which simplified the microscale observations of evolving damage. The fixture was mounted in an Instron ElectroPuls E1000 load frame equipped with a 1 kN load cell. The evolution of damage in pristine and notched samples was observed in real-time using a custom traveling microscope equipped with a manual Mitutoyo objective lens turret, Mitutoyo Infinity corrected long working distance 10× and 50× objective lenses, and a Point Grey Grasshopper GRAS-50S5M-C 5MP resolution camera. During testing, the 10× and 50× lenses were used interchangeably. Typically, the 10× lens was used to acquire full-field displacement and strain fields across approximately four plies in the region of maximum tension. The 50× objective lens, which clearly showed fiber/matrix interfaces, was used to image initiation and evolution of microscale fracture. The microscope was attached to a manual 3-axis translation stage and mounted to the base on the E1000 load frame. Acquisition of force, displacement, and image data was done using Correlated Solutions VicSnap software at a rate of 1 Hz. All specimens were loaded in displacement control at a rate of 0.15 mm/min.

25.3 Results and Discussion

The transverse strength of pristine 3 PB specimens was calculated using Euler-Bernoulli beam theory based on a critical failure load corresponding to macroscopic fracture of the entire sample. After testing 36 pristine specimens, the average Y_T was 143.10 MPa with a coefficient of variation of 8.34%, which is consistent with previously reported values for the same material system [2]. Surprisingly, out of all specimens imaged on the 2–3 plane using the traveling microscope, approximately 75% exhibited some amount of microscale fracture well before macroscopic failure. Figures 25.2 and 25.3 show typical microscale fracture on the 2–3 plane viewed with the 10× lens. Figure 25.2 shows the entire frame captured by the microscope, and Fig. 25.3 zooms in on four areas in the microstructure that exhibited fracture prior to macroscopic specimen failure. In Fig. 25.3, the three rows correspond to the microstructure before testing (top row), at the first evidence of crack formation (middle row), and right before macroscopic specimen failure (bottom row). For this particular specimen, the four cracks formed at loads ranging from 16–71% of ultimate failure load. It is interesting to note that the crack in area 2 is within 23 μm to the path of the crack that ultimately broke the specimen (highlighted in red in Fig. 25.2), but the two cracks did not interact. Although this particular specimen broke directly over the center roller, this was not the case for every specimen. Many of the specimens tested broke near but not directly over the center roller. This suggests that location of transverse fracture is dictated by the competition between ply level stresses (due to far-field 3 PB loading), local stress concentrations (caused by random distribution of fibers), and variability in local material strength properties.

The strengths of the PFIB notched specimens were calculated using the same expression used to calculate the strength of the pristine specimens while ignoring the presence of the notches. A summary of the results collected from this study is shown in Fig. 25.4. In this figure, calculated strength is plotted against PFIBed notch size. The x-axis ranges from less than one fiber diameter, which is highlighted in pink on the left-hand side, to greater than one ply thickness, highlighted in green on the right-hand side of the plot. The strength of the pristine specimens (blue square in Fig. 25.4) is plotted on the far left with error bars encompassing the entire range of the pristine strength data. The experimental data (green triangles in Fig. 25.4) were fitted with a power-law trend line to visualize what strength would be predicted for a notch with zero

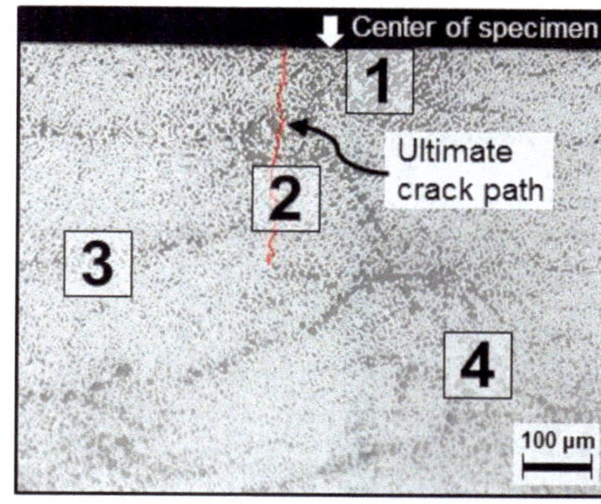

Fig. 25.2 Single frame of microstructure on the tensile side directly over the center roller taken at 10× magnification with crack locations labeled

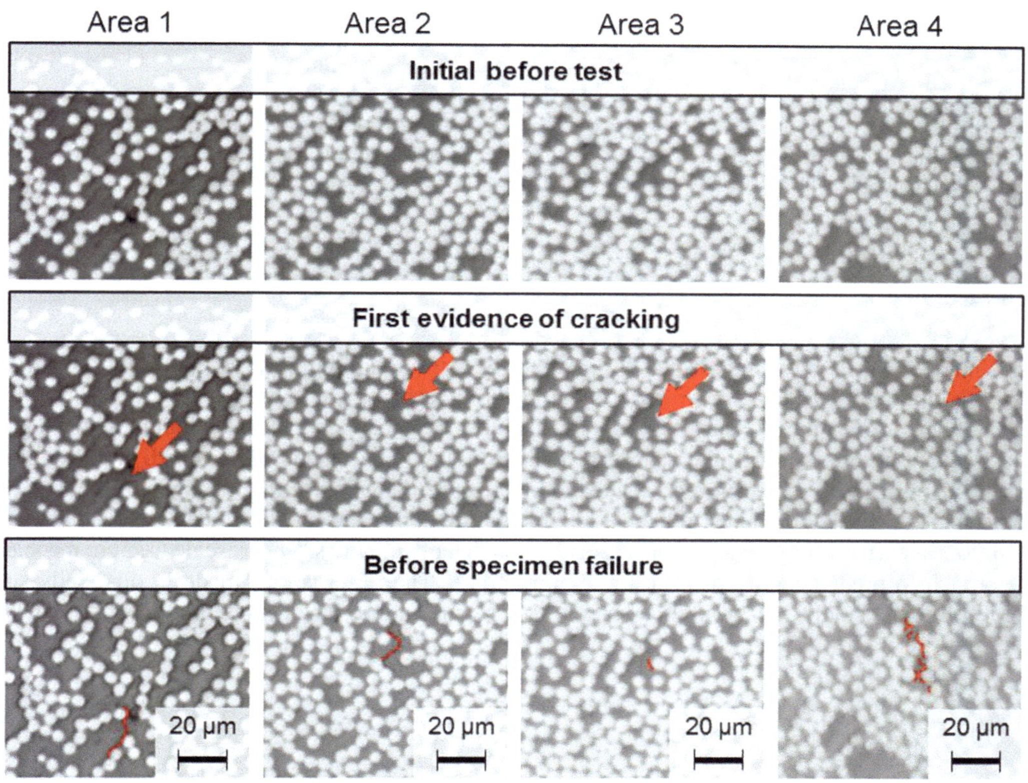

Fig. 25.3 Micrographs taken at 10× magnification of four areas with progressive crack formation indicated with red arrows. Top row: Pristine microstructure taken before testing. Center row: First evidence of cracking. Bottom row: Final stage of crack growth before ultimate specimen failure, cracks outlined in red

length. Examining the linear fit in Fig. 25.4, it appears that as the notch lengths approach 5–10 μm, or the size of one to two fibers, the estimated strength approaches that of pristine specimens. This would indicate that flaws smaller than the smallest microstructural feature do not influence the apparent Y_T. A linear-elastic fracture mechanics (LEFM) solution for an orthotropic three-point bend configuration with a width-wise crack was used to estimate strengths of specimens over the range of notches presented in Fig. 25.4. Intralaminar mode I fracture toughness of 200 J/m^2 was used in all calculations [5]. The strength values predicted using LEFM are plotted with a yellow line in Fig. 25.4. As seen in this figure, the LEFM solution clearly over-predicts the transverse strength for notch sizes that are less than 1.5 times a single-ply thickness.

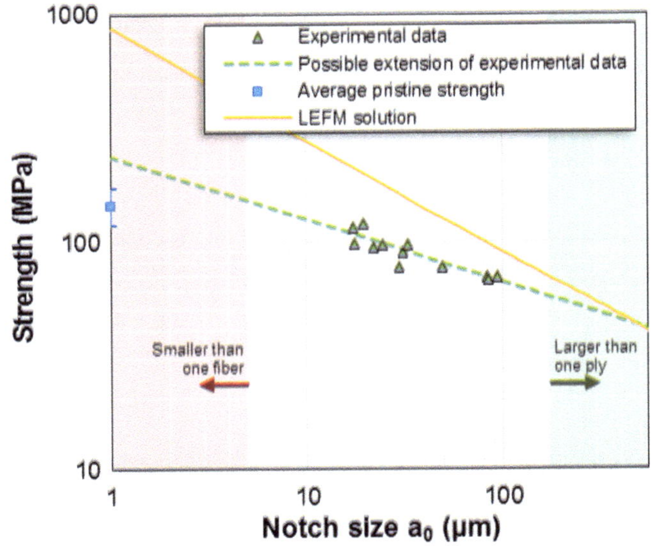

Fig. 25.4 Plot of experimentally obtained notch vs. measured strength (green markers) overlaid with a plot of the LEFM solution for a cracked three-point bend configuration (yellow line). The transition between fracture at a length scale smaller than a single fiber, and larger than one ply are highlighted

25.4 Conclusions

The observations made in this study suggest that Y_T may not be a true material property, but rather, a structural property dependent on the distribution of fibers in the matrix, local material variations, and pre-existing flaws. Optical observations made *in situ* reveal that cracks on the 2–3 surface initiate well below the specimen's ultimate failure load, which may indicate that Y_T is not a true measure of microscale transverse tensile failure. The results of the PFIB notched specimens study provide insight into the transition between microscale and macroscale damage in FRPs. As the size of a crack decreases to approximately the size of one fiber diameter, the local microstructure dominates and a strength-based approach to predicting failure may be appropriate. As the crack increases in size to about one ply thickness, macroscale fracture mechanics dominate, and a LEFM approach to predict failure may be more appropriate. The observations made in this study suggest that the analytical LEFM solutions should not be used to predict transverse failure when modeling cracks smaller than one ply thickness.

References

1. O'Brien, T.K., Salpekar, S.A.: Scale Effects on the Transverse Tensile Strength of Graphite Epoxy Composites Rep. NASA TM-107637. NASA, Hampton (1992)
2. O'Brien, T.K., Chawan, A.D., DeMarco, K., Paris, I.L.: Influence of Specimen Preparation and Specimen Size on Composite Transverse Tensile Strength and Scatter Rep. TM-2001-211030. NASA, Hampton (2001)
3. O'Brien, T.K., Chawan, A.D., Krueger, R., Paris, I.L.: Transverse Tension Fatigue Life Characterization Through Flexure Testing of Composite Materials Rep. NASA TM-2001-211035. NASA, Hampton (2001)
4. Adams, D.F., King, T.R., Blackketter, D.M.: Evaluation of the transverse flexure test method for composite materials. Compos. Sci. Technol. **39**, 341–353 (1990)
5. Czabaj, M.W., Ratcliffe, J.G.: Comparison of intralaminar and interlaminar mode I fracture toughnesses of a unidirectional IM7/8552 carbon/epoxy composite. Compos. Sci. Technol. **89**, 15–23 (2013)

Chapter 26
Optimization of Kerf Quality During CO_2 Laser Cutting of Titanium Alloy Sheet Ti-6Al-4V and Pure Titanium Ti

B. El Aoud, M. Boujelbene, E. Bayraktar, and S. Ben Salem

Abstract CO_2 laser cutting is an advanced processing technology, which can, according to the computer-aided design graphics, cut a variety of shapes in the surfaces of many metallic sheets. Laser cutting of various materials including the Titanium alloy Ti-6Al-4V and pure Titanium Ti is carried out to assess the kerf width size variation along the cut section. This work aims to analyze the effect of laser power, cutting speed, and gas pressure on the kerf quality of Ti-6Al-4V alloy and Ti with CO_2 laser cutting process. The kerf width size is formulated and predicted using the lump parameter analysis and it is measured from the experiments. The influence of laser output power Pu and laser cutting speed V and pressure nitrogen assisting gas p on the kerf width size variation is analyzed using the analytical tools including scanning electron and optical microscopes. The quality of laser cut kerf mainly depends on appropriate selection of process parameters. Uniform kerf with minimum kerf width is always demand. It has been found that the kerf width during CO_2 laser process is not uniform along the length of cut. A considerable improvement in kerf quality has been achieved.

Keywords CO_2 laser cutting · Kerf quality · Titanium alloys Ti-6Al-4V · Pure titanium Ti · And optimization

26.1 Introduction

Sheet-metal cutting is the single largest, in terms of sales, global industrial laser application. CO_2 lasers dominate this application due to their good-quality beam combined to high output power. It is estimated that more than 40,000 cutting machines using CO_2 lasers have been installed worldwide. However, aluminum alloys, typical engineering materials in different industries such as automotive or aerospace industry, are not extensively cut by lasers [1]. Laser cutting involves with high temperature processing of materials including solid phase heating, melting, and evaporation. In the case of metallic materials processing, assisting gas is used to reduce the oxidation reactions in the cutting section. Since the oxidation reactions gives rise to excessive heating in the cutting section via high temperature exothermic reactions, the resulting section suffers from cutting asperities such as sideways burning, dross attachments, and thermal erosion [2]. The proper controlling of the laser cutting process through appropriate selection of the cutting parameters minimizes the defect sites along the cut sections. However, further investigations are needed for net shaping of the materials, which involve with high thermal conductivities and low fracture toughness such as titanium alloys and alumina [3]. Laser cutting of thick sections offers considerable advantages over the conventional techniques due to precision of operation, short processing time, and low cost. The physical processes involved in laser cutting of thick sections are complicated and significantly influences the end product quality. Laser parameters, in particular laser output power, focus on setting of focusing lens, cutting speed, assisting gas, and its pressure influence the physical processes in the cutting section. In this case, controlling the affecting parameters results in improved cutting quality. Consequently, investigation into affecting parameters in laser cutting process is necessary to improve the end product quality [4–6]. Laser cutting widely used in metallic and non-metallic materials such as polymer, composites or ceramics [7–9]. Most authors were focusing on the power intensity and cutting speed.

B. El Aoud
Supmeca/Paris School of Mechanical and Manufacturing Engineering, Paris, France

University of Tunis El Manar, ENIT, Ecole Nationale d'Ingénieurs de Tunis, Le Belvedere, Tunisia

M. Boujelbene · S. Ben Salem
University of Tunis El Manar, ENIT, École Nationale d'Ingénieurs de Tunis, Tunis, Tunisia

E. Bayraktar (✉)
Supmeca-Paris School of Mechanical and Manufacturing Engineering, Paris, France
e-mail: bayraktar@supmeca.fr

P. R. Thakre et al. (eds.), *Mechanics of Composite, Hybrid and Multifunctional Materials, Volume 5*, Conference Proceedings of the Society for Experimental Mechanics Series, https://doi.org/10.1007/978-3-319-95510-0_26

Table 26.1 Chemical composition of Ti-6Al-4 V (%)

Ti	Al	V	Fe	C	N	H	O
Base	6	4	0.3	0.08	0.05	0.01	0.2

Table 26.2 Mechanical and physical properties of pure Titanium Ti

Mechanical break (MPA)	345
Elasticity limit 0,2 (MPA)	275
Elongation %	20
Hardness	160 HB/30
Normal modulus of elasticity (GPA)	103
Tangential modulus of elasticity (GPA)	40

Laser cutting of advance ultra-high strength steel 22MnB5 by Abdul Fattah et al. [10] resulted in combined effect of laser power and cutting speed on kerf width formation. Most study show that Increment of cutting speed reduced kerf width formation and HAZ region, whereas the higher laser power produced inverse results [11, 12]. Laser cutting process and the assessment of cutting parameters on kerf size and geometry are important aspects of the quality evaluation of the end product. Optimization studies on laser cutting offers improved process control and securing of the end product quality. Modelling and optimization study for the assessment of the cut quality of a thin aluminum-alloy was carried out by Sharma and Yadava [13]. They introduced the entropy measurement methodology for the calculation of weight corresponding to each quality characteristic. Ghany and Newishy [14] have observed the variation of kerf width with cutting speed, laser power, and type of gas and pressure as above during experimental study of Nd: YAG laser cutting of 1.2 mm-thick austenitic stainless steel sheet. During his experimental investigation, found that kerf width increases with increasing laser power and decreasing the cutting speed. He also observed that oxygen or air gives wider kerf while use of inert gas gives the narrow kerf. They have also found that on increasing pulse frequency the kerf width decreases. A CO_2 laser cutting of 3 mm thick of Titanium alloy sheet grade 5 Ti-6Al-4V is investigated by B. El AOUD et al. [15]. Results indicate that the thickness of the Heat affected Zone increases with the evolution of laser power and decreases with the increase of cutting speed, and the optimum cutting condition was found to be 1 kW for the laser power, 2400 mm/min for the cutting speed and 2 bars for gas pressure. Also, it is underlined that the kerf width is mainly influenced by laser power and cutting speed.

In the present study, laser cutting of composites consisting of Titanium alloy (Ti-6Al-4V) and pure Titanium (Ti) are carried out. The kerf width variation due to the laser output power parameters is examined. The study is extended firstly to investigate the kerf width variation along the cut edges with the cutting parameters such as laser power, cutting speed and gas pressure and secondly to introduce the use of Taguchi method to optimize kerf variation. The resulting cut sections are examined using the scanning electron microscope (SEM) and optical microscope.

26.2 Experimental Work

The CO_2 laser from laser machine type 4000 TLF TURBO is used to irradiate the workpiece surface. Nitrogen emerging from a conical nozzle and co-axially with the laser beam is utilized. The workpieces accommodated are Ti-6Al-4V and pure Ti with their elemental composition are given in Tables 26.1 and 26.2 respectively. The sheet dimensions were 20 mm × 15 mm with thickness of 3 mm for Titanium alloy Ti Ti-6Al-4V and 12 mm × 8 mm with thickness of 3 mm for pure Titanium Ti.

26.3 Design of Experiment

The experiment was designed based on a three levels of parameters. Laser power ($Pu = 1$–4 kW), cutting speed ($V = 480$–2400 mm/min) and gas pressure ($p = 2$–14 bars). In order to achieve best cutting quality, Taguchi's experimental design, an efficient plan, was used for conducting experiments. According to Table 26.3, L_{18} orthogonal array are used to reduce number of the experiments.

Table 26.3 Experimental multi-performance results

EXP. No	*Pu (kW)*	*V (mm/min)*	*p (bar)*	*Kw [Ti6Al4V] (μm)*	*Kw [Ti] (μm)*
1	2	480	2	623.10	574.78
2	2	480	8	586.20	517.20
3	2	480	14	607.42	562.02
4	2	1440	2	582.01	542.53
5	2	1440	8	534.71	483.65
6	2	1440	14	571.63	532.79
7	2	2400	2	545.23	487.01
8	2	2400	8	474.68	471.57
9	2	2400	14	483.16	519.77
10	3	480	2	654.60	599.07
11	3	480	8	591.65	554.77
12	3	480	14	743.64	631.40
13	3	1440	2	615.28	571.48
14	3	1440	8	555.43	546.55
15	3	1440	14	586.27	586.65
16	3	2400	2	559.62	512.07
17	3	2400	8	476.97	530.42
18	3	2400	14	499.95	575.47

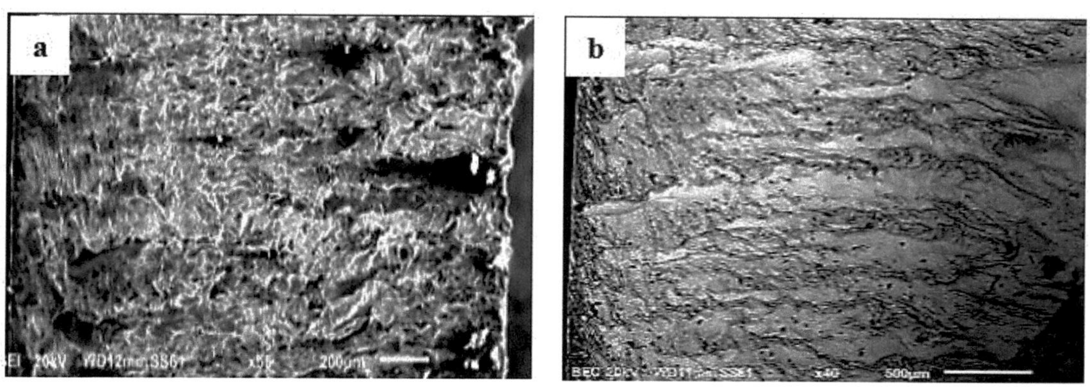

Fig. 26.1 SEM micrographs of laser cut kerf surfaces at laser power $Pu = 3$ kW, the cutting speed $V = 480$ mm/min and the gas pressure $p = 2$ bars: (**a**) Ti-6Al-4 V alloy and (**b**) pure Titanium

26.4 Results and Discussion

CO_2 Laser gas assisted cutting of titanium alloy (Ti-6Al-4V) and pure Titanium (Ti), is carried out. The effect of laser output power and cutting speed on the kerf width size variation is analyzed.

Figures 26.1 and 26.2 show SEM micrographs of laser cut kerf surfaces of the workpieces for two different conditions of laser output power and cutting speeds. The kerf surfaces appear to be almost similar for metallic materials; in which case, some striations with flow of liquid metal are observed. The rapid solidification of the liquid metal at the surface, due to the convection cooling effect of the high pressure assisting gas, gives rise to the formation of the cast layer at the kerf surface.

In the case of high power or low speed cutting process Fig. 26.1a, b, the cracks are partially extended on the kerf surface unlike the low power and high speed cutting process. This behavior is attributed to the thickness of the melt cast layer, which increases with the laser output power or the low cutting speeds.

In the case of high speed cutting or low power Fig. 26.2a, b, the cast layer formed at the kerf surface remains thin while resulting in attainment of high temperature gradients. This situation results in the crack network formation on the kerf surface [3].

Moreover, in the case of titanium alloy, the striation depth remains large at the kerf surface. This is associated with the high oxidation potential of titanium, which results in sideways burning and deep molten layer formation at the kerf surface.

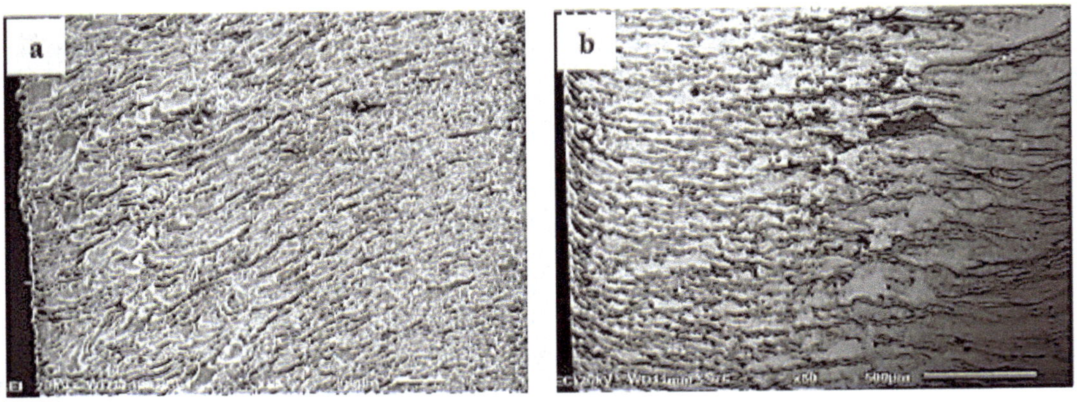

Fig. 26.2 SEM micrographs of laser cut kerf surfaces at laser power $Pu = 2$ kW, the cutting speed $V = 2400$ mm/min and the gas pressure $p = 2$ bars: (**a**) Titanium alloy Ti-6Al-4 V alloy, (**b**) pure Titanium Ti

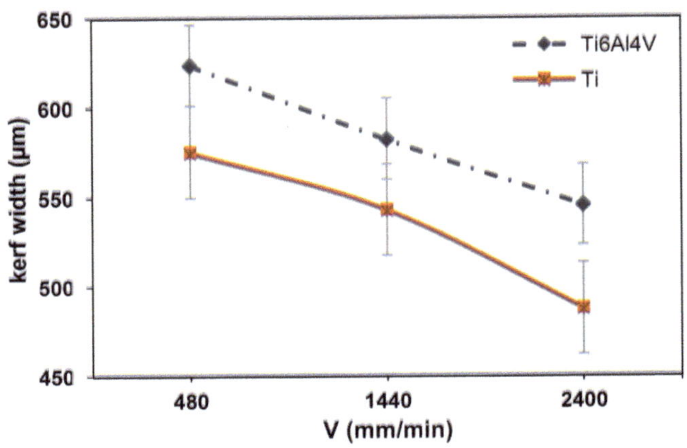

Fig. 26.3 Kerf width size variation with the laser cutting speed for Ti-6Al-4V and pure Titanium Ti at $Pu = 2$ kW and $p = 2$ bars

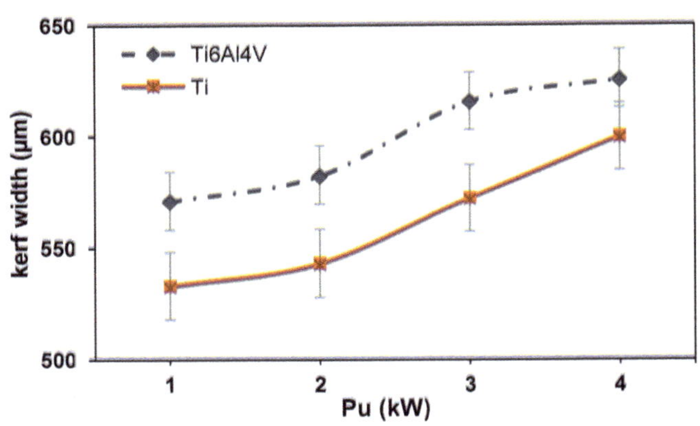

Fig. 26.4 Kerf width size variation with the laser power for Ti-6Al-4V and pure Titanium Ti at $V = 1440$ mm/min and $p = 2$ bars

It should be noted that high temperature exothermic reactions results in excessive energy generation contributing to the laser beam energy in the cutting section [16]. This causes deep striation patterns at the surface [17].

Figures 26.3 and 26.4 show the effect of two output laser parameters; cutting speed (V) and Laser power (Pu) on the kerf width variation for two different materials; Titanium alloy (Ti-6Al-4V) and pure Titanium (Ti).

It is evident from Fig. 26.3 that the kerf width size decreases as the cutting speed increases. This is associated with the formation of large molten layer due to the excessive energy provided via ectothermic reactions in the cut section. In fact, when using slow cutting speed more heat would be introduced to the specimen and then more materials will be melted and ejected causing the kerf to increase.

According to Fig. 26.4, kerf width size variation demonstrates the opposite behavior to that for the laser output cutting speed. In fact, increasing laser output power increases the kerf width size, which is more pronounced for the titanium alloy. This is due to the high temperature exothermic reactions, which provide excess energy in the cutting section. Since the assisting gas used is nitrogen, the presence of the oxygen remains low in the cutting section. However, high oxygen affinity of titanium alloy gives rise to the oxidation reactions taking place locally in the cutting section. Therefore, localized thermal erosion is resulted at the kerf surface while causing the deep striation formations. In addition, this gives rise to the formation of varying kerf width size along the cut edges. Consequently, kerf width size increases with increasing output Laser power.

26.5 Kerf Width Optimization

Taguchi method of robust parameter design is an offline statistical quality control technique in which the level of controllable factors or input process parameters are so chosen to nullify the variation in responses due to uncontrollable or noise factors.

In Taguchi method, the experimental values of quality characteristics are used to compute the quality loss values for each quality characteristic in all experimental runs. In the present case the smaller value of kerf width is desired, therefore the S/N ratio for smaller-the-better case will be used which is given below in equation form [18].

$$\frac{\mathrm{S}}{\mathrm{N}} = -10\log\frac{1}{n}\left(\sum y^2\right) \tag{26.1}$$

Where n is the number of observations and y the observed data or each type of the characteristics. S/N ratios obtained from this equation are given in Table 26.4.

As shown in Table 26.4 and according to the Taguchi method, the optimum cutting conditions is found as S/N equal to −53.53 and − 53.47 for kerf width *Kw* of Titanium alloy Ti-6Al-4V and *Kw* of pure Titanium Ti respectively. Thus, the optimum cutting conditions which were the laser power of 2 kW, the cutting speed of 2400 mm/min and the gas pressure of 8 bars.

The interpretations can be made according to the level values of *Pu*, *V*, and *p* factors obtained for kerf width *Kw* of Ti-6Al-4V given in Table 26.5 and for *Kw* of pure Titanium Ti given in Table 26.6. The different values of S/N ratio between maximum and minimum are (main effect) also shown in these tables. Figures 26.5 and 26.6 show the graphic of the level values given in Tables 26.5 and 26.6 respectively.

The cutting speed and the gas pressure are two factors that have the highest difference between values 1.94 and 0.94 respectively for Titanium alloy and 0.91 and 0.81 respectively for pure Titanium Ti. Based on the Taguchi prediction the

Table 26.4 S/N Ratios values

EXP. No	*Pu (kW)*	*V (mm/min)*	*p (bar)*	*[Ti6Al4V] S/N ratios*	*[Ti] S/N ratios*
1	2	480	2	−55.89	−55.19
2	2	480	8	−55.36	−54.27
3	2	480	14	−55.67	−54.99
4	2	1440	2	−55.30	−54.69
5	2	1440	8	−54.56	−53.69
6	2	1440	14	−55.14	−54.53
7	2	2400	2	−54.73	−53.75
8	2	2400	8	−53.53	−53.47
9	2	2400	14	−53.68	−54.32
10	3	480	2	−56.32	−55.55
11	3	480	8	−55.44	−54.88
12	3	480	14	−57.43	−56.01
13	3	1440	2	−55.78	−55.14
14	3	1440	8	−54.90	−54.75
15	3	1440	14	−55.36	−55.37
16	3	2400	2	−54.96	−54.19
17	3	2400	8	−53.57	−54.49
18	3	2400	14	−53.98	−55.20

Table 26.5 S/N response table for *Kw* [Ti-6Al-4 V] factor

Level	*Pu (kW)*	*V(mm/min)*	*p(bar)*
1	−54.87	−56.02	−55.50
2	−55.30	−55.17	−54.56
3		−54.07	−55.21
Delta	0.43	1.94	0.94

Table 26.6 S/N response table for kerf width *Kw* [Ti] factor

Level	*Pu (kW)*	*V(mm/min)*	*p(bar)*
1	−54.32	−55.15	−54.75
2	−55.06	−54.70	−54.26
3		−54.24	−55.07
Delta	0.74	0.91	0.81

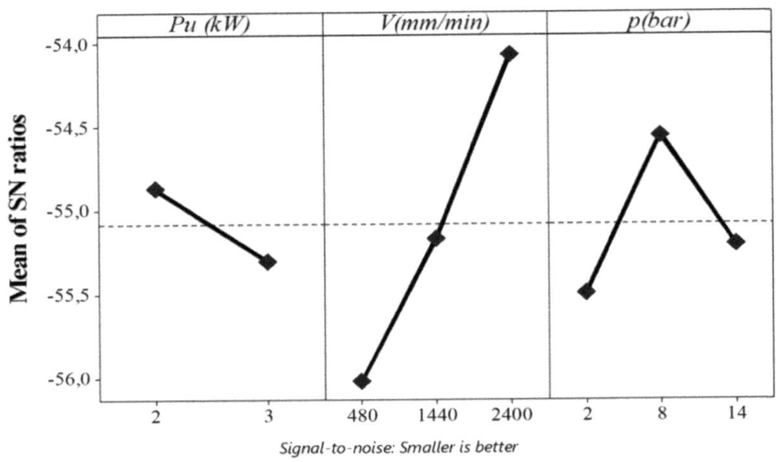

Fig. 26.5 The graphic of mean of S/N ratios for kerf width *Kw* [Ti-6Al-4V]

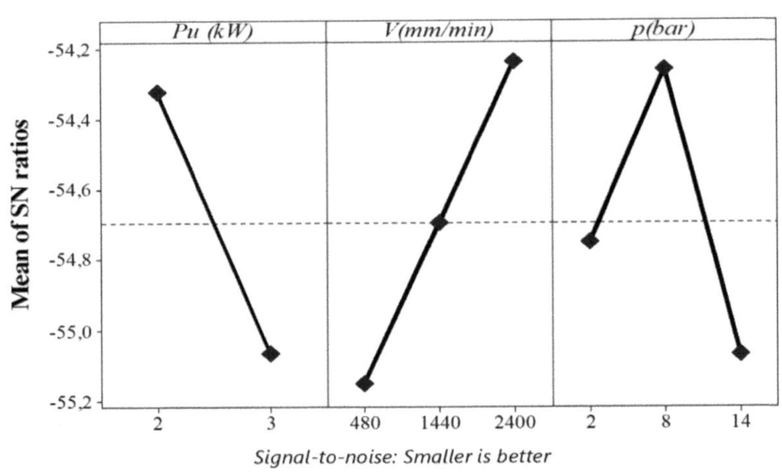

Fig. 26.6 The graphic of mean of S/N ratios for kerf width *Kw* of pure Titanium [Ti]

larger different between value of S/N ratio will have a more significant effect on Kerf width size. According to Table 26.5, the first level of laser power, the third level of cutting speed and the second level of gas pressure are higher. Consequently, the optimum cutting conditions determined under the same conditions for the experiments to be conducted will be 2 kW for laser power, 2400 mm/min for cutting speed and 8 bars for gas pressure for Ti-6Al-4V. For pure Titanium the same results are showed, hence, the optimum cutting conditions determined will be 2 kW for laser power, 2400 mm/min for cutting speed and 8 bars for gas pressure (see Table 26.6).

26.6 Conclusion

The present research consist of parametric study and analysis of influence of cutting parameters in CO_2 laser cutting of Titanium alloy (Ti-6Al-4V) and pure Titanium (Ti). The cutting parameters selected during experiments are laser power (*Pu*), cutting speed (*V*) and gas pressure (*p*), while the response parameter taken is the kerf width variation.

Basing on the results of conducted experimental research and their analysis, following conclusions were drawn:

- In the case of high power or low speed cutting process, the cracks are partially extended on the kerf.
- In the case of high speed cutting or low power, the cast layer formed at the kerf surface remains thin.
- In the case of titanium alloy, the striation depth remains large at the kerf surface.
- For both materials; Titanium alloy Ti-6Al-4V and pure Titanium Ti, the kerf width size decreases as the cutting speed increases.
- For both materials; Titanium alloy Ti-6Al-4V and pure Titanium Ti, the kerf width size increases with increasing output Laser power.
- The kerf width is proportional to laser power and inversely proportional to cutting speed.
- Kerf width Kw [Ti-6Al-4V] and Kw [Ti], S/N ratios are found as a result of experiments conducted according to the L_{18} orthogonal array. The maximum value was found by using the S/N ratio equation of "the smaller-the better," the maximum S/N ratio yielded optimum cutting parameters.
- The optimum cutting conditions for Titanium alloy Ti-6Al-4V and pure Titanium Ti determined will be 2 kW for laser power, 2400 mm/min for cutting speed and 8 bars for gas pressure.

References

1. Riveiro, A., Quintero, F., Lusquinos, F., Comesana, R., Pou, J.: Parametric investigation of CO2 laser cutting of 2024-T3 alloy. J. Mater. Process. Technol. **210**, 1138–1152 (2010)
2. Arif, A.F.M., Abdul Aleem, B.J.J.: Laser cutting of rectangular blanks in thick sheet steel: effect of cutting speed on thermal stresses. J. Mater. Eng. Perform. **19**(2), 177–184 (2010)
3. Yilbas, B.S., Shaukat, M.M., Ashraf, F.: Laser cutting of various materials: kerf width size analysis and life cycle assessment of cutting process. Opt. Laser Technol. **93**, 67–73 (2017)
4. Hernandez-Castaneda, J.C., Sezer, H.K., Li, L.: The effect of moisture content in fiber laser cutting of pine wood. Opt. Lasers Eng. **1139**, 49–52 (2011)
5. Klotzbach, A., Hauser, M., Beyer, E.: Cutting of carbon fiber reinforced polymers using highly brilliant laser beam sources. Phys. Procedia. **572**, 7–12 (2011)
6. Yilbas, B.S.: Laser cutting of thick sheet metals: effects of cutting parameters on kerf size variations. J. Mater. Process. Technol. **201**, 285–290 (2008)
7. Choudhury, I.A., Shirley, S.: Laser cutting of polymeric materials: an experimental investigation. Opt. Laser Technol. **42**(3), 503–508 (2010)
8. Shyha, I.: An investigation into CO_2 laser trimming of CFRP and GFRP composites. Procedia Eng. **63**, 931–937 (2013)
9. Riveiro, A., Mejías, A., Soto, R., Quintero, F., delVa, J.l., Boutinguiza, M., Lusquiños, F., Pardo, J.: CO_2 laser cutting of natural granite. Opt. Laser Technol. **76**, 19–28 (2016)
10. Tahir, A.F.M., NurAqida, S.: An investigation of laser cutting quality of 22MnB5 ultra high strength steel using response surface methodology. Opt. Laser Technol. **92**, 142–149 (2017)
11. Eltawahni, H.A., Hagino, M., Benyounis, K.Y., Inoue, T., Olabi, A.G.: Effect of CO_2 laser cutting process parameters on edge quality and operating cost of AISI316L. Opt. Laser Technol. **44**(4), 1068–1082 (2012)
12. Rajaram, N., Sheikh-Ahmad, J., Cheraghi, S.: CO_2 laser cut quality of 4130 steel. Int. J. Mach. Tools Manuf. **43**(4), 351–358 (2003)
13. Sharma, A., Yadava, A.V.: Modelling and optimization of cut quality during pulsed Nd: YAG laser cutting of thin Al-alloy sheet for straight profile. Opt. Laser Technol. **44**(1), 159–168 (2012)
14. Ghany, K.A., Newishy, M.: Cutting of 1.2 mm thick austenitic stainless steel sheet using pulsed and CW Nd: YAG laser. J. Mater. Process. Technol. **168**, 438–447 (2005)
15. El Aoud, B., Boujelbene, M., Bayraktar, E., Ben Salem, S., Miskioglu, I.: Studying Effect of CO2 Laser Cutting Parameters of Titanium Alloy on Heat Affected Zone and Kerf Width Using the Taguchi Method. Mechanics of Composite and Multi-functional Materials, Volume 6, Conference Proceedings of the Society for Experimental Mechanics Series. Springer International Publishing AG (2018)
16. Keles, O., Oner, U.: A study of the laser cutting process: influence of laser power and cutting speed on cut quality. Lasers Eng. **20**, 319–327 (2010)
17. Yilbas, B.S., Akhtar, S.S., Karatas, C.: Laser hole cutting into Ti-6Al-4V alloy and thermal stress analysis. Int. J. Adv. Manuf. Technol. **59**(9), 997–1008 (2012)
18. Shetty, R., Pai, R.B., Rao, S.S., Nayak, R.: Taguchi's technique in machining of metal matrix composites. J. Braz. Soc. Mech. Sci. Eng. **31**(1), 12–20 (2009)

Chapter 27
A Study of the Surface Integrity of Titanium Alloy Ti-6Al-4V in the Abrasive Water Jet Machining Process

M. Douiri, M. Boujelbene, E. Bayraktar, and S. Ben Salem

Abstract The abrasive water jet (AWJ) cutting technique is one the most rapidly improving technological methods of cutting materials. In this cutting technique, a thin, high velocity water jet accelerates abrasive particles that are directed through an abrasive water jet nozzle at the material to be cut.

Using the abrasive water jet machining process, this work investigated the effect of machining conditions, specially the cutting speed and the material thickness, on the integrity surface (including surface microstructure alterations and surface roughness) of work piece of the Titanium alloy Ti-6Al-4V. The GMT garnet was used as an abrasive material with 80 mesh. Surface integrity is defined as the inherent or enhanced condition of the surface produced in a machining or other surface generating operation. It is very important to examine the influence of cutting conditions on surface roughness and microstructure. Photographs of cut surfaces were analyzed and roughness parameters were measured in different locations across depth of cut. Differences between surface geometric structures and measured roughness parameter values obtained were presented and discussed.

Keywords AWJ · Ti-6Al-4V · Microstructure · Surface roughness · Cutting speed · Material thickness

27.1 Introduction

Presently, abrasive water jet cutting is amongst the most rapidly developing and intensely researched methods of material cutting. It belongs to the spectrum of non-conventional manufacturing methods and is widely used for cutting a wide array of construction materials, including metals, fiberglass, rubber, stone and plastics. Other advantages include small cutting forces and low cutting zone temperatures [1, 2]. Advantages of abrasive water jet cutting include the ability to cut almost all materials, such as titanium alloys Ti-6Al-4V which have been widely used in industries, especially aerospace, energy and medical industries, due to their good mechanical and chemical properties [3–5]. However, titanium alloys are typically difficult-to-cut materials [6, 7]. That is why, after cutting of Ti-6Al-4V, the surface quality, and surface roughness must receive more attention [7–9].

A considerable number of studies have investigated the effects of cutting velocity, spreading distance, water pressure, abrasive grain size and other factors on the surface roughness [10–12]. Thus, it is necessary to have a deeper knowledge of the optimal conditions of operation, which will allow us to ensure a good surface roughness.

In recent years, a large amount of research effort has been made to understand the Abrasive Water Jet process and improve its cutting performance such as the depth of cut and surface finish for various materials [13]. One of the earliest studies was carried out by Douiri et al. [14]. They used granite samples for their experimental studies and investigated the effect of process parameters on rock cutting. It was found that entraining of abrasive particles increase the cutting capability of water jet and increases of water jet pressure allow obtaining deeper or material thickness.

M. Douiri
Supmeca-Paris, School of Mechanical and Manufacturing Engineering, St-Ouen, Paris, France

University of Tunis El Manar, ENIT, Ecole Nationale d'Ingénieurs de Tunis, Le Belvedere, Tunisia

M. Boujelbene · S. Ben Salem
University of Tunis El Manar, ENIT, École Nationale d'Ingénieurs de Tunis, Tunis, Tunisia

E. Bayraktar (✉)
Supmeca-Paris, School of Mechanical and Manufacturing Engineering, St-Ouen, Paris, France
e-mail: bayraktar@supmeca.fr

P. R. Thakre et al. (eds.), *Mechanics of Composite, Hybrid and Multifunctional Materials, Volume 5*, Conference Proceedings of the Society for Experimental Mechanics Series, https://doi.org/10.1007/978-3-319-95510-0_27

Miranda and Quintino [15] experimentally studied the effect of material properties on cutting performances using calcareous stones. Liu and Chen [16] conducted an experimental study to define the effects of AWJ process parameters on cutting mechanisms and performances of granite.

While machining ductile materials like aluminium, the effect of feed rate was observed to be significant in affecting the roughness of cut surface, while the kerf characteristics were affected by the traverse speed [17, 18].

The taper on the cut surface was found to be affected by the ingredients of the abrasive slurry. Further, the presence of a polymer in the slurry could improve the metal removal rate MRR [19]. While machining the fibre-reinforced plastics, jet traverse rate was observed to affect the quality of machined surface [20]. Three distinct regions (an initial damage region, a smooth cutting region and a rough cutting region) were visualized *along the kerf* wall produced by AWJ cut process [21]. Generally, garnet was used as abrasive in machining operations but colemanite powder could also be used for obtaining better cut characteristics [22]. The multiple process inputs in AWJC and factor interactions need a thorough study to enable complete quality control of the process. Accordingly, it is very clear that information concerning surface integrity of the machined surface region will be very valuable in the design and manufacture of parts. Efforts have been made by several researchers in the past few decades to study the relationships between the machining process conditions, the nature of the surface alterations produced and their effect on product's functional performance.

Yuvaraj and Pradeep Kumar [23] experimentally are deduced that an increasing wear rate by increasing the cutting speed used was noticed. Their scanning electron microscopy examinations showed that plastic deformation and surface fracture are the dominant mechanisms during the abrasive water jet process.

Based on the recent studies, a several factors can be identified that affect the surface condition obtained even the grains of abrasive used. Thus contaminations on the depth of cut of the machined part have been noticed, this is at the origin of the abrasive grates used. Traces of metallurgical agitation of the samples were characterized by plastic deformation and damage in the form of plastic flow extending parallel to the direction of water slip [23, 24].

This work investigated the effect of machining conditions, using the GMT garnet as an abrasive material with 80 meshes, on surface integrity of workpiece of the Titanium alloy Ti-6Al-4V alloy machined using three different cutting speeds. Topography of the surface obtained was studied based on measurement of roughness depth of cut and the microstructure of the sub-machined surface was observed using an optical microscope in order to investigate the metallurgical deformation. Photographs of cut surfaces were taken with scanning electronic microscope at the difference of the surface geometric structures and local deformation. The various results of surface was analyzed.

27.2 Materials and Experimental Set-Up

The workpiece materials used in all the experiments were taken from a sheet of the titanium alloy Ti–6Al–4V. The chemical composition of the Ti–6Al–4V alloy is presented in Table 27.1.

The cutting operations were carried out under the following conditions mentioned in Table 27.2.

To ensure the measurement results, 3 measurements are taken for each sample in order to avoid recovery errors. The Taguchi method and L9 Orthogonal Array were used to reduce number of the experiments. The design of experiments (DOE) and measured Ra values are done with Minitab 17 program.

Table 27.1 Chemical composition of Ti-6Al-4 V (%)

Ti	Al	V	Fe	C	N	H	O
Base	6	4	0.3	0.08	0.05	0.01	0.2

Table 27.2 Experimental design levels used

Parameters	Symbol	Unit	−1	0	+1
Cutting speed	V	mm/min	200	300	400
Thickness material	t	mm	2	4	6

27.3 Results and Discussions

27.3.1 Microstructure of Cut Edge

The observation by scanning electron microscope (SEM) shows that the surface cut by the water jet machining is divided into two parts, the top zone, and the bottom zone as shown in Fig. 27.1a, this confirms the results of many researches which discussed this phenomenon done by Begic-Hajdarevic et al. 2014 [3] .The first texture, the top zone was located at the beginning of the cut and was characterized by the smooth surface. The second texture are presented in Fig. 27.1b. It was located at the bottom where the zone possesses a surface morphology with irregular undulations Fig. 27.1.

Surface roughness obtained in AWJ cutting application is determined by striations looked as a plastic deformation, caused by jet at the lower region of the cut surface. Typical cut front geometry and surface profile at lower region of the cut of Ti-6Al-4V alloy sample is presented. In the first hind, we can deduce the same deduction of many researches us Löschner et al. [2] so that the quality deteriorates as the material thickness and the cutting speed increases. As a consequence of the deeper penetration of the jet into the depth of workpiece, there is a loss of its energy due to the interaction with workpiece material, mutual collisions of abrasive particles, etc. That results in rougher surface characteristics at the lower region of the cut surface.

The scanning electron microscope observation, shows a voids can be observed on the abrasive water jet cut surfaces of titanium alloy, as shown in Fig. 27.2 .As indicated in the Fig. 27.2a, b, the observation by scanning electron micro-scope shows that the increase the cutting speed reduces the surface quality of the machined part. Further the increasing in cutting speed increase the voids and the lip formation;

The SEM images of the AWJ cut surfaces in Fig. 27.2b shows a wear tracks and lip formation observed in the AWJ cut surfaces, when the cutting speed in the terms of 400 mm/min were used. This is confirms that the increase in the cutting speed reduces the surface quality;

Despite, the decrease in cutting speed shows a severe contamination on the machined surface. It was occurred as a result of the fractured abrasive particles getting embedded in the kerf wall top cutting regions. The high impulse of the abrasive particles that exist with the decreasing of the cutting speed causes the long distribution of the abrasive particles in the depth

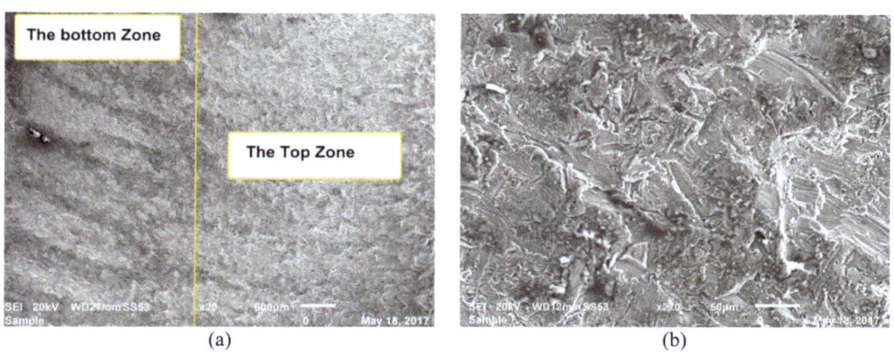

Fig. 27.1 Cut edge microstructure at V = 400 mm/min and t = 4 mm (**a**) General view, (**b**) Enlargement of the Bottom Zone

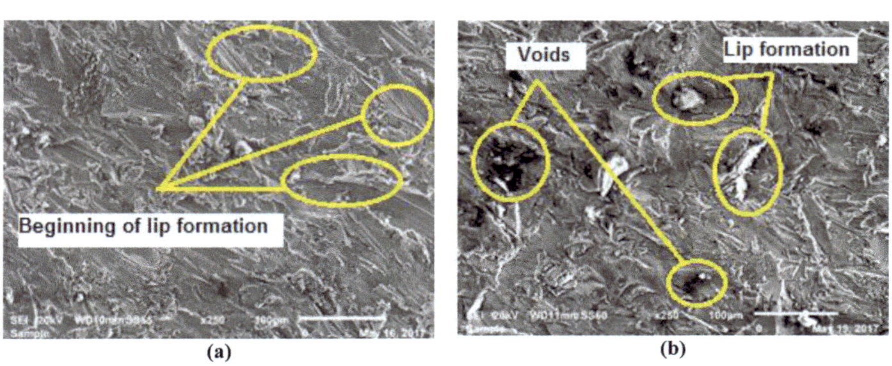

Fig. 27.2 The SEM photograph of a AWJ cut surface at t = 6 mm: (**a**) V = 200 mm/min, (**b**) V = 400 mm/min

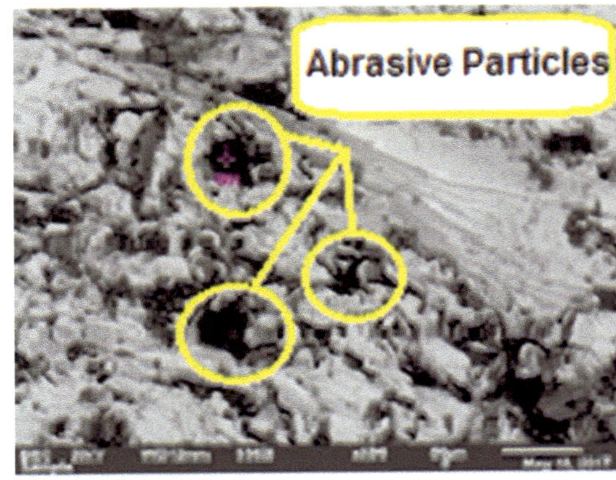

Fig. 27.3 Embedment of abrasive particles

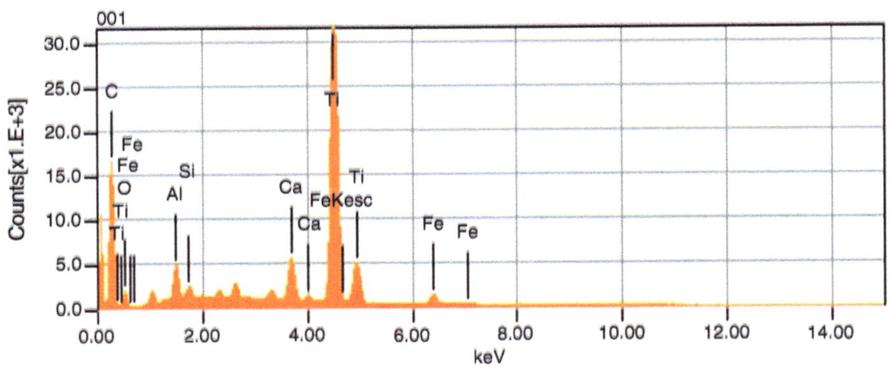

Fig. 27.4 Energy dispersive spectroscopy analysis

cutting. This happens because at higher cutting speed, lesser number of abrasive particles impinges on cutting surface due to smaller interaction time between abrasive jet and workpiece. At lower speeds, more number of abrasive particles impinges on cutting surface due to more interaction time between abrasive jet and workpiece and, therefore, better surface finish is observed. The increasing of cutting speed decrease the severe contamination, which was analyzed as an abrasive particles which was getting embedded in the cutting surface wall top cutting regions.

Figures 27.3 and 27.4 shows the results of energy dispersive spectroscopy analysis. This confirms the results of many researches which discussed this phenomenon done by Yuvaraj and PradeepKumar [23].

27.3.2 Surface Roughness Analysis

Surface roughness of the cutting surface is defined as the inherent irregularities of the workpiece form circular holes affected by interaction abrasive water jet /workpiece. The parameter used in this study to characterize surface roughness is the arithmetic mean heights of surface (Sa), which is given in Eq. (27.1). In addition, the output process parameters were obtained using a 3D optical profilometer surface treated by software program MountainsMap. The layout of:

$$Sa = \frac{1}{a} \iint a \; |Z \; (x, y)| \, dx \, dy \tag{27.1}$$

Where "a" is the surface taken for observation and "z" the ordinate of the surface roughness.

There are many parameters used in the literature and industry related to surface roughness. Among the two dimensional (2D) surface roughness parameters, the most popular is average roughness. It is quoted as Ra (see Eq. (27.2)). Mathematically, Ra is the arithmetic value of the departure of the profile from centerline along sampling length:

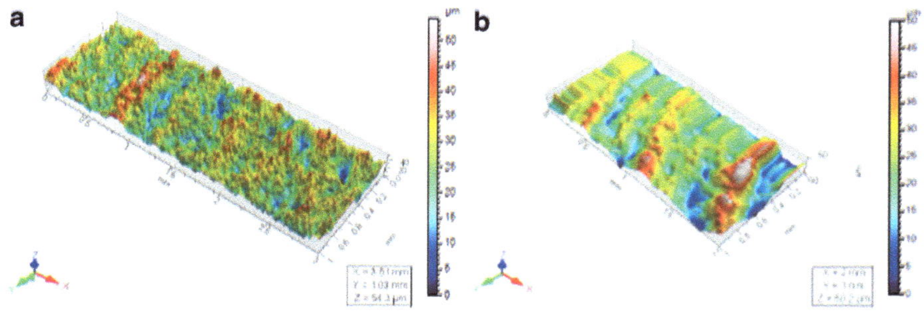

Fig. 27.5 3D surface topography: (a) V = 300 mm/min; (b) V = 400 mm/min

Fig. 27.6 The effect of the workpiece thickness (t), on the Roughness (Ra), and Cutting Speed (V) fixed

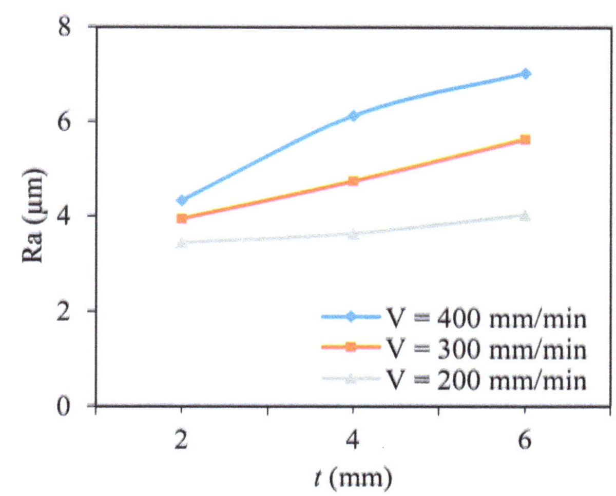

$$Ra = \frac{1}{a}\int_0^a |Z(x)|\,\mathrm{d}x \tag{27.2}$$

A considerable number of studies have investigated the effects of cutting velocity, spreading distance, water pressure, abrasive grain size and other factors on the surface roughness. Thus, it is necessary to have a deeper knowledge of the optimal conditions of operation, which will allow us to ensure a good surface roughness.

In our study, the 3D topography study show in Fig. 27.5 that an increases the surface roughness is noticed with the increasing of cutting speed. This is confirms the first part study of microscopic observation.

The 3D topography shows us a variation in the surface quality, in Fig. 27.5: An increase in the cutting speed caused an increase in the surface roughness, in particular at the bottom of the cut this is occurred with our deduction of the earliest SEM observation.

In order to demonstrate the shift effect on the average roughness value of the part to be cut, roughness measurements were taken at the sample media with different thicknesses (t): 2 mm, 4 mm and 6 mm. The study of the shifting effect on the surface roughness during abrasive water jet cutting measured in Fig. 27.6 . Increasing the material thick value involves increasing the value of Ra. A smaller variation in surface roughness occurred at the lower cutting speed value. It is found that the surface roughness increases by increasing the cutting speed and the material thick. Furthermore, it can be observed that the roughness Ra changes slightly over the entire depth of the cut surface at low displacement speeds. Starting at 300 mm/min, the increase of Ra is very considerable and is almost equal to 50% of the variation;

27.4 Analyzing and Evaluating Results of the Experiments Using Taguchi Method

The Taguchi method is an experimental design technique which is useful in reducing the number of experiments dramatically by using orthogonal arrays. Taguchi creates a standard orthogonal array to accommodate this requirement. Taguchi used the signal-to-noise (S/N) ratio as the quality characteristic of choice. S/N ratio is used as a measurable value instead of standard

Table 27.3 S/N Ratios values

Experimental number	V (mm/min)	t (mm)	R_a (μm)	S/N ratios
1	200	2	3.4	−10.629
2	200	4	3.6	−11.126
3	200	6	4.0	−12.041
4	300	2	3.9	−11.821
5	300	4	4.5	−13.064
6	300	6	5.6	−14.963
7	400	2	4.3	−12.669
8	400	4	6.1	−15.706
9	400	6	7.0	−16.902

Table 27.4 S/N response table for Ra factor

Level	V (mm/min)	t (mm)
1	−11.27	−11.71
2	−13.28	−13.30
3	−15.09	−14.64
Delta	**3.83**	**2.93**

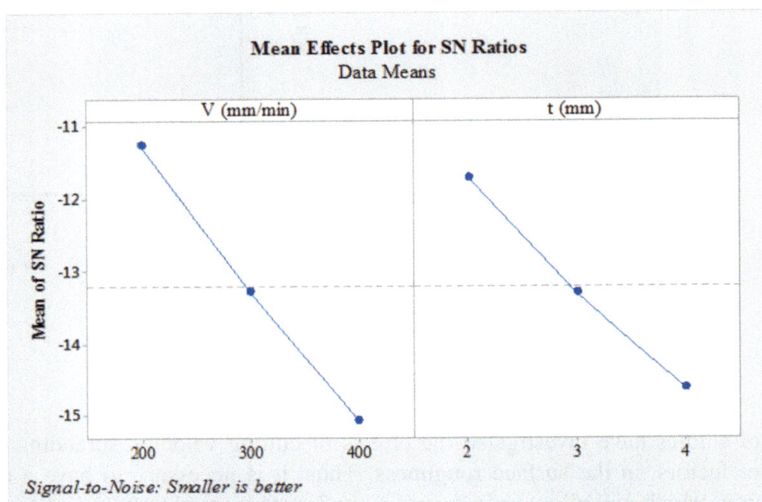

Fig. 27.7 The graphic of mean of S/N ratios for Ra

deviation because as the mean decreases, the standard deviation also decreases and vice versa. In the study, we use the S/N ratio us "smaller is the better" characteristics by Eq. (27.3), when the characteristic is continuous.

Smaller is the better characteristic,

$$\frac{\mathrm{S}}{\mathrm{N}} = -10\log\frac{1}{n}\left(\sum y^2\right) \tag{27.3}$$

Where y is the average of observed data, $\frac{\mathrm{S}}{\mathrm{N}}$ the variation of y, (n) the number of observations, and (y) the observed data or each type of the characteristics. S/N ratios and level values were calculated by using Eq. (27.3), "the smaller-the better" in the MINITAB 17 Program. S/N ratios obtained from this equation are given in Table 27.3.

As shown in Table 27.3 and according to the Taguchi method, the optimum cutting conditions was found as S/N equal to −10.629.Thus, the optimum cutting conditions which were the cutting speed of 200 mm/min, and the thickness material of 2 mm.

The interpretations can be made according to the level values of (*V*), and (*t*) factors obtained for Ra given in Table 27.4. The different values of S/N ratio between maximum and minimum are (main effect) also shown in this table. Figure 27.7 shows the graphic of the level values given in Table 27.4.

The cutting speed and thickness material are two factors that have the highest difference between values 3.83 and 2.39 respectively. Based on the Taguchi prediction that the larger different between value of S/N ratio will have a more significant effect on surface roughness. According to Table 27.4, the third level of cutting speed, the first level of thickness material are higher. Consequently, the optimum cutting conditions determined under the same conditions for the experiments to be conducted will be 400 mm/min for cutting speed and 2 mm for thickness material.

27.5 Conclusion

Surface integrity studies of 3D surface topography, 2D surface roughness profiles, and micrographs have been carried out on AWJ cutting of Titanium alloy Ti-4V-6Al through variation of cutting speed and thickness material. The major conclusions are given below:

- The SEM analysis confirms the presences of wear tracks by the AWJ cutting surfaces which is increasing with the increasing of cutting speed. Further, a very small number of abrasive particles were embedded in the machined surface wall cut surface as necessary.
- An increase of the cutting speed always has the same effect to increase the surface roughness value.
- Surface roughness parameters Ra values increase with the increasing of the cutting speed and the thickness material, which increase the plastic deformation in the lower part of the workpiece;
- The better surface integrity was achieved through a lower cutting speed value and uniform pattern, inconsequential presence of the abrasive contamination in the cut surfaces.

References

1. Ramprasad, Upadhyay, G., Hassan, K.: Optimization MRR of stainless steel 403 in abrasive water jet machining using Anova and Taguchi method. J. Eng. Res. Appl. **5**, 86–91 (2015)
2. Löschnera, P., Jarosza, K., Niesłonya, P.: Investigation of the effect of cutting speed on surface quality in abrasive water jet cutting of 316L stainless steel. Procedia Eng. **149**, 276–282 (2016)
3. Begic-Hajdarevic, B., Cekic, A., Mehmedovic, M., Djelmic, A.: Experimental study on surface roughness in abrasive water jet cutting. Procedia Eng. **100**, 394–399 (2015)
4. Krajcarz, D.: Comparison metal water jet cutting with laser and plasma cutting. Procedia Eng. **69**, 838–843 (2014)
5. Daymi, A., Boujelbene, M., Bayraktar, E., Ben Amara, A., Katundi, D.: Influence of feed rate on surface integrity of titanium alloy in high speed milling. Adv. Mater. Res. **264**, 1228–1233 (2011)
6. Cojbasic, Z., Petkovic, D., Shamshirband, S., Wen Tong, C., Sudheer, C., Jankovic, P., Ducic, N., Baralic, J.: Surface roughness prediction by extreme learning machine constructed with abrasive water jet. Precis. Eng. **43**, 86–92 (2016)
7. Hreha, P., Radvanska, A., Hloch, S., Perzel, V., Krolczyk, G., Monkova, K.: Determination of vibration frequency depending on abrasive mass flow rate during abrasive water jet cutting. Int. J. Adv. Manuf. Technol. **77**(1–4), 763–774 (2015)
8. Lehocka, D., Klich, J., Foldyna, J., Hloch, S., Krolczyk, J.B., Carach, J., Krolczyk, G.M.: Copper alloys disintegration using pulsating water jet. Measurement. **82**, 375–383 (2016)
9. Boujelbene, M., Ezzdini, S., Elboughdiri, N., Ben Salem, W., Youssef, W.: Investigation on the surface roughness of the high steel material after wire electrical discharge machining (WEDM) process. Int. J. Adv. Appl. Sci. **4**(6), 130–136 (2017)
10. Kovacevic, R.: Surface texture in abrasive waterjet cutting. J. Manuf. Syst. **10**, 32–40 (1991)
11. Akkurt, A., Kulekci, M.K., Seker, U., Ercan, F.: Effect of feed rate on surface roughness in abrasive water jet cutting applications. J. Mater. Process. Technol. **147**, 389–396 (2004)
12. Axinte, D.A., Srinivasu, D.S., Kong, M.C., Butler-Smith, P.W.: Abrasive water jet cutting of polycrystalline diamond; a preliminary investigation. Int J Mach Tool Manu. **49**, 797–803 (2009)
13. Valicek, J., Drzik, M., Ohlidal, M., Madr, V., Hlavac, L.M.: Optical method for surface analyses and their utilization for abrasive liquid jet automation. In: Proceedings of the American Waterjet Conference (WJTA), Minneapolis, pp. 1–11 (2001)
14. Douiri, M., Boujelbene, M., Ben Salem, S., Bayraktar, E.: Effects of the Abrasive Water Jet Machining Process Parameters on the Surface Roughness of the Titanium Alloy Ti-6Al-4V, Conference AMPT (2017)
15. Miranda, M.R., Quintino, L.: Microstructural study of material removal mechanisms observed in abrasive waterjet cutting of calcareous stones. Mater. Charact. **54**, 370–377 (2005)
16. Liu, Y., Chen, X.: A study on the abrasive water jet cutting for granite. Key Eng. Mater. **257–258**, 527–532 (2004)
17. Akkurt, A., Kulekci, M.K., Seker, U., Ercan, F.: Effect of feed rate on surface roughness in abrasive waterjet cutting applications. J. Mater. Proc. Technol. **147**(3), 389–396 (2004)
18. Ma, C., Deam, R.T.: A correlation for predicting the kerf profile from abrasive water jet cutting. Exp. Thermal Fluid Sci. **30**(4), 337–343 (2006)

19. Mahabalesh, P.: A study of taper angles and material removal rates of drilled holes in the abrasive water jet machining process. J. Mater. Proc. Technol. **189**(1–3), 292–295 (2007)
20. Azmir, M.A., Ahsan, A.K., Rahmah, A.: Effect of abrasive water jet machining parameters on aramid fibre reinforced plastics composite. Int. J. Mater. Form. **2**(1), 37–44 (2009)
21. Ay, M., Caydas, U., Hascalik, A.: Effect of traverse speed on abrasive waterjet machining of age hardened Inconel 718 nickel-based super alloy. Mater. Manuf. Process. **25**(10), 1160–1165 (2010)
22. Cosansu, C., Cogun, C.: An investigation on use of colemanite powder as abrasive in abrasive waterjet cutting (AWJC). J. Mech. Sci. Technol. **26**(8), 2371–2380 (2012)
23. Yuvaraj, N., Pradeep Kumar, M.: Cutting of aluminum alloy with abrasive water jet and cryogenic assisted abrasive water jet: a comparative study of the surface integrity approach. Wear. **362–363**, 18–32 (2016)
24. Shetty, R., Pai, R.B., Rao, S.S., Nayak, R.: Taguchi's technique in machining of metal matrix composites. J. Braz. Soc. Mech. Sci. Eng. **31**(1), 12–20 (2009)

Chapter 28
Process Reliability of Abrasive Water Jet to Cut Shapes of the Titanium Alloy Ti-6Al-4V

M. Douiri, M. Boujelbene, E. Bayraktar, and S. Ben Salem

Abstract Titanium alloys have been widely used in industries, especially aerospace, energy and medical industries, due to their good mechanical and chemical properties. However, titanium alloys are typically difficult-to-cut materials. That is why, in industries we are using many process to cut this alloys. In fact, several studies have made a comparisons between the various machining processes currently used in industries to cut Titanium alloys.

Remains the abrasive water jet process is the most answered in terms of quality quantity and reliability. In this cutting technique, a thin, high velocity water jet accelerates abrasive particles that are directed through an abrasive water jet nozzle at the material to be cut. Advantages of abrasive water jet cutting include the ability to cut almost all materials, no thermal distortion, and high flexibility, small cutting forces and being environmentally friendly. The mechanism and rate of material removal during AWJ cutting depends both on the type of abrasive and on a range of process parameters.

The presented work aims at studying the behavior machinability in Ti-6AL-4V alloys using the GMTas an abrasive material with 80 meshes. Photographs of cut shapes were taken with a 2D machining which show the impact of parameters conditions on surface geometric in different location. A comparative measurement of the kerf and the precision of the angle was taken with a profile projectors, and the defects of cut was discussed.

In order to increase the relibility of the abrasive water jet process, and to anticipate an estimation of the machinability of the material, a mathematical model of the Kerf width has been put in place, which aims at optimization of cutting parameters by minimizing the Kerf.

Keywords Abrasive Water Jet Cutting · Titanium alloy Ti-6Al-4V · Cutting speed · Material thickness · Surface morphology · Kerf

28.1 Introduction

The technology abrasive water jet cutting (AWJC) has been upgraded significantly to meet the necessity of the modern industries. AWJC is used in difficult-to hard material for material removal by several manufacturing industries. AWJC process provides a high-quality cutting with a great surface finish and therefore it is advantageous in the manufacturing of several components in automotive, aerospace, surgical equipment's and defense sector industries owing to its characteristics. AWJC process utilizes the mixture of water and abrasive to erode material from the target surface. The AWJC process has several machining parameters, i.e., water pressure, transverse speed, abrasive flow rate, standoff distance, abrasive grit size, etc., which affect the performance parameters (i.e., surface quality, kerf top width, taper angle and material removal rate). To obtain the optimum parameter setting for machining process is utmost important to achieve desired quality. Few researchers have reported the influence of the process parameters on the performance characteristics of the AWJC process [1–10].

Sreenivasa Rao et al. [11] studied the effect water pressure, traverse speed and stand-off distance on mild steel in abrasive water jet machining. In this research they have taken one output parameter i.e. surface roughness. They have cut the mild

M. Douiri
Supmeca-Paris School of Mechanical and Manufacturing Engineering, St-Ouen, Paris, France

University of Tunis El Manar, ENIT, Ecole Nationale d'Ingénieurs de Tunis, Le Belvedere, Tunisia

M. Boujelbene · S. Ben Salem
University of Tunis El Manar, ENIT, École Nationale d'Ingénieurs de Tunis, Tunis, Tunisia

E. Bayraktar (✉)
Supmeca-Paris School of Mechanical and Manufacturing Engineering, St-Ouen, Paris, France
e-mail: bayraktar@supmeca.fr

P. R. Thakre et al. (eds.), *Mechanics of Composite, Hybrid and Multifunctional Materials, Volume 5*, Conference Proceedings of the Society for Experimental Mechanics Series, https://doi.org/10.1007/978-3-319-95510-0_28

steel at different water pressure, traverse speed and stand-off distance with using the Taguchi method L9. By analyzed they have found that the optimum value of the different parameters water pressure, traverse speed and stand-off distance, at which the surface roughness will be decrease [11].

Kamlesh H. Thakkar et al. [12] studied the effect of traverse speed, stand-off distance and abrasive flow rate on abrasive water jet machining of mild steel 250. They have cut the mild steel at different traverse, stand-off distance and abrasive flow rate by using the Taguchi method L9. By analyzed they have found that the optimum selection of the three basic parameters, i.e., traverse speed, abrasive flow rate and standoff distance is at the minimum surface roughness and maximum metal removal rate. The effect of each of these parameters is studied while keeping the other parameters considered in this study as constant. Chithirai Pon Selvan et al. [13] studied the different parameters in abrasive water jet machining of aluminium with dimensions of the sheet 150 × 100 × 60 mm. They have cut the aluminium with the variable parameters water pressure, traverse speed, abrasive flow rate and stand-off distance using Taguchi method L9 and the garnet 80 mesh is used in the form of abrasive. By analyzed they have found that the effect of surface roughness with respect to these four variable parameters. As the jet pressure increases, the surface becomes smoother [13].

Boud et al. [14] have studied in abrasive water jet machining of aluminum 7475 with taking parameters traverse speed, stand-off distance and jet pressure. They have taken aluminium plate with thickness 3 mm to find the effect of material removal rate and fatigue life [14]. They have find out that the error bars were determined by the result of the four samples. By analyzed in the graph the number of cycles where the machined surface has failed is approximately 46,000, as shown by the red line, however for the machined surface the failure was approximately between 39,000 and 49,000. And that results in a reduction of material removal rate as the water jet has to overcome the residual compressive stress.

Wang [15] studied about the changing different parameters such as water pressure, abrasive flow rate, traverse speed and stand-off distance. To analysis the effect of depth of cut at changing these parameters. In this research they have cutting of the borosilicate glass with changing these parameters with using the Taguchi method L27 and find out that Testing of the estimated depth of cut model and validation of the optimum result indicate their practical adequacy. Finally, SEM study of the cut wall gives a possible material erosion mechanism of amorphous material [15]. John Kechagias et al. [16] studied for quality characterization of abrasive water jet machining of TRIP sheet steels with dimension 10 × 10 cm^2 by using the Taguchi method L18. They have find out that signal to noise (S/N) ratio represents the response of the data observed in the Taguchi design of experiments. Both, the arithmetic mean roughness and the mean kerf width was characterized as "the smaller the better" quality characteristics since lower values are desirable [16].

Zhongbo Yue et al. [17] studied the material removal rate effect in the abrasive water jet machining process with changing in the input parameters such as water pressure, jet feed speed, abrasive mass flow rate and nozzle tilted angle. They obtained that if cutting of the alumina ceramic then increase the MRR with the increasing water pressure and mass flow rate [17]. Jun Wang [18] studied the effect of depth of cut in multi-pass cutting and single pass cutting in the abrasive water jet machining. Jun Wang have cut the alumina ceramic in both cutting form with four controlled parameters i.e. water pressure, nozzle traverse speed, nozzle stand-off distance and abrasive flow rate [18]. He found that in single pass cutting nozzle oscillation cutting creates a scanning cutting action by the particles that not only reduces the particle interference, but also clears the target surface and in the multi-pass cutting the number of passes may be not endlessly increased to increase the total depth of cut for thick materials.

The presented work aims at studying the behavior machinability in Ti-6AL-4V alloys using the GMT as an abrasive material with 80 meshes. Photographs of cut shapes were taken with a 2D machining which show the impact of parameters conditions on surface geometric in different location. A comparative study and measurement of the kerf width and the precision was taken with a profile projectors, and the defects of cut was discussed. This study shows that an increasing of the defects of cut by increasing the cutting speed used was noticed.

28.2 Experimental Setup

28.2.1 Material Work

The workpiece materials used in all the experiments were taken from a sheet of the titanium alloy Ti–6Al–4V for three thickness. The chemical composition of the Ti–6Al–4V alloy is presented in Table 28.1.

28.2.2 Worpiece to Cut

The workpiece material Ti-6AL-4V is used to be cut in this form with the water jet process (Fig. 28.1);

28.2.3 Experiments Machine

The experiments were carried out on an NC 3015 EB abrasive water jet cutting system with a KMT Streamline TM system. Ultra-high pressure pumps SL-V 50 capable of providing a maximum water pressure of 600 MPa. In this experimental study, there are 6 constant parameters throughout the Abrasive water Jet Machine AWJM process to cutting a sheet of the titanium alloy Ti–6Al–4V (Fig. 28.1).

The Table 28.2 show the six constant technical parameters.

28.2.4 Variables Parameters and Factors Studied

For this study, the cutting speed (V) and the material thickness (t) to be machined were selected as a variable process parameter (Table 28.3).

Table 28.1 Chemical composition of the titanium alloy

Element	Ti	Al	V	Fe	C	N	H	O
%	Balance	6	4	0.3	0.08	0.05	0.01	0.2

Fig. 28.1 Form of the workpiece

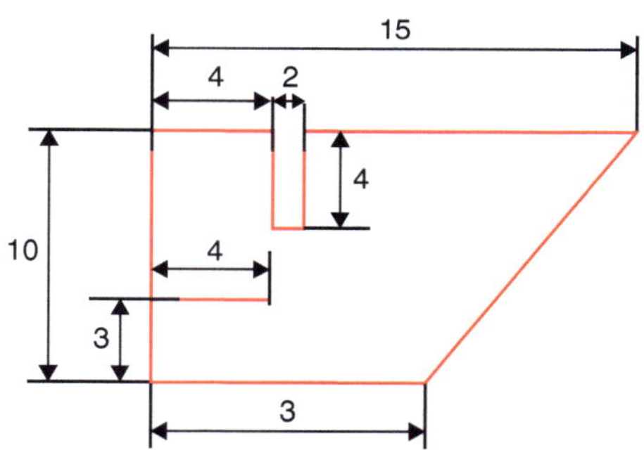

Table 28.2 Constant technological parameters

Technical Parameters	Value
Cuttinglength	150 mm
Water pressure	415 MPa
Type of abrasive	Mineral
Density of the abrasive	0.5 g/l
Abrasive grain size	80 g
Nozzlediameter	0.76 mm

Table 28.3 Design structure of experiment parameters and levels

Control Parameter	Min	Inter	Max	Out Put
Cutting speed V (m/min)	200	300	400	Width of the kerf
Thickness material t (mm)	2	4	6	

28.3 Surface Morphology

It consists of the geometric characteristics of the external cut surface, such as surface texture, surface roughness, surface morphology and visual examination. The following characterization techniques constitute the evaluation of the surface topography of the Abrasive Water Jet cutting or machining.

Figure 28.2 shows cutting surface view of samples machined at traverse speed of 200–400 mm/min, respectively. Microstructural evaluation of the cutting surfaces of samples revealed three distinct zones which were identified as:

- *Zone 1*, a initial damage region IDR, which is cutting zone at shallow angles of attack;
- *Zone 2*, a smooth cutting region, SCR, witch is cutting zone at large angles of attack;
- *Zone 3*, a rough cutting region (RCR), which is the jet upward deflection zone.

The surface morphology in different regions of cutting surface is generated from the instantaneous penetration of abrasive waterjet (Fig. 28.2). It is expected that these regions would change with an increasing interaction of the abrasive water jet with the material, i.e., increased overlap at any region of cut.

As can be seen from the Fig. 28.2, as the cutting speed or traverse speed increases, the number of particles impinging on a given exposed target area decreases, which in turn reduces the *zone 1*, initial damage region width slightly.

Scanning electron microscopy analysis of the cutting surface texture revealed that the mechanism of material removal was a combination of scooping induced ductile shear and ploughing actions of the abrasive particles (Fig. 28.2). In a previous study, it was claimed that the cutting mechanism in the *zone 1*, (initial damage region) and zone 2, (smooth cutting region) could be considered as cutting wear and deformation wear, respectively, while in the *zone 3*, (rough cutting region), the cutting process is considered as being controlled by erosive wear at large particle attack angles (Fig. 28.2).

The degree of plastic deformation increases from the top to the bottom of the cutting surfaces. Typical microstructures from these regions are shown in Fig. 28.2. In the *zone 1*, of the cutting surface, abrasive particles have a sufficient level of kinetic energy to destruct the material. This small damaged region, *zone 1*, is characterized by a small rounded corner at the top edge of the workpiece due to the plastic deformation of material caused by the initial AWJ bombardment. As the abrasive particles penetrate into the material, some of the energy is used in eroding the material in the *zone 2*, smooth cutting region, and the stream loses kinetic energy.

The degree of deflection increases with increasing traverse speed or cutting speed *V*. We suggest that striation formation is a result of reduction in the jet energy. According to other reported studies, Four major sources contribute to the striation formation simultaneously, i.e., the nature of the step formation cutting process, the waterjet dynamic characteristics, the rigidity and the system vibration of the machine and the microstructure of the material.

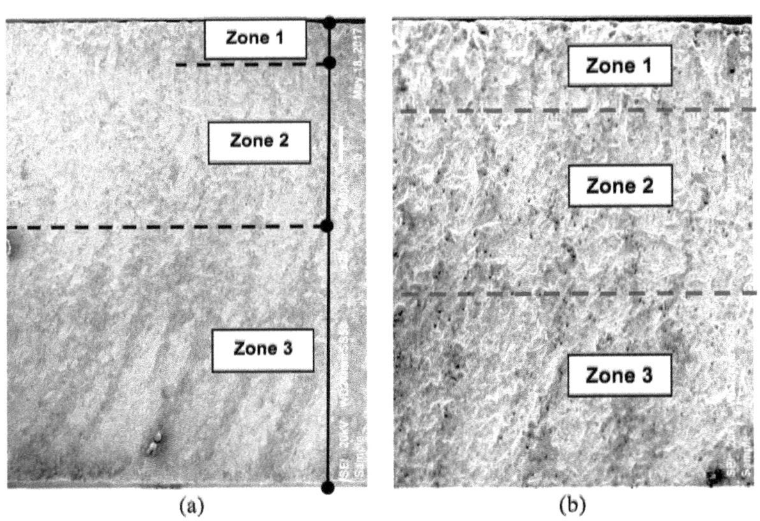

Fig. 28.2 Photograph cut surfaces obtained after AWJ cutting; (**a**) Thickness $t = 6$ mm, (**b**) Thickness $t = 4$ mm

28.4 Kerf Width Analysis

The generation of the kerf is an important phenomenon in abrasive water jet machining and is to be modeled analytically. Figure 28.3 show the cut produced by an AWJC in different sections, known as a kerf, which calculated by the formula (28.1). The top of the kerf is wider than the bottom of the kerf. Generally, in practice, the top width of the kerf is superior to the diameter of the AWJ. The top and bottom width of cut (W_{top} and W_{bottom} respectively), were measured with a stereoscopic microscope.

$$Kw = (\,Top\ width + Bottom\ width)\,/2 \tag{28.1}$$

Table 28.4 and Figs. 28.3, 28.4, and 28.5 shows the variation of kerf width with cutting speed and the thickness material, the mass flow of abrasive grain are kept constant;

In Fig. 28.4, the kerf width is found to increase with the increase in thickness material. However, the kerf decreases with the increasing of cutting speed. As the thickness material is increased, the interaction time between the abrasive water jet and the metal decreases. This results in the increase in the kerf.

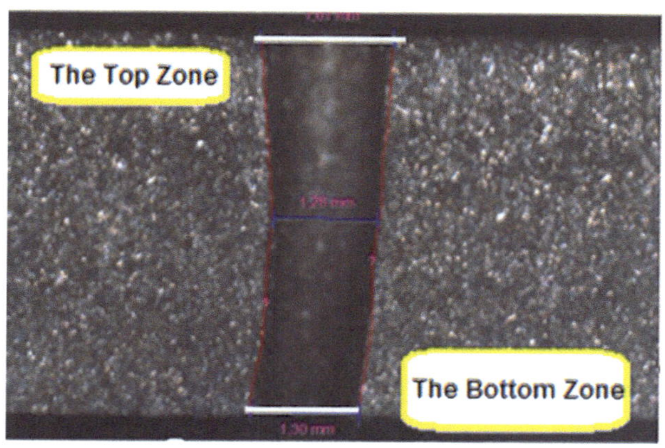

Fig. 28.3 Form of the Kerf to thickness material equal to 6 mm, V = 300 mm/min

Table 28.4 Variation of Kerf width with cutting speed at $p = 2$ bar

Cutting speed V(mm/min)	$t = 2\ mm$	$t = 4\ mm$	$t = 6\ mm$
	Kerf width (mm)		
200	1.55	1.76	1.85
300	1.50	1.65	1.80
400	1.32	1.55	1.60

Fig. 28.4 Kerf as function of the material thickness (t) and the cutting speed V

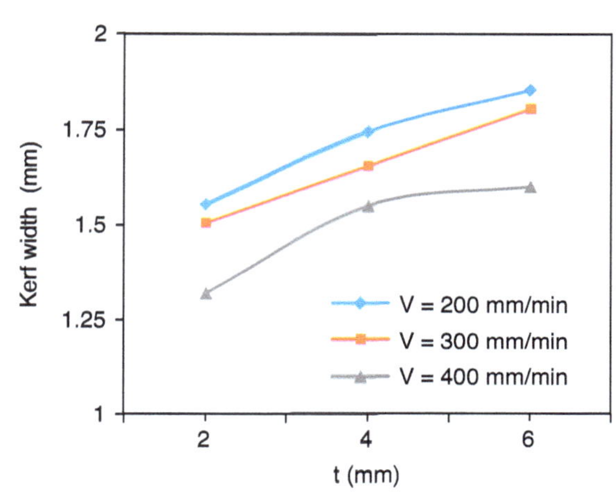

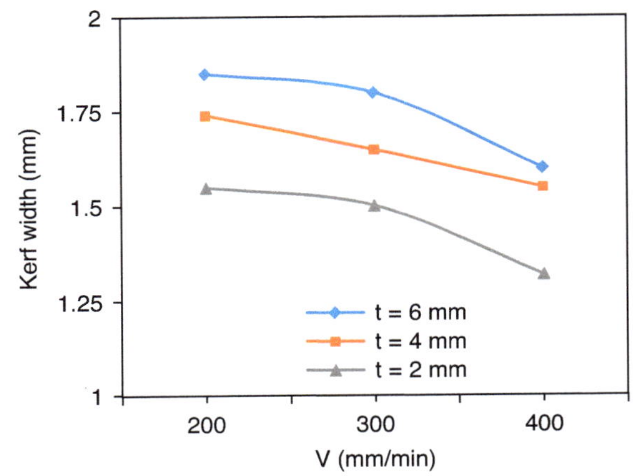

Fig. 28.5 Evolution of the Kerf as function of the material thickness (t) and cutting speed V

If the cutting speed is made very higher in the order of 400 mm/min and the thickness of the workpiece material is made very small in the order of 2 mm, the interaction between abrasive water jet and the metal may not be sufficient to initiate the kerf.

Figure 28.5 presents the variation of the kerf width with the cutting speed. It is seen that though the kerf width decreases with the decrease of the material thickness. The traverse speed has a great effect on the width kerf. By increasing the depth of cut or the material thickness and the traverse speed, the kerf width decreases (Fig. 28.5). As shown in Fig. 28.5, the kerf width decreases when the traverse speed increases, and when the thickness material decreases. This confirms the results of many researches which discussed this phenomenon.

28.5 Results Kerf Analysis Using Taguchi Method

The goal of this research was to produce minimum arithmetic mean heights of width kerf (K_W) of the friction stir welded joint. A smaller K_W values represent the best response. In order to evaluate influence of input parameters (V and t) on response (K_W); the performance indicator correspondent for this study is the S/N ratio with a lower-the-better characteristic developed by Taguchi (Tables 28.5, 28.6, and 28.7). The ratio is given by the Eq. (28.3).

$$\frac{S}{N} = -10\log\left(\frac{1}{n}\sum y^2\right) \tag{28.2}$$

Whereas the design of experiment based on Taguchi's L9 Orthogonal Array method is shown in Table 28.5. The obtained values of responses are then compared with predicted values of regression equations. Minitab 17 version statistical software is used to generate regression equations and for analysis of obtained data Taguchi Method is used.

The influence of cutting parameters on the Kerf width was studied using the statistical method; Analysis of variance "ANOVA". A mathematical model of the *kerf width* was determined using the Taguchi method.

Based on these results, we calculated the mean effects and the total of each parameter and their interaction which allows us to determine the ANOVA Table 28.6.

In Table 28.6 Df stands for Degree of Freedom, SS is the Sum of Squares and Show and the mean Squares MS, Eqs. (28.4) to (28.6).

$$MS = \frac{SS}{Df} \tag{28.3}$$

$$F_{test} = \frac{MS}{MS_{error}} \tag{28.4}$$

$$F_{theoritical} = F_{(Df, Df_{error})} \tag{28.5}$$

Table 28.5 Experimental conditions and results

	Original Number		Value Code			
N°	V (mm/min)	t (mm)	A	B	The Kerf width (mm)	S/N Kerf width (mm)
1	200	2	−1	−1	1.55	−3.806
2	200	4	−1	1	1.76	−5.105
3	200	6	−1	+1	1.85	−5.343
4	300	2	1	−1	1.50	−3.521
5	300	4	1	1	1.65	−4.349
6	300	6	1	+1	1.80	−5.105
7	400	2	+1	−1	1.32	−2.411
8	400	4	+1	1	1.55	−3.806
9	400	6	+1	+1	1.60	−4.082

Table 28.6 ANOVA kerf width results for average roughness

Source	Df	Sum of squares SS	Mean Square MS	F_{test}		$F_{theoretical}$
A	2	648.96	324.48	14.30	≥	11.3
B	2	1234.24	617.12	27.19	≥	11.3
Errors	4	90.80	22.70			
Total	8	1974.01				

Table 28.7 S/N response table for Kerf factor

Level	Cutting Speed (mm/min)	Thickness material (mm)
1	−4.752	−3.247
2	−4.326	−4.421
3	−3.434	−4.844
Delta	1.318	1.597

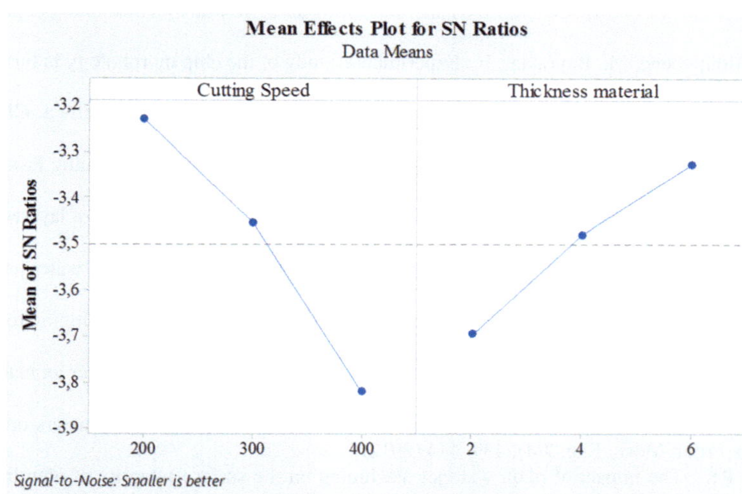

Fig. 28.6 The graphic of mean of S/N ratios for K_w

F is determined from the Fisher-Snedecor table for $\alpha = 5\%$.

According to Table 28.6, F_{test} (A) and F_{test} (B) are superior to $F_{theoretical}$ in contrast to their interaction. Thus, only the parameters A and B have a significant effect on the quality surface. The thickness is greater than the thickness of material to be machined because F_{test} (B) > F_{test} (A) (Fig. 28.6).

$$\text{Kerf} = \exp\left(Y'\right) \tag{28.6}$$

$$Y' = 0.52 – 0.00075\, V + 0.045\, t \tag{28.7}$$

28.6 Conclusions

In this study, the profiles of machined surfaces, kerf geometries and microstructural features of the machined surfaces in terms of cutting speed or traverse speed in abrasive water jet AWJ-machined Titanium alloy Ti–6Al–4V were investigated experimentally. Summarizing the main features of the results, the following conclusions may be drawn:

- Microstructural evaluation of the cutting surfaces of samples revealed two distinct zones which were identified as:
  - *Zone 1*, an initial damage region or zone 1, which is cutting zone at shallow angles of attack,
  - *Zone 2*, a smooth cutting region, which is cutting zone at large angles of attack,
  - *Zone 3*, a rough cutting region, which is the jet upward deflection zone.
- As the cutting speed increases, the AWJ cuts narrower kerf widths with a greater kerf taper ratio. This is because the traverse speed or cutting speed of abrasive waterjet allows fewer abrasives to strike on the jet target and hence generates a narrower slot.
- The kerf taper ratio differs for increasing cutting speed in the abrasive water jet cutting process.
- The size of the Three zone is depend of the AWJ cutting conditions and the parameters design.

References

1. Wang, J., Liu, H.: Profile cutting on alumina ceramics by abrasive waterjet. Part 1: experimental investigation. Proc. Inst. Mech. Eng. Part C: J. Mech. Eng. Sci. **220**, 703–714 (2006)
2. Wang, J.: Predictive depth of jet penetration models for abrasive waterjet cutting of alumina ceramics. Int. J. Mech. Sci. **49**, 306–316 (2007)
3. Boujelbene, M., Abellard, P., Bayraktar, E., Torbaty, S.: Study of the milling strategy on the tool life and the surface quality for knee prostheses. J. Achiev. Mater. Manuf. Eng. **31**(2), 610–615 (2008)
4. Azmir, M.A., Ahsan, A.K.: A study of abrasive water jet machining process on glass/epoxy composite laminate. J. Mater. Process. Technol. **209**, 6168–6173 (2009)
5. Khan, A.A., Haque, M.M.: Performance of different abrasive materials during abrasive water jet machining of glass. J. Mater. Process. Technol. **191**, 404–407 (2007)
6. Mhamdi, M.B., Salem, S.B., Boujelbene, M., Bayraktar, E.: Experimental study of the chip morphology in turning hardened AISI D2 steel. J. Mech. Sci. Technol. **27**(11), 3451–3461 (2013)
7. Matsumura, T., Muramatsu, T., Fueki, S.: Abrasive water jet machining of glass with stagnation effect. CIRP Ann. Manuf. Technol. **60**, 355–358 (2011)
8. Douiri, M., Boujelbene, M., Bayraktar, E., Ben Salem, S.: Effects of the Abrasive Water Jet Machining Process Parameters on the Surface Roughness of the Titanium Alloy Ti-6Al-4V, Conference AMPT, Decembre (2017)
9. Shanmugam, D.K., Masood, S.H.: An investigation on kerf characteristics in abrasive water jet cutting of layered composites. J. Mater. Process. Technol. **209**, 3887–3893 (2008)
10. Akkurt, A., Kulekci, M.K., Seker, U., Ercan, F.: Effect of feed rate on surface roughness in abrasive water jet cutting applications. J. Mater. Process. Technol. **147**, 389–396 (2004)
11. Sreenivasa Rao, M., Ravinder, S., Seshu Kumar, A.: Parametric optimization of abrasive waterjet machining for mild steel: Taguchi approach. Int. Conf. Adv. Mech. Sci. **2**, 28–30 (2014)
12. Thakkar, K.H., Prajapati, V.M., Thakkar, S.A.: A machinability study of mild steel using abrasive water jet machining technology. Int. J. Eng. Res. Appl. (IJERA). **3**(3), 1063–1066 (2013)
13. Chithirai Pon Selvan, M., Mohana Sundara Raju, N., Sachidananda, H.K.: Effects of process parameters on surface roughness in abrasive waterjet cutting of aluminium. Front. Mech. Eng. **7**(4), 439–444 (2012)
14. Boud, F., Loo, L.F., Kinnell, P.K.: The impact of plain waterjet machining on the surface integrity of aluminium 7475. Procedia CIRP. **13**, 382–386 (2014)
15. Wang, J.: Abrasive waterjet machining of polymer matrix composites: cutting performance erosive analysis and predictive models. Int. J. Adv. Manuf. Technol. **15**, 757–768 (1999)
16. Kechagias, J., Petropoulos, G., Vaxevanidis, N.: Application of Taguchi design for quality characterization of abrasive water jet machining of TRIP sheet steels. Int. J. Adv. Manuf. Technol. **62**, 635–643 (2012)
17. Yue, Z., Huang, C., Zhu, H., Wang, J., Yao, P., Liu, Z.W.: Optimization of machining parameters in the abrasive water jet turning of alumina ceramic based on the response surface methodology. Int. J. Adv. Manuf. Technol. **71**, 2107–2114 (2014)
18. Wang, J.: Depth of cut models for multipass abrasive waterjet cutting of alumina ceramics with nozzle oscillation. Front. Mech. Eng. China. **5**(1), 19–32 (2010)

Chapter 29
Optimization of the High Energy Milling Process of Chips of a Stainless Steel Using the Response Surface Modeling

Fábio Gatamorta, C. S. P. Mendonça, M. M. Junqueira, E. Bayraktar, B. G. Andrade, M. de L. M. Melo, and G. Silva

Abstract The present paper discusses response surface methodology as an efficient approach for predictive model building and optimization of the high energy milling process from the chip duplex stainless steel with a carbide vanadium addition willing to obtain the smaller particle size. The process parameters studied was milling time of 10 h and 50 h. Rotations were performed at 250 and 350 rpm. Vanadium carbide was added from 0% to 3% in weight, and a mass/ball ratio of 1/10 and 1/20. An analysis of particle size and a scanning electronic microscopy were used to measure and characterize particle size. With addition of carbide in milling process resulted on a reduction of particle size compared to the material without carbide added around 66%. The results predicted using factorial regression model showed high values of regression coefficients ($R^2 = 0.952$) indicating good agreement with experimental data. The minimum value of particle size was obtained for following optimal conditions: rotation of 325 rpm, time of 42 h, ball/mass 18:1 and carbide 2, 67%wt. The particle size of fabricated powders after 50 h of milling with 3% vanadium carbide addition was about 186 times lower than that the initial chips.

Keywords Duplex stainless steel · Chip · Response surface methodology · High energy milling · Particle size; factorial planning · Carbides · Time · Rotation · Relation mass/ball

29.1 Introduction

Duplex stainless steels (DSS) are characterized by an outstanding combination of strength, toughness and corrosion resistance [1, 2]. Powder metallurgy stainless steel components constitute an important and growing segment of the Powder Metallurgy (PM) industry [3, 4]. The reuse or recycling of materials avoids the materials waste in the environment, reduces the natural resources extraction and reduces the energy input in the manufacture of numerous parts types and the natural resources extraction [5].

To improve the mechanical properties of steels produced by powder metallurgy, the creation of a metal-ceramic composite with carbide addition as reinforcement is advantageous [6, 7]. Carbides are compounds formed by carbon and metal atoms. The metal carbides present important properties, such stability at high temperature, great hardness, and good electric and thermal conductivity [8].

Shashanka and Chaira 2015 who developed a duplex stainless steel in a Fritsch planetary mill in the time of 40 h and with a mill speed rotation of 300 rpm [9]. Yonekubo, 2014, reused a super-duplex stainless steel through the milling of chips coming of a machining process and produced the powder in a planetary ball mill, in the mass/sphere relationship of 1/16, during the time of 100 h, at a rotation of 300 rpm [10]. Kuffner et al., 2015 in their studies made the high-energy milling of chips of a SAE 52100 steel, with 0%, 1% and 3% niobium carbide addition in a planetary mill during milling times of 5, 10, 15 and 20 h at a milling speed of 350 rpm and a mass/ball of 1:10, found greater efficiency in the milling process with the addition of carbide [7].

F. Gatamorta
University of Campinas, São Paulo, Brazil

C. S. P. Mendonça · M. M. Junqueira · B. G. Andrade · M. d. L. M. Melo · G. Silva
University of Itajubá, Itajubá, Minas Gerais, Brazil

E. Bayraktar (✉)
Supmeca-Paris, School of Mechanical and Manufacturing Engineering, Paris, France
e-mail: bayraktar@supmeca.fr

P. R. Thakre et al. (eds.), *Mechanics of Composite, Hybrid and Multifunctional Materials, Volume 5*, Conference Proceedings of the Society for Experimental Mechanics Series, https://doi.org/10.1007/978-3-319-95510-0_29

Factorial designs are widely used to investigate the effects of experimental factors and the interactions between those factors, that is, how the effect of one factor varies with the level of the other factors in a response. This paper presents an investigation on the effect of milling parameters such as time, speed rotation, relation mass/ball and percentage of carbide vanadium in the process. Analysis of variance (ANOVA) is the one that determinates the significant factors and the smaller particle size.

29.2 Experimental Conditions

29.2.1 Materials and Process Used in this Work

In this work, the UNS S31803 duplex stainless steel has the following chemical composition: 22,3% Cr; 5,44% Ni; 2,44% Mo; 0,02% C, 0,160 N and Fe bal. The raw material for the research was obtained through of the machining step at low speed and without the use of lubricants to avoid contamination by oil-soluble.

29.2.2 Experimental Procedure

The milling was made using high energy ball milling in a planetary ball mill in an inert argon atmosphere to avoid oxidation of the powders. The parameters that were analyzed were the milling time, the relation of mass/ball of the sphere, and finally the amount of vanadium carbide that was added at the time of milling. The significance of factors and their interactions has been evaluated by the analysis of variance (ANOVA) using the Minitab program.

The values of rotation, milling time, mass/ball relation and percentage of vanadium carbide were varied. The values of these levels are shown in Table 29.1. Based on the specification of these variables, the sequencing of the experiments was then defined based on the Central Composite Design (CCD), containing four parameters (k) at two levels ($2^k = 2^4 = 16$), eight axial points ($2^{k-1} = 8$) and seven central points, resulting in 30 experiments. The CCD is an experimental arrangement of the Response Surface Methodology that incorporates to the factorial experiments a set of central points and a set of extra levels, called axial points. These values were chosen based on the literature and, mainly, empirical tests performed in high energy milling. The regression model has been represented for each response and optimization of the process has been carried out to the minimum particle size.

29.2.3 Structural Characterization

The characterization of the duplex stainless steel milled powder was realized using a Carl Zeiss EVO MA15 scanning electron microscope. In the secondary electron (SE) mode, the particle size variation and morphology of powder were analyzed. Using the back-scatter electron (BSD) and energy dispersive x-ray (EDS and Mapping) modes, the distribution of a VC produced by the Hermann C. Starck Company was evaluated in the matrix. Particle size distribution was performed in a Microtrac Bluewave S3500 equipment to analyze of powder size with increasing milling time.

To describe the average proportion of the alloy chip sizes, a stereoscope was used. In this work was used chips of duplex stainless steel UNS S31803 as starting point to the milling process with and without vanadium carbide presence with initial size between 5 and 15 μm.

Table 29.1 Real levels of the factors being studied

	−1	1
Time (hours)	10	50
Rotation (rpm)	250	350
Mass/ball relation	1/10	1/20
Carbide percentage (%)	0	3.0

29.3 Results and Discussion

The initial characterization of the scraps duplex stainless steel UNS S31803 is shown in Fig. 29.1. It can be observed that the duplex steel scraps from the machining process showed a mean size of 8000 μm (Fig. 29.1).

The initial characterization in Fig. 29.2 shows the micrographs of vanadium carbides. It is feasible to note that there are agglomerates with an average size between 15 and 40 μm composed of small particles upon larger particles and smaller particles with near nanometer sizes.

An analysis was performed to identify the significant influence of the parameters as well as their interactions on the milling process. For this, the analysis of variance (ANOVA) was performed through the Minitab software, which results are shown in Table 29.2. By analyzing the results, it is noticed that, considering a 95% confidence interval, the values in bold in the table are significant on the response. Analyzing the value p (descriptive level) is a statistic used to synthesize the result of a hypothesis test, for a level of significance ($\alpha = 0.05$). The values in bold in the Table 29.2 are significant.

In addition, the model presented a square correlation adjusted coefficient of R^2 (adjustment) of 90.37%, fitting the statistical model very well. The mathematical model coded by 2^4 factorials can be given by Eq. (29.1).

$$\begin{aligned} R = X_0 + X_1 \cdot A + X_2 \cdot B + X_3 \cdot C + X_4 \cdot D + X_5 \cdot A^2 + X_6 \cdot B^2 + X_7 \cdot C^2 + X_8 \cdot D^2 \\ + X_9 \cdot A \cdot B + X_{10} \cdot A \cdot C + X_{11} \cdot A \cdot D + X_{12} \cdot B \cdot C + X_{13} \cdot B \cdot D + X_{14} \cdot C \cdot D \end{aligned} \quad (29.1)$$

In that R is the mean particle size, X0 is the global mean, Xi represents the other regression coefficients and A, B, C, D represents the rotation, milling time, mass/ball ratio and percentage of carbide respectively. Substituting the respective values of Xi in Eq. (29.1) we obtain:

Fig. 29.1 Scraps duplex stainless steel UNS S31803

Fig. 29.2 Morphology of vanadium carbide powder

Table 29.2 Analysis of variance for medium size (μm)

Factor	Effect	Coefficient	SE Coef	T	p_value
Constant		88.17	3.15	**27.97**	**0.000**
Rotation	−101.37	−50.68	2.29	**−22.14**	**0.000**
Time	−151.22	−75.61	2.44	**−30.93**	**0.000**
Relation	−79.95	−39.98	2.44	**−16.35**	**0.000**
Carbide	−131.89	−65.94	2.44	**−26.97**	**0.000**
Rotation* rotation	72.34	36.17	6.20	**5.83**	**0.000**
Time*time	96.32	48.16	7.98	**6.03**	**0.000**
Relation*relation	30.01	15.00	7.98	1.88	0.090
Carbide *carbide	83.98	41.99	7.98	**5.26**	**0.000**
Rotation* time	5.87	2.93	2.45	**1.20**	**0.259**
Rotation*relation	28.53	14.26	2.45	**5.82**	**0.000**
Rotation*carbide	−4.52	−2.26	2.45	−0.92	0.378
Time*relation	8.84	4.42	2.45	1.80	0.102
Time* carbide	47.57	23.79	2.45	**9.71**	**0,000**
Relation*carbide	−15.92	−7.96	2.45	**−3.25**	**0.009**

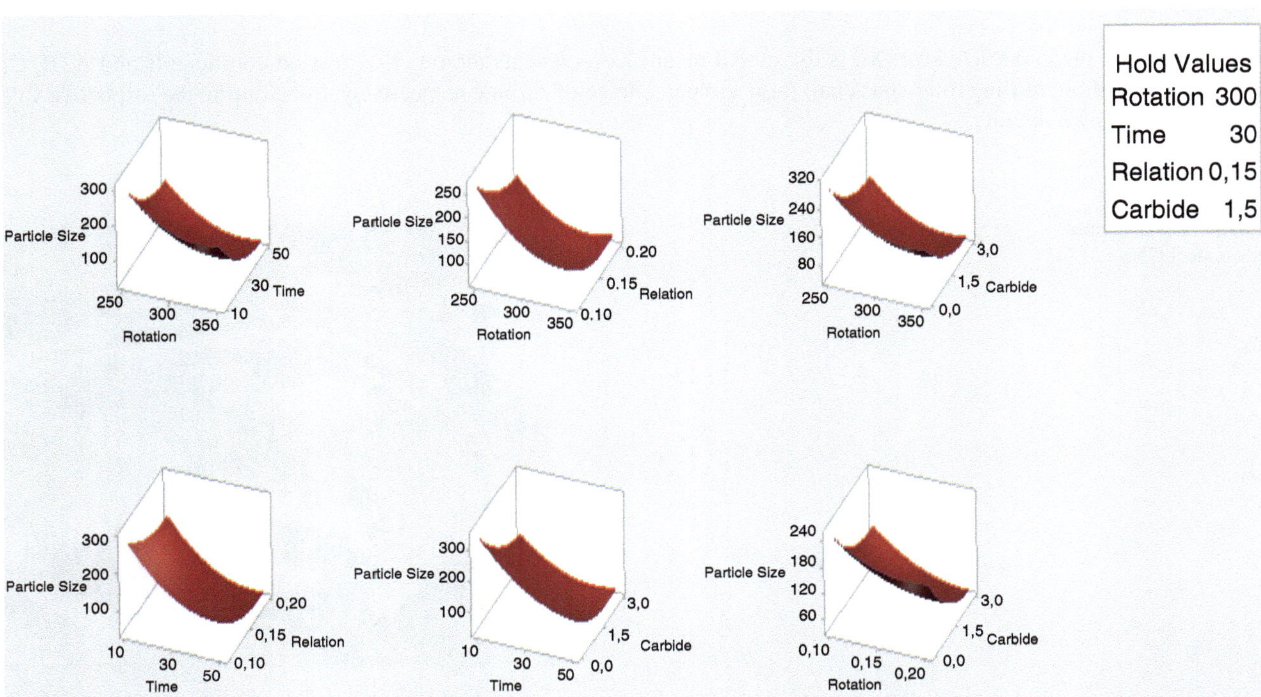

Fig. 29.3 Response surface curve for particle size

$$R = 2580 - 10.59 \cdot A - 13.74 \cdot B - 4285 \cdot C - 98.8 \cdot D + 0.015 \cdot A^2 + 0.012 \cdot B^2 + 6002 \cdot C^2 + 18.66 \cdot D^2 + 0.0029 \cdot A \cdot B + 5.704 \cdot A \cdot C + 0.0301 \cdot A \cdot D + 4.42 \cdot B \cdot C + 0.793 \cdot B \cdot D - 106.2 \cdot C \cdot D \quad (29.2)$$

Minitab 17 was used to plot the response surface curves (Fig. 29.3), which relates the variables (rotation, milling time, mass/ball ratio and percentage of carbide) with the particle size variable, in these figures two variables are held constant, and by varying the levels of factors in the values maximum (+1) and minimum (−1). Results showed that the variables influenced in the milling process and the reduction of particle size. This means that the correlation between the variables of the milling process, the effects together, aid in the milling process.

a) b)

Fig. 29.4 (**a**) Particles morphology after the high energy milling process**:** Milling condition: 300 rpm, mass / ball ratio of 1/20, milling time of 50 h, addition of 3% VC**,** and also (**b**) Particles morphology after the high energy milling optimizer process

Table 29.3 Analysis of response optimizing for particle size (μm)

Response optimization	
Rotation (rpm)	324.8 rpm
Time (hours)	41.7 h
Ball/mass ration	18:1
Carbide percentage (%)	2.67
Particle size (μm)	13.53
Composite desirability	1.00

For milling for the rotation of 300 rpm, with the time of 50 h and mass / ball ratio of 1/20 and with addition of 3% of CV (Fig. 29.4), it is observed that the particles have acquired an irregular morphology with the size of particles between 30 μm and 150 μm, and the mean size is around 56 μm.

For the lower values of rotation, milling time, mass/ball ratio and carbide addition, the obtained particle size is greater than 300 μm. For surface response curves, the surface obtained is concave, the tendency of concave profile is ideal to find a point of less value for the particles, since it converges to a point of minimum. The lower values for particle size are obtained for intermediate regions, the smaller particle sizes obtained in the milling process are between 2.5% and 3% carbide regions, for mass / ball ratio values between 1/15 and 1/20, for times greater than 40 h and rotation at the highest level. However, with increasing milling time occurs in the process a cold welding due to collisions between the particles and the balls, leads to an increase in particle size.

In order to obtain optimization of the high energy milling process was used Minitab tool called "*Response Optimizer*", the established values are shown in Table 29.3. In order to obtain the smallest particle size the adjustment values are:

To validate the optimal point, a confirmation experiment was performed. Values were adjusted as indicated by "Response Optimizer". After the milling process, that the material was transformed in particles with irregular morphology, with size ranging from 40 to 100 μm. The size obtained for the particles was less than 186 times the initial chip as observed in Fig. 29.4b.

29.4 Conclusion

The present article presented a study to optimize the milling process of UNS S31803 steel chip with carbide addition. It can be concluded that varying the conditions of rotation, milling time, mass / ball ratio and percentage of carbide, have different effects on milling. In order to obtain the smallest particle size, after milling, the values must be varied at the highest levels. The particle size after the process of optimization for the time of 42 h of milling, ball/mass 18:1 and carbide percentage of 2.67 weight was about 186 times lower than that of the initial chips. Therefore, the high energy milling process can be used as an alternative for the reuse of the chip of a duplex stainless steel.

Acknowledgements The authors acknowledge the support of FAPEMIG, FAPESP CAPES, UNICAMP and UNIFEI.

References

1. Gunn, R.N.: Duplex Stainless Steels- Microstructures, Properties and Applications. Woodhead Publishing, Cambridge (2003)
2. Matias, J.V.S., Tavares, S.S.M., Pardal, J.M., Ribeiro, R.S.A.: Mat. Res. **20**, 1–5 (2017)
3. Martin, F., Garcia, C., Blanco, Y., Rodriguez-Mendez, M.L.: Mater. Sci. Eng. A. **642**, 360–365 (2015)
4. Klar, E., Samal, P.K.: Powder Metallurgy Stainless Steels: Processing, Microstructures, and Properties. ASM International, Materials Park (2007)
5. Delforge, D.Y.M., Ferreira, I., Silva, C.G.R., Paris, E.A.G., Marcelo, A.L., Novaes, R.H.: Rev. Esc. Minas. **60**, 95–100 (2007)
6. Trueman, A.R., Schweinsberg, D.P., Hope, G.A.: Corros. Sci. **41**, 1377–1389 (1999)
7. Kuffner, B.H.B., Diogo, W.S., Amancio, D.A., Rodrigues, G., Silva, G.: Rev. Esc. Minas. **68**, 295–300 (2015)
8. Gubernat, A., Zych, Ł.: J. Eur. Ceram. Soc. **34**, 2885–2894 (2014)
8. Kosolapova, T.Y.: Carbides Properties, Production and Applications. Springer Publishing, New York (1971)
9. Shashanka, R., Chaira, D.: Development of nano-structured duplex and ferritic stainless steels by pulerisette planetary milling followed by pressureless sintering. Mater. Charact. **99**, 220–229 (2015)
10. Yonekubo, A. E.: Caracterização Microestrutural do aço inoxidável Superduplex UNS S32520 (UR 52N*) Processado por moagem de alta energia. Dissertação de Mestrado, Universidade Estadual de Ponta Grossa- Ponta Grossa (2010)

Chapter 30
Iron Contents on Recycle Aluminum and Influence on Mechanical Properties

Claudomiro Alves, Bruna Vilas Boas, and Fábio Gatamorta

Abstract This study aims to investigate the influence of iron content on the mechanical property of the recycle AA356 aluminium alloy. The recycle Alloys enriched with iron were prepared by casting and samples were chemically analyzed by mass spectrometry to confirm their composition. The alloy tixoability was evaluated by thermodynamic simulation via Thermocalc®, and then other alloys samples were prepared by partial melting. The samples were analysed metallurgically and mechanically. Through microstructural analysis, it was noted the presence of Fe-rich intermetallic in the alloys enriched with iron, which were not found in the alloy as received. The presence of these intermetallic contribute to improve the mechanical strength.

Keywords Aluminum alloys · Mechanical properties · Tixoability · Recycling · Composite design

30.1 Introduction

Aluminum alloys have several advantages in applications that require low specific weight and high thermal and electrical conductivity. The aluminum alloy AA356 (Al-7Si 0.3 Mg) is widely used in the manufacture of automotive components and aerospace industries due to its physical and mechanical properties [1]. The presence of some elements modifies these properties because of inclusions formation, intermetallic phases, micro porosities, etc. For the best combination of ductility and strength, single-phase structures are the most appropriate. However, they are difficult to achieve in Al due to its limited solubility [2].

Fe is a secondary element which can be presented as a contamination result. It influences the alloy characteristics since it favors the intermetallic compounds formation [3], especially with Fe-rich intermetallic which contributes improving the mechanical strength. The Fe content between 0.1% and 0.4% are indicated in applications by gravity casting and between 0.8% and 1.2% is beneficial in high press die casting. For high levels of Fe, the strength and ductility decrease due to the presence of intermetallic in platelets form (β-Al5FeSi) which also provide an increase in porosity, thereby decreasing the mechanical properties [4].

30.2 Experimental Procedure

For the production of the samples three classes of scrap were used: mixed plates, mixed profiles and cables that were separated in portions of 1000 g and had their measured mass in a precision electronic scale. This procedure was performed to ensure that the difference between the mean values of the samples were statistically within the accepted deviation values.

To determine the pouring temperature, the reference values obtained from the literature were used. The superheating chosen was approximately 10% to avoid excessive oxidation and the solidification conditions chosen were that of conventional solidification with ingot molds preheated to 200 °C. For the production of the specimens of the cast material,

C. Alves (✉) · F. Gatamorta
University of Campinas, São Paulo, Brazil
e-mail: brunavb@fem.unicamp.br; fabiog@fem.unicamp.br

B. V. Boas
Paulista University, Jundiaí, São Paulo, Brazil
e-mail: miro@fem.unicamp.br

P. R. Thakre et al. (eds.), *Mechanics of Composite, Hybrid and Multifunctional Materials, Volume 5*, Conference Proceedings of the Society for Experimental Mechanics Series, https://doi.org/10.1007/978-3-319-95510-0_30

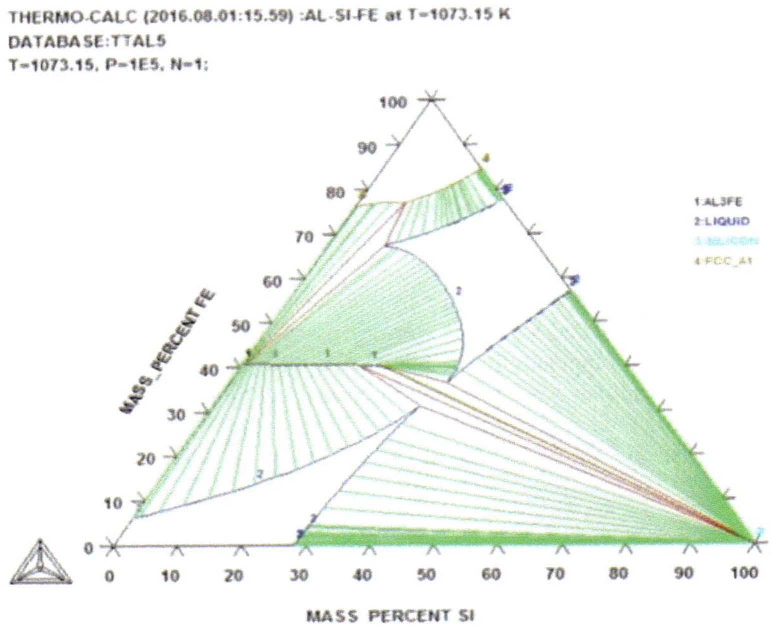

Fig. 30.1 Ternary diagram Al-Si-Fe obtained by thermodynamic simulation carried out in Thermocalc® from AA 356 alloy

1045 steel (ingot) molds were used. The ingot molds and the scrap samples were heated in an oven at 150 °C for total removal of moisture that could present a risk of accidents during pouring.

For the material production, the alloy AA356 was placed in a SiC crucible that was heated to 720 °C until the liquid state. Subsequently, the liquid was mechanically stirred via rod SAE 1045 steel to enrich the alloy with Fe by diffusion mechanisms. It was used different times (5, 10, 15 and 25 min) aiming residual levels of disparate iron. The casting occurred with 10% overheating in 1045 steel molding and remained in it until the room temperature. Thermodynamic simulation was carried out through Thermocalc® software (Fig. 30.1).

The cast alloys were sectioned and samples of each alloy were submitted to chemical analysis through atomic emission spectrometry (AES). For microstructural characterization, the samples were prepared according metallographic patterns and etched with 1% HF to reveal the microstructure. Micrographic images were acquired using a microscope and used for porosity analysis by *ImageJ*. Microhardness measurements were performed using a Vickers Hardness Tester with a load of 0.3 kg applied for 10 s to mechanical characterization of the samples. The results were submitted to Tukey test (with a 95% confidence interval) to check the influence of the Fe content in the hardness values and porosity of the samples.

For the mechanical characterization of the materials samples of each of the ingots were cut into the groups. From the samples, cylindrical specimens were made and standard tensile tests were performed. The test specimens conformed to ASTM E8 / E8M-15a and the test speed was set at 5 mm/s with a limit of 100 KN of force, and maximum deformation until rupture. The mechanical strength, yield stress and elongation were determined according to the same standard.

30.3 Results and Discussion

Table 30.1 shows the chemical composition of the obtained alloys and the stirring time.

Figure 30.2 shows the microstructures obtained for AA 356 alloy with different iron contents. It can be noted the presence of Al5FeSi acicular lamellae. These compounds are favored if there is a low Fe: Mn ratio, and their formation depend on the amount of iron present in the chemical composition. This iron compound reduces the alloy fluidity and impairs the material workability since it has high hardness. The high hardness values found in the AA356–2,0% Fe alloy are justified by the greater amount of Al5FeSi.

Table 30.2 shows the variation in the micro hardness and porosity as a function of iron content. Figure 30.3 shows the Tukey's test results for the hardness and porosity of the AA 356 alloy with different iron contents. The result shows that only at levels greater than 2.0% Fe the porosity and hardness values have a significant increase.

Table 30.1 Chemical composition of the alloys

Sample	Stirring Time(min)	Al	Si	Cu	Mg	Mn	Fe
AA 356	–	92,30	6,74	0,130	0,219	0,112	0,280
AA 356-0,3Fe	5	92	7	0,137	0,356	0,109	0,341
AA 356-0,7Fe	10	92,384	6,27	0,119	0,235	0,106	0,680
AA 356-1,2Fe	15	91,47	6,63	0,125	0,253	0,120	1199
AA 356-2,0Fe	25	90,73	6,44	0,122	0,227	0,121	1,96

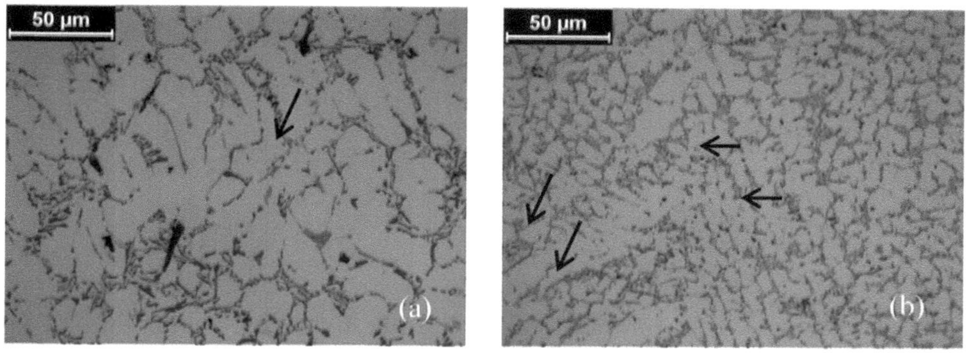

Fig. 30.2 Longitudinal microstructures. Presence of Al5FeSi acicular lamellae in the alloy microstructure (**a**) AA356–0,7%Fe (**b**) AA356–2,0%Fe

Table 30.2 Microhardness and porosity as a function of iron content

Type	AA356	0,3%Fe	0,7%Fe	1,2%Fe	2,0%Fe
Porosity (%)	2,14 ± 0,1	1,93 ± 0,3	2,07 ± 0,4	1,81 ± 0,2	3,5 ± 0,9
Micro hardness (HV)	71,8 ± 4,0	73,6 ± 3, 9	72,5 ± 3,9	71,5 ± 2,3	86,8 ± 3,5

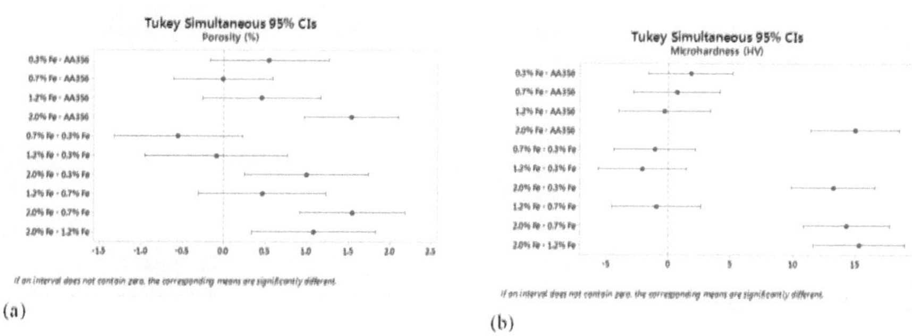

Fig. 30.3 Turkey Analysis for (**a**) porosity and (**b**) hardness in the alloy AA 356 with different iron contents

The results of tensile tests were introduced in the form of a graph of mean values through the software Origin8®. For a better visualization of the numerical values the results were also showed in the form of a table with the mean values of the test for mechanical strength, elongation and rupture stress.

Figure 30.4 shows the average curves obtained in the stress versus strain tests performed on the samples obtained within the groups in the off scouring classes used in the work. From the graph, it can be seen that the highest resistance value is observed in the cable class. This characteristic can be associated to the high values of alloying elements among such as Silicon that is an element that, in the Aluminum, promotes the increase of the mechanical resistance. The group that presents the lowest resistance is the class of profiles.

The graph also shows high deformation value within the mixed profile class. This characteristic is probably associated with the low values of alloying elements that increase the plasticity of the material due to the lower presence of phases in the material.

The mean values of Maximum Stress, Maximum Stress and Stress Ratio are shown in Table 30.3. The values found were compared with the mean values of the ranges observed in the literature.

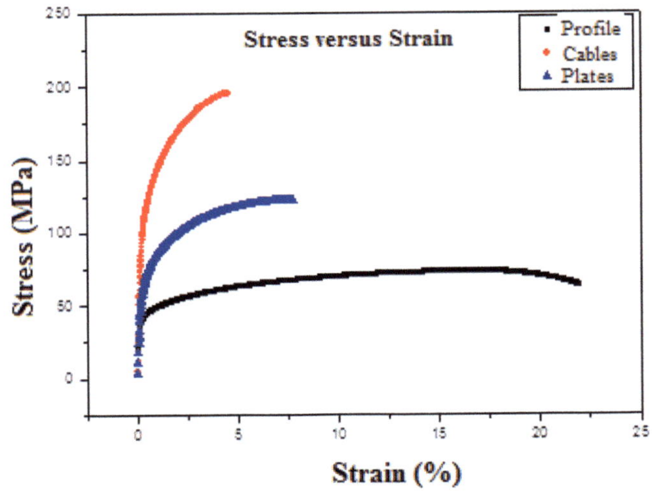

Fig. 30.4 Average curves obtained with the tensile test

Table 30.3 Values of mechanical properties obtained from the stress versus strain tests found in the off scouring classes used for the work

Type	Maximum stress (MPa)	Rupture stress (MPa)	Maximum deformation (%)
Cables	193	110	4
Plate	145	80	7,5
Profiles	73	52	22

In the cable class, the Maximum Stress and Rupture Stress values are compatible with the values of the 3000 series alloys (Mn series alloys) showing, however, values of deformation much lower than the average values of the alloys in this series. In the mixed plate class the maximum tensile strength and tensile strength values are slightly higher than the values for the 1000 series alloys and also have the lower deformation values. In the mixed profile class the Maximum Stress and Rupture Stress values are close to the values of the commercially pure aluminum also showing the proportional deformation values [5].

From the data obtained in the tensile test, a factorial analysis was performed in the MiniTab software to verify the influence of the iron content on the mechanical properties of the samples tested.

Figure 30.5 shows the factorial analysis graphs for the main effects in the maximum stress values. It is possible to observe from the graph that the percentage of iron doesn't follow the performance of the change of classes. For increasing iron percentage values from 0.13%, an increase in values was observed to 0.16%. This increase was already observed previously and was explained by the presence of Mn (present in the composition of the plates class) which assists in the formation of α-Fe intermetallic which are less damaging than the β-Fe intermetallic phase because they have a shape being preferable to cast structures. From values of 0.16% occurs the decrease of values for maximum stress and it can be explained by the absence of Mn in the composition of the mixed profile class.

30.4 Conclusions

The variation presented in the final chemical composition of the recycled materials is related to the variation in the mechanical properties in each of the classes and the values found can be associated in the series provided by the norms;

From the results found, a change in microstructural morphology, mechanical properties and porosity was observed for Fe percentages higher than 1.2%. The presence of low Mn values of the alloy inhibits the formation of Chinese script (Fe-Mn) 3Si2Al15 and the rate of solidification inhibits the presence of the sludge phase.

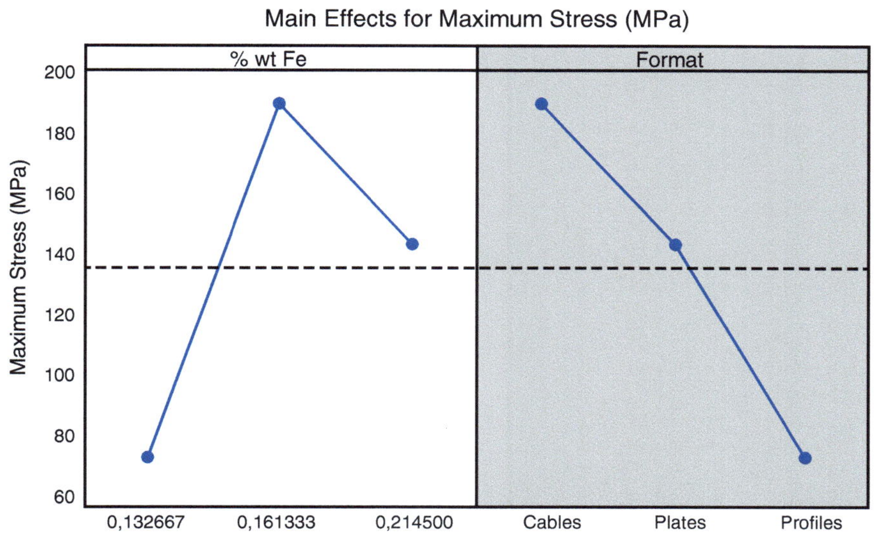

Fig. 30.5 Graphs of factorial analysis of the main effects in the values of maximum tension for the parameters format (class) chemical composition of iron

References

1. Hernández-Paz, J.F., Paray, F., Gruzleski, J.E.: Natural Aging and Heat Treatment of A356 Aluminium Alloy. AFS Transaction (2004)
2. Molinari, M.: Ligas de alumínio para fundição e influência dos elementos de liga e impurezas, Joinville (1997)
3. Fuoco, R.: Processos de produção de peças fundidas em ligas de alumínio. IPT, São Paulo (2001)
4. Samuel, A.M., Samuel, F.H.: Effect of Iron and Cooling Rate on Tensile Properties of B319.2 Alloys in Non-Modified and Sr-Modified Conditions. AFS Transactions (2004)
5. ASM International Handbook Committee: Properties and Selection: Nonferrous Alloys and Special-Purpose Materials, vol. 2. 10 ed., (1990)

Chapter 31
Experimental Comparison of the Microstructure and Surface Roughness in CO_2 Laser Cutting of the Titanium Alloy Ti–6Al–4V and the Pure Titanium Ti

B. El Aoud, M. Boujelbene, E. Bayraktar, S. Ben Salem, and A. Boudjemline

Abstract Titanium and its alloys have over the years proven themselves to be technically superior and cost effective materials for a wide range of applications spanning the industries of aerospace, industrial, marine, and even commercial products. In this paper, the surface integrity of titanium alloys Ti-6Al-4V and pure titanium Ti, cut with high power CO_2 laser is determined and the results are presented and discussed. The aim of this experimental investigation is to compare the microstructure and the cut surface quality of both titanium based on the variation of three different cutting parameters namely, laser power (*Pu*), cutting speed (*V*) and gas pressure (*p*).

The microstructure analysis is evaluated by investigating the cut edge microstructure and the cut section roughness for both Titanium alloy Ti-6Al-4V and pure Titanium Ti. An experimental analysis is carried out to identify the main effects and interactions of these parameters. Optical microscopy and scanning electron microscopy (SEM) are carried out to examine the cutting defects.

Keywords CO_2 laser cutting · Experimental comparison · Surface roughness · Cut microstructure · Ti-6Al-4V · Pure titanium

31.1 Introduction

Laser cutting can be a promising tool in the machining of titanium alloy parts because it has more advantages such as the small heat affected zone, cutting accuracy, contactless processing and short time machining [1]. Laser, is a tool of exceptional precision and power, offers several advantages over conventional cutting methods due to precision of operation, short processing time, and low cost. The physical processes involved in laser cutting of thick sections are complicated and significantly influences the end product quality [2]. Titanium finds extensive use in vacuum, aerospace and medical applications. Cutting of titanium sheets is one of the primary requirements in the fabrication of most of the components [3]. Ti–6Al–4V is the most commonly used alloy; it is an extensively employed material for medicine, aerospace and

B. El Aoud
Supmeca/Paris School of Mechanical and Manufacturing Engineering, Paris, France

University of Tunis El Manar, ENIT, Ecole Nationale d'Ingénieurs de Tunis, Tunis, Tunisia

M. Boujelbene · S. Ben Salem
University of Tunis El Manar, ENIT, École Nationale d'Ingénieurs de Tunis, Tunis, Tunisia

E. Bayraktar (✉)
Supmeca-Paris School of Mechanical and Manufacturing Engineering, Paris, France
e-mail: bayraktar@supmeca.fr

A. Boudjemline
University of Hail, College of Engineering, Hail, Kingdom of Saudi Arabia

P. R. Thakre et al. (eds.), *Mechanics of Composite, Hybrid and Multifunctional Materials, Volume 5*, Conference Proceedings of the Society for Experimental Mechanics Series, https://doi.org/10.1007/978-3-319-95510-0_31

other components that operate at high temperatures and in corrosive as well as hostile environments [4]. Titanium and its alloys have over the years proven themselves to be technically superior and cost effective materials for a wide range of applications spanning the industries of aerospace, industrial, marine, and even commercial products, this is because of their good combination of mechanical properties to include excellent specific strength, stiffness, immune to corrosion in sea water environment, good erosion resistance in environments spanning a range of aggressiveness, and importantly their acceptable mechanical properties at elevated temperatures coupled with an intrinsic capability to withstand and safely function at elevated temperatures [5]. Laser parameters, in particular laser output power, focus on setting of focusing lens, cutting speed, assisting gas, and its pressure influence the physical processes in the cutting section. In this case, controlling the affecting parameters results in improved cutting quality. Consequently, investigation into affecting parameters in laser cutting process is necessary to improve the end product quality [6]. Most noticeably and worthy of noting is that titanium metal is allotropic and has a hexagonal close packed (HCP) {α}crystal structure at low temperatures and a body-centered cubic {β} structure at temperatures above 885 °C. Overall, the structural titanium alloys have been classified into three categories, denoted as: alpha, alpha plus beta, and beta subtract phase alloys. The alpha phase alloys of titanium, which are categorized as commercially pure titanium, are relatively weak in strength but offer a combination of good corrosion resistance, good weldability, creep resistance, receptive to heat treatment coupled with ease of processing and fabrication [7]. Laser cutting can be used to cut a variety of different materials, numerous researches have been done on laser cutting of mild steel, alloy steels, nickel alloys, aluminum and even very hard materials difficult to machine otherwise, such as CBN (cubic boron nitride) [8]. Miraoui et al. [9] were studied experimentally the effect of laser beam diameter on cut edge of steel plates. They showed that laser cutting and its input parameter (laser beam diameter) have a thermal effect on the surface microstructure and on the micro hardness beneath the cut surface. Yilbas et al. [10] were predicted, using finite element method, the thermal stress field developed in the cutting region of Kevlar laminates obtained by laser cutting. Apart from laser power and cutting speed, processing mode affected the quality of cut. Cutting of 4130 steel at higher duty cycles produced wider kerf and thicker HAZ region whereas by setting pulse frequency at more than 100 Hz, a better cutting quality was obtained [11]. At high cutting speed, pulse mode led to rough surface. In addition, pulse mode cutting produced straight kerf and uniform surface compared to CW [12]. Working at CW and pulsed mode produced different cutting quality and it's critical to understand the combined effect of processing mode, cutting speed and laser power on Ultra High Strength Steel (UHSS) blanks that are widely used today in automotive industry. Most work reviewed in the literature considers only one or two characteristic properties of the laser cut surface to describe quality [13, 14]. Surface roughness and size of heat affected zone are often used to describe laser cut quality. The effect of laser power, cutting speed and nitrogen assist gas pressure on the cut quality in laser cutting analyzed in [15–17]. The predicted temperature distribution is correlated with the HAZ and a critical temperature range corresponding to the maximum depth of the HAZ is identified using a combination of metallography, hardness testing, and thermal modeling. Parametric studies showed that the depth and width of the heat affected zone increased with an increase in the laser power and decreased with an increase of the laser spot size and the laser scan speed [18].

The purpose of this article is to investigate CO_2 laser micro-cutting of Titanium alloy (Ti-6Al-4V) and pure Titanium (Ti). In this paper, the effect of three input parameters: laser power (*Pu*), cutting speed (*V*) and gas pressure (*p*) on the cut edge microstructure and the surface quality (R_a) behavior, taken under experimental investigation have been studied.

31.2 Experimental Research

All experiments were performed using 4000 TLF TURBO CO_2 laser machine with TEM_{01} processing mode. All parameters were precisely controlled. The machine was capable of using nitrogen as an assist gas at right angle and coaxially to the laser beam. By trial experiments it was decided to use nitrogen as inert gas.

The workpieces accommodated are firstly Titanium alloy (Ti-6Al-4V) secondly pure Titanium (Ti) with their elemental composition given in Table 31.1 and with their mechanical and physical properties cited in Table 31.2.

In experiments, number of fixed and variable parameters are involved. The fixed parameters were focal length, nozzle diameter, nozzle tip distance and material thickness whereas the variable parameters were Laser Power ($Pu = 1, 2, 3, 4$ *kW*), Cutting Speed ($V = 480, 1440, 2400$ *mm/min*) and Gas Pressure ($p = 2, 8, 14$ *bars*).

Table 31.1 Chemical composition of Ti-6Al-4V and Pure Ti (%)

Chemical elements	Titanium alloy (Ti-6Al-4V)	Pure Titanium (Ti)
Ti	Base	100
Al	6	–
V	4	–
Fe	0.3	–
C	0.08	–
N	0.05	–
H	0.01	–
O	0.2	–

Table 31.2 Mechanical and physical properties of Ti-6Al-4V and Pure Ti

	Pure Titanium (Ti)	Titanium alloy (Ti-6Al-4V)
Elastic limit 0,2 (MPA)	275	700
Elongation %	20	10
Brinell hardness	160 HB/30	–
Modulus of elasticity (GPA)	103	120
Modulus of rigidity (GPA)	40	–

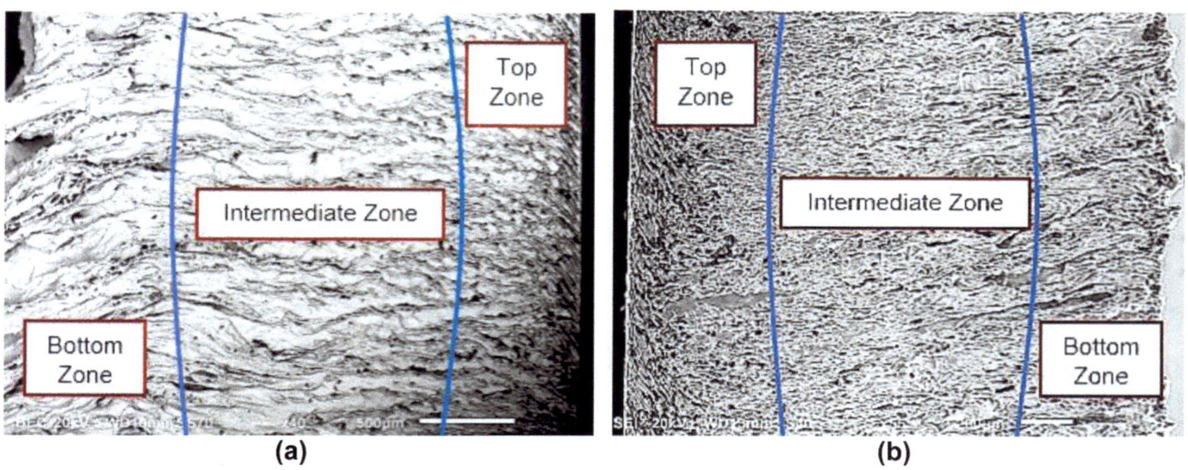

Fig. 31.1 SEM of general view of cut edge at laser power $Pu = 2$ kW, the cutting speed $V = 1440$ mm/min and the gas pressure $p = 8$ bars: (**a**) Titanium alloy Ti-6Al-4V, (**b**) pure Titanium Ti

31.3 Experimental Results

31.3.1 Cut Edge Microstructure Analysis for Titanium Alloy Ti-6Al-4V and Pure Titanium Ti

After laser beam cutting operations, the machined specimens were analyzed with the help of scanning electron microscope (SEM). As the laser beam process cuts the material with the help of thermal effect and due to this it may change the microstructure of machined surface. For both materials, Titanium alloy (Ti-6Al-4V) and pure Titanium (Ti), Laser cut surface, shown in Fig. 31.1a, b, can be broadly categorized into three zones; narrow top zone, intermediate zone and bottom zone and this confirms the results of research done by Shanjin et al. [19].

The top zone is the initialization zone of laser cutting and it's seems the same for Titanium alloy (Ti-6Al-4V) and pure Titanium (Ti). It involves formation of depressed melt pool caused by the very high gas thrust on the surface of metal sheet. Progressive movement of laser beam gives rise to movement of the depressed melt pool thus leading to the formation of very fine striations in this top zone.

However, the intermediate zone represents a small difference between Titanium alloy (Ti-6Al-4V) and pure Titanium (Ti). This difference can be clearly observed from the enlargement of this zone (See Fig. 31.2a, b).

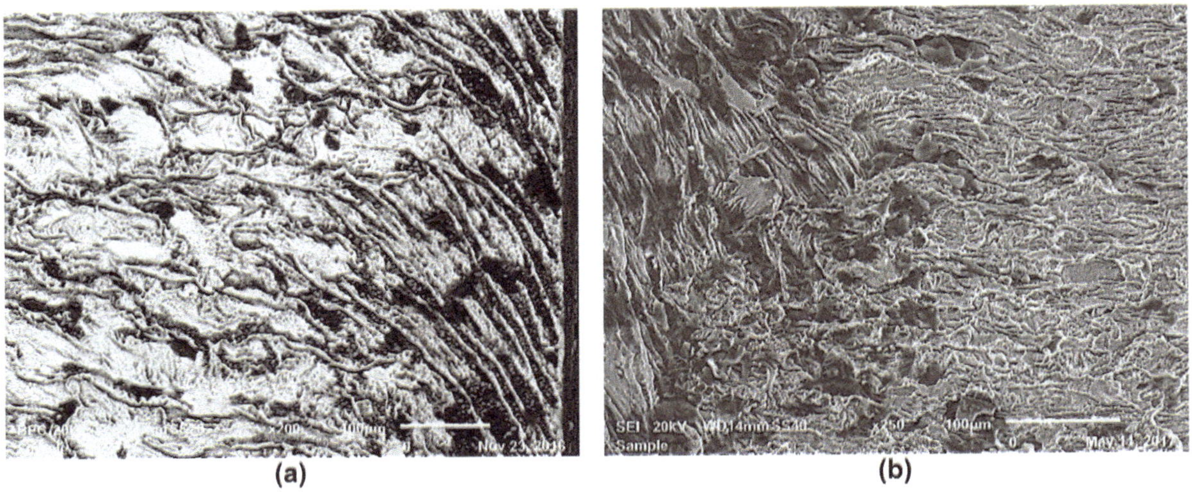

Fig. 31.2 Enlargement of the intermediate zone at laser power $Pu = 2$ kW, the cutting speed $V = 1440$ mm/min and the gas pressure $p = 8$ bars: (**a**) Titanium alloy Ti-6Al-4V, (**b**) pure Titanium Ti

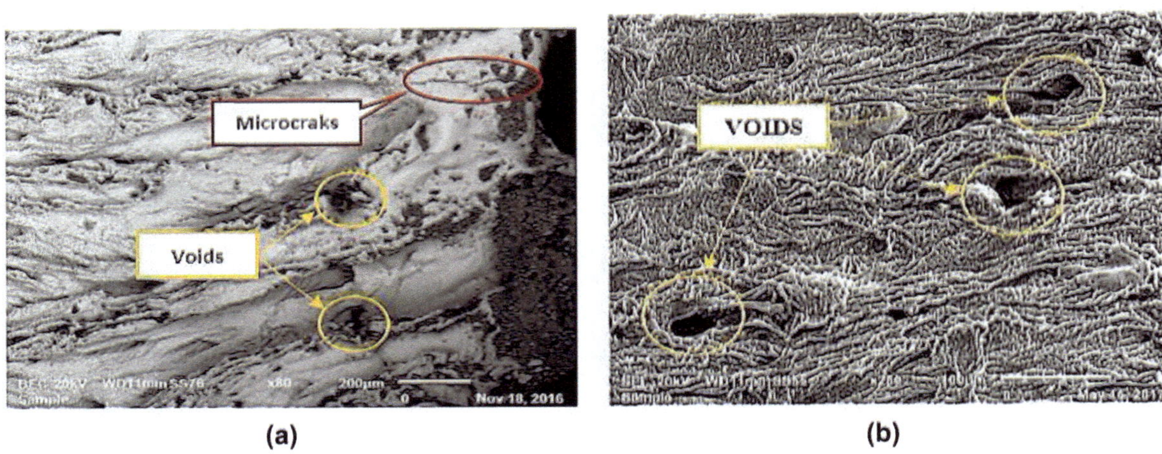

Fig. 31.3 Cut edge microcracks and voids (**a**) Titanium alloy Ti-6Al-4V, (**b**) pure Titanium Ti

In fact, for Titanium alloy (Ti-6Al-4V), in the laser material interactive region, the surface material is heated to a relatively high temperature between the melting point and the beta-transformation point. As the laser beam moves away, the surface material under-takes a rapid cooling due to the cooling effect of the gas flow and the heat conduction to the matrix. As the temperature drops down to below the beta-transformation point, non-diffusion phase transformation occurs leading to metastable acicular martensite grains.

For pure Titanium (Ti), the intermediate zone where there is the evaporation and ejection of liquid debris along with vapors. No transverse striations are observed on the cut surface made with CO_2 laser operation. In addition Limited laser interaction time and intermittent cooling stages in pulsed mode cutting did not result in boiling of the material in the intermediate zone, and hence transverse striations are not observed.

Also the bottom zone has almost the same morphological aspect for both materials. It possesses a surface morphology with irregular undulations. Added it's the zone where the material evaporation ceases as laser beam moves past the concerned region.

The CO_2 laser cut surfaces of Titanium alloy (Ti-6Al-4V) and pure Titanium (Ti) carried signatures of voids, produced as a result of the flow of high pressure gas through converging nozzle tip [3]. Microcracks can be clearly observed on laser cut surfaces of titanium alloy, as shown in Fig. 31.3a, b.

The formation of micocraks and voids due to the tension forces on the laser cut surface and the brittleness of the reaction products, namely titanium oxide and titanium nitride.

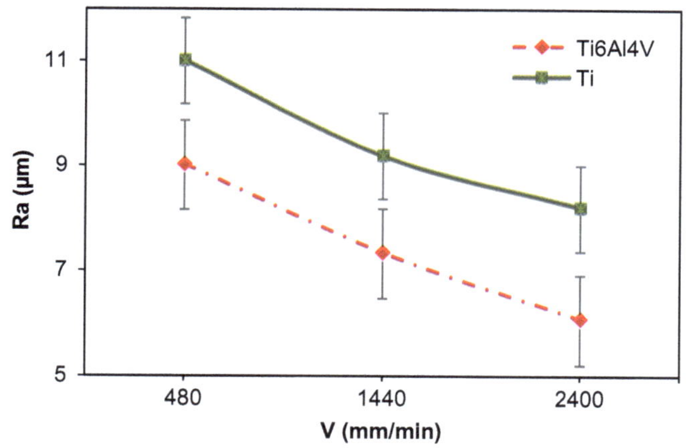

Fig. 31.4 Surface Roughness variation with the laser cutting speed for Titanium alloy Ti-6Al-4V and pure Titanium Ti at $Pu = 3$ kW and $p = 2$ bars

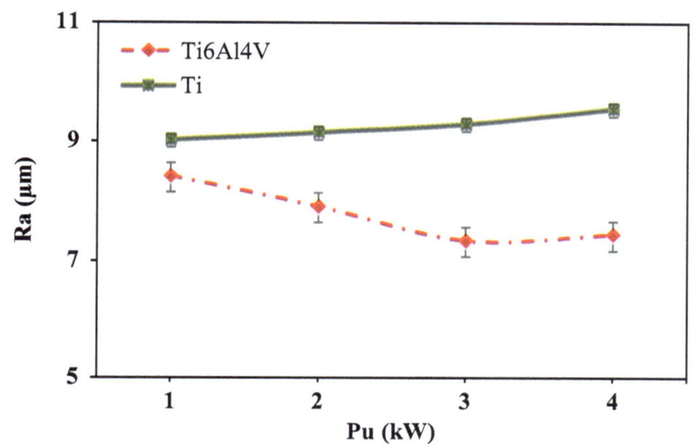

Fig. 31.5 Surface Roughness variation with the laser power for Titanium alloy Ti-6Al-4V and pure Titanium Ti at $V = 1440$ mm/min and $p = 2$ bars

31.3.2 Roughness Analysis for Titanium Alloy Ti-6Al-4V and Pure Titanium Ti

Laser cutting operations generally produce regular patterns in the cut surface, known as striations. The severity of frequency and amplitude of these striations has a direct impact on surface quality. Roughness is defined as minor and periodically repeated disorders on the surface of materials, excluding shape and undulation faults. The characteristic morphology of a surface that has undergone laser cutting is due to the enormous amount of heat generated by the discharges, which causes melting and vaporization of the material, followed by rapid cooling. There are many parameters used in the literature and industry related to surface roughness. Among the surface roughness parameters, the most popular is average roughness R_a.

Figures 31.4 and 31.5 show the effect of two output laser parameters; cutting speed (*V*) and Laser power (*Pu*) on the surface roughness variation for two different materials; Titanium alloy (Ti-6Al-4V) and pure Titanium (Ti).It is clearly observed that surface roughness value decreases as the cutting speed increases for Titanium alloy (Ti-6Al-4V) as well as pure Titanium (Ti). In fact, when the cutting speed is less it results in a damaged surface but if the cutting speed is very high, the surface quality is better because it results in insufficient exposure of the work to the laser beam and may not initiate the cutting process at all. On the other hand, the behavior of the roughness following the effect of the laser power is not the same for Titanium alloy (Ti-6Al-4V) and pure Titanium (Ti). It can be seen in Fig. 31.5 that for Titanium alloy (Ti-6Al-4V), R_a decreases as Laser power increases, which is in agreement with the results found by Miraoui et al. [20]. But, for pure Titanium (Ti), it can be seen from Fig. 31.5 that the increase in Laser power produces a slight increase in average roughness. So we can say that the variation of the roughness is almost constant and the explanation of this result is not very significant.

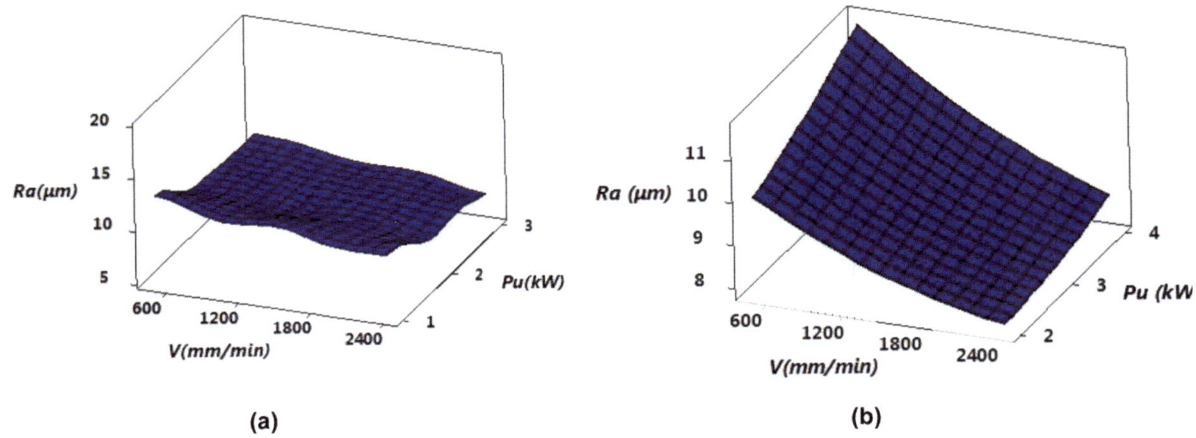

Fig. 31.6 Effect of *V* and *Pu* on R_a (**a**) Titanium alloy Ti-6Al-4V, (**b**) pure Titanium Ti

The same relations can be shown in Fig. 31.6a, b. The combination of laser power and cutting speed give the effects plots which the influences of cutting speed (*V*) and laser power (*Pu*) on the surface roughness parameters R_a while the gas pressure (*p*) is kept constant.

31.4 Conclusion

Laser cutting of titanium alloy sheet is a complex thermal process. This paper presents CO_2 laser cutting of titanium alloy sheet (Ti-6Al-4V) and pure Titanium (Ti).This article investigates the effect of different laser cutting parameters like Laser power (*Pu*) and Cutting speed (V) on such quality factors including surface roughness and cut edge microstructure.

Scanning electron microscope (SEM) is carried out to study the microstructure of cut edge. Nitrogen-assisted laser cutting of titanium alloy leads to micro-cracks and reduced surface quality due to the embrittlement of oxygen and nitrogen on the laser cut surfaces.

The average roughness of the cut surface of the specimens was measured by using roughness tester.

It is found that:

- For both materials, Titanium alloy (Ti-6Al-4V) and pure Titanium (Ti), Laser cut surface is categorized into three zones: narrow top zone, intermediate zone and bottom zone.
- Each of the three different zones is characterized by its own specific microstructure.
- Laser power (*Pu*) and cutting speed (*V*) have major effect on surface roughness for Ti-6Al-4V as well as Pure Ti.
- For Titanium alloy (Ti-6Al-4V), average surface roughness (R_a) is inversely proportional to Laser power and cutting speed. R_a decreases with the increase of these two parameters.
- Improved surface roughness for Titanium alloy (Ti-6Al-4V), can be obtained at higher laser power and higher cutting speed.
- For pure Titanium (Ti), average surface roughness (R_a) is proportional to Laser power and inversely proportional to cutting speed. R_a decreases with the increase of cutting speed and it increases as Laser power increases.
- Improved surface roughness for pure Titanium (Ti), can be obtained at lower laser power and higher cutting speed.

References

1. Hernandez-Castaneda, J.C., Sezer, H.K., Li, L.: The effect of moisture content in fiber laser cutting of pine wood. Opt. Lasers Eng. **1139**, 49–52 (2011)
2. Klotzbach, A., Hauser, M., Beyer, E.: Cutting of carbon fiber reinforced polymers using highly brilliant laser beam sources. Phys. Procedia. **572**, 7–12 (2011)
3. RaoB, T., Kaul, R., Tiwari, P., Nath, A.K.: Inert gas cutting of titanium sheet with pulsed mode CO2 laser. Opt. Lasers.n Eng. **43**, 1330–1348 (2005)

4. Daymi, A., Boujelbene, M., Linares, J.M., Bayraktar, E., Ben, A.A.: Influence of workpiece inclination angle on the surface roughness in ball end milling of the titanium alloy Ti–6Al–4V. J. Achiev. Mater. Manuf. Eng. **35**(1), 79–86 (2009)
5. Poondla, N., Srivatsan, T.S., Patnaik, A., Petraroli, M.: A study of the microstructure and hardness of two titanium alloys: commercially pure and Ti–6Al–4V. J. Alloys Compd. **486**, 162–167 (2009)
6. Mei, Z., Hongfei, D., Jing, L., Ning, R., He, Y., Haomin, J., Keshan, D., Xinpin, C.: A method for establishing the plastic constitutive relationship of the weld bead and heat-affected zone of welded tubes based on the rule of mixtures and a microhardness test. Mater. Sci. Eng. **527**, 2864–2874 (2010)
7. Hascalik, A., Ay, M.: CO2 laser cut quality of Inconel 718 nickel based super alloy. Opt. Laser Technol. **48**, 554–564 (2013)
8. Mao, C., Sun, X., Huang, H., Kang, C., Zhang, M., Wu, Y.: Characteristics and removal mechanism in laser cutting of cBN–WC–10Co composites. J. Mater. Process. Technol. **230**, 42–49 (2016)
9. Miraoui, I., Zaied, M., Boujelbene, M.: Effect of laser beam diameter on cut edge of steel plates obtained by laser machining. Appl. Mech. Mater. **467**, 227–232 (2014)
10. Yilbas, B.S., Akhtar, S.S.: Laser cutting of Kevlar laminates and thermal stress formed at cutting sections. Opt. Lasers Eng. **50**, 204–209 (2013)
11. Riveiro, A., Quintero, F., Lusquiños, F., Comesaña, R., Pou, J.: Effects of processing parameters on laser cutting of aluminium–copper alloys using off-axial supersonic nozzles. Appl. Surf. Sci. **257**(12), 5393–5397 (2011)
12. Mullick, S., Madhukar, Y.K., Roy, S., Nath, A.K.: Performance optimization of water jet assisted under water laser cutting of AISI 304 stainless steel sheet. Opt. Lasers Eng. **83**, 32–47 (2016)
13. Anicic, O., Jović, S., Skrijelj, H., Nedić, B.: Prediction of laser cutting heat affected zone by extreme learning machine. Opt. Lasers. Eng. **88**, 1–4 (2017)
14. El Aoud, B., Boujelbene, M., Bayraktar, E., Ben, S.S., Miskioglu, I.: Studying effect of CO2 laser cutting parameters of titanium alloy on heat affected zone and Kerf width using the Taguchi method. Springer International Publishing AG (2018), Mechanics of Composite and Multi-functional Materials, vol. 6, Conference Proceedings of the Society for Experimental Mechanics Series
15. Jarosz, K., Löschner, P., Niesłony, P.: Effect of cutting speed on surface quality and heat-affected zone in laser cutting of 316L stainless steel. Proc. Eng. **149**, 155–162 (2016)
16. Roy, S., Zhao, J., Shrotriya, P., Sundararajan, S.: Effect of laser treatment parameters on surface modification and tribological behavior of AISI 8620 steel. Tribol. Int. **112**, 94–102 (2017)
17. Yang, T., Tomus, D., Rometsch, P., Wu, X.: Influences of processing parameters on surface roughness of Hastelloy X produced by selective laser melting. Addit. Manuf. **13**, 103–112 (2017)
18. Petković, D., Nikolić, V., Milovančević, M., Lazov, L.: Estimation of the most influential factors on the laser cutting process heat affected zone (HAZ) by adaptive neuro-fuzzy technique. Infrared Phys. Technol. **77**, 12–15 (2016)
19. Lv, S., Wang, Y.: An investigation of pulsed laser cutting of titanium alloy sheet. Opt. Lasers Eng. **44**, 1067–1077 (2006)
20. Miraoui, I., Boujelbene, M., Bayraktar, E.: Analysis of roughness and heat affected zone of steel plates obtained by laser cutting. Adv. Mater. Res. **974**, 169–173 (2014)

Chapter 32
Influence of Crumb Rubber Reinforcement on the Properties of Medium Density Fiberboard

Libin K. Babu, Kunal Mishra, and Raman P. Singh

Abstract Recycling of scrap tires in the form of crumb rubber is one way to address the environmental concern of tire disposal. In this investigation we have utilized crumb rubber to reinforce medium density fiberboard (MDF). For the purpose of comparison, a baseline MDF material is prepared by mixing polypropylene pellets and wood flour in the ratio of 1:2 by weight using a double helical screw mixer at 165 °C. The crumb rubber reinforced MDF (CR–MDF) material is prepared by adding crumb rubber particles of a size between 1.2 to 1.4 mm, into to the polypropylene pellets and wood flour mixture. Subsequently, the prepared samples are molded into the desired dimension using a hot press. The bulk density of CR–MDF materials is found to be 19% higher than the baseline MDF samples. This indicates that crumb rubber assists in the consolidation process. Prepared MDF and CR–MDF specimens are mechanically characterized as per ASTM D 1037. Compressive modulus of CR–MDF shows an increase of 200%, as compared to baseline MDF. With the addition of crumb rubber, flexural modulus and flexural strength of the MDF samples are improved by 120% and 75%, respectively. Scanning electron microscopy (SEM) on failed CR–MDF samples indicates extensive tearing of crumb rubber particles which suggests that particle bridging is the dominant strengthening mechanism. In addition, water immersion experiments show that the presence of crumb rubber improves the resistance to moisture absorption. Finally, Fourier Transform Infrared Spectroscopy (FTIR) of CR–MDF samples indicates sulfur–sulfur bonds and the CN group from the nitrile component of the crumb rubber.

Keywords Recycled composite · Crumb rubber · Medium density fiber-board · Mechanical properties · Moisture absorption

32.1 Introduction

Every year there are more than one billion tires that are discarded as waste [1]. The three dimensional cross-linked structure makes the chemical decomposition of rubber challenging. Therefore, less than 1% of the raw materials are typically reclaimed from waste tires [2]. The increasing number of waste tires discarded each year is rapidly reducing the available sites for disposal in the form of land-fill, which has caused significant environmental concern. Most of these tires eventually get burnt and used as fuel, but this method emits harmful gases into the atmosphere. Therefore, there is an ongoing need for recycling and reclamation of rubber from scrap tires. Some promising options to meet challenges of tire disposal are: (i) use of tire rubber in asphalt concrete mixture, (ii) incineration of reclaimed rubber for production of steam and (iii) reuse of ground tire rubber in composites as filler material [3–5].

One such method of recycling scrap tires is re-using them in the form of crumb rubber. Crumb rubber is a granular form of tire rubber obtained by shredding, grinding, and magnetically separating the scrap tire rubber [3]. Crumb rubber obtained from waste tires is classified into four groups based on average size: *coarse-* 9.5 to 6.3 mm; *mid range-* 2 to 1 mm; *fine-* 0.4 to 0.2 mm; and *superfine-* 0.15 to 0.075 mm. The size of crumb rubber is designated utilizing a mesh through which it is passed during its production. On an average, 10 to 12 pounds of crumb rubber is generated from a single waste tire [4].

Numerous efforts have been made to use crumb rubber as filler in construction materials like asphalt and concrete [6–9]. Studies have demonstrated that using crumb rubber as a filler in structural materials is an effective method to recycle waste tires [10, 11]. In addition to overcoming the hazards posed on the environment, the inclusion of crumb rubber particles as filler reduces the production cost of materials by approximately 30% [12]. These studies also indicate that, although the compressive and flexural strength of concrete material is lowered by addition of crumb rubber particles due to the lack of

L. K. Babu (✉) · K. Mishra
School of Mechanical and Aerospace Engineering, Oklahoma State University, Stillwater, OK, USA
e-mail: libin.babu@okstate.edu; raman.singh@okstate.edu

R. P. Singh
College of Engineering, Architecture & Technology, State University, Stillwater, OK, USA

P. R. Thakre et al. (eds.), *Mechanics of Composite, Hybrid and Multifunctional Materials, Volume 5*, Conference Proceedings of the Society for Experimental Mechanics Series, https://doi.org/10.1007/978-3-319-95510-0_32

Fig. 32.1 Prepared samples. (**a**) MDF sample. (**b**) Crumb rubber reinforced MDF sample

bonding, the crumb rubber particles in concrete are found to increase the toughness, crack resistance, shock wave absorption, noise level reduction, and also the flexibility of concrete materials [6, 13]. Promising mechanical features and enhanced performance by the use of crumb rubber in concrete have been reported in all the studies. In this study, we extended the incorporation of crumb rubber in to another structural material, namely medium density fiberboard (MDF) material.

32.2 Materials and Methodology

Baseline MDF material were prepared by mixing polypropylene with wood flour. Polypropylene pellets acquired were from PolyOne (Avon Lake, OH) and wood flour was obtained from System Three Resins, Inc. (Auburn, WA). The two materials were mixed in the ratio of 1:2 by weight in a double helical screw mixer (C.W. Brabender, South Hackensack, NJ) at 165 °C, for a duration of 30 min. This temperature is close to the melting point of polypropylene and the selected duration was enough to obtain a visually homogeneous mixture. The mixture heated to 250 °C and subsequently compacted in aluminum molds using moderate pressure. This resulted in blocks that could then be used for physical and mechanical characterization. The crumb rubber reinforced medium density fiberboard (CR–MDF) materials were prepared in a similar manner, but included the addition of 20 wt. % of 1.2 mm sized crumb rubber particles. Figure 32.1 shows images of typical baseline MDF and the CR–MDF materials prepared using this process.

In order to analyze the changes in properties of MDF due to the presence of CR, several tests were conducted, including mechanical characterizations as per ASTM standards. Additionally, the moisture absorption by both batch of samples were analyzed to determine the impact of CR in the mixture.

32.2.1 Density Measurements

The buoyancy method was used to determine the bulk density of samples. First, the dry mass of the sample, m_1, was recorded. Thereafter, the apparent mass of the sample immersed in water, m_2, was noted using a density determination kit (Sartorius YDK 01, Sartorius AG, Goettingen, Germany). Temperature of water is a factor that affects the density of the liquid causing buoyancy. Hence, the temperature of water was recorded to correct for variations in the density, ρ_{fl}, as a function of temperature. The bulk density of the material, ρ_b was evaluated using Eq. (32.1)

$$\rho_b = \frac{m_1}{m_1 - m_2}\rho_{fl} \tag{32.1}$$

32.2.2 Compression Experiments

The rectangular blocks of samples were machined to obtain cylindrical samples of diameter 25.4 mm in accordance with ASTM D–1037 [14], shown in Fig. 32.2.

Compression experiments were carried out on a universal testing machine (INSTRON 5567, Norwood, MA). Three samples of each type were tested to obtain the average value of the properties. The load was applied continuously through the

Fig. 32.2 Cylindrical samples for compression tests

duration of the test at a uniform rate of 0.5 mm/min till the specimen failure, as per ASTM D–1037 [14]. Load–displacement data was recorded continuously through the duration of the test, and the compression modulus was evaluated using Eq. (32.2),

$$E_{compression} = \frac{l_g}{bd}\frac{\Delta P}{\Delta y} \tag{32.2}$$

where l_g was the distance between the gage points in mm and $\Delta P/\Delta y$ was the slope of the straight line portion of the load–displacement curve in N/mm.

32.2.3 *Flexural Experiments*

Flexural samples were sectioned from the rectangular blocks according to ASTM D–1037 standard [14] with dimensions $3.175 \times 51 \times 127.2$ mm. The samples were then loaded on the universal testing machine (INSTRON 5567, Norwood, MA). The span of the test was set as 76.2 mm, and the specimen was loaded continuously at a rate of motion N, of 1.524 mm/min, calculated using Eq. (32.3),

$$N = \frac{zL^2}{6d} \tag{32.3}$$

where, z was the outer fiber strain rate of 0.005 mm/mm/min as specified by the ASTM standard. L and d were the span in mm and thickness of the specimen in mm, respectively. Load and deflection data were continuously recorded during the test, and its values at maximum load were noted. The flexural strength was calculated as,

$$\sigma_{fM} = \frac{3P_{max}L}{2bd^2} \tag{32.4}$$

where, P_{max} was the maximum load in N and b was the width of the specimen measured in dry conditions in mm. The flexural modulus was calculated as,

$$E_{bend} = \frac{L^3}{4bd^3}\frac{\Delta P}{\Delta y} \tag{32.5}$$

where, $\Delta P/\Delta y$ was the slope of the straight line portion of the load–deflection curve in N/mm.

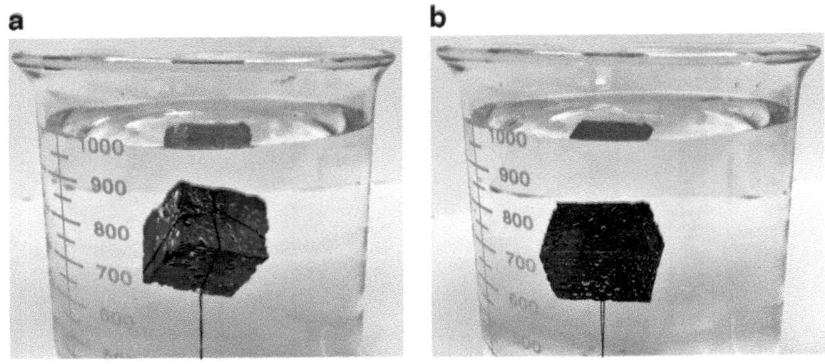

Fig. 32.3 Moisture absorption test apparatus. (**a**) MDF sample. (**b**) CR–MDF sample

32.2.4 *Moisture Absorption Tests*

Moisture absorption tests were conducted according to ASTM D–1037 [14] to determine the water absorption characteristics of the two types of materials. Samples 25.4 × 25.4 × 25.4 mm in size were cut for the test, which corresponds to reduced dimensions with respect to ASTM standard. This was used to provide ease with sample preparation and the test process. Method A: 2–Plus–22h submersion in water test was conducted, as shown in Fig. 32.3, and weights of all the samples were recorded as a function of immersion duration. Distilled water was used for each analysis. The samples were suspended in water using a string attached to the bottom of the beaker. Before being weighed the samples were suspended to drain for approximately 10 min and then excess surface water was removed. Extended water absorption tests up to 48 h were done for few samples for further study. After submersion, the samples were further dried in an oven till 103 °C to determine the moisture content as per ASTM D- 4442 [15].

32.2.5 *Fourier Transform Infrared Spectroscopy*

Absorbance mode FTIR spectroscopy was performed using an FTIR spectrometer (Varian 680-IR, Agilent Technologies, Santa Clara, CA, USA). Liquid nitrogen was used to make the specimens brittle, and to obtain the necessary powdered sample mixture. Background to the spectra was acquired using potassium bromide (KBr). Approximately 5 mg of the powdered specimen was mixed in circa 95 mg of KBr using a pestle and mortar. Spectra peaks ranging between wavenumbers 4000 and 400 cm^{-1} was generated and characteristic IR chart was used to interpret the spectra for each specimen. The influence of noise in the spectra generated was minimized through multiple scans.

32.3 Results and Discussion

32.3.1 *Density*

The average density of baseline MDF was evaluated to be 0.85 g/cm^3. This measurement in keeping with commercial grade MDF which has an average density of 0.75 g/cm^3. The average density of CR–MDF samples was found to be 1.01 g/cm^3. This denotes that addition of CR makes the composite material relatively more dense by 18.8%.

32.3.2 *Mechanical Properties*

Figure 32.4 shows the load–deflection curves for the different samples subjected to compression tests. It is observed that the maximum load that was recorded before failure is significantly higher for CR–MDF samples. Table 32.1 indicates the average compression modulus for the two materials considered. It can be noted that the compression modulus of CR–MDF

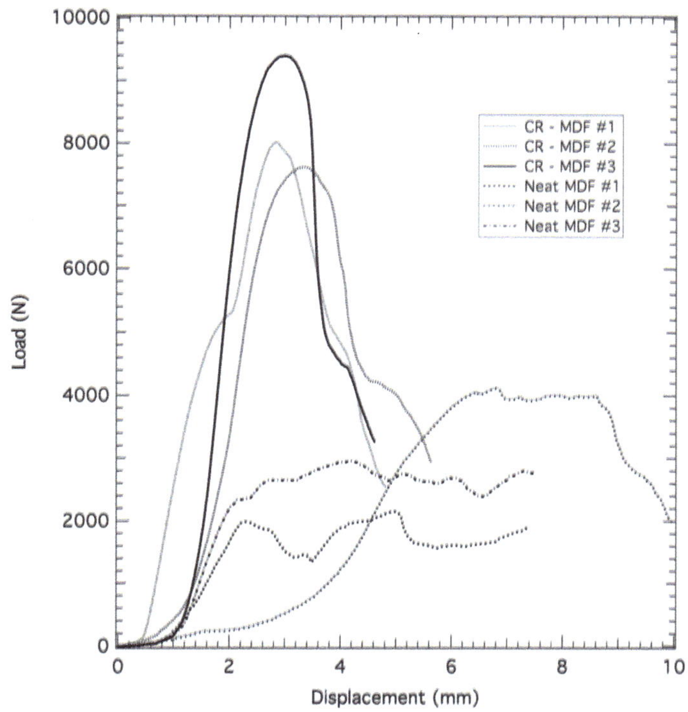

Fig. 32.4 Load–deflection curves from compression tests

Table 32.1 Average compression modulus of the samples

Sample	Average compression modulus (in MPa)
MDF	0.7 ± 0.3
CR–MDF	2.4 ± 0.9

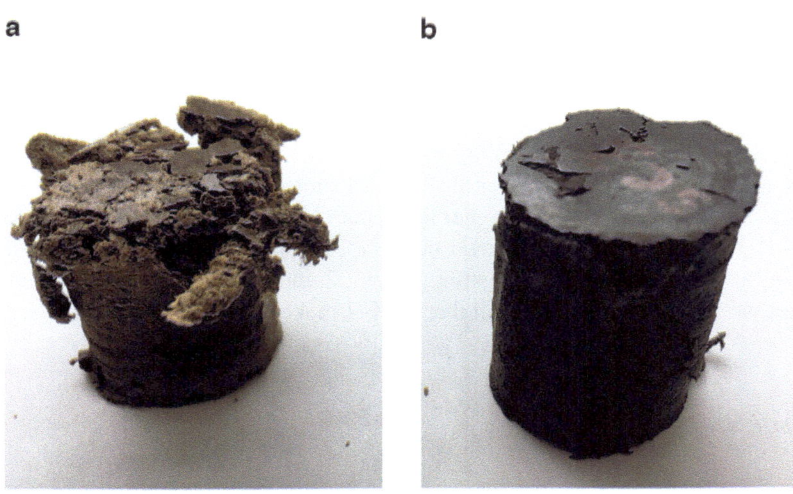

Fig. 32.5 Deformed compression test samples. (**a**) MDF sample. (**b**) Crumb rubber reinforced MDF sample

is approximately 3.5 times higher than MDF samples. As expected, these results contrast the observations regarding the previously noted compressive behavior of CR added concrete or bitumen. This increase in the compressive modulus of CR–MDF samples is due to better binding between the constituents of the mixture.

Figure 32.5a, b shows the deformation due to compression tests in MDF and CR–MDF samples, respectively. CR–MDF underwent less deformation compared to MDF samples, and it can be attributed to the higher density of CR–MDF samples. It corroborates the thought that crumb rubber promotes binding. Better binding crumb rubber in MDF provided higher resistance against the applied load that eventually led to enhancement in the modulus values.

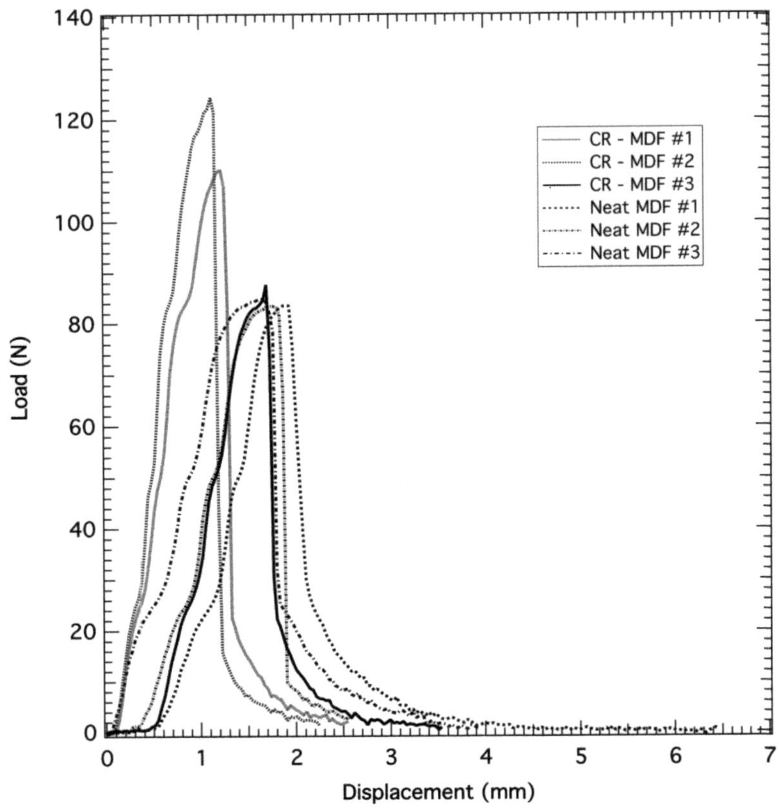

Fig. 32.6 Load–deflection curves from flexural tests

Table 32.2 Average flexural properties of the samples

Sample	Average flexural modulus (in MPa)	Average flexural strength (in kPa)
MDF	3.2 ± 0.5	16.2 ± 6.7
CR–MDF	7.1 ± 2.1	28.3 ± 10.8

Figure 32.6 shows the load–displacement graphs for the samples subjected to flexural testing. Similar to the compression data, the CR–MDF samples supported considerably higher loads than MDF samples. Flexural modulus and flexural strength evaluated using ASTM D–1037 standard [14] are shown in Table 32.2. It can be observed that the reinforcement of MDF using CR increases its flexural modulus and flexural strength by 2.2 and 1.75 times, respectively. It is very similar to the result obtained from compression experiments denoting that the CR makes the MDF structure more compact and cohesive.

32.3.3 Moisture Absorption

Table 32.3 provides the gravimetric measurements that were recorded for the two batch of samples using ASTM D–1037. It can be noted that the presence of CR in MDF considerably reduced the amount of water absorbed by MDF for the same duration of the test. In comparison to MDF samples, the rate of water absorption is found to be gradual for CR–MDF samples.

Using ASTM D–4442, it was found that MDF samples exhibited an average moisture content of approximately 4% with a deviation of 2.3%. However, CR–MDF showed only 0.77% of average moisture content, with a deviation of 0.002%. It can be inferred that inclusion of CR has helped MDF to resist moisture absorption. From the test data, it was understood that CR reduces the porosity within the structure, compared to the neat MDF specimens.

Table 32.3 Weights of samples recorded at different intervals

	Weight of sample (g)			
Sample	MDF #1	MDF #2	CR–MDF #1	CR–MDF #2
0 h	11.464	13.512	10.006	12.937
2 h	11.605	13.579	10.292	12.995
24 h	11.782	13.679	10.570	13.085
48 h	–	–	10.692	13.143
Oven dried	11.509	13.573	10.119	13.041

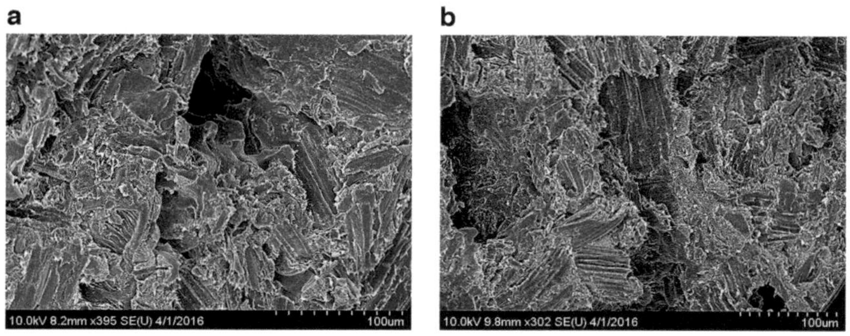

Fig. 32.7 Scanning electron images of failed samples. (**a**) MDF sample. (**b**) Crumb rubber reinforced MDF sample

32.3.4 SEM Analysis

Scanning electron microscope (SEM) images of the deformed surfaces of both batches were obtained using field emission scanning electron microscopy (Hitachi S-4800 FESEM, Dallas, TX). Figure 32.7a indicates that wood surface has majorly smooth cleavage upon failure due to the various tests. However, Fig. 32.7b reveals that crumb reinforced MDF samples underwent extensive tearing during shear failure. It can be understood that through particle bridging mechanism, crumb rubber helps to improve the properties of MDF sample

32.3.5 Fourier Transform Infrared Analysis

Figure 32.8 shows the FTIR spectra of MDF and CR modified MDF samples. O–H bonds of hemicelluloses and cellulose present in MDF sample are shown by the peaks around 3000 and 3600 cm^{-1}. Also, the peaks between 1630 and 1780 cm^{-1} are the absorption bands related to stretching vibration of C = O in acetyl groups present in hemicelluloses [15]. Spectra of crumb rubber reinforced MDF has some new peaks due to the presence of crumb rubber in the structure. This was indicated by new peaks formed at 2350 cm^{-1} by the CN group from the nitrile component of the CR. Also, the presence of carbon black of the phenolic group (OH) stretching, present in CR was observed around 3600 cm^{-1} [16]. The change in peaks in the region 1520 and 1700 cm^{-1} denoted the stretching of the lactone ring (C = O and C = C). CR reinforced MDF also showed few peaks in the 550 to 400 cm^{-1} region due to the presence of sulfur– sulfur bonds, seen in crumb rubber [16].

32.4 Conclusion

Medium density fiber (MDF) samples were made using polypropylene and wood flour. Crumb rubber of average particle size, 1.2 mm was added to the mixture by 20% by weight. A mechanical mixer was employed to prepare the samples. Density of MDF was found to be 0.85 g/cm^3 and CR–MDF as 1.01 g/cm^3, implying that addition of crumb rubber increased the density of MDF samples by 18.8%. Both types of samples were cut according to ASTM standards for different mechanical tests. Compression and flexural tests indicated that crumb rubber improved the mechanical properties such as compression modulus, flexural modulus and flexural strength of MDF samples. The compression modulus and flexural modulus increased

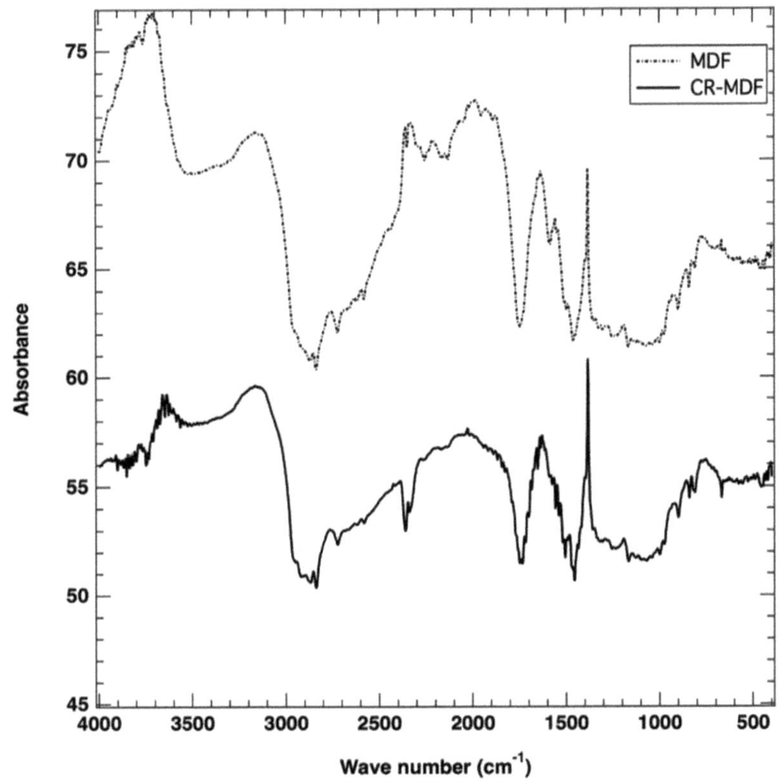

Fig. 32.8 FTIR spectra of MDF and CR reinforced MDF

by about 200 and 120%, respectively. Also, flexural strength was higher by approximately 75%. From SEM images of failed MDF and CR reinforced MDF samples, it was understood that crumb rubber enhanced the properties of neat MDF, through particle bridging mechanism. Moisture content in CR–MDF specimens was lower than MDF by approximately 80%. FTIR spectra indicated the presence of CN group from the nitrile part of the crumb rubber. The aim of this work was to evaluate the changes in specific mechanical properties of MDF by addition of crumb rubber. This study will be helpful to ascertain the potential use of crumb rubber as a suitable additive in the manufacturing process of medium density fiberboard that has significant demand in different applications.

Acknowledgements The authors gratefully acknowledge the support provided for this research by Oklahoma State University—Technology Business Development Program (TBDP) through the research grant (AA-1-55079).

References

1. Siddique R., Naik, T.R.: Properties of concrete containing scrap-tire rubber – an overview. Waste Manag. **24**(6), 563–569 (2004)
2. Adhikari B., De D., Maiti S.: Reclamation and recycling of waste rubber. Prog. Polym. Sci. (Oxford) **25**, 909 (2000)
3. Ganjian E., Khorami M., Maghsoudi, A.A.: Scrap-tyre-rubber replacement for aggregate and filler in concrete. Constr. Build. Mater. **23**, 1828–1836 (2009)
4. Smith, F.G., Daniels, E.J., Teotia, A.P.S.: Testing and evaluating commercial applications of new surface-treated rubber technology utilizing waste tires. Resour. Conserv. Recycl. **15**, 133–144 (1995)
5. Amari T., Themelis, N.J., Wernick, I.K.: Resource recovery from used rubber tires. Res. Policy **25**, 179–188 (1999)
6. Khaloo, A.R., Dehestani M., Rahmatabadi P.: Mechanical properties of concrete containing a high volume of tire-rubber particles. Waste Manag. (New York, N.Y.) **28**, 2472–2482 (2008)
7. Sukontasukkul P., Chaikaew, C.: Properties of concrete pedestrian block mixed with crumb rubber. Constr. Build. Mater. **20**, 450–457 (2006)
8. Yilmaz A., Degirmenci N.: Possibility of using waste tire rubber and fly ash with Portland cement as construction materials. Waste Manag. **29**, 1541 (2009)
9. Batayneh, M.K., Marie I., Asi, I.: Promoting the use of crumb rubber concrete in developing countries. Waste Manag. **28**, 2171 (2008)

10. Thomas, B.S., Gupta, R.C.: A comprehensive review on the applications of waste tire rubber in cement concrete Renew. Sust. Energ. Rev. **54**, 1323–1333 (2016)
11. Navarro, F.J., Partal P., Martínez-Boza, F., Gallegos C.: Influence of crumb rubber concentration on the rheological behavior of a crumb rubber modified bitumen. Energy Fuels **19**, 1984–1990 (2005)
12. Fukumori K., Matsushita M., Okamoto H., Sato N., Suzuki, Y., Takeuchi, K.: Recycling technology of tire rubber. JSAE Rev. **23**, 259–264 (2002)
13. Fattuhi, N.I., Clark, L.A.: Cement-based materials containing shredded scrap truck tyre rubber. Constr. Build. Mater. **10**, 229–236 (1996)
14. ASTM.: ASTM D1037-13 standard test methods for evaluating properties of wood-base fiber and particle. Technical report (2013)
15. ASTM. ASTM D4442-16: Standard test methods for direct moisture content measurement of wood and wood- based materials. ASTM International (2015)
16. Yu, G.X., Li, Z.M., Zhou, X.L., Li, C.L.: Crumb rubber-modified asphalt: microwave treatment effects. Pet. Sci. Technol. **29**, 411–417 (2011)

Chapter 33
Sub-components of Wind Turbine Blades: Proof of a Novel Trailing Edge Testing Concept

Malo Rosemeier, Alexandros Antoniou, and Catherine Lester

Abstract Wind turbine rotor blade sub-component testing (SCT) confines the structural validation to design critical blade parts. Unlike full-scale blade testing, SCT can be adjusted to replicate the stress state of the local structure closer to field conditions and thus augment towards increasing the structural reliability. One of the blade regions often subjected to fatigue loads is the trailing edge bond line. In this work, the proof of concept of a novel trailing edge sub-component test is presented. An outboard specimen of 3 m length was cut out of a 34 m wind turbine rotor blade. The sub-component was installed on a customized test rig and loaded statically with an hydraulic actuator. The imposed strain field along the specimen length is found in a good agreement to the strain distribution in the corresponding area developed during the static full-scale blade test. The experimental setup was simulated by analytical and finite element models. Data recordings with electrical strain gauges and a four-camera digital image correlation system were obtained to validate the predicted structural response of the specimen.

Keywords Experimental validation · Digital image correlation · Finite element analysis · Certification · Out-of-plane displacement

33.1 Introduction

In the past decades rotor blade designs of wind turbines have been thoroughly investigated by structural testing. International standards describe the validation process of the design assumptions through full-scale blade testing [11]. Therefore, blades are tested under static and fatigue loads. The static loads represent the design envelope of extreme load cases, while the cyclic loads correspond to damage equivalent fatigue loads of at least 20 years of operation. A cyclic full-scale blade test in resonance excitation, however, is reaching its limits when it is used to investigate the structural response under combined flap- and lead-lag loading scenarios subjected to specific stress ratios [17].

As a supplement to the full-scale blade experiment, testing of representative blade elements and sub-components is presented as a design validation method for rotor blades in widely accepted guidelines [9]. Generic elements and structures have been developed for this purpose, e. g. [1, 13, 19]. In the present research, however, a sub-component is considered as a structural part cut directly out of a wind turbine rotor blade [12]; it therefore corresponds to the blade full-scale.

Berring et al. [2] worked on the determination of the bend-twist effect and tested an outboard blade segment. Finite element (FE) simulations of the test were conducted using shell and volume elements to obtain a reasonable representation of the torsional behavior of the specimen [7]. Furthermore, Ridzewski [15] proposed a setup for testing the leading edge of a wind turbine blade. Lahuerta et al. [14] presented experimental results of a wind turbine blade segment test with clamped edges focusing on the structural performance of the trailing edge. With this configuration it is possible to introduce the applied bending moment of a cross-section as in a full-scale test. Moreover, Branner et al. [8] presented a similar test concept, but cutting the blade along the spar cap. In that context a segment/sub-component test configuration with flexible edges was introduced by Rosemeier et al. [18]. The aim was to impose a strain distribution along and across the sub-structure similar to the corresponding full-scale blade test.

Analytical and numerical results of this sub-component test concept are presented in this work and further compared to deformation measurements recorded with electrical strain gauges and a digital image correlation system.

M. Rosemeier (✉) · A. Antoniou · C. Lester
Division Structural Components, Fraunhofer IWES, Fraunhofer Institute for Wind Energy Systems, Bremerhaven, Germany
e-mail: malo.rosemeier@iwes.fraunhofer.de

P. R. Thakre et al. (eds.), *Mechanics of Composite, Hybrid and Multifunctional Materials, Volume 5*, Conference Proceedings of the Society for Experimental Mechanics Series, https://doi.org/10.1007/978-3-319-95510-0_33

33.2 Methodology

33.2.1 Testing Concept

The proposed test concept is designed with a single actuator (Fig. 33.1) allowing the flexible adjustment of combined static and fatigue loading scenarios for blade sub-components, which are either cut out of a full-scale blade or manufactured separately.

The specimen is glued into load frames at its span-wise edges, which are mounted to ball joints R and T. Joint T is mounted to a stiff wall and joint R is mounted to a vertical beam that is hinged on a strong floor at joint B. The actuator is mounted between another stiff wall (joint C) and the vertical beam (joint D) allowing for a flexible positioning along the height, which makes it possible to operate the actuator in its most efficient position w. r. t. force and displacement. The concept allows the adjustment of the following parameters: (a) the distance between the axis through ball joints R and T (RT-axis), and the line of centroids z_c' defines the bending moment distribution M along the specimen due to an eccentrically introduced force F, and (b) the superposition of the resulting normal force distribution N and M defines the target strain distribution across the cross-section. This concept is described in detail by Rosemeier et al. [16, 18]. The shape of the bending moment distribution depends on the shape of the line of centroids z_c', which can be influenced by the shape of the specimens cut along the span or by addition of material or springs connected in parallel to the specimen.

33.2.2 Models

The structural response of the ball-joint sub-component test concept is analyzed using the example of a 34 m blade design. The trailing edge sub-component of approximately 3 m length was virtually cut out, such that the target cross-sections in the middle of the specimen is at 71% blade length. The segment was further cut in span-wise direction such that only the area of interest remains, i.e. the trailing edge cell including one main shear web. An analytical beam model (AM) on the basis of Euler-Bernoulli beam theory [10] was assembled with cross-section properties determined by the Beam Cross Section Analysis Software BECAS [3–6]. Additionally, finite element shell models (FEM) of the full-scale blade test and the sub-component test setup were implemented in ANSYS APDL [20]. The blade parameterization and input generation for the models was conducted using workflows of the FUSED-Wind framework [22].

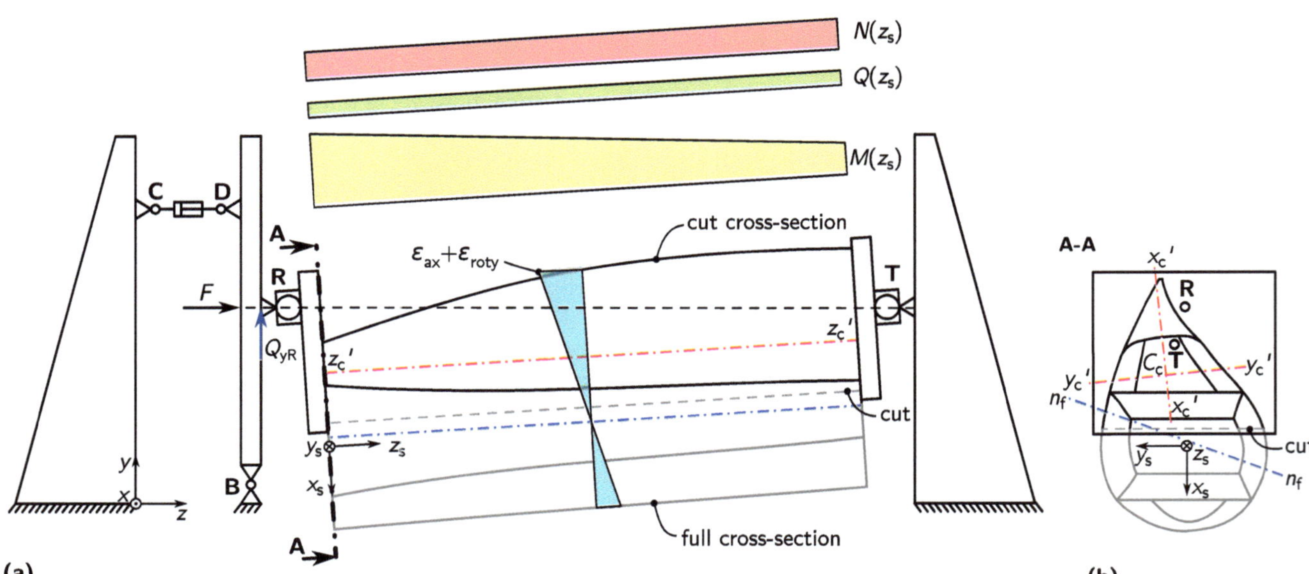

Fig. 33.1 Side view of the setup (a): a segment cut cross-section (black) of a full cross-section (grey) is cut out and clamped between ball joint R and T. Ball joint T is connected to a strong wall. Ball joint R is connected to vertical beam, which is mounted to a strong floor at hinge B. An adjustable actuator connects the beam at joint D with another stiff wall at joint C. The front view into plane A-A is shown in (b) [18]

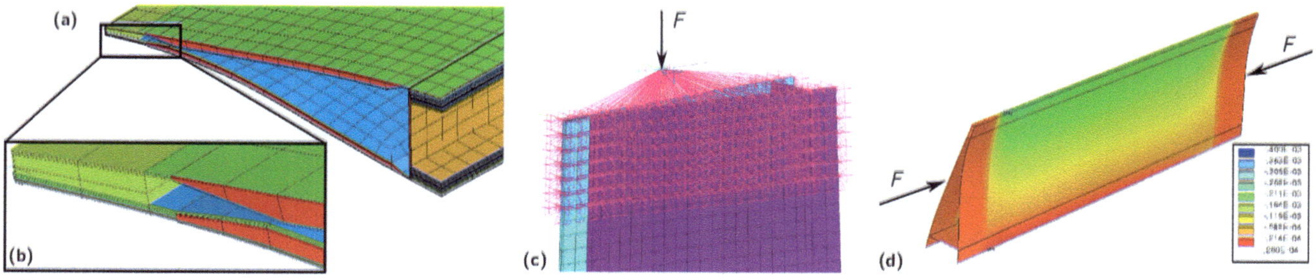

Fig. 33.2 Finite element model showing extruded SHELL281 elements of the tested 3 m section (**a**), a zoom into the trailing edge bond-line modelled with SOLID186 elements (**b**), the load frame representation by CERIG elements and the load introduction via a master node (**c**) and its structural strain response (**d**). The color scale shows the longitudinal strain ε_z along the specimen at 20 % load level of the final test load of the leading-to-trailing edge (LTT) load case

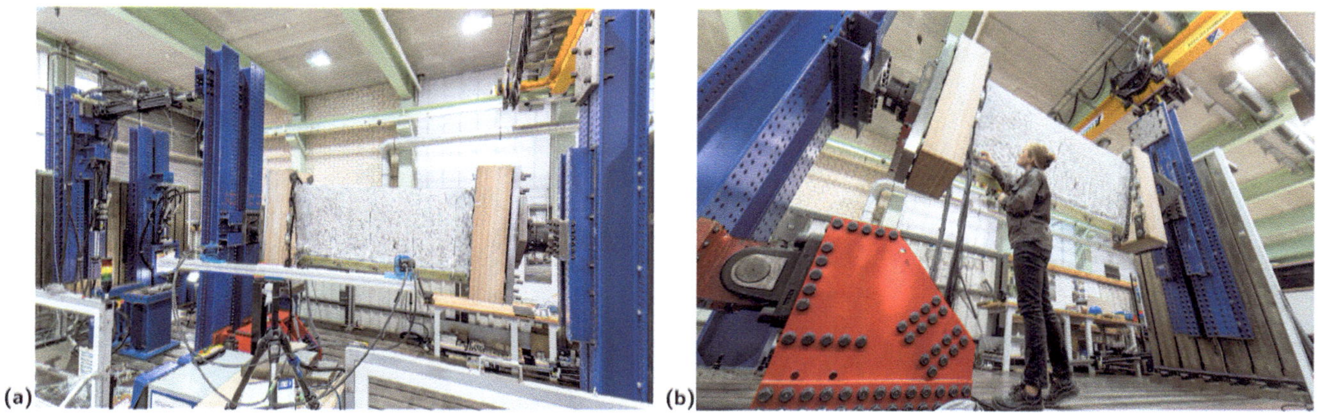

Fig. 33.3 Test rig overview (**a**) and specimen view (**b**). The hydraulic actuator is shown on the left introducing the loading into a hinged vertical beam. The specimen is mounted via ball joints between the vertical beam and a strong wall on the right. In the front two of the four DIC cameras are seen facing the suction side of the specimen

The positions of the load introduction points at the specimen edges (Fig. 33.2d) were determined using the analytical model of the full and the cut cross-section. The structural response of the static full-scale leading-to-trailing edge (LTT) load case was chosen as reference.

33.2.3 Experimental

33.2.3.1 Setup

The test concept described in Sec. 33.2.1 was set up on a strong floor with two adjustable strong walls (Fig. 33.3). The hydraulic actuator was used in displacement control and had a nominal load of 210 kN with a maximum stroke displacement range of 600 mm. A benefit of this setup is that relatively small variations in specimen length can be compensated by the stroke displacement. The ball joints used were designed for a nominal load of 400 kN. The gravity load constrains the 3 m long specimen's rotational degree of freedom about its length axis. While mounting the specimen, it finds its stable position depending on the position of both joints R and T within the cross-section.

33.2.3.2 Specimen Preparation

The specimen was cut along the spar cap and ground at the load introduction zones (Fig. 33.4a) both to improve the adhesion with the load frames and enhance visual inspections during testing (Fig. 33.4b). Wooden frames were milled according to the outer and inner surface of the specimen. In a casting process (Fig. 33.4c) the frames were glued to the specimen with an epoxy system. The curing of the epoxy resin was performed approximately for 12 h at 65 °C.

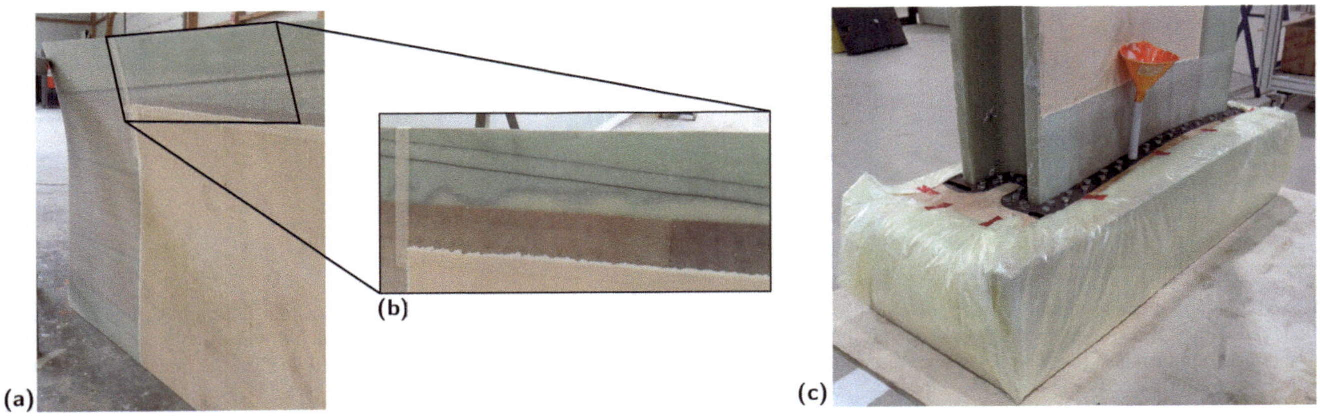

Fig. 33.4 Ground of the paint at the load introduction zones (**a**) as well as on the trailing edge bond line (**b**). Root and tip ends of the specimen were glued into wooden load frame using a casting process

Fig. 33.5 Specimen coordinate system in suction side view. Root end is on the left and tip and on the right hand side

33.2.3.3 Measurement Equipment

Strain Gauges

Strain gauges (SG) were applied according to the coordinate system as shown in Fig. 33.5, where the relative arc-length coordinate s starts at the trailing edge of the pressure side surface going towards the leading edge and ending at the trailing edge of the suction side surface. The s-coordinate is parallel to the load frames and the z-coordinate is perpendicular to the load frames pointing from the root towards the tip end. The load frames were mounted such that they were positioned parallel to the virtual root coordinate system of the full-scale blade.

Digital Image Correlation

In addition to strain gauge measurements, a stereo digital image correlation (DIC) system measured the specimen movement [23], providing out-of-plane displacement and full-field strain of the entire specimen surface. A black and white stochastic speckle pattern was painted on both specimen surfaces (Fig. 33.6a). Flags were placed on the wood frame visible by all cameras, such that the cameras were coupled into a single, global coordinate system during post-processing.

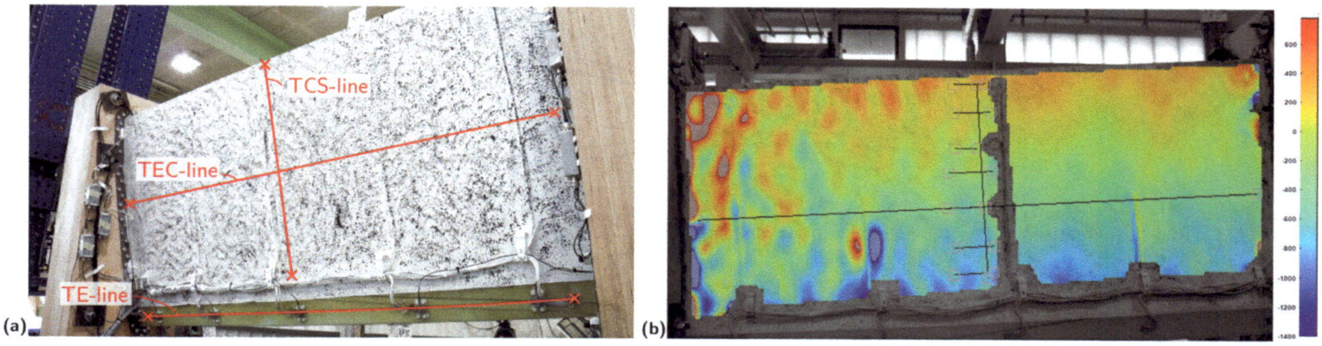

Fig. 33.6 Speckle pattern for the digital image correlation (DIC) system shown on the suction side surface (**a**) and virtual strain gauge positions (**b**). Three reference markers are positioned at the trailing edge and spar cap edge of the specimen to define the common coordinate system for suction and pressure side. Post-processing lines are defined along the target cross-section (TCS-line) and along the middle of the trailing edge cell (TEC-line). Additionally, the trailing edge electrical strain gauge line (TE-line) is highlighted. The color scale in (**b**) shows the longitudinal strains ε_z

Two pairs of 2.8 megapixel cameras were installed to record the movement of each panel separately - two cameras on the SS and two on the PS. The use of two stereo DIC setups and a global coordinate system provided the ability to measure out-of-plane displacements of both panels relative to the other.

Virtual strain gauges placed next to the electrical strain gauges compared the TCS-line (Fig. 33.6b).

33.3 Experimental Results and Model Validation

33.3.1 Geometry Comparison

The comparison of the finite element model (FEM) with the measured geometry of the specimen shows that the FEM is thicker compared to the actual geometry (Fig. 33.7).

33.3.2 Strain Response Along the Trailing Edge

The trailing edge specimen was loaded in a configuration aiming to reach the 100 % strain level of the leading-towards-trailing edge (LTT) full-scale blade test load case [21]. The trend of the longitudinal strains measured with strain gauges (SG) at the trailing edge pressure (PS) and suction side (SS) (TE-line at $s \approx 50$ mm, see Fig. 33.6a) are in a good agreement with the measured values of the LTT full-scale test (Fig. 33.8a). The slope of the strain distribution along the trailing edge was successfully replicated. The discrepancy close to the load frames is caused from the stiffening of the support wooden blocks. Moreover, the back to back values at both sides of the TE compare very well to each other. At the target cross-section of the specimen they move apart indicating a bending deformation of an S-shape.

The analytical (AM) and finite element models (FEM) are validated against the experimental, longitudinal strains along the specimen length. The strain data were derived from SGs. The longitudinal strain along the trailing edge bond-line is benchmarked (Fig. 33.8a). In general, it can be seen that the experimental strain level in both full-scale test (FST) and ball-joint sub-component test concept (BJ-SCT) is lower compared to the FEM, but for both experiments the strain is at the same level. The relative strain trend between suction (SS) and pressure side (PS), the pathway of the FEM and experiment BJ-SCT is similar: Along the length the strain level on the PS is getting lower than on the SS, finding its peak value at the target cross-section at $\approx$24 m blade length. Strain peaks were recorded next to both load frames as also prognosed by the FE model The peaks are caused by the load frame constraints on to the local Poisson's ratio of the structure.

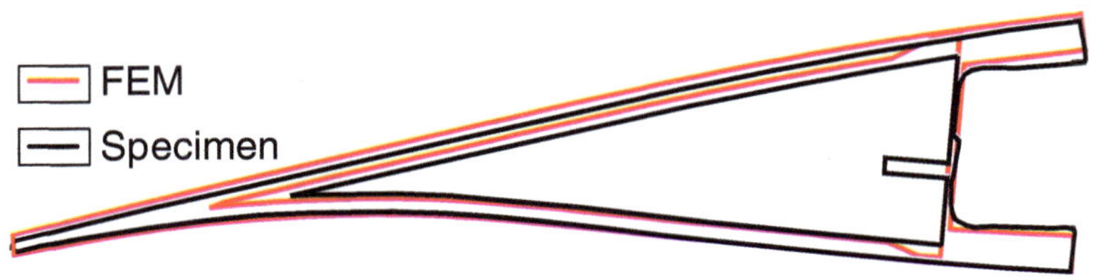

Fig. 33.7 Geometry comparison of the finite element model and actual blade geometry at the tip end of the sub-component specimen

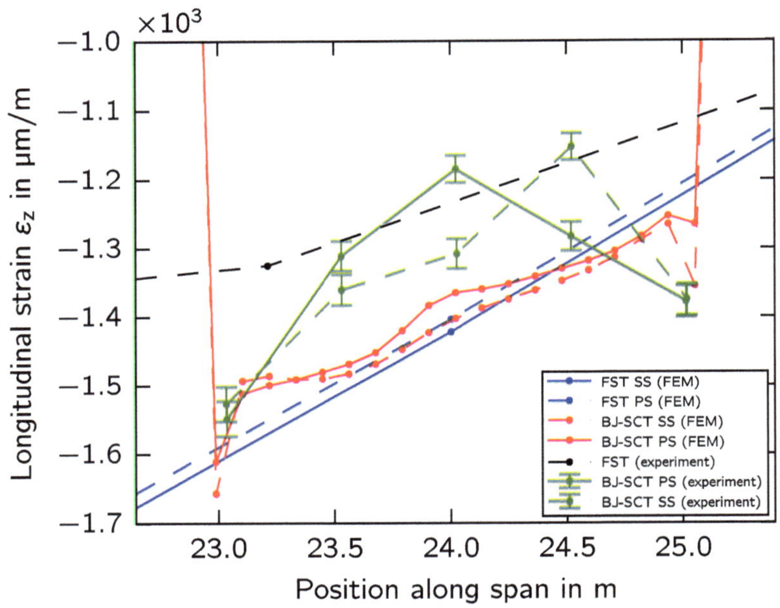

Fig. 33.8 Validation of the finite element (FEM) and analytical (AM) model of the full-scale blade (FST) and sub-component ball joint concept (BJ-SCT). The models are benchmarked with the full-scale experiment [21] and the sub-component experiments using the full-scale leading-to-trailing edge (LTT) load case at 100 % load level. The longitudinal strain ε_z is compared for both pressure and suction side along the trailing edge bond line (**a**) and along the target cross-section surface coordinate s (**b**)

33.3.3 Strain Response Across the Target Cross-section

The longitudinal strain across the cross-section coordinate s (Fig. 33.9a) is observed at the target cross-section in the middle of the specimen at about 24 m blade length. The strain level of the BJ-SCT experiment is in a good agreement with the FEM and AM results. The experimental strain level on the SS, however, is lower than the level observed in the FEM and AM. The slightly increased strain dimple at $s = 0.83$ correlates well to the trend of the convex strain distribution in the BJ-SCT FE model.

The longitudinal strain along the surface of the target cross-section (TCS-line) in the middle of the specimen was also monitored with SGs (Fig. 33.9a). The recordings are compared to the digital image correlation (DIC) system results at discrete points on the TCS-line. It can be seen that both measurement methods show a fair agreement regarding absolute values but they both show the same trend. The DIC strains, particularly on the PS, have a slight offset compared to the SGs, which is also the case for the mid arc-length range on the SS. The discrepancies can be explained through a spatial offset of the DIC virtual strain gauges. Moreover, the electrical strain gauges measure the local deformation and the DIC strains are a result of a smoothing algorithm over a relatively long gauge length compared to the SG length.

Furthermore, the transversal strains were extracted from SG and DIC measurements along the TCS-line (Fig. 33.9b). As in the longitudinal DIC strain results, an offset towards a lower strain level is observed on the PS compared to FEM and two SG measurement points.

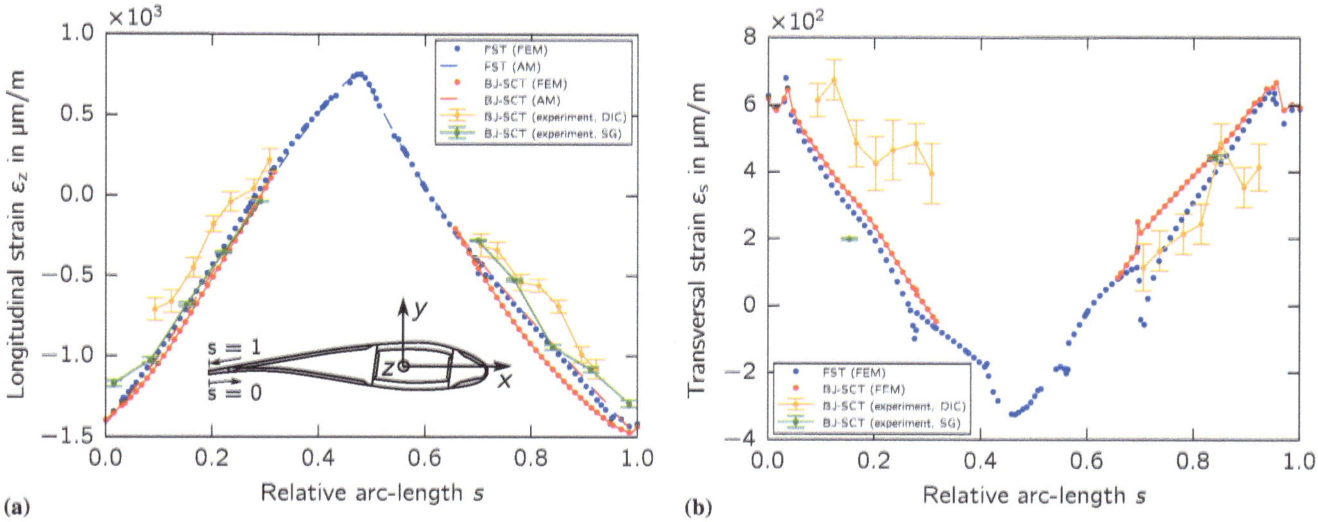

Fig. 33.9 Validation of the finite element (FEM) and analytical (AM) model with sub-component ball joint concept (BJ-SCT). The models are benchmarked with the sub-component experiment using the full-scale leading-to-trailing edge (LTT) load case at 100 % load level. The longitudinal (**a**) and transversal (**b**) strain ε_z is compared along the target cross-section surface coordinate s

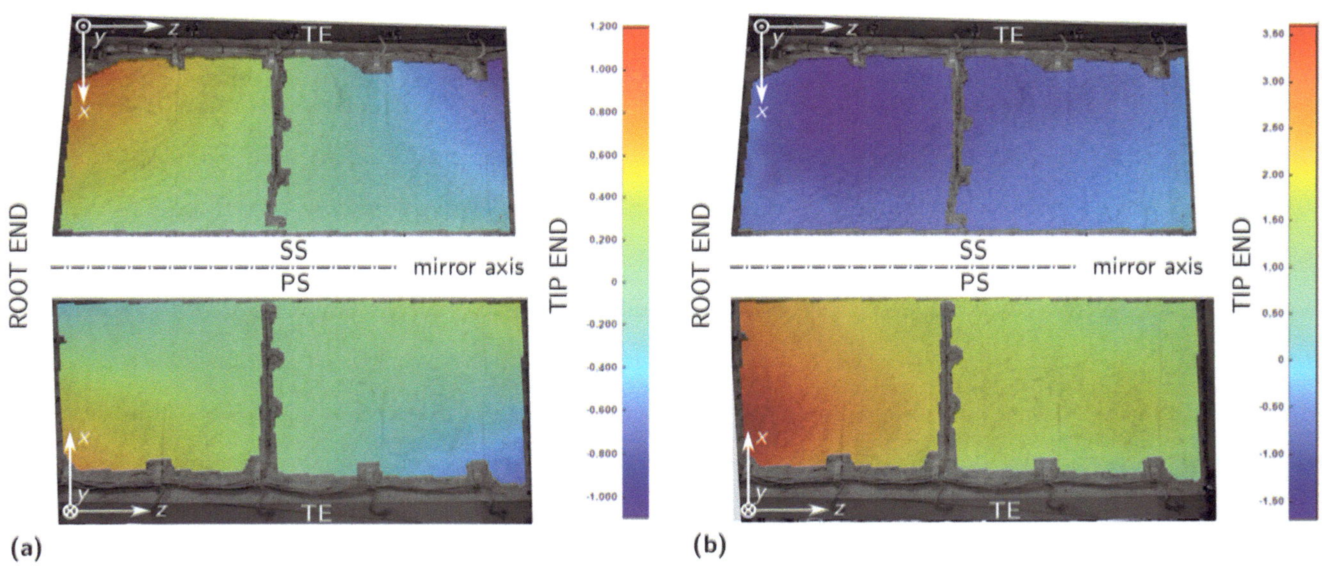

Fig. 33.10 In-plane displacement u_z with rigid body motion removed (**a**) and out-of-plane displacements u_y (**b**) of the suction side and pressure side surface. The displacement field was extracted at 100 % test load level of the LTT load case

33.3.4 Displacement Response

In the displacements field, u_z indicates the lead-lag bending compression of the specimen (Fig. 33.10a). The compression is slightly larger on the SS compared to the PS indicating a flap-wise bending.

The out-of-plane displacement u_y along the TCS-line and along the TEC-line shows that the PS surface moves towards the SS surface, whereas the SS surface moves towards the PS surface. The whole displacement field is shown in Fig. 33.10b.

33.4 Conclusions

The structural strain response is compared to the sub-component test configuration to assess the applicability of the proposed test setup. The validation of analytical and finite element models (FEM) shows good agreement with trends seen in experimental full-scale and sub-component test results.

An advanced four camera digital image correlation (DIC) system was measuring the global deformations of the specimen. The derived DIC strain results were correlating within acceptable deviations from the electrical strain gauge recordings.

The actual blade is more flexible than the FEM since the strains along the trailing edge in the full-scale and the sub-component test are lower. This fact is supported by the geometry comparison indicating that the FEM is thicker than the actual specimen. Assuming the same structural layup and material properties, the second moment of area of the cross-section leads to a larger stiffness in the FEM. A feedback of the actual geometry into the FEM is required for a model improvement.

Acknowledgements We acknowledge the support of the European Commissions Seventh Framework Programme within the IRPWind project (609795) and the support within the Future Concept Fatigue Strength of Rotor Blades project granted by the German Federal Ministry for Economic Affairs and Energy (BMWi) (0325939) and the Senator for Health, Environment and Consumer Protection of the Free Hanseatic City of Bremen within the ERDF programme Bremen 2014-2020 (201/PF_IWES_Zukunftskonzept_Betriebsfestigkeit_Rotorbltter_Phase I).

References

1. Andersen, S., Bach, P., Bonne, W., Kensche, C., Lilholt, H., Lystrup, A., Sys, W.: Fatigue of materials and components for wind turbine rotor blades. Technical report, German Aerospace Center (DLR) (1996)
2. Berring, P., Branner, K., Berggreen, C., Knudsen, H.W.: Torsional performance of wind turbine blades-part 1: experimental investigation. In: Proceedings of the 16th International Conference on Composite Materials, vol. 43. Japan Society for Composite Materials, Kyoto, Japan (2007)
3. Blasques, J.P.: Multi-material topology optimization of laminated composite beams with eigenfrequency constraints. Compos. Struct. **111**, 45–55 (2014)
4. Blasques, J.P., Stolpe, M.: Multi-material topology optimization of laminated composite beam cross sections. Compos. Struct. **94**(11), 3278–3289 (2012)
5. Blasques, J., Bitsche, R.: An efficient and accurate method for computation of energy release rates in beam structures with longitudinal cracks. Eng. Fract. Mech. **133**, 56–69 (2015)
6. Blasques, J., Bitsche, R., Fedorov, V., Lazarov, B.: Accuracy of an efficient framework for structural analysis of wind turbine blades. Wind Energy. **119**(9), 1603–1621 (2015)
7. Branner, K., Berring, P., Berggreen, C., Knudsen, H.W.: Torsional performance of wind turbine blades–Part II: numerical validation. In: Proceedings of the International Conference on Composite Materials (ICCM-16) (2007)
8. Branner, K., Berring, P., Haselbach, P.: Subcomponent testing of trailing edge panels in wind turbine blades. In: Proceedings of 17th European Conference on Composite Materials (2016)
9. DNV GL AS: DNVGL-ST-0376 – Rotor blades for wind turbines (2015). https://rules.dnvgl.com/docs/pdf/DNVGL/ST/2015-12/DNVGL-ST-0376.pdf
10. Euler, L.: Methodus inveniendi lineas curvas maximi minimive propietate gaudentes, Additamentum I, De curvis elasticis (1744)
11. IEC: IEC 61400-23 – wind turbines part 23: full-scale structural testing of rotor blades. International Electrotechnical Commission (2012)
12. IEC: IEC 61400-5 – wind turbines part 5: wind turbine blades (under review). International Electrotechnical Commission (2017)
13. Kühlmeier, L.: Buckling of wind turbine rotor blades: analysis, design and experimental validation (2007)
14. Lahuerta, F., de Ruiter, M.J., Espinosa, L., Koorn, N., Smissaert, D.: Assessment of wind turbine blade trailing edge failure with sub-component tests. In: Proceedings of 21st International Conference on Composite Materials (2017)
15. Ridzewski, J.: Fatigue of composite materials and structures. Presentation at Haus der Technik, Essen (2013)
16. Rosemeier, M., Massart, P., Antoniou, A.: Tailoring the design of a trailing edge sub- component test. In: Presented at 3rd Annual IRPWind Conference in Amsterdam, The Netherlands, 19–20th Sept 2016. https://doi.org/10.5281/zenodo.153837.
17. Rosemeier, M., Basters, G., Antoniou, A.: Benefits of sub-component over full-scale blade testing elaborated on a trailing edge bond line design validation. Wind Energy Science Discussions **2017**, 1–14 (2017). https://doi.org/10.5194/wes-2017-35. https://www.wind-energ-sci-discuss.net/wes-2017-35/
18. Rosemeier, M., Bätge, M., Antoniou, A.: A novel single actuator test setup for combined loading of wind turbine rotor blade sub-components. In: Proceedings of the 2nd International Symposium on Multiscale Experimental Mechanics: Multiscale Fatigue in Lyngby, Denmark (2017)
19. Sayer, F., Antoniou, A., van Wingerde, A.: Investigation of structural bond lines in wind turbine blades by sub-component tests. Int. J. Adhes. Adhes. **37**, 129–135 (2012)
20. Swanson, J.A.: ANSYS Mechanical APDL, version 15.0 (2014)
21. Thomsen, C.L., Eisenberg, Y.P.: Blade test SSP34#2 edgewise and flapwise final static test. Technical report, Risø National Laboratory, Denmark (2003)
22. Zahle, F., Réthoré, P.E., Graf, P., Dykes, K., Ning, A.: Fused-wind v0.1.0 (2015). https://doi.org/10.5281/zenodo.13899
23. Sutton MA, Orteu JJ, Schreier HW (2009) Image correlation for shape, motion and deformation measurements. Springer Science & Business Media, Berlin

Chapter 34
Toughening Mechanisms on Recycled Rubber Modified Epoxy Based Composites Reinforced with Alumina Fibers

A. B. Irez, I. Miskioglu, and E. Bayraktar

Abstract Environmental and economic concerns has been a motivation for the material manufacturers to produce new, robust, lightweight and cost-effective materials. Therefore, aeronautic and automobile industries are investigating multifunctional composite materials that can meet their expectations. In this regard, polymer based composites can have high specific strength with the contribution of resistant fillers. As a polymer, epoxy is considered an appropriate matrix, which can be modified by different agents. First option is the addition of hard particles and second is that the inclusion of thermoplastics or the addition of elastomeric materials. Since, epoxy exhibits high brittleness due to its highly cross-linked nature, modifiers less rigid than the matrix can serve as unique tougheners enhancing the ductility. In this study, a good combination of both hard and soft modifiers is used. In addition to mechanical characteristics, using of fresh clean scrap EPDM rubbers adds an economic and environmental value to this study. Also, due to its favorable structural characteristics such as interlocking effects of fibers, addition of alumina fibers (AF) ensures desired mechanical properties in case of a homogeneous distribution. This paper primarily explains the mechanical behavior as well as damage mechanisms of epoxy-fresh scrap rubber composites. The mechanical and physical properties of these composite systems are studied in the present work. Dynamic Mechanical Analysis (DMA) analyses were carried out to determine thermal-mechanical properties. Three-point bending and fracture toughness tests were realized with single edge notched beam (SENB) and smooth specimens. Finally, scanning electron microscope (SEM) was used to observe fracture surfaces and the microstructure.

Keywords Three-point bending · Epoxy – rubber · Alumina fibers · SEM · Fracture toughness

34.1 Introduction

In the last decades, the companies in aeronautic and automobile industries are conducting their projects by considering some fundamental restrictions to be able to compete with their rivals. First restriction is the decreasing of the total cost of the vehicles and second one is the reduction of the total mass of the vehicle to lower the CO_2 emissions and fuel consumption. Therefore, the engineers in these companies are in search of light weight and low-cost materials to manufacture certain components of the vehicles. In this regard, the use of recycled materials brings some advantages to the material manufacturers in terms of economic and environmental aspects. Hence, there is a trend of manufacturing some components of automobiles and aircrafts by using recycled materials. Also, with regards to second restriction light weight robust materials should be designed. In this respect, polymer-based composites by using recycled elements may bring solutions to these restrictions.

As a thermoset polymer, epoxy is widely used in many different applications due to its favourable properties such as decent chemical resistance, durable mechanical properties, and ease in modification versatility as well as its inherent low viscosity and volatility [1–6]. Nevertheless, it also has some unenviable properties such as brittleness because of highly cross-linked network structure. Thus, researchers have been working for a while to improve the brittleness of epoxies in order to extend their applications [4–12]. To toughen epoxy resins, secondary phase particles can be incorporated. These secondary phase particles can be dispersed over the matrix and they can either be soft fillers such as thermoplastic particles,

A.B. Irez
CentraleSupélec, Université Paris-Saclay, Gif-sur-Yvette, France

I. Miskioglu
Michigan Technology University, Engineering Mechanics Department, Houghton, MI, USA

E. Bayraktar (✉)
Supmeca-Paris, School of Mechanical and Manufacturing Engineering, Paris, France
e-mail: bayraktar@supmeca.fr

P. R. Thakre et al. (eds.), *Mechanics of Composite, Hybrid and Multifunctional Materials, Volume 5*, Conference Proceedings of the Society for Experimental Mechanics Series, https://doi.org/10.1007/978-3-319-95510-0_34

rubber or rigid fillers [8–14]. In this respect, to manufacture various cost-effective consumer products, recycled rubbers can be used as soft fillers.

Blending the recycled rubber with epoxy enables its shaping to beneficial products at a low cost. This idea can be achieved by mixing powdered scrap rubber with epoxy as well as necessary additives. The manufacture of recycled rubber modified composites is realized by means of continuous processes, such as injection molding or pultrusion of thermosets. Another important point is that addition of rubber to plastics also promotes some key properties of plastics especially impact resistance and 'good feel' behavior.

Second, the modification of epoxies also done by the incorporation of some inorganic–particulate fillers such as silica [6, 7] and alumina [8–11]. They generally improve the material properties such as fracture toughness [12, 15–23, 25], resin stiffness [14–19] and wear resistance [24, 25]. In this study alumina was used as a rigid filler. In particular, alumina has good thermal conductivity, inertness to most acids and alkalis, high adsorption capacity, thermal stability and electrical insulation, and so on. Also, it is inexpensive, non-toxic and highly abrasive [16–18, 26]. More specifically, in this study alumina has been used in fibre form and it has advantages compared to particle form.

This work presents processing of recycled rubbers blended with epoxy resin to create novel composites in an economic way. Main objective of this research is to determine the mechanical properties and toughening mechanisms of these composites. During this study, physical properties such as glass transition temperature were determined with dynamical mechanical analysis (DMA). Then, bending tests were realized with single edge notched beam (SENB) and smooth specimens. At the end, fracture surfaces were observed by means of scanning electron microscopy (SEM) to study the toughening and damage mechanisms.

34.2 Experimental Procedure

34.2.1 Material Processing

Recycled rubber modified epoxy based composites were prepared according to following procedure. First Bisphenol A-type Huntsman™ Araldite DBF epoxy resin was heated to 65 °C in order to decrease its viscosity. Then the alumina fibers were added to epoxy resin. This blend was mixed by using a high shear stirrer for 1 h at 65 °C to homogenize the mixture. The temperature of the mixture was controlled by a digital thermometer during the mixing process. Next, EPDM rubbers were added gradually to the mixture and mixing of this blend was continued for an additional 2 h, again at 65 ° C to prevent agglomerations. The recycled rubber powders are obtained from the unused scrap parts obtained during sportive equipment manufacturing. It means that they are fresh, clean and completely free of other materials that is present in ground tire. After mixing procedure, the mixture was cooled down to room temperature and the hardener was added. Because, addition of hardener at higher temperatures results in rapid solidification which may lead to uncompleted cross-linking reaction in some regions of the composite. When the mixture cooled down to room temperature, the hardener (Huntsman™ HY-952) was added regarding the stoichiometric ratio and this final compound was mixed for 15 min. Afterwards, the mixture was degassed in a vacuum chamber at room temperature for about 30 min. Then, the mixture was poured into silicon molds which have been cleaned previously and molds were left at room temperature for 24 h to complete the curing.

After bulk composites were manufactured, test specimens were prepared by machining of the bulks according to test standards (Table 34.1).

34.2.2 Experimental Procedure

Dynamic properties, storage modulus (E') and mechanical loss angle tangent (tan δ) of the epoxy based composites were investigated using a Dynamic Mechanical Thermal Analyser Q800 system (TA Instruments). The data were obtained at a

Table 34.1 Composition of the EPDM rubber modified epoxy based composites

LRAL composites	Alumina fiber content			
Rubber content	0%	5%	7.50%	10%
20%	LR20	LR2AL5	LR2AL7.5	LR2AL10

frequency of 1 Hz, 0.1% strain in the temperature range from −80 °C to 150 °C using a heating rate of 3 °C/min in single cantilever bending mode. The dimensions of the investigated samples were as follows: width 10 mm, length 30 mm, and the thickness were 4 mm.

Microstructure and fracture surface damage analyses have been carried out by means of optical (OM) and scanning electron microscopy (SEM). SEM observation was realized on the fracture surface of the tested specimens with Scope/JSM-6010LA Jeol® electron microscope.

As mechanical characterization, there point bending tests (3 PB) have been carried out according to the ASTM D790 standards. Deflection of the specimen was measured by the crosshead position and crosshead speed was selected as 1 mm/min. Flexural strength and strain were obtained from the test results. The test specimen was placed in the Instron 5569 bending test fixture and force was applied until it fractured. At least five specimens for each composition were used and results were given with standard deviations.

In addition, fractural properties such as plain strain fracture toughness (K_{Ic}) and critical strain energy release rate (G_{Ic}) were investigated with SENB specimens and the tests were realized according to ASTM D5045 standard. Notches were introduced by tapping a fresh razor blade.

34.3 Results and Discussions

34.3.1 Microstructure of the Composites and Surface Hardness

After manufacturing of the composites, one of the compositions, here LR2AL10, was sectioned and polished. Then the microstructure was observed to see the distribution of the modifiers. This microscopical observation is given in Fig. 34.1. In the figure, circular particles are determined as alumina fibers and other irregular shaped particles are the recycled EPDM rubbers according to EDS analysis. The microstructure of compositions showed a considerably homogenous distribution of the reinforcements in the microstructure. However, some small local agglomerations of rubber particles were observed in the microstructure due to the mixture process; homogenous distributions of the reinforcements should be improved by longer mixing process. Also, the diameters of the alumina fibers showed a variation in the interval of 10–100 μm.

After microstructural observation, density and surface hardness of the composites were determined by a pycnometer and Shore D durometer respectively. The density results are given in Table 34.2.

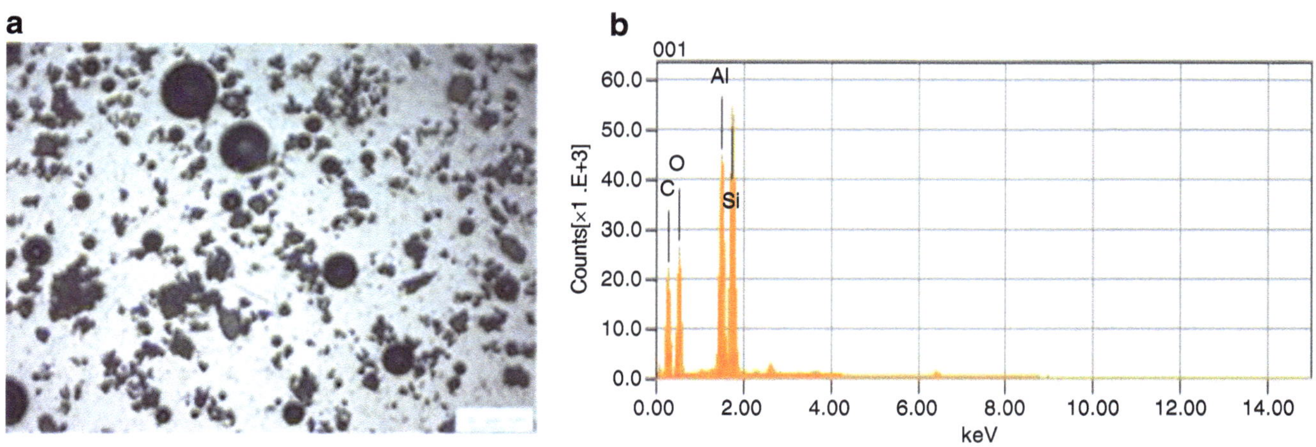

Fig. 34.1 (**a**) Microstructural observation of LR2AL10 composite and (**b**) EDS analysis for this composition

Table 34.2 Density of the manufactured composites

Composition name	Density (g/cm^3)
LR20	1,121
LR2AL5	1,167
LR2AL7,5	1,175
LR2AL10	1,121

Table 34.3 Shore D hardness of the manufactured composites

Composition name	Shore D
LR20	72,8
LR2AL5	75,4
LR2AL7,5	76,8
LR2AL10	75,2

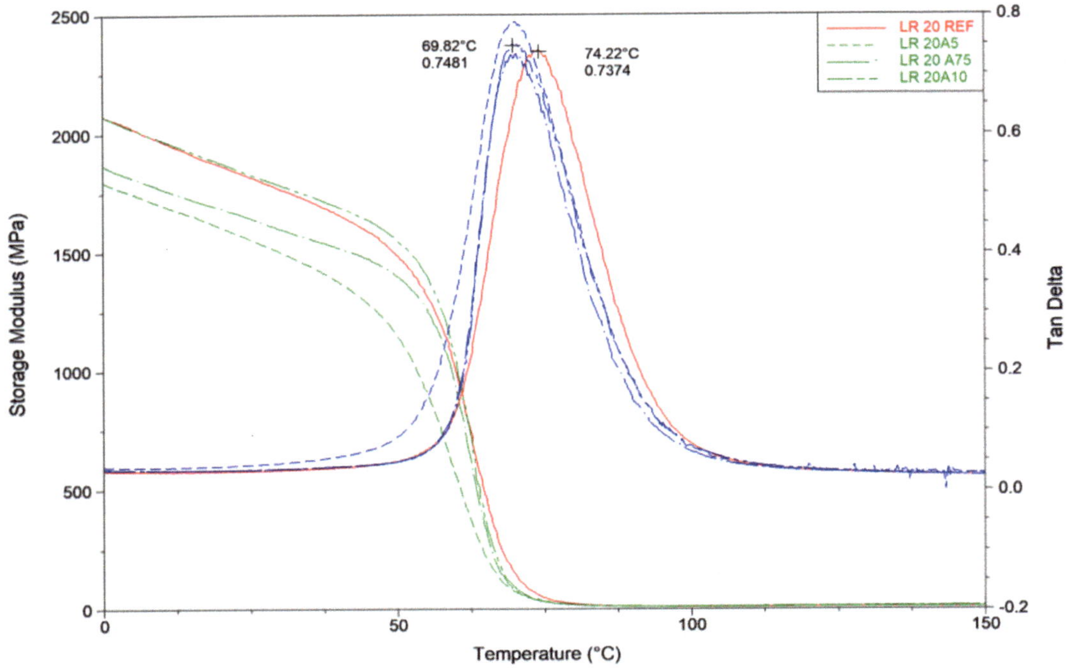

Fig. 34.2 Evolution of storage modulus and tanδ as a function of the temperature for the LRAL

Densities of the manufactured composites indicated a rising trend depending on the increasing alumina fiber content as expected. After density measurements, surface hardness measurements are presented in Table 34.3.

Shore D hardness results showed that rubber content decrease the surface hardness whereas alumina fibers followed an opposite tendency.

Following to fundamental physical properties, thermomechanical of the composites were evaluated by means of DMA analysis.

34.3.2 Dynamic Mechanical Analysis

After microstructural observations, glass transition temperature (T_g) of the composites was determined by means of DMA tests. Also, from these analyses, the viscoelastic characteristics of the composites can be evaluated. Results of DMA analyses were given in Fig. 34.2.

Figure 34.2 demonstrates the evolution of the storage modulus and tan δ of LR2AL group composites depending on temperature for different AF contents. From this figure it is commented that alumina fibers increase energy storing capability of the composites. On the other hand, loss modulus did not follow the same tendecy. Rising AF content manifested itself in different level of heat dissipation from the material. Therefore, one can comment that, alumina fibers improved the elastic response of the rubber modified epoxy based composites. Since, high rigidity of AFs forces the composites in to show elastic behaviour. In addition, T_g of these composites were determined around 70 °C and it was seen that there is not a direct correlation between the T_g and alumina content. Besides, lower E' values under T_g gives an indication about more flexible structure of the LR2AL5 composites. It means that alumina fiber content decreases the dynamical flexibility of the composites. This DMA analysis is important for the manufactured composites. Since, the determined glass transition temperature can be thought as a temperature limit for the structural applications. Because mechanical resistance of these composites drops drastically after T_g and this should be considered during design process.

34.3.3 Bending Tests and Fracture Toughness Determination

Three-Point Bending (SEN-3 PB) tests have been carried out for each different type of composites. In tensile test, failed areas do not stay included in the specimen gauge length. Therefore, the bend test is more appropriate with this type of composites.

Fracture toughness is a property which describes the ability of a material containing a crack to resist the fracture, and is one of the most important properties of any material for many design applications.

Flexural stress is calculated during three-point bending test according to Eq. 34.1:

$$\sigma = \frac{3\ P\ l}{2\ b\ h^2} \tag{34.1}$$

In this formula, l is the span length, P is the maximal bending load, b and h are the specimen thickness and depth, respectively.

Flexural strain, ε_f, was determined according to the Eq. 34.2:

$$\varepsilon_f = \frac{6Dh}{l^2} \tag{34.2}$$

D is the maximum deflection at the center of the specimen.

E_B is the modulus of elasticity in bending and it is expressed with Eq. 34.3 as follows:

$$E_B = \frac{l^3 m}{4bh^3} \tag{34.3}$$

m is the tangent of the initial straight portion of the force-displacement curve.

The mode I fracture toughness, K_{Ic}, was determined by testing of the SENB specimens and K_{Ic} was calculated according to the Eq. 34.4:

$$K_{Ic} = \frac{F}{B\ w^{1/2}} f(x); \quad x = \frac{a}{W}\ , 0 < \frac{a}{W} < 1 \tag{34.4}$$

F is the maximum force from the load-elongation trace; B is the thickness of the specimen; W is the width and "a" is the total notch length.

$f(x)$ is the geometry correction factor and is expressed with the Eq. 34.5 as follows:

$$f(x) = 6(x)^{0.5} \left\{ \frac{\left[1.99 - x(1-x)\left(2.15 - 3.93x + 2.7x^2\right)\right]}{(1+2x)(1-x)^{1.5}} \right\} \tag{34.5}$$

Critical strain energy release rate (fracture energy) G_{Ic} was calculated using Eq. 34.4:

$$G_{Ic} = \frac{{K_{Ic}}^2}{E} \tag{34.6}$$

where E is the elasticity modulus for plane stress approach examined for thin specimens.

Table 34.4 indicates the mechanical properties in bending mode. Fracture toughness values were also presented in the table.

From Table 34.4 it is commented that flexural strength of these composites decreases by the increasing content of alumina fiber. However, flexural moduli were generally improved by the addition alumina fibers. Also, strain at break values did not follow a constant tendency. These unstable results may derive from the inhomogeneous distribution of the fibers or rubbers. Because of the adhesion forces alumina fibers may create clusters and they can influence the material's properties. On the other hand, for fracture toughness K_{Ic} values and fracture energy (G_{Ic}) was improved by the addition of alumina fibers. In

Table 34.4 Comparison of mechanical properties of LRAL specimens

Composition name	Flexural stress (MPa)	Flexural modulus (MPa)	Strain at break (ε %)	K_{Ic} (MPa m$^{1/2}$)	G_{Ic} (kJ/m^2)
LR20	48,34 ± 2,05	1149,64 ± 41	4,6 ± 0,4	0,991 ± 0,998	0,879 ± 0,127
LR2AL5	40,17 ± 1,86	1220,77 ± 76	3,5 ± 0,17	1159 ± 0,043	1,106 ± 0,081
LR2AL7,5	40,06 ± 2,40	1143,54 ± 84	4,3 ± 1,2	1186 ± 0,028	1,237 ± 0,059
LR2AL10	40,65 ± 1,82	1302,47 ± 166	3,4 ± 0,3	1103 ± 0,027	0,936 ± 0,045

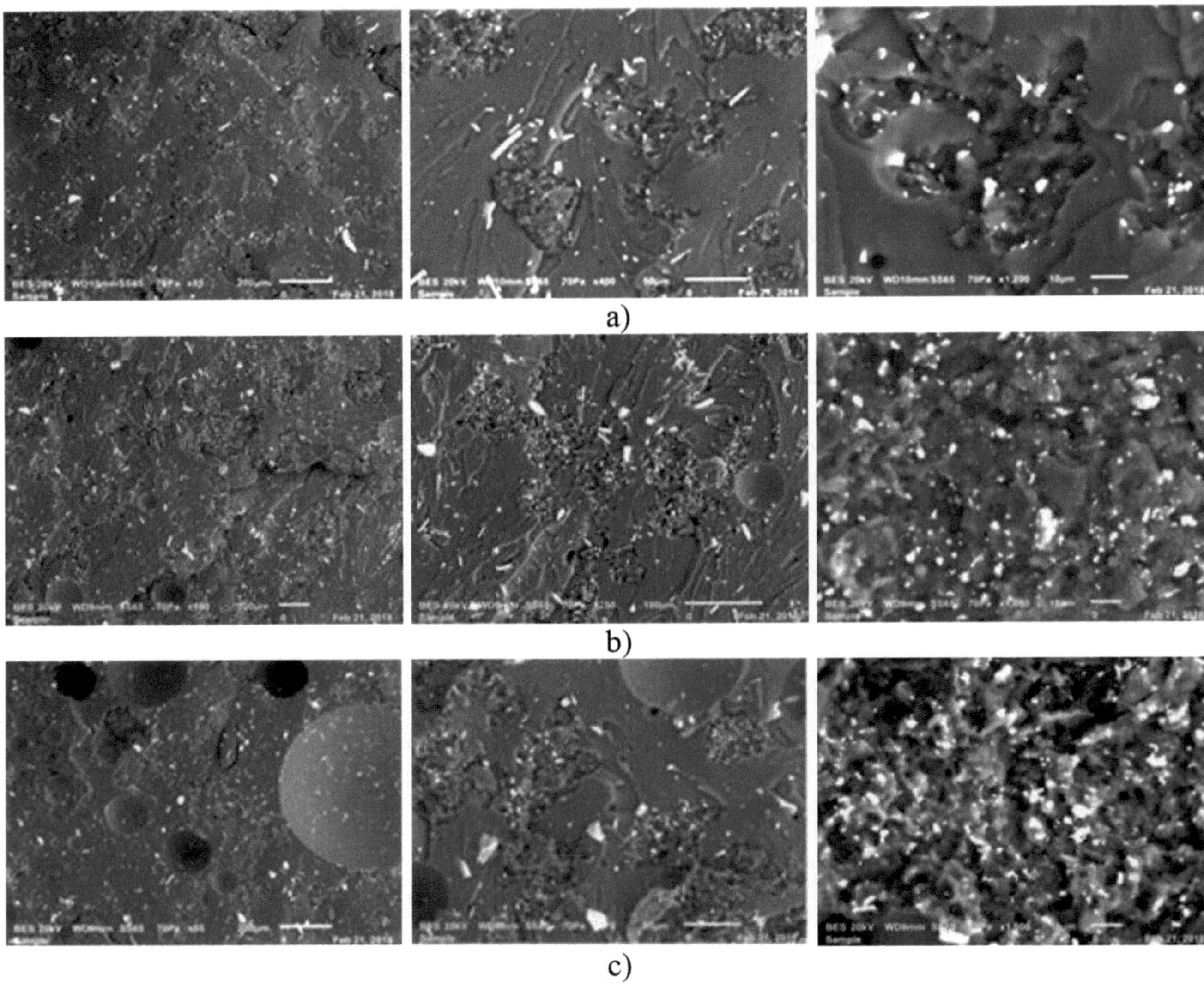

Fig. 34.3 Fracture surface observation after 3 PB tests for the (**a**) LR2AL5 (**b**) LR2AL7,5 (**c**) LR2AL10 specimens

general, hard particles as alumina may stop the crack propagation by crack blunting and crack pinning mechanisms. Also, rubbers may bring some toughening mechanisms such as shear band formation close to rubber particles, debonding and tearing of rubber particles as well as void formation and shear yielding. However, SEM observations on fracture surfaces is required to identify toughening mechanisms. In this regard, SEM observation was conducted in these ternary composites and the images were given in Fig. 34.3.

Figure 34.3 shows the fracture surfaces of LRAL composites in different magnifications. As first impression, it can be commented that rough fracture surfaces give a sign about the good adhesion between matrix and reinforcements. Also, the fibers which are perpendicular to fracture surface created a link (crack bridging) between two sides of the crack and it has a positive effect against to crack opening. In addition, tortuous structure observed in higher magnifications, is a result of crack deflection mechanisms. In this case, increased fracture area dissipates fracture energy from the matrix and filler interfaces which increases the fracture toughness of these composites.

34.4 Conclusion

In this study, microstructure observation indicated a considerably homogeneous distribution of the fillers. Also, different particle size of the fillers was seen from the OM images. After microstructural observation, glass transition temperature of the composites was determined by means of DMA analyses. Also, DMA analyses showed that alumina fibers increase energy storing capability of the composites. In addition, alumina fibers generally increased the flexural modulus of the composites and the main contribution of alumina was seen in fracture toughness.

After mechanical tests, fracture surfaces were observed by means of SEM to identify toughening mechanisms. Crack deflection and crack blunting with the contribution of fiber bridging were thought as the main toughening mechanisms. Improvement in fracture toughness is obtained as a combination of these mechanisms for the composites in consideration. As a result, these composites can be utilized in the automotive industry for the manufacturing of some components under the hood such as air filter box.

References

1. McGrath, L.M., Parnas, R.S., King, S.H., Schroeder, J.L., Fischer, D.A., Lenhart, J.L.: Investigation of the thermal, mechanical, and fracture properties of alumina–epoxy composites. Polymer. **49**, 999–1014 (2008)
2. Zaimova, D., Bayraktar, E., Miskioglu, I., Dishovsky, N., Katundi, D.: Epoxy resin reinforced scrap rubber composites used for aeronautical applications, SEM-2013, Annual Conference – Experimental and Applied Mechanics, The Westin Lombard Yorktown Center Lombard, IL, June 3–5, 2013, 1, 1, 1–9, ed. SEM/Springer-Link (2013)
3. Wetzel, B., Rosso, P., Haupert, F., Friedrich, K.: Epoxy nanocomposites – fracture and toughening mechanisms. Eng. Fract. Mech. **73**, 375–398 (2006)
4. Zaimova, D., Bayraktar, E., Miskioglu, I.: Design and manufacturing of new elastomeric composites: mechanical properties, chemical and physical analysis Int. J. Compos. Part B Elsevier USA 1, 1, 1–12, November 2016, on line (2017)
5. Garg, A.C., Mai, Y.-W.: Failure mechanisms in toughened epoxy resins—a review. Compos. Sci. Tech. **31**, 179–223 (1988). https://doi.org/10.1016/0266-3538(88)90009-7
6. Lee, J., Yee, A.F.: Inorganic particle toughening I: micro-mechanical deformations in the fracture of glass bead filled epoxies. Polymer. **42**, 577–588 (2001). https://doi.org/10.1016/S0032-3861(00)00397-9
7. Irez, A.B., Bayraktar, E., Miskioglu, I.: Mechanical characterization of epoxy – scrap rubber based composites reinforced with alumina fibers. In: Mechanics of Composite and Multi-functional Materials, vol. 6, pp. 59–70. Springer, Cham (2017)
8. Zaimova, D., Bayraktar, E., Miskioglu, I., Dishovsky, N.: Wear resistance of elastomeric based composites by continuous multi-cycle indentation used in manufacturing engineering. Adv. Mater. Res. **939**, 106–113 (2014., Trans Tech Publications, Switzerland). https://doi.org/10.4028/www.scientific.net/AMR.939.106
9. Nakamura, Y., Yamaguchi, M., Okubo, M., Matsumoto, T.: Effects of particle size on mechanical and impact properties of epoxy resin filled with spherical silica. J Appl Polym Sci. **45**, 1281–1289 (1992)
10. Okazaki, M., Murota, M., Kawaguchi, Y., Tsubokawa, N.: Curing of epoxy resin by ultrafine silica modified by grafting of hyperbranched polyamidoamine using dendrimer synthesis methodology. J. Appl. Polym. Sci. **80**, 573–579 (2001). https://doi.org/10.1002/1097-4628(20010425)80:4<573::AID-APP1132>3.0.CO;2-E
11. Zaimova, D., Bayraktar, E., Katundi, D., Dishovsky, N.: Study of the influence of SiC and Al2O3 as reinforcement elements in elastomeric matrix composites, JAMME-2013, International-Annual Conference of WAMME(Academy of Science), Krakow-Poland, 23–26 June 2013, 1, 1, 1–9, ed. WAMME-JAMME, (2013), ISSN 1734-8412
12. Wetzel, B., Haupert, F., Qiu, Z.M.: Epoxy nanocomposites with high mechanical and tribological performance. Compos. Sci. Technol. **63**, 2055–2067 (2003). https://doi.org/10.1016/S0266-3538(03)00115-5
13. Bayraktar, E., Miskioglu, I., Zaimova, D.: Low-cost production of epoxy matrix composites reinforced with scarp rubber, boron, glass bubbles and alumina. Mech. Compos. Multifunct. Mater. Springer- USA. **7**(1), 163–172., ed. Kristin B. Zimmerman, Springer-USA (2015)
14. Zee, R.H., Huang, Y.H., Chen, J.J., Jang, B.Z.: Properties and processing characteristics of dielectric-filled epoxy resins. Polym Compos. **10**, 205–214 (1989). https://doi.org/10.1002/pc.750100402
15. Ferreira, L.M.P., Miskioglu, I., Bayraktar, E., Katundi, D.: Mechanical and tribological properties of scrap rubber reinforced with Al2O3 fiber, aluminium and TiO2, Conference proceedings of the society for experimental mechanics series, vol. 7, pp. 37–44. Springer (2016). https://doi.org/10.1007/978-3-319-41766-0

16. Zaimova, D., Bayraktar, E., Tan, M-J., Miskioglu, I., Katundi, D.: Design of multifunctional energetic structural composites: a preliminary study on an epoxy-rubber matrix with exothermic mixture reinforcements, SEM-2013, Annual Conference – Experimental and Applied Mechanics, The Westin Lombard Yorktown Center Lombard, IL, USA, June 3–5, 2013, 1, 1, 1–8, ed. SEM/Springer-Link (2013)
17. Jin Kim, D., Hyun Kang, P., Chang, N.Y.: Characterization of mechanical properties of γAl2O3 dispersed epoxy resin cured by γ-ray radiation. J Appl Polym Sci. **91**, 1898–1903 (2004). https://doi.org/10.1002/app.13250
18. Arayasantiparb, D., McKnight, S., Libera, M.: Compositional variation within the epoxy/adherend interphase. J. Adhes. Sci. Technol. **15**, 1463–1484 (2001). https://doi.org/10.1163/156856101753213312
19. Landel, R.F., Nielsen, L.E.: Mechanical Properties of Polymers and Composites, 2nd edn. CRC Press, New York (1993)
20. Irez, A.B., Miskioglu, I., Bayraktar, E.: Mechanical characterization of epoxy – scrap rubber based composites reinforced with nano graphene. Springer Link, Mech. Compos. Multi-funct. Mat. **6**, 45–58 (2017., ISBN 978-3-319-63,408-1). https://doi.org/10.1007/978-3-319-63,408-1
21. Rothon, R.: Particulate-filled Polymer Composites. iSmithers Rapra Publishing, Shawbury (2003)
22. Rothon, R.N.: Mineral Fillers in Thermoplastics: Filler Manufacture and Characterisation. Mineral Fillers in Thermoplastics I, pp. 67–107. Springer, Berlin, Heidelberg (1999). https://doi.org/10.1007/3-540-69220-7_2
23. Branch, M.: Preparation of nano-scale α-Al2O3 powder by the sol-gel method. Ceram. Silic. **55**, 378–383 (2011)
24. Amirshaghaghi, A., Kokabi, M.: Tailoring size of α-Al2O3 n nano powders via polymeric gel-net method. Iran. Polym. J. **19**, 615–624 (2010)
25. Zhang, Z., Lei, H.: Preparation of α-alumina/polymethacrylic acid composite abrasive and its CMP performance on glass substrate. Microelectron. Eng. **85**, 714–720 (2008)
26. Irez, A.B., Miskioglu, I., Bayraktar, E.: Mechanical characterization of epoxy: scrap rubber based composites reinforced with nanoparticles. Springer Link, Mech. Compos. Multi-funct. Mat. **6**, 33–44 (2017., ISBN 978-3-319-63,408-1). https://doi.org/10.1007/978-3-319-63,408-1

Chapter 35
Toughening Mechanisms on Recycled Rubber Modified Epoxy Based Composites Reinforced with Graphene Nanoplatelets

A. B. Irez, I. Miskioglu, and E. Bayraktar

Abstract Recycling is a subject undergoing intense study in terms of sustainable development. In every area, recycling is strongly encouraged by governments due to the international agreements on environmental issues. For example, many industrial manufacturers have a tendency to find clean and cost-efficient solutions by utilizing recycled materials to produce new components. In this regard, rubbers have very wide usage in aeronautic and automotive industries both in structural and in interior body components. Rubbers are also used to modify brittle polymer components in the existence of hard, resistant fillers. In this study, fresh scrap EPDM rubbers are used to manufacture novel composites by modifying epoxy resin with the inclusion of graphene nano platelets (GnPs). Rather than micro sized particles, nanoparticles have high surface area, which means that a low content of these nanoparticles may enhance the material's properties more efficiently. Besides, due to its superior structural, thermal and physical characteristics, addition of graphene promises improved mechanical properties if they can be dispersed homogeneously. This paper is focused on the fracture characteristics and toughening mechanisms of epoxy – fresh scrap rubber composites. The mechanical and physical properties of these composite systems are studied in the present work. Mechanical properties are examined by means of three-point bending tests with smooth and single edge notched beam specimens (SENB). Also, nano indentation tests were realized to see the creep compliance and viscoelastic properties. Finally, scanning electron microscope (SEM) was used to observe the fracture surfaces and the microstructure.

Keywords Toughness · Nano indentation · Epoxy – rubber composites · Graphene nanoplatelets · SEM

35.1 Introduction

Currently recycling of materials is attracting a considerable attention in all around the world due to the environmental and economic concerns. Amongst these materials, rubbers are having a significant place thanks to their broad application areas in many industries. In order to improve material's properties, rubbers are generally vulcanized with sulphur under a well-defined pressure and temperature. However, non-used vulcanized scrap rubber and tires after completing their life cycle, is problematic for landfills. Also, in the context of landfills, due to the impermeability and shape of discarded tires, they hold water for a long time and this creates a habitat for mosquito larvas as well as other animals such as rodents and snakes [1, 2]. These may carry infections namely malaria, cephalitis, dengue and chikungunya. Besides, if these piles burst into flames, it is quite difficult to extinguish them [3]. Last but not least, some additives of discarded tires in landfills such as colorants, stabilizers, flame retardants and plasticizers may leach from the bulk of the tires to the soil. Hence, they can harm or kill some beneficial bacterial colony in the soil [4]. For this reason, in this study our first goal is to use these scrap rubbers to manufacture novel composites by blending them with a matrix for various industrial applications without causing excessive degradation in the mechanical properties of the matrix. In this regard, blending the recycled rubber with a material that can flow at high temperatures enables its shaping to beneficial products at a low cost. This idea can be achieved by mixing powdered scrap rubber with polymers as well as other necessary reinforcements.

A. B. Irez
CentraleSupélec, Université Paris-Saclay, Gif-sur-Yvette, France

I. Miskioglu
Michigan Technology University, Engineering Mechanics Department, Houghton, MI, USA

E. Bayraktar (✉)
Supmeca-Paris, School of Mechanical and Manufacturing Engineering, Paris, France
e-mail: bayraktar@supmeca.fr

P. R. Thakre et al. (eds.), *Mechanics of Composite, Hybrid and Multifunctional Materials, Volume 5*, Conference Proceedings of the Society for Experimental Mechanics Series, https://doi.org/10.1007/978-3-319-95510-0_35

Here in this study, the above-mentioned polymer or in other words the matrix was chosen as epoxy thanks to its low cost, high specific strength and stiffness, chemical resistance, ease of process and environmental stability [5]. Second, as a reinforcing agent, the modification of epoxies is also done by the incorporation of some inorganic fillers. In the last decades nano sized materials have become very popular compared to ordinary materials thanks to their advantageous properties in terms of their electrical, mechanical, thermal and magnetic features [6]. In this regard, epoxies reinforced with nano fillers has shown promising improvements in comparison with neat epoxy [7, 8].

Nanoparticles have high surface area, which means that a low content of these nanoparticles may enhance the materials properties [9–11]. In this respect, nano graphene has unique mechanical, thermal and electrical properties. It has a two-dimensional honeycomb structure coming from a single-atom-thick sheet of sp^2 bonded carbon atoms and this makes it a novel promising nano scale filler for polymer nanocomposites. Its outstanding properties can be given as specific elasticity modulus (1 TPa) and mechanical strength (130 GPa), thermal conductivity (5000 W/mK) and high electrical conductivity (up to 6000 S/cm) [12–14]. In this study, graphene nano platelets (GnPs) were used. Simple process and abundance of inexpensive naturally existing graphene precursor (graphite) lead to economical produce of graphene in large quantities. Thus, outstanding balance between properties and cost make GnPs ideal reinforcing nano filler for polymer matrices [15]. However, due to the synthesis procedure of graphene, interlinear van der Waal forces facilitate the agglomeration of the graphene sheets. As a result, during its mixing with polymers it is difficult to provide a homogeneous dispersion [16]. This situation should be considered during the manufacturing process.

This work presents processing of recycled rubbers blended with epoxy resin and GnPs to create novel composites in economic way. In this research the mechanical properties and toughening mechanisms of these newly manufactured composites were studied. During this study, mechanical properties were investigated by three point bending tests by using notched and smooth specimens. After that, fracture surfaces were observed by means of scanning electron microscopy (SEM) to study the toughening and damage mechanisms. Then, creep compliance and viscoelastic behaviour were investigated by nano-indentation tests.

35.2 Experimental Procedure

35.2.1 Composite Manufacturing

Recycled rubber modified epoxy-based composites were prepared according to following procedure. Huntsman™ Araldite DBF epoxy was preheated at 65 °C in order to decrease its viscosity. Then, GnPs were added to epoxy resin at predetermined wt%. This blend was mixed by using a magnetic stirrer at 1000 rpm for 1 h at 65 ° C to homogenize the mixture and during the mixing the temperature was controlled by a digital thermometer. Then, recycled EPDM rubbers were gradually added to this mixture and this blend was mixed for two more hours again at 65 ° C to prevent the agglomerations of rubbers and to provide a better wetting of rubbers. These recycled rubber powders are obtained from the unused scrap parts during sportive equipment manufacturing. It means that they are fresh clean and completely different from ground tire. Following to mixing procedure, this mixture was cooled down to 25 ° C and the hardener was added. The addition of hardener at higher temperatures is not desirable since rapid solidification occurs which lead to uncompleted cross-linking reaction in some regions of the composite. After that, when the mixture cooled down to room temperature, the hardener was added according to the stoichiometric ratio of the epoxy content and this final compound was mixed for 15 min. Afterwards, the mixture was degassed in a vacuum chamber at room temperature for 30 min. Then, the mixture was poured into silicon molds and molds were left at room temperature for 24 h to complete the curing. After bulk composites were manufactured, test specimens were prepared by machining of the bulks according to test standards. Table 35.1 presents the different compositions of the manufactured composites.

Table 35.1 Composition of the GnPs reinforced liquid epoxy-EPDM rubber based composites

	Graphene nano platelets content (wt. %)			
LRG composites	0%	0.5%	1.0%	1.5%
20 wt. % rubber	LR20	LR2G0.5	LR2G1	LR2G1.5

35.2.2 Experimental Procedures

Microstructure and fracture surface damage analyses have been carried out by means of optical (OM) and scanning electron microscopy (SEM). SEM observation was realized on fracture surface of the tested specimens with Scope/JSM-6010LA Jeol® electron microscope.

There point bending tests (3PB) have been carried out according to the ASTM D790 standards. Deflection of the specimen was measured by the crosshead position and crosshead speed was selected as 1 mm/min. Flexural strength and strain were obtained from the test results. The test specimen was placed in the Instron 5569 bending fixture and force was applied until it fractured. At least five specimens for each composition were used and standard deviation and average values were given in results.

In addition, fracture properties such as plain strain fracture toughness (K_{Ic}) and critical strain energy release rate (G_{Ic}) were investigated with SENB specimens and the tests were realized according to ASTM D5045 standard. Notches were introduced by tapping with a fresh razor blade.

Creep tests using a nanoindenter were performed on the two of the manufactured compositions to see the time dependent behaviour. On each sample 20 indents were performed on a 5 × 4 grid with a Berkovich indenter. The indents were spaced 50 μm along the 5-indent side and 75 μm along the 4-indent side. The load was increased at a rate of 1 mN/s to the max load and kept at the maximum load for 500 s then unloaded.

35.3 Results and Discussions

35.3.1 Microstructure of the Composites and Surface Hardness

After the manufacturing of the composites their densities were measured with a pycnometer and the results were given in Table 35.2. According to the results, density of the composites were not affected significantly by the GnPs content.

After measuring the densities, microstructure of the composites was observed by means of optical microscopy after sectioning and polishing procedures. The microstructure of LR2G1 is given in Fig. 35.1. According to this figure,

Table 35.2 Density of the manufactured composites

Composition name	Density (g/cm^3)
LR20	1,121
LR2G0,5	1,13
LR2G1	1,11
LR2G1,5	1,10

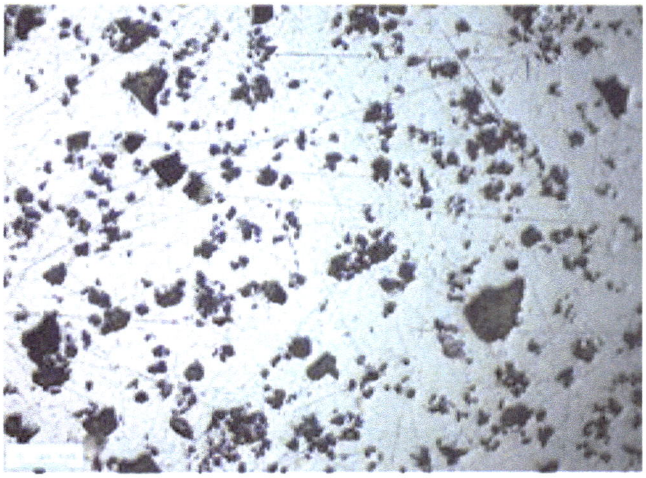

Fig. 35.1 Microstructural observation of LR2G1 composite

Table 35.3 Shore D hardness of the manufactured composites

Composition name	Shore D
LR20	72,8
LR2G0,5	75
LR2G1	75,8
LR2G1,5	73

considerably homogeneous distribution of rubbers can be observed. However, some larger rubber particles were observed and GnPs cannot be detected in this magnification.

After microstructural observation, surface hardness of the composites was determined by a Shore D durometer and results are given in Table 35.3.

Shore D hardness results showed that GnP content increased the surface hardness compared to only rubber modified composites.

35.3.2 *Bending Tests and Fracture Toughness Determination*

Three-point bending (3PB) tests have been carried out for each different type of composites. Bending tests provide a practical determination of the strength of the materials. After extracting beam type specimens from the composite bulks, mechanical properties were determined in bending mode.

Flexural stress is calculated during three-point bending tests according to the Eq. 35.1:

$$\sigma = \frac{3\,P\,l}{2\,b\,h^2} \tag{35.1}$$

In this formula, l is the span length, P is the maximal bending load, b and h are the specimen thickness and depth, respectively.

Flexural strain, ε_f, was determined according to Eq. 35.2:

$$\varepsilon_f = \frac{6Dh}{l^2} \tag{35.2}$$

D is the maximum deflection at midspan of the specimen.

E_B is the modulus of elasticity in bending and it is expressed with Eq. 35.3 as follows:

$$E_B = \frac{l^3 m}{4bh^3} \tag{35.3}$$

m is the tangent of the initial straight portion of the force-displacement curve.

After fundamental mechanical calculations, fracture toughness of the composites was determined. Fracture toughness is a property which describes the ability of a material containing a crack to resist the fracture, and is one of the most important properties of any material for many design applications.

The mode I fracture toughness, K_{Ic}, was determined by testing of the SENB specimens and K_{Ic} was calculated according to Eq. 35.4:

$$K_{Ic} = \frac{F}{B\ w^{1/2}}\,f(x);\quad x = \frac{a}{W}\ ,0 < \frac{a}{W} < 1 \tag{35.4}$$

F is the maximum force from the load-elongation trace; B is the thickness of the specimen; W is the width and a is the total notch length.

$f(x)$ is the geometry correction factor and is expressed in Eq. 35.5 as follows:

$$f(x) = 6(x)^{0.5}\left\{\frac{\left[1.99 - x\,(1-x)\left(2.15 - 3.93x + 2.7x^2\right)\right]}{(1+2x)\,(1-x)^{1.5}}\right\} \tag{35.5}$$

Table 35.4 Comparison of mechanical properties of LRG specimens

Composition name	Flexural stress (MPa)	Flexural modulus (MPa)	Strain at break (ε %)	K_{Ic} (MPa $m^{1/2}$)	G_{Ic} (kJ/m^2)
LR20	48,34 ± 2,05	1149,64 ± 41	4,6 ± 0,4	0,991 ± 0,098	0,879 ± 0,127
LR2G0,5	47,94 ± 0,25	1417,84 ± 13,69	3,5 ± 0,1	1,172 ± 0,048	0,974 ± 0,078
LR2G1	48,49 ± 1,01	1474,58 ± 19,79	3,4 ± 0,1	1,001 ± 0,03	0,681 ± 0,04
LR2G1,5	48,61 ± 0,74	1537,54 ± 28,97	3,3 ± 0,1	1,066 ± 0,101	0,76 ± 0,141

Critical strain energy release rate (fracture energy) G_{Ic} was calculated using the expression in Eq. 35.6:

$$G_{Ic} = \frac{{K_{Ic}}^2}{E} \tag{35.6}$$

where E is the elasticity modulus for plane stress approach examined for thin specimens.

Table 35.4 indicates the measured mechanical properties in bending mode. Fracture toughness values are also presented in this table.

From Table 35.4 main improvement was clearly seen in the flexural moduli. GnPs are increasing the flexural modulus of the composites as expected. This improvement is due to the high modulus of GnPs. However, similar increase was not observed in the strength to the same extent as well as strain at break. Even, strain at break lowered compared to only rubber consisting composites. This situation is possibly arising from the lowered mobility of the polymer chains of epoxy resin. Hard particles may restrict deformation capability of the polymer. This situation manifests itself in the descending strain values. On the other hand, fracture toughness was improved by GnP addition. There are various toughening mechanisms in GnP and rubber modified epoxies. Addition of rubber may bring some mechanisms such as shear band formation close to rubber particles, debonding and tearing of rubber particles as well as void formation and shear yielding [17]. In addition, GnPs are also contributing to toughening of epoxies due to matrix shear banding, crack deflection and other various mechanisms. Nevertheless, the identification of these mechanisms requires SEM observation on fracture surfaces. In this regard, SEM observation was conducted in these ternary composites and the images are given in Fig. 35.2.

First, fracture surfaces show fracture propagation lines and it is clear that there is a strong adhesion between matrix and reinforcements. Homogeneously distributed white small particles are detected as GnPs from EDS analysis. In addition, local rough surfaces indicate increased fracture areas in these composites. This situation is an indicator for the crack deflection mechanism.

Another important point is that, when the propagating crack encounters GnPs, it creates twisted crack pathways as shown in the high magnification images in the right column of Fig. 35.2. In these regions, cracks may interact with GnPs and by doing so, the fracture energy can be dissipated. On the other hand, the interface contact between the rubbers and epoxy may be reduced due to large rubbery domains, probably due to the clustering of the nano reinforcements having lower surface area to volume ratio. This may result in the rubber particles to be pulled out because of crack propagation along the interface between recycled rubbers and epoxy showing ductile fracture onthe rubber side [15, 18].

Once and for all, greater crack areas on rubber and GnP modified epoxies can be reasoned as a combination of matrix shear banding, crack deflection, layer breakage and delamination of GnP layer. This ends up with the higher fracture of recycled rubber – GnP modified ternary composites [19].

35.3.3 *Time Dependent Behaviour by Means of Nano-indentation*

During each test, data collected was used to calculate the creep compliance and the stress exponent defined in Eq. 35.7 [20]:

$$\varepsilon(t) = \sigma_0 J(t) \tag{35.7}$$

where σ_0 is the constant stress applied and $J(t)$ *is* calculated using Eq. 35.8.

$$J(t) = A(t)/(1 - \nu)\, P_0 \tan\theta \tag{35.8}$$

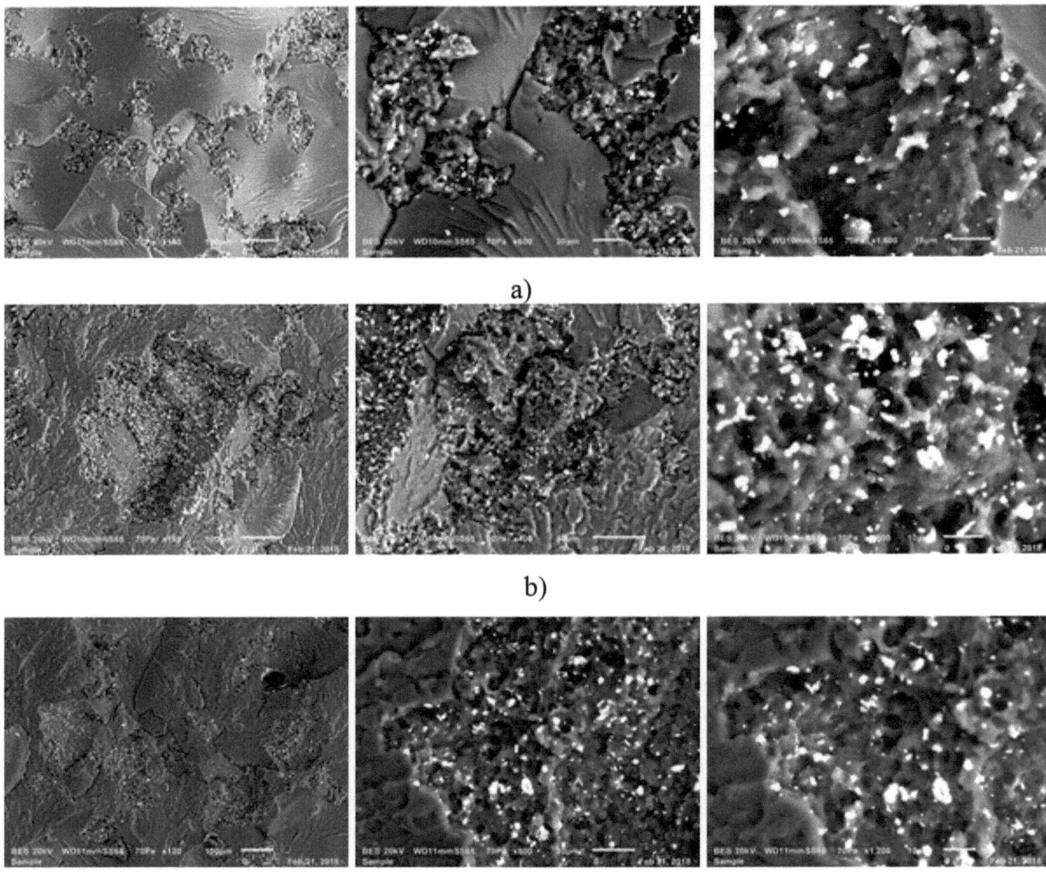

Fig. 35.2 Fracture surface observation after 3 PB tests for the LRG specimens (**a**) LR2G0,5 (**b**) LR2G1 (**c**) LR2G1,5

In Eq. 35.8 $A(t)$ is the contact area, P_0 is the constant applied load, θ is the effective cone angle which is 70.3° for a Berkovich indenter. The Poisson's ratio ν is assumed to be 0.3. This approach considers how the contact area under the Berkovich tip changes while displacement into the surface alters.

The strain versus time behavior during creep is characterized by a high strain rate $\dot{\varepsilon} = d\varepsilon/dt$ in the first stage of creep and then in the secondary, steady state stage of creep, the strain rate is given in Eq. 35.9

$$\dot{\varepsilon} = K\sigma^n \tag{35.9}$$

where K is a constant and n is the stress exponent. The strain rate is determined in the software and in turn n is obtained from the log-log plot of strain rate versus stress in the secondary stage of creep.

According to Fig. 35.3, it is seen that creep compliance is dependent to GnP content for high loading values. On the other hand, creep compliance is not very sensitive to GnP content at 20 mN of loading.

According to Fig. 35.3 it is seen that creep compliance is higher for the composites without GnP addition which means that GnPs lower the capacity to flow of the composites in return to an abruptly applied stress.

The results may exhibit differences compared to linear viscoelastic models. This may arise from the non-consideration of tip-specimen adhesion in models as regards to experimental approaches.

The average indentation modulus and stress exponent values were obtained from the 20 indentations performed under 20 mN and 50 mN constant loads and they are presented in Fig. 35.4a, b with error bars showing ± one standard deviation.

From Fig. 35.4b it is seen that, addition of GnPs results in a drop in the stress exponent which shows that only rubber modified compositions showing more viscoelastic behaviour. Different researchers have shown that for elasticity modulus determined by nanoindentation is generally higher than that determined by macroscopic tensile tests for polymers and polymer-based composites. This inequality is thought because of pile up of material around the contact impression and

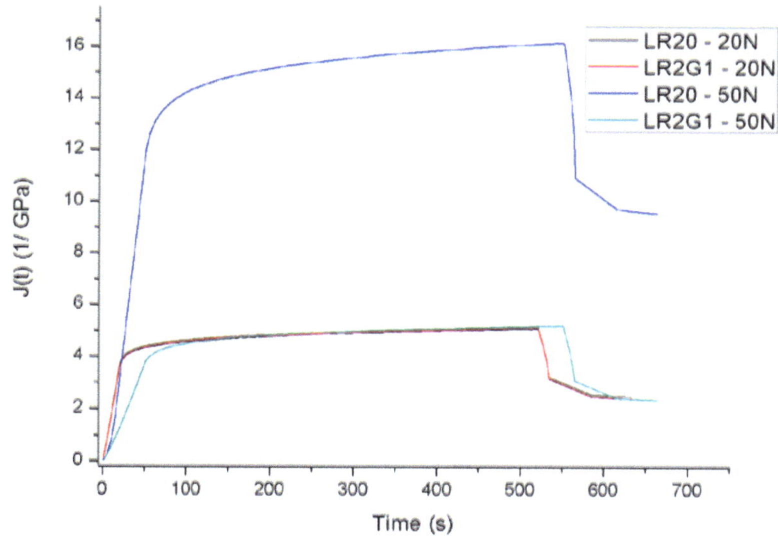

Fig. 35.3 Creep Compliance curves for specimens LR20, LR2G1, under 20 mN and 50 mN load

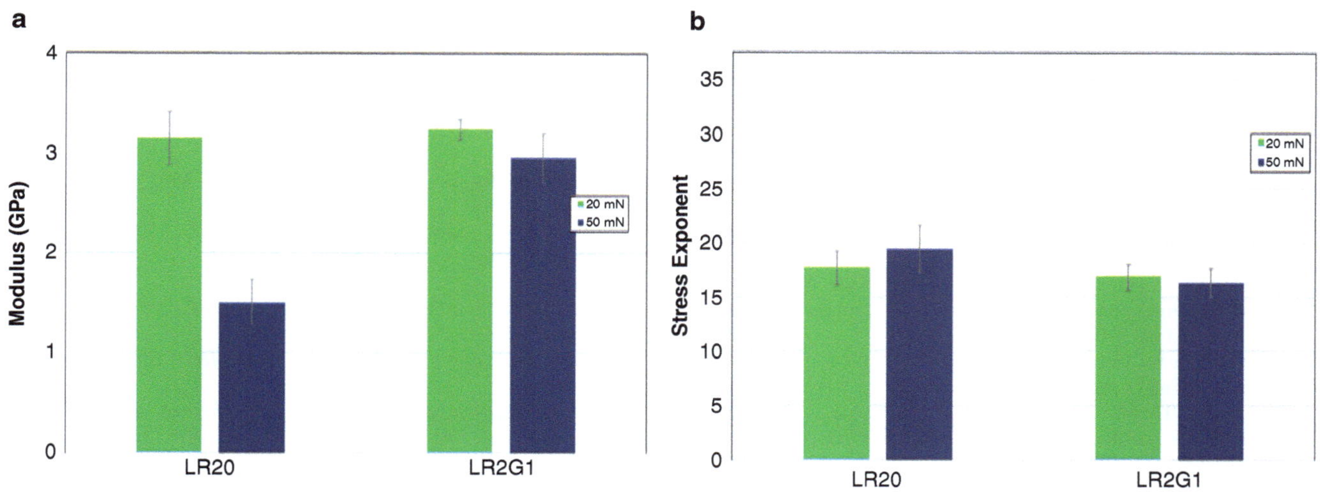

Fig. 35.4 (a) Indentation modulus (b) Stress exponent for the LR20&LR2G1 composites under 20 mN and 50 mN

viscoelasticity of the polymer and polymer based composites, which are not taken into consideration during the modulus determination by the Oliver-Pharr method [20–23].

35.4 Conclusion

In this study, nano graphene reinforced rubber modified epoxy based composites have been successfully designed. These composites were manufactured and tested under laboratory conditions, with the long term goal of using in the automotive and aeronautical applications. It seems feasible that they can be utilized in different components such as suspension pads and radiator baffles. Microstructure observation has indicated that microstructure was relatively homogeneous even if small amount of agglomerations of rubber particles was found in certain areas; this is completely related the mixing conditions that may be improved further to distribute particles more homogeneously in the matrix. In terms of mechanical test results, main improvement was seen in the flexural modulus whereas the deformation capability of the composites was lowered by the increasing content of the GnPs. Also, fracture toughness of the composites was ameliorated by GnPs addition. This situation is possibly due to f the crack deflection, layer breakage, and separation/delamination mechanisms thanks to GnPs. In particular, increased fracture area is a sign of the crack deflection mechanisms. Fracture surface observation by means of SEM helped us to identify these mechanisms. Nanoindentation results indicated that addition of GnPs degrades the viscoelastic behavior of the composites in general.

Acknowledgments Authors acknowledge to Dr. H-A. ALHAS-Airbus Space-London/UK for supporting this project. During our collaboration when he was in Airbus – Helicopter in Paris, he has supported and participated partially to discussion for application of GnPs to epoxy resin modified rubber based composited.

References

1. Adhikari, B., De, D., Maiti, S.: Reclamation and recycling of waste rubber. Prog. Polym. Sci. **25**, 909–948 (2000). https://doi.org/10.1016/S0079-6700(00)00020-4
2. Fiksel, J., Bakshi, B.R., Baral, A., Guerra, E., Dequervain, B.: Comparative life cycle assessment of beneficial applications for scrap tires. Clean Techn. Environ. Policy. **13**, 19–35 (2011). https://doi.org/10.1007/s10098-010-0289-1
3. Fang, Y., Zhan, M., Wang, Y.: The status of recycling of waste rubber. Mater. Des. **22**, 123–128 (2001). https://doi.org/10.1016/S0261-3069(00)00052-2
4. De, S.K., Isayev, A., Khait, K.: Rubber Recycling. CRC Press, Boca Raton (2005)
5. May, C.: Epoxy Resins: Chemistry and Technology, 2nd edn. CRC Press, Boca Raton (1987)
6. Wang, Q.: The effect of nano size fillers on electrical performance of epoxy resin. Phd. University of Southampton (2012)
7. Ochi, M., Takahashi, R.: Phase structure and thermomechanical properties of primary and tertiary amine-cured epoxy/silica hybrids. J. Polym. Sci. B Polym. Phys. **39**, 1071–1084 (2001). https://doi.org/10.1002/polb.1084
8. Okazaki, M., Murota, M., Kawaguchi, Y., Tsubokawa, N.: Curing of epoxy resin by ultrafine silica modified by grafting of hyperbranched polyamidoamine using dendrimer synthesis methodology. J. Appl. Polym. Sci. **80**, 573–579 (2001). https://doi.org/10.1002/1097-4628(20010425)80:4<573::AID-APP1132>3.0.CO;2-E
9. Wichmann, M.H.G., Schulte, K., Wagner, H.D.: On nanocomposite toughness. Compos. Sci. Technol. **68**, 329–331 (2008). https://doi.org/10.1016/j.compscitech.2007.06.027
10. Liang, Y.L., Pearson, R.A.: Toughening mechanisms in epoxy–silica nanocomposites (ESNs). Polymer. **50**, 4895–4905 (2009). https://doi.org/10.1016/j.polymer.2009.08.014
11. Srivastava, I., Koratkar, N.: Fatigue and fracture toughness of epoxy nanocomposites. JOM. **62**, 50–57 (2010). https://doi.org/10.1007/s11837-010-0032-8
12. Novoselov, K.S., Geim, A.K.: The rise of graphene. Nat. Mater. **6**, 183–191 (2007)
13. Superior thermal conductivity of single-layer graphene – Nano Letters (ACS Publications): https://pubs.acs.org/doi/abs/10.1021/nl0731872. Accessed 21 Feb 2018
14. Effective Elastic Mechanical Properties of Single Layer Graphene Sheets – IOPscience: http://iopscience.iop.org/article/10.1088/0957-4484/20/6/065709/meta. Accessed 21 Feb 2018
15. Wang, F., Drzal, L.T., Qin, Y., Huang, Z.: Enhancement of fracture toughness, mechanical and thermal properties of rubber/epoxy composites by incorporation of graphene nanoplatelets. Compos. A: Appl. Sci. Manuf. **87**, 10–22 (2016). https://doi.org/10.1016/j.compositesa.2016.04.009
16. Pathak, A.K., Borah, M., Gupta, A., Yokozeki, T., Dhakate, S.R.: Improved mechanical properties of carbon fiber/graphene oxide-epoxy hybrid composites. Compos. Sci. Technol. **135**, 28–38 (2016). https://doi.org/10.1016/j.compscitech.2016.09.007
17. The Effect of Graphene and Natural Rubber Content on Mechanical and Electrical Conductivity Properties of Epoxy/Natural Rubber/Graphene Conductive Materials. https://www.scientific.net/MSF.888.209. Accessed 21 Feb 2018
18. Ladani, R.B., Wu, S., Kinloch, A.J., Ghorbani, K., Zhang, J., Mouritz, A.P., et al.: Multifunctional properties of epoxy nanocomposites reinforced by aligned nanoscale carbon. Mater. Des. **94**, 554–564 (2016). https://doi.org/10.1016/j.matdes.2016.01.052
19. Kuo, W.-S., Tai, N.-H., Chang, T.-W.: Deformation and fracture in graphene nanosheets. Compos. A: Appl. Sci. Manuf. **51**, 56–61 (2013). https://doi.org/10.1016/j.compositesa.2013.03.020
20. Irez, A.B., Miskioglu, I., Bayraktar, E.: Mechanical characterization of epoxy–scrap rubber based composites reinforced with nano graphene. In: Mechanics of Composite and Multi-functional Materials, vol. 6, pp. 45–57. Springer, Cham (2018)
21. Tranchida, D., et al.: Mechanical characterization of polymers on a nanometer scale through nanoindentation. A study on pile-up and viscoelasticity. Macromolecules. **40**(4), 1259–1267 (2007)
22. Lagoudas, D.C., Thakre, P.R., Amine Benzerga, A.: Nanoindentation of cnt reinforced epoxy nanocomposites. In: Fracture of Nano and Engineering Materials and Structures, pp. 649–650. Springer, Dordrecht (2006)
23. Tranchida, D., et al.: Accurately evaluating Young's modulus of polymers through nanoindentations: A phenomenological correction factor to the Oliver and Pharr procedure. Appl. Phys. Lett. **89**(17), 171905 (2006)

Chapter 36
Damage Accumulation in CMCs

B. Swaminathan, J. D. Kiser, A. S. Almansour, K. Sevener, and S. Daly

Abstract Silicon carbide/ silicon carbide ceramic matrix composites (SiC/SiC CMCs) are a subgroup of structural ceramics that are well-suited for high-temperature aerospace applications, due to their low weight, excellent creep resistance, damage tolerance, and high specific strength. Damage initiation and accumulation in CMCs depends on a number of factors, including constituent architecture, thermo-mechanical loading parameters, and environmental conditions. However, much is still unknown about the impact of these factors and the interactions between them. In order to accurately predict life of these advanced composites, it is critical to understand the relationships between damage initiation/accumulation and key microstructural features, and to characterize which early damage mechanisms subsequently lead to crack coalescence and macroscopic failure. In this study, SiC/SiC minicomposites, which are representative of a fundamental element of the macroscopic CMC, are characterized via an experimental approach combining acoustic emission (AE) with microscale deformation tracking via digital image correlation inside a scanning electron microscope (SEM-DIC) in order to examine early damage initiation at room temperature below the proportional limit. This dual technique enables the capture of multi-modal data on early damage mechanisms in CMCs at both the subsurface and surface levels. This research provides the foundation for characterization of damage in SiC/SiC CMCs under more complex testing conditions.

Keywords Ceramic matrix composite · DIC · AE · Matrix cracking · Damage accumulation

36.1 Introduction

Silicon carbide/ silicon carbide ceramic matrix composites (SiC/SiC CMCs) are a subgroup of structural ceramics that are candidate materials for extreme environment applications due to their low weight, creep resistance, high specific strength, damage tolerance, and oxidation resistance at temperatures exceeding 2000 °F (1093 °C). CMCs can endure high loads for extended times at elevated temperatures (exceeding the capabilities of metals) and can be laid up into complex shapes, providing greater design flexibility. Characterizing and understanding the degradation of SiC/SiC CMCs under the stresses and the temperatures encountered during operation depends on combining novel experimental techniques with the appropriate modeling. Improved fundamental understanding of damage initiation and the effects of the environment on mechanical properties is needed for reductions in component mass and to ensure reliable operation in the elevated temperature and extreme environment conditions associated with aerospace and space hardware.

Recent studies of damage accumulation in SiC/SiC CMCs have incorporated sophisticated damage detection techniques. Morscher et al. developed the acoustic emission technique as a means of monitoring and quantifying matrix cracking and other damage mechanisms in CMCs as they relate to time, stress, and environment [1, 2]. Acoustic emission (AE) is a damage monitoring technique that captures sound waveforms produced from the AE energy released from CMC damage events. AE only captures dynamic processes; only active features such as crack initiation and growth are detected, such that developing

B. Swaminathan (✉)
Department of Materials Science, University of California, Santa Barbara, Engineering II Building, Santa Barbara, CA, USA
e-mail: bswaminathan@umail.ucsb.edu

J. D. Kiser · A. S. Almansour
NASA Glenn Research Center (GRC), Cleveland, OH, USA

K. Sevener
Department of Materials Science and Engineering, University of Michigan, Ann Arbor, MI, USA

S. Daly
Department of Mechanical Engineering, University of California, Santa Barbara, Engineering II Building, Santa Barbara, CA, USA

P. R. Thakre et al. (eds.), *Mechanics of Composite, Hybrid and Multifunctional Materials, Volume 5*, Conference Proceedings of the Society for Experimental Mechanics Series, https://doi.org/10.1007/978-3-319-95510-0_36

and stagnant features can be discerned [2]. This approach has enabled new insights into surface and subsurface cracking behavior in SiC/SiC CMCs.

Another technique for damage characterization is a combination of scanning electron microscopy (SEM) and distortion-corrected digital image correlation (DIC) in order to obtain full-field quantitative deformation measurements at the length scale of the constituent architecture. DIC is a non-contact optical technique that maps subsets of tracking markers on a specimen surface during loading to capture full-field displacements, from which strains can be computed. As DIC is length-scale independent, it can be used accurately at the microscale. This technique was originally conceived by Sutton et al. [3, 4] and later developed by the Daly group [5, 6]. Tracy et al. recently used this approach to capture full-field microscale data of CMC damage and cracking at 800 °C under vacuum and explore statistical characterization of architecture-damage interactions [6].

36.2 Experimental Method and Materials

In this presentation, an experimental approach to detect and characterize damage in SiC/SiC CMCs will be discussed. The goal of this work is to understand the relationships between microstructural features and early stress-dependent changes below the proportional limit, utilizing samples that are representative of a fundamental element of the macroscopic CMC. This approach combines acoustic emission with microscale deformation tracking via digital image correlation inside a scanning electron microscope in order to examine early damage accumulation (at room temperature). Combining these two experimental approaches enables the capture of multi-modal data on early damage mechanisms in CMCs at both the surface and subsurface levels.

Studying the microscale effects of composite structure on damage initiation in-SEM requires statistically significant full-field microscale deformations, which is difficult due to the low strain levels exhibited by CMCs. This is overcome by combining distortion corrected DIC with nanoparticles for tracking to acquire the necessary high accuracy and spatial resolution data. A self-assembled gold nanoparticle (Au NP) pattern, is used to track damage. Using a technique developed by Kammers et al., 2-D arrays of nanoparticles are produced that are ideal for the surface patterning need for SEM-DIC: they provide dense coverage, secure bonding of NPs to the surface, dynamic feature size, and functionality in multiscale experiments. The NPs facilitate imaging at fields of view (FOVs) small enough to resolve low strains indicative of deformation in CMCs [7].

This approach is leveraged to study Hi-Nicalon Type S ™ SiC fiber-based minicomposites. These minicomposites consist of a single fiber tow of 500 filaments that is coated with a thin Boron Nitride (BN) interphase, and then encased in an overlayer of chemical vapor infiltrated SiC matrix. This talk will cover how the AE + SEM-DIC approach can be used and combined with a large data framework in the mechanical testing of SiC/SiC minicomposites to study the evolution of damage and its relation to constituent architecture. This research provides the foundation for characterization of damage in SiC/SiC CMCs under more complex testing conditions.

Acknowledgments This research is supported by the NASA Space Technology Research Fellowship program.

References

1. Morscher, G.N.: Modal acoustic emission of damage accumulation in a woven SiC/SiC composite. Compos. Sci. Technol. **59**(5), 687–697 (1999)
2. Morscher, G.N., Godin, N.: Use of acoustic emission for ceramic matrix composites. In: Bansal, N.P., Lamon, J. (eds.) CMCs: Materials, modeling, and technology, pp. 569–590. Wiley, Hoboken (2015)
3. Sutton, M.A., Li, N., Joy, D.C., Reynolds, A.P., Li, X.: Scanning Electron microscopy for quantitative small and large deformation measurements part I: SEM imaging at magnifications from 200 to 10,000. Exp. Mech. **47**(6), 775–787 (2007)
4. Sutton, M.A., Li, N., Garcia, D., Cornille, N., Orteu, J.J., Mcneil, S.R., Reynolds, A.P.: Scanning electron microscopy for quantitative small and large deformation measurements part II: experimental validation for magnifications from 200 to 10,000. Exp. Mech. **47**(6), 789–804 (2007)
5. Kammers, A.D., Daly, S.: Digital image correlation under scanning electron microscopy: methodology and validation. Exp. Mech. **53**(9), 1743–1761 (2013)
6. Tracy, J., Daly, S., Sevener, K.: Multiscale damage characterization in continuous fiber ceramic composites using digital image correlation. J. Mater. Sci. **50**(15), 5286–5299 (2015)
7. Kammers, A.D., Daly, S.: Self-assembled nanoparticle surface patterning for improved digital image correlation in a scanning electron microscope. Exp. Mech. **53**(8), 1333–1341 (2013)

Chapter 37
Investigating Intralaminar Crack Growth in Biaxially Stressed Composites

Jordan French, Jessica Christensen, and Michael M. Czabaj

Abstract Permeability concerns of fiber-reinforced composites has delayed their implementation into next-generation space launch vehicle structures. Namely composite cryotanks, which could help achieve significant weight savings, are not being implemented due to a deficient understanding of the mechanisms that lead to the generation of through-thickness crack networks and permeation paths. A novel in-plane cruciform test method was developed to investigate intralaminar crack growth under biaxial load using ex situ X-ray computed tomography (CT). Specimens were tested at predefined load levels, determined from a progressive damage model, then imaged with X-ray CT. 3-dimensional representations of the specimen after each load interval were reconstructed to capture damage accumulation as a function of load. The CT data were then used to establish initial hypotheses of crack initiation and growth in biaxially stressed composites.

Keywords Biaxial testing · Progressive damage · Intralaminar crack growth

To reduce the weight and cost of future space-launch vehicles, tape-laminate composites are being investigated for use in cryogenic propellant tanks (cryotanks) [1]. However, recent development programs have concluded that tape-laminate composites are not ready for this application due to the deficient understanding of ply-level damage caused by cryobiaxial loading [2]. Cryobiaxial loading occurs from internal pressurization of the vehicle's cryotank coupled with thermal strains induced by the cryogenic propellant's low temperature. Such loading scenario can cause initiation and evolution of intralaminar crack networks, which may lead to uncontrolled permeation of the cryogenic propellant and failure of the entire cryotank.

This works presents a combined numerical and experimental effort to progressively image the initiation and evolution of damage in tape laminate composites subjected to biaxial loading. To this end, a new in-plane cruciform specimen has been developed to include several key features that make it suitable for investigation of ply-level damage. These features include: (1) an ability to apply a wide range of biaxial load ratios, (2) a relatively simple specimen geometry that allows for in-situ and ex-situ observation of 3D damage evolution using high-resolution X-ray CT, (3) an ability to test different of stacking sequences, and (4) an ability to cool the gage region to LN2 temperatures through conduction or cryogen submersion. Despite these advantages, most of the cruciform geometries that have been developed to date appear to fail away from the gage region. In recent years, an extensive effort has taken place to modify the cruciform geometry and ensure failure in the gage region and away from specimen corners, but no study has attempted to develop a cruciform to investigate intralaminar damage evolution exclusively [3, 4]. To this end, a newly developed cruciform geometry is proposed which incorporates aluminum doublers, compound corner radii, and a semi-hemispherical gage region (see Fig. 37.1). These unique features are meant to minimize damage accumulation caused by the inherent stress concentrations at the 90-degree transitions between adjacent loading arms and ensure centralized damage formation.

Previous work by the authors indicated that cruciform specimens are highly sensitive to geometric features and manufacturing-induced flaws [5, 6]. In consideration of this, the Air Force Research Lab's high-fidelity progressive damage modeling tool, BSAM, was used to validate the specimen geometry, and to ensure a centralized crack network evolves during biaxial loading. An in-depth description of BSAM can be found in a publication by Iarve et al. [7].

Similar to what has been done previously in [5, 6], BSAM predictions were used to identify load intervals at which damage first initiated or evolved to a new state. These results were used to guide a series of progressive damage tests using a custom-built table-top load frame shown in Fig. 37.2 and described in [6]. During progressive testing, strains in the gage region were monitored using a 3D digital image correlation system. After each increment, the specimen was unloaded,

J. French (✉) · J. Christensen · M. M. Czabaj
Department of Mechanical Engineering, University of Utah, Salt Lake City, UT, USA
e-mail: J.French@utah.edu

P. R. Thakre et al. (eds.), *Mechanics of Composite, Hybrid and Multifunctional Materials, Volume 5*, Conference Proceedings of the Society for Experimental Mechanics Series, https://doi.org/10.1007/978-3-319-95510-0_37

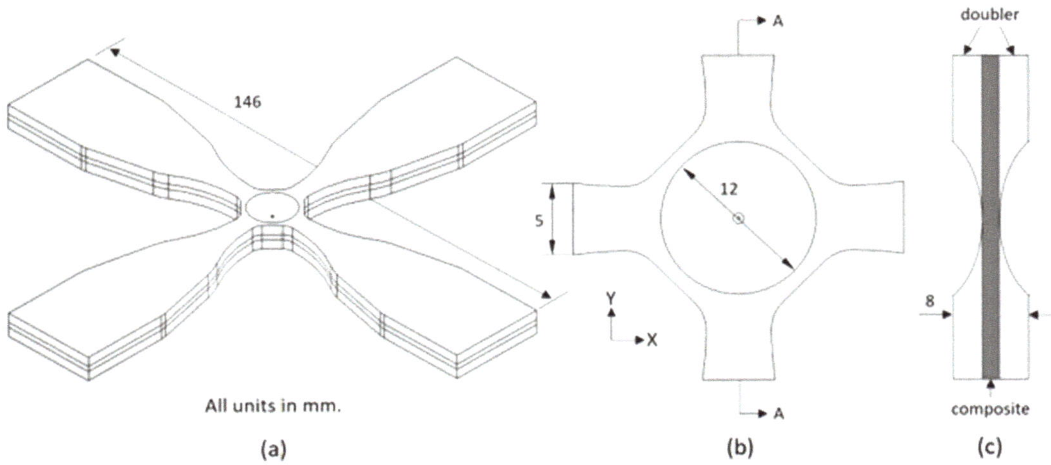

Fig. 37.1 (**a**) Isometric view of proposed cruciform design. (**b**) Detail view of cruciform gage region. (**c**) Cross sectional view of cruciform gage region with aluminum doublers bonded to the composite specimen

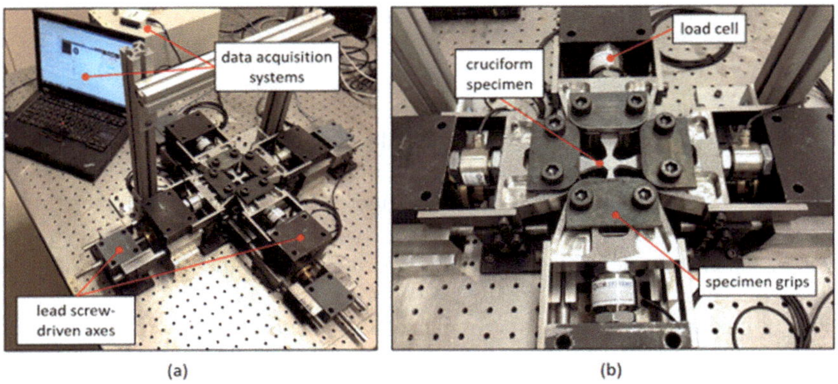

Fig. 37.2 (**a**) Experimental setup of the biaxial progressive damage tests. (**b**) Detail view of specimen grips and load cells

brought to room temperature, and subsequently imaged using X-ray CT to record crack location, density, and size. Results from numerical simulations and progressive damage tests are used to propose mechanisms through which transverse cracks initiate and evolve in composites subjected to biaxial loading.

Acknowledgements This work was supported by a NASA Space Technology Research Fellowship.

References

1. NASA.: NASA Technology Roadmaps – TA 12: Materials Structures, Mechanical Systems, and Manufacturing. The office of the chief Technologist @ NASA HQ (2015)
2. Niedermeyer, M.: X-33 LH2 Tank Failure Investigation Findings. NASA Marshall Space Flight Center, NTRS-20010020398 (2000)
3. Ohtake, Y., et al.: Geometry determination of cruciform-type specimen and biaxial tensile test of C/C composites. Key Eng. Mater. **164**, 151–154 (1999)
4. Kumazawa, H., Takatoya, T.: Biaxial strength investigation of CFRP composite laminates by using cruciform specimens. In: Proceedings of the 17th International Conference on Composite Materials (2009)
5. French, J., Rapking, D., Mollenhauer, D., Czabaj, M.: Development of a novel in-plane tension-tension biaxial cruciform specimen. In: Proceedings of the 31st Technical Conference for the American Society of Composites (2016)
6. French, J., et al.: Numerical and experimental design of a cruciform specimen for a novel cryobiaxial experiment. In: Proceedings of the 32nd Technical Conference for the American Society of Composites (2017)
7. Iarve, E., Mollenhauer, D.: Chapter 9: Mesh-independent matrix cracking and delamination modeling in advanced composite materials. In: Camanho, P., Hallett, S. (eds.) Numerical Modeling of Failure in Advanced Composite Materials, 1st edn (2015)

Chapter 38
Determination of Stress Free Temperature in Composite Laminates for Residual Stress Modeling

Brian T. Werner, Helena Jin, and Timothy M. Briggs

Abstract As the complexity of composite laminates rises, the use of hybrid structures and multi-directional laminates, large operating temperature ranges, the process induced residual stresses become a significant factor in the design. In order to properly model the initial stress state of a structure, a solid understanding of the stress free temperature, the temperature at which the initial crosslinks are formed, as well as the contribution of cure shrinkage, must be measured. Many in industry have moved towards using complex cure kinetics models with the assistance of commercial software packages such as COMPRO. However, in this study a simplified residual stress model using the coefficient of thermal expansion (CTE) mismatch and change in temperature from the stress free temperature are used. The limits of this simplified model can only be adequately tested using an accurate measure of the stress free temperature. Only once that is determined can the validity of the simplified model be determined. Various methods were used in this study to test for the stress free temperature and their results are used to validate each method. Two approaches were taken, both involving either cobonded carbon fiber reinforced polymer (CFRP) or glass fiber reinforced polymer (GFRP) to aluminum. The first method used a composite-aluminum plate which was allowed to warp due to the residual stress. The other involved producing a geometrical stable hybrid composite-aluminum cylinder which was then cut open to allow it to spring in. Both methods placed the specimens within an environmental chamber and tracked the residual stress induced deformation as the temperature was ramped beyond the stress free temperature. Both methods revealed a similar stress free temperature that could then be used in future cure modeling simulations.

Keywords Residual Stress · Stress Free Temperature · Composites · Digital Image Correlation · Bonded joint

As the complexity of composite laminates rises, the use of hybrid structures and multi-directional laminates, large operating temperature ranges, the process induced residual stresses become a significant factor in the design [1, 2]. In order to properly model the initial stress state of a structure, a solid understanding of the stress free temperature, the temperature at which the initial crosslinks are formed, as well as the contribution of cure shrinkage, must be measured. Many in industry have moved towards using complex cure kinetics models with the assistance of commercial software packages such as COMPRO [3–5]. However, in this study a simplified residual stress model using the coefficient of thermal expansion (CTE) mismatch and change in temperature from the stress free temperature are used [4, 6]. The limits of this simplified model can only be adequately tested using an accurate measure of the stress free temperature. Only once that is determined can the validity of the simplified model be determined. Various methods were used in this study to test for the stress free temperature and their results are used to validate each method. Two approaches were taken, both involving either cobonded carbon fiber reinforced polymer (CFRP) or glass fiber reinforced polymer (GFRP) to aluminum. The first method used a composite-aluminum plate which was allowed to warp due to the residual stress. The other involved producing a geometrical stable hybrid composite-aluminum cylinder which was then cut open to allow it to spring in. Both methods placed the specimens within an environmental chamber and tracked the residual stress induced deformation as the temperature was ramped beyond the stress free temperature. Both methods revealed a similar stress free temperature that could then be used in future cure modeling simulations.

Due to the CTE mismatch between the composite and aluminum plate, the residual stresses that were developed at temperatures other than the initial curing temperature will lead to the warpage change of the hybrid structure. The initial stress-free state can be identified as the plate turns to flat. In this work, the hybrid plate was set up free standing inside a thermal chamber with a transparent glass door. A 3D-digital image correlation (DIC) system was applied to measure the

B. T. Werner (✉) · H. Jin · T. M. Briggs
Sandia National Laboratories, Livermore, CA, USA
e-mail: btwerne@sandia.gov

P. R. Thakre et al. (eds.), *Mechanics of Composite, Hybrid and Multifunctional Materials, Volume 5*, Conference Proceedings of the Society for Experimental Mechanics Series, https://doi.org/10.1007/978-3-319-95510-0_38

full-field displacement of the plate as it undergoes the temperature change. Figure 38.1a shows the experimental set-up for heating/cooling the plate and measuring their warpage and (b) shows the specimen with DIC patterns.

(a) (b)

Fig. 38.1 (**a**) Experimental setup for measuring plate warpage; (**b**) Specimen with DIC speckle patterns

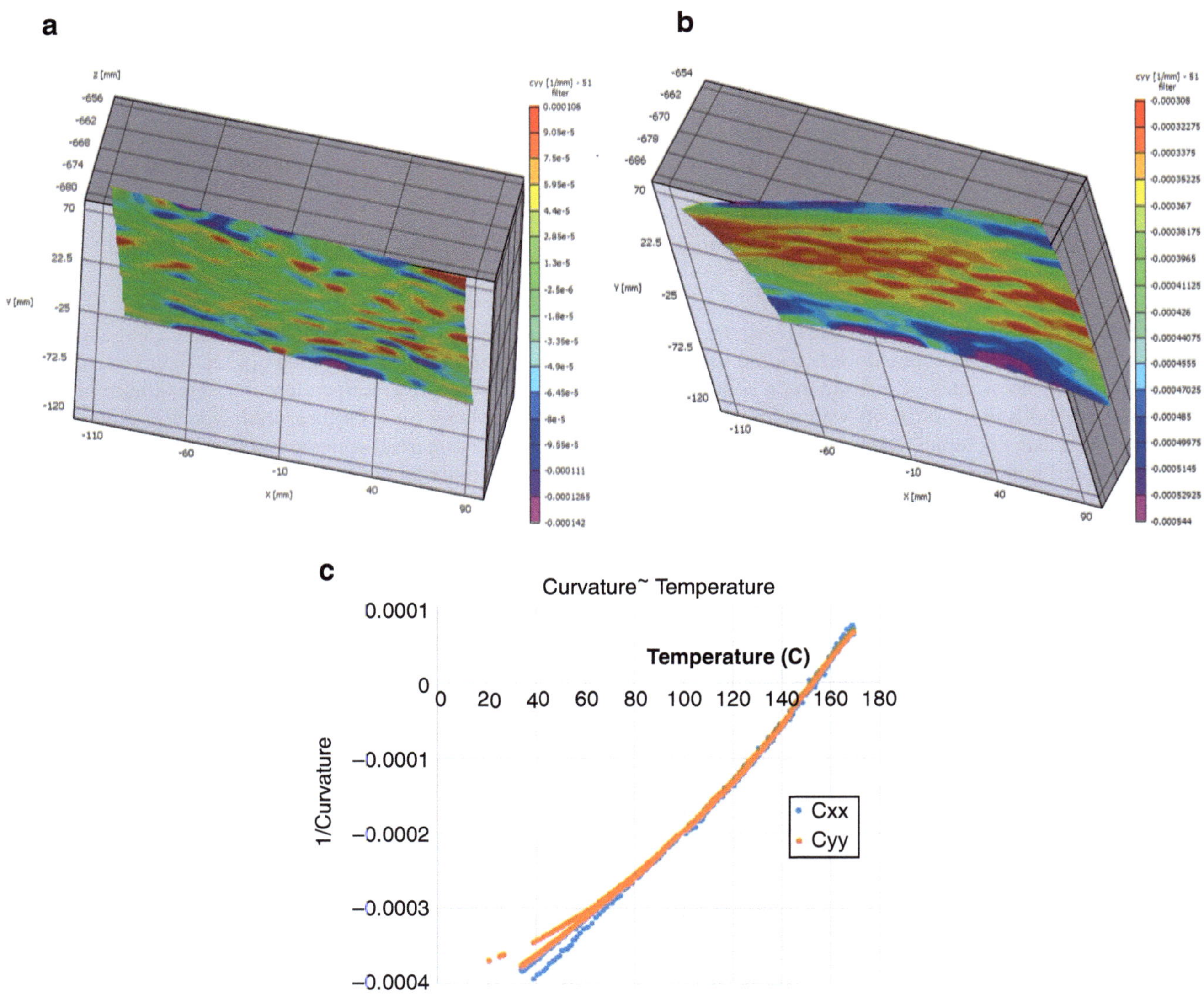

Fig. 38.2 (**a**) Curvature of warped plate at room temperature, (**b**) Curvature of flat plate at heated stress-free temperature, (**c**) Plate curvature history with temperature

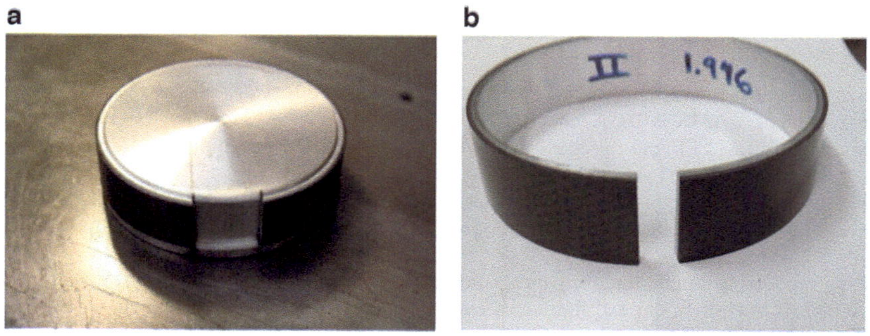

Fig. 38.3 Split ring (**a**) in cutting jig and (**b**) removed from cutting jig with resulting spring in

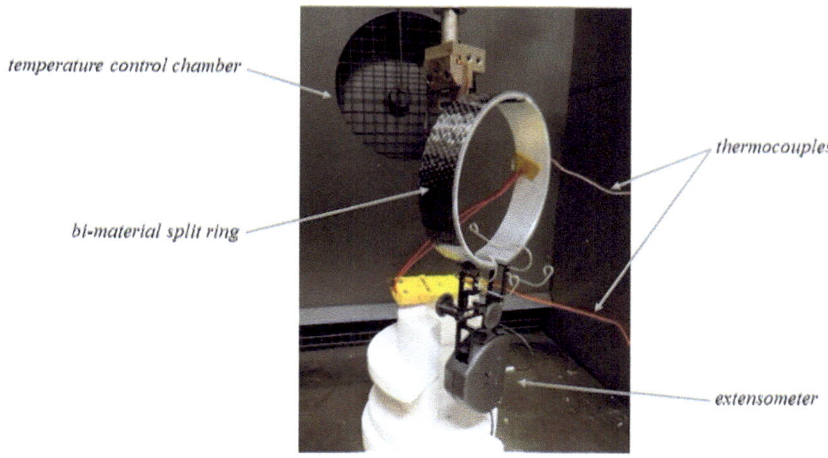

Fig. 38.4 Split ring test setup

The hybrid plate was heated up steadily at 1 °C per minute inside the thermal chamber till 180 °C and then cool down to room temperature steadily. The heating rate was chosen slow enough to reach uniform temperature among the chamber, aluminum and composite plates. The warpage of the plates was changing as the stress due to the coefficient of thermal expansion mismatch between the composite and aluminum plates was varying with temperature. The 3D-DIC system recorded the shapes of the plate as it went through the temperature cycle. The digital image correlation technique was then applied to these images to calculate the curvature of the plate. Figure 38.2a, b show the curvature of warped plate at the room temperature and the flat stress-free status at heated temperature. Figure 38.2c shows an example of the CFRP 0°/Al plate curvature history with the temperature. The inverse of both curvatures $1/C_{xx}$ and $1/C_{yy}$ crosses the temperature axis at 152 °C, which stands for the stress-free status when the plate is flat.

The split ring approach used a cobonded composite-aluminum cylinder which was cut into 25 mm wide sections. Prior removing a sector, two scribe lines were made on either side of the ply overlap region. The distance between these scribe lines was measured at room temperature using calipers. Then using a custom jig, an approximately 25 mm sector was removed which included the overlap region (Fig. 38.3). Following this removal, the split ring sprung in to provide a much smaller gap than was removed. An extensometer was then affixed with each of the blades resting on the scribe lines and the temperature was ramped to 180 °C as the extensometer tracked the opening of the split ring's gap (Fig. 38.4). Once the distance between the scribe lines was measured to be equal to the distance measured at room temperature, while factoring in the global expansion of the hybrid structure, the stress free temperature was measured.

Acknowledgement Sandia National Laboratories is a multimission laboratory managed and operated by National Technology and Engineering Solutions of Sandia LLC, a wholly owned subsidiary of Honeywell International Inc. for the U.S. Department of Energy's National Nuclear Security Administration under contract DE-NA0003525.

References

1. Nelson, S., Werner, B.: Experimental and computational investigations into the effect of process induced stresses on the mode I fracture toughness of composite materials. In: CAMX 2016. Anaheim, CA (2016)
2. Werner, B.T., Nelson, S.M., Briggs, T.M.: Effect of process induced stresses on measurement of frp strain energy release rates. In: Society for Experimental Mechanics Annual Conference. Indianapolis, IN (2017)
3. Volk, B.L., et al.: Predicting the failure of polymer matrix composites as a function of processing cycle. In: CAMX 2014. Orlando, FL (2014)
4. Volk, B.L., et al.: Evaluation of process modeling methodologies for out-of-autoclave polymer matrix composites. In: CAMX 2016. Anaheim, CA (2016)
5. Brockman, R.A., et al.: OMC Processing simulation using an elastic-viscoplastic model. In: SAMPE 2015. Baltimore, MD (2015)
6. Hanson, A.A., et al.: Experimental measurement and finite element modeling of residual stresses in simple composite structures. In: CAMX 2016. Anaheim, CA (2016)

Chapter 39
Calibration of a Simple Rate Dependent Elastic-Plastic Constitutive Model for a Toughened Carbon Epoxy Composite System

Brian T. Werner and Joseph D. Schaefer

Abstract The concept of progressive failure modeling is an ongoing concern within the composite community. A common approach is to employ a building block approach where constitutive material properties lead to lamina level predictions which then lead to laminate predictions and then up to structural predictions. There are advantages to such an approach, developments can be made within each step and the whole workflow can be updated. However, advancements made at higher length scales can be hampered by insufficient modeling at lower length scales. This can make industry wide evaluations of methodologies more complicated. For instance, significant advances have been made in recent years to strain rate independent failure theories on the lamina level. However, since the Northwestern Theory is stress dependent, for adequate use in a progressive damage model, a similarly robust constitutive model must also be employed to calculate these lamina level stresses. An improper constitutive model could easily cause a valid failure model to produce incorrect results. Also, any global strain rate applied to a multi-directional laminate will produce a spectrum of local lamina level strain rates so it is important for the constitutive law to account for strain rate dependent deformation.

Keywords Composites · Plasticity · Constitutive modeling · Failure · Calibration

The concept of progressive failure modeling is an ongoing concern within the composite community. A common approach is to employ a building block approach [1] where constitutive material properties lead to lamina level predictions which then lead to laminate predictions and then up to structural predictions. There are advantages to such an approach, developments can be made within each step and the whole workflow can be updated. However, advancements made at higher length scales can be hampered by insufficient modeling at lower length scales [2]. This can make industry wide evaluations of methodologies more complicated [3]. For instance, significant advances have been made in recent years to strain rate independent failure theories on the lamina level [4–6]. However, since the Northwestern Theory is stress dependent, for adequate use in a progressive damage model, a similarly robust constitutive model must also be employed to calculate these lamina level stresses. An improper constitutive model could easily cause a valid failure model to produce incorrect results [7, 8]. Also, any global strain rate applied to a multi-directional laminate will produce a spectrum of local lamina level strain rates so it is important for the constitutive law to account for strain rate dependent deformation.

In this study, an elastic viscoplastic model developed by Daniel, et al. [9] for a brittle carbon/epoxy, AS4/3501-6, was calibrated for a more ductile toughened carbon/epoxy system, IM7/8552. The model was developed in a similar manner to C.T. Sun's one-parameter constitutive law for unidirectional composite lamina [10–14]. Daniel's model however includes a second parameter which can account for loading direction based on the sign of the hydrostatic stress component. This becomes important when moving to multi-directional laminates where the applied stress can be of opposite sign of the lamina stress. Since this model was developed for a brittle resin system, it may not be sufficient for more ductile material systems. Tension loads typically result in fairly brittle failure within unidirectional lamina; thus, the bulk of the characterization in this effort was performed in compression.

B. T. Werner (✉)
Sandia National Laboratories, Livermore, CA, USA
e-mail: btwerne@sandia.gov

J. D. Schaefer
Boeing Research and Technology, St. Louis, MO, USA

P. R. Thakre et al. (eds.), *Mechanics of Composite, Hybrid and Multifunctional Materials, Volume 5*, Conference Proceedings of the Society for Experimental Mechanics Series, https://doi.org/10.1007/978-3-319-95510-0_39

The initial characterization focused on obtaining the elastic properties for the lamina. All of the testing was performed at a quasi-static strain rate (10^{-4}) with unidirectional on-axis tension testing providing the in-plane Young's moduli and Poisson's ratio (E_1, E_2, ν_{12}) and 10° off-axis testing providing the in-plane shear modulus, G_{12}. Thicker unidirectional panels were manufactured to allow for on- and off-axis compression testing. Specimens were cut at 15°, 30°, 45°, 60°, 75°, and 90° to the fiber direction, 1-axis. These specimens were then loaded in compression using a 50-kip servo-hydraulic load frame at strain rates of 10^{-4} and 1 s^{-1}. The specimens were 7.5 mm × 7.5 mm × 25 mm. Each specimen was bonded to steel loading blocks and held with hydraulic wedge grips. The axial strain was recorded using 120-ohm resistance strain gages with an excitation voltage of 2 V. For the dynamic testing, a split-Hopkinson pressure bar was used to obtain strain rates up to 500 s^{-1}. While the strain rate varied for each off-axis angle due to differences in mechanical impedance, all of the dynamic tests fell within a range of 250–500 s^{-1}. Since the strain rate effect on strength and constitutive behavior is typically measured on a logarithmic scale, this was deemed acceptable. Achieving higher strain rates was not possible due to the required length of the specimens. For each off-axis specimen there had to be fibers that reached from one transverse edge of the specimen to the other. This would allow the specimen to fully deform in shear without influence from the loading faces of the bars. Bing and Sun also showed that for smaller angles, friction with the bar faces provided a major influence on the specimen's behavior [15]. For this reason, each specimen's loading faces were coated with graphite and bonded to steel discs made of the same diameter and material as the split-Hopkinson pressure bars' incident and transmission bars. These discs were also coated with graphite to allow for a low friction interface.

The calibration of the model for each strain rate was straight forward. The 90° compression and tension data was used to determine the hydrostatic parameter. Using the proposed potential function, the parameter was varied until the nonlinear portion of the data for the tension and compression lined up. The remaining deviatoric parameter was then fit by using the 15°, 45°, and 90° compression data. This parameter was fit by varying it until the data from the three tests converged into one master curve on the effective stress-effective plastic strain plot. Then, the master curve was fit using a power law. This provided the model fit for each individual strain rate. The 30°, 60°, and 75° data was left out of the model calibration as the resulting fits could be used for validation of the model. By plotting the relationship of the power law parameters against strain rate, the model could then be used for any strain rate between the bounds of the quasi-static and dynamic strain rates. Such a model fit could then be used in the framework of a more complex, progressive damage model (Figs. 39.1 and 39.2).

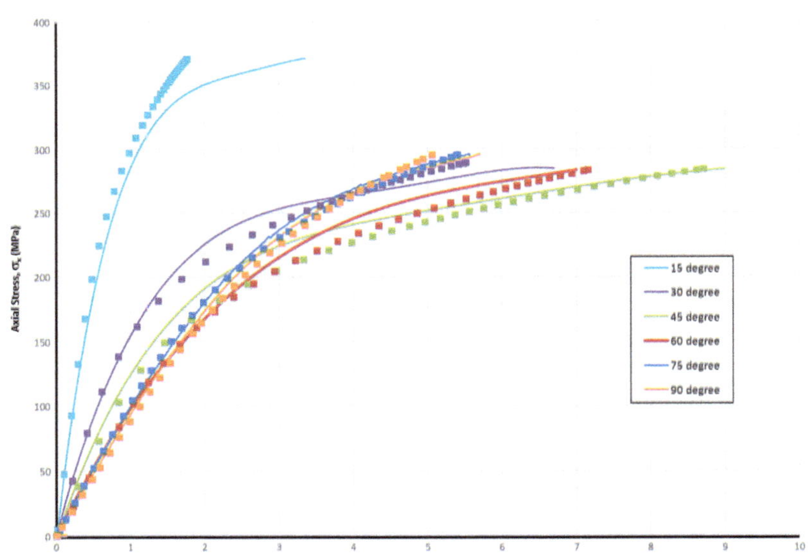

Fig. 39.1 Experimental data and model fits under quasi-static loading conditions for IM7/8552

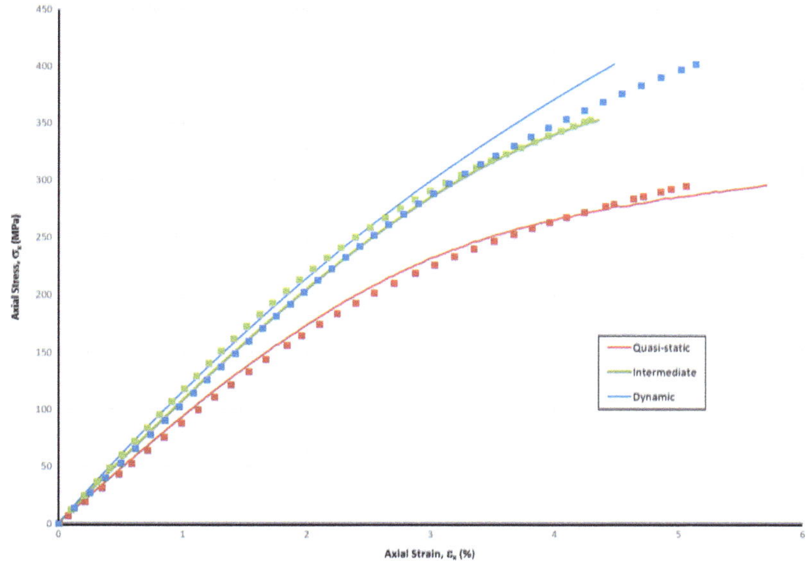

Fig. 39.2 Resulting model fits for 90° compression data across the three strain rates tested

Acknowledgement Sandia National Laboratories is a multimission laboratory managed and operated by National Technology and Engineering Solutions of Sandia LLC, a wholly owned subsidiary of Honeywell International Inc. for the U.S. Department of Energy's National Nuclear Security Administration under contract DE-NA0003525.

References

1. The Composite Materials Handbook – MIL 17: Polymer matrix composites: Materials usage, design and analysis. Technical Standards Service (1997)
2. Razi, H., et al.: Rapid integration of new analysis methods in production. In: American Soceity for Composites: Thirty-First Technical Conference. Williamsburg, VA (2016)
3. Pankow, M., Justusson, B., Schaefer, J. D.: Impact and CSAI analysis and validation using high fidelity test and inspection data. In: Proceedings of the American Society for Composites. 32nd Annual Technical Conference. West Lafayette, IN, USA (2017)
4. Daniel, I.M., et al.: Interfiber/interlaminar failure of composites under multi-axial states of stress. Compos. Sci. Technol. **69**(6), 764–771 (2009)
5. Daniel, I.M., Werner, B.T., Fenner, J.S.: Strain-rate-dependent failure criteria for composites. Compos. Sci. Technol. **71**(3), 357–364 (2011)
6. Schaefer, J.D., Werner, B.T., Daniel, I.M.: Strain-rate-dependent failure of a toughened matrix composite. Exp. Mech. **54**(6), 1111–1120 (2014)
7. Wanthal, S., et al.: Verification and validation process for progressive damage and failure analysis methods in the NASA Advanced Composites Consortium. 32nd American Society for Composites Conference, West Lafayette (2017)
8. Hyder, I., et al.: Assessment of intralaminar progressive damage and failure analysis methods using an efficient evaluation framework. In: The American Society for Composites (2017)
9. Daniel, I.M., et al.: Characterization and constitutive modeling of composite materials under static and dynamic loading. AIAA J. **49**(8), 1658–1664 (2011)
10. Chen, J.L., Sun, C.T.: A New Plasticity Theory for Anisotropic Fiber Composites. Proceedings of the second international symposium on composite materials and structures, pp. 664–671 (1992)
11. Gates, T.S., Sun, C.T.: An Elastic Viscoplastic Constitutive Theory for Fiber Reinforced Thermoplastic Composites. Aiaa/Asme/Asce/Ahs/Asc 30th Structures, structural dynamics and materials conference, Pts 1-4, pp. 845–851 (1989)
12. Gates, T.S., Sun, C.T.: Elastic viscoplastic constitutive model for fiber reinforced thermoplastic composites. AIAA J. **29**(3), 457–463 (1991)
13. Sun, C.T., Chen, J.L.: A simple flow rule for characterizing nonlinear behavior of fiber composites. J. Compos. Mater. **23**(10), 1009–1020 (1989)
14. Tsai, J., Sun, C.T.: Constitutive model for high strain rate response of polymeric composites. Compos. Sci. Technol. **62**(10–11), 1289–1297 (2002)
15. Bing, Q.D., Sun, C.T.: Modeling and testing strain rate-dependent compressive strength of carbon/epoxy composites. Compos. Sci. Technol. **65**(15–16), 2481–2491 (2005)

Chapter 40
Imaging the Life-Cycle of CMCs Using High-Resolution X-Ray Computed Tomography

Peter J. Creveling, Noel LeBaron, and Michael W. Czabaj

Abstract In this study, the entire life cycle of a ceramic matrix composite (CMC) manufactured using polymer infiltration and pyrolysis (PIP) is imaged using high-resolution X-ray micro-computed tomography (μCT). The entire PIP process is imaged *ex situ* to capture the evolution of voids and shrinkage cracks during laminate densification. After manufacturing, two specimens are extracted from the CMC laminate, and subsequently tested to failure in flexure and tension at 1000°C. Gray-scale image segmentation methods are used to quantify the evolution of porosity within the microstructure during PIP processing. X-ray μCT image results from *in situ* testing are qualitatively examined to quantify presence of individual fiber breaks, fiber pull-outs, matrix cracking, fiber-matrix decohesion, to name a few. These results are used to motivate development of new algorithms for segmentation of microstructural features from X-ray μCT data sets.

Keywords Ceramic matrix composites · Silicon-carbide/silicon-carbide · X-ray micro-computed tomography

40.1 Introduction

Fiber-reinforced ceramic matrix composites (CMCs) are seeing increased use in the aerospace industry due to their high specific strength and stiffness, and ability to withstand prolonged exposures to extremely high temperatures and oxidative environments. The material properties of CMCs make them ideal for use in high-efficiency gas turbines, aircraft engines, and hypersonic re-entry space vehicles [1]. Despite the increasing use of CMCs, the fundamental and quantitative links governing the physics of their processing, the resulting stochastic microstructure, and the long-term structural performance remain largely undiscovered [1–6]. To this end, the aim of this research is to use high-resolution X-ray micro-computed tomography (μCT) to image the entire life-cycle of a silicon-carbide/silicon-carbide CMC, starting with initial formation of the microstructure during manufacturing, and ending with ultimate failure during mechanical loading in realistic operational conditions.

40.2 Experimental Methods

In this study, X-ray μCT imaging was performed on a CMC sample manufactured by multi-step polymer infiltration and pyrolysis (PIP). The composite system consisted of 8HS satin weave CG Nicalon-fiber fabric infiltrated with SMP-10 resin, and which was formed into a 60 mm × 30 mm × 1.5 mm laminate. The laminate was manufactured by C-staging and pyrolyzing the uncured resin at 1000 ° C, followed by five additional PIP steps until the laminate was deemed fully densified. At the end of each manufacturing step, the composite sample was imaged *ex situ* in two distinct regions of interest (ROI). Imaging was performed using a Zeiss Versa-520 X-ray CT 3D microscope at the Air Force Research Laboratory (Dayton, OH) with a spatial resolution of 1.21 μm/voxel. After PIP manufacturing, two "matchstick" specimens, each containing one of the ROI, were extracted from the laminate and subsequently tested in flexure and tension at 1000 ° C. Prior to testing, each sample was notched in the center of the gage region to promote initiation of damage within the ROIs. The tensile and flexure testing was performed *in situ* at the Advanced Light Source (ALS) beamline 8.3.2 (Berkeley, CA). All imaging was done in near-real time with spatial resolution of 1.26 μm/voxel.

P. J. Creveling · N. LeBaron · M. W. Czabaj (✉)
Department of Mechanical Engineering, University of Utah, Salt Lake City, UT, USA
e-mail: m.czabaj@utah.edu

P. R. Thakre et al. (eds.), *Mechanics of Composite, Hybrid and Multifunctional Materials, Volume 5*, Conference Proceedings of the Society for Experimental Mechanics Series, https://doi.org/10.1007/978-3-319-95510-0_40

40.3 Results and Discussion

The tomographic images collected throughout the manufacturing process can be used to draw qualitative and quantitative observations about the evolution of microstructure during the entire PIP process. Figure 40.1 shows an example of the microstructure evolution observed within a single fiber-tow. The cross-sectional images in Fig. 40.1 depict the evolution of shrinkage cracks during pyrolysis, as well as, the reduction of porosity as a result of multi-step re-infiltration and pyrolysis. To quantify the evolution of the porosity in 3D, the tomographic image stacks from each ROI were segmented using gray-scale thresholding in Avizo Lite 9.4.0. From the 3D segmentation results for each processing step, qualitative changes in porosity are shown in Fig. 40.2. These images help visualize the increase in porosity after initial pyrolysis, and the decrease in porosity after each new PIP step.

The void volume percentage within one of the ROIs is plotted as a function of each processing step in Fig. 40.3. The largest increase in porosity occurs between C-staging to the initial pyrolysis. The largest decrease in porosity occurs during PIP #1–2, after which porosity reaches a stead state value for PIP #3–5. Examining the evolution of voids throughout the entire manufacturing process, two types of voids are observed, including: (1) persistent voids that form during C-staging and remain unfilled, and (2) voids that form during C-staging and/or pyrolysis that are filled during PIP #1–5. The 3D image data suggests that persistent voids remained unfilled due to blockage of pathways leading to the free surfaces occurring during the first two infiltrations.

Preliminary results shown in Fig. 40.4 highlight the damage initiating from the notch within the ROI before (a) and after (b) flexure testing. The observed damage includes individual fiber breaks, fiber pullout, matrix cracking, and crack-tip

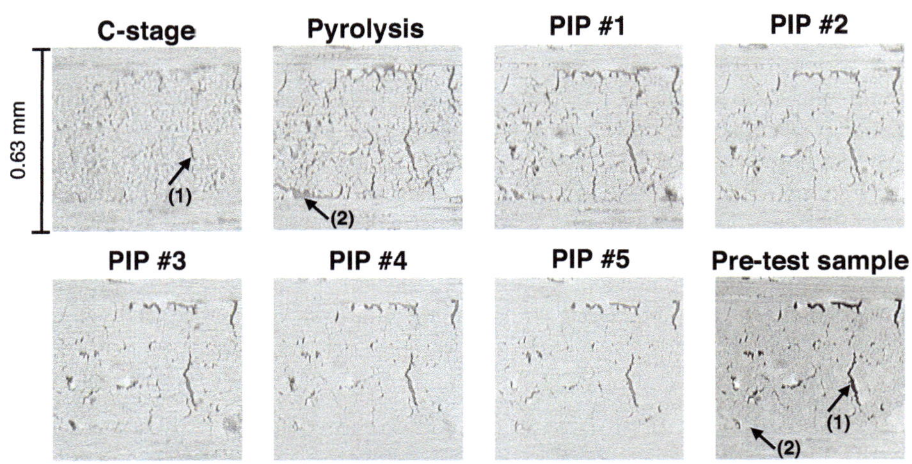

Fig. 40.1 Cross-sectional images collected *ex situ* of the microstructure evolving throughout the entire PIP process. Circular shapes correspond to individual fiber cross-sections, while dark regions correspond to shrinkage cracks and voids. Arrow (1) indicates a persistent shrinkage crack formed during consolidation. Arrow (2) indicates a void formed during pyrolysis that is later filled during multi-step re-infiltration

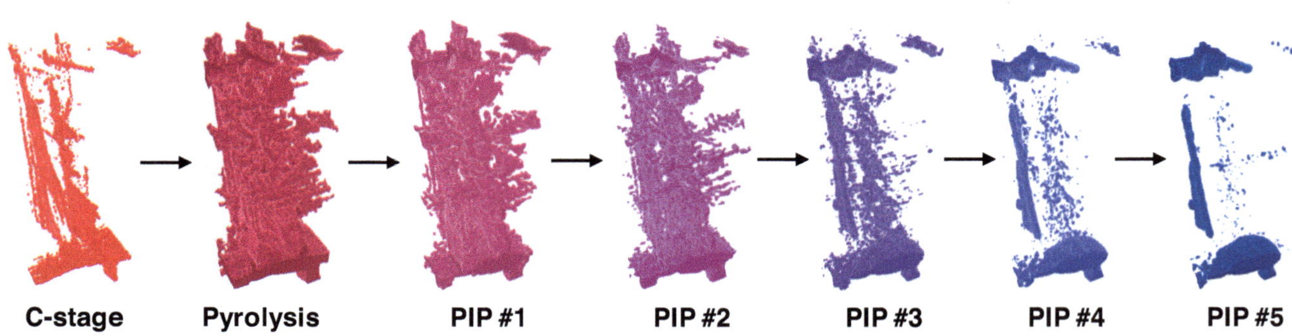

Fig. 40.2 3D segmentation of a void network throughout the entire PIP process.

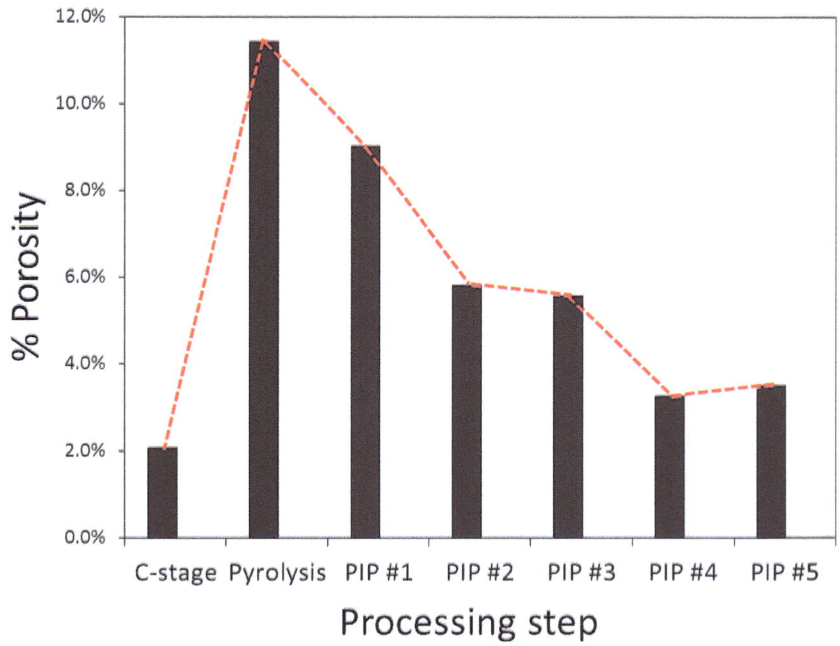

Fig. 40.3 Change of void volume at each step of the PIP process. The data used in this plot was extracted from a small 0.63 mm × 0.63 mm × 0.63 mm sub-volume within one of the ROIs

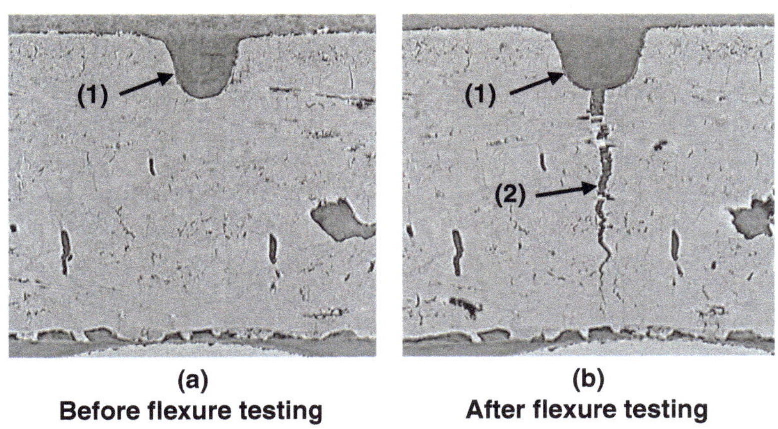

Fig. 40.4 Reconstruction of *in situ* tomographic data before (**a**) and after (**b**) flexure testing. Arrow (1) highlights the notched region, and (2) highlights the damage within the microstructure resulting from mechanical loading in three-point flexure

displacements. The inherently 3D features observed in these images provide a wealth of previously unobtainable information, and efforts are currently under to develop manual, semi-manual, and automatic algorithms for quantification of these rich data sets.

40.4 Conclusion

This study presents results from high-resolution tomographic imaging of CMC composites obtained throughout the PIP manufacturing process. In addition, *in situ* X-ray CT imaging was used to investigate failure of the samples manufactured via PIP. Image segmentation was applied in order to quantify, in 3D, the change in net porosity during the multi-step manufacturing process. Future work includes of new segmentation methods in order to accurately quantify shrinkage cracks, pore networks, stochastic locations of individual fibers in 3D, fiber breaks, and fiber pullout. Once completed, this study is expected to provide new insights that will help improve the structural performance of current and future CMCs manufactured via PIP.

References

1. Larson, N.M., Zok, F.W.: In-situ 3D visualization of composite microstructure during polymer-to-ceramic conversion. Acta Mater. **144**, 579–589 (2018)
2. Bale, H.A., Haboub, A., et al.: Real-time quantitative imaging of failure events in ultrahigh-temperature materials under load at temperatures above 1600C. Nat. Mater. **12**(1), 40–46 (2013)
3. Marshall, D.B., Cox, B.N.: Integral textile ceramic structures. Annu. Rev. Mat. Res. **38**, 435–443 (2008)
4. Bale, H., Blacklock, M., et al.: Characterizing three-dimensional textile ceramic composites using synchrotron X-ray computed micro-tomography. J. Am. Ceram. Soc. **95**(1), 392–402 (2012)
5. Cox, B.N., Bale, H.A., et al.: Stochastic virtual tests for high-temperature ceramic matrix composites. Annu. Rev. Mat. Res. **44**, 479–529 (2014)
6. Haboub, A., Bale, H.A., et al.: Tensile testing of materials at high temperatures above 1700 C with in situ synchrotron X-ray micro-tomography. Rev. Sci. Instrum. **85**(8), 083702 (2014)

Chapter 41
Effect of Process Induced Residual Stress on Interlaminar Fracture Toughness on Hybrid Composites

Brian T. Werner, Kevin Nelson, and Ciji Nelson

Abstract Delaminations are of great concern to any fiber reinforced polymer composite (FRPC) structure. In order to develop the most efficient structure, designers may incorporate hybrid composites to either mitigate the weaknesses in one material or take advantage #of the strengths of another. When these hybrid structures are used at service temperatures outside of the cure temperature, residual stresses can develop at the dissimilar interfaces. These residual stresses impact the initial stress state at the crack tip of any flaw in the structure and govern whether microcracks, or other defects, grow into large scale delaminations. Recent experiments have shown that for certain hybrid layups which are used to determine the strain energy release rate, G, there may be significant temperature dependence on the apparent toughness. While Nairn and Yokozeki believe that this effect may solely be attributed to the release of stored strain energy in the specimen as the crack grows, others point to a change in the inherent mode mixity of the test, like in the classic interface crack between two elastic layers solution given by Suo and Hutchinson. When a crack is formed at the interface of two dissimilar materials, while the external loading, in the case of a double cantilever beam (DCB), is pure mode I, the stress field at the crack tip produces a mixed-mode failure. Perhaps a change in apparent toughness with temperature can be the result of an increase in mode mixity. This study serves to investigate whether the residual stress formed at the bimaterial interface produces a noticeable shift in the strain energy release rate-mode mixing curve.

Keywords Fracture · Hybrid composites · Residual stress · Non-destructive evaluation · Mixed-mode

Delaminations are of great concern to any fiber reinforced polymer composite (FRPC) structure. In order to develop the most efficient structure, designers may incorporate hybrid composites to either mitigate the weaknesses in one material or take advantage #of the strengths of another. When these hybrid structures are used at service temperatures outside of the cure temperature, residual stresses can develop at the dissimilar interfaces. These residual stresses impact the initial stress state at the crack tip of any flaw in the structure and govern whether microcracks, or other defects, grow into large scale delaminations. Recent experiments have shown that for certain hybrid layups which are used to determine the strain energy release rate, G, there may be significant temperature dependence on the apparent toughness [1–4]. While Nairn and Yokozeki believe that this effect may solely be attributed to the release of stored strain energy in the specimen as the crack grows [1, 3], others point to a change in the inherent mode mixity of the test, like in the classic interface crack between two elastic layers solution given by Suo and Hutchinson [5]. When a crack is formed at the interface of two dissimilar materials, while the external loading, in the case of a double cantilever beam (DCB), is pure mode I, the stress field at the crack tip produces a mixed-mode failure. Perhaps a change in apparent toughness with temperature can be the result of an increase in mode mixity. This study serves to investigate whether the residual stress formed at the bimaterial interface produces a noticeable shift in the strain energy release rate-mode mixing curve.

The panels under investigation are carbon fiber reinforced polymer (CFRP) co-cured with glass reinforced polymer (GFRP). Each is an 8-harness satin weave prepreg with the same matrix epoxy. Each panel was designed so the bending stiffness of each CFRP and GFRP beam was the same, this resulted in a layup of $[0/90]_{5s}{}^{carbon}/[0/90]_{10s}{}^{glass}$. The size of each panel cured was 400 mm × 500 mm and a 12.7 μm thick strip of Teflon film was inserted into both 400 mm edges toserve as a precrack. The DCB, end-notched flexure (ENF), and mixed-mode bending (MMB) tests were performed across a

B. T. Werner (✉) · K. Nelson
Sandia National Laboratories, Livermore, CA, USA
e-mail: btwerne@sandia.gov

C. Nelson
Sandia National Laboratories, Albuquerque, NM, USA

P. R. Thakre et al. (eds.), *Mechanics of Composite, Hybrid and Multifunctional Materials, Volume 5*, Conference Proceedings of the Society for Experimental Mechanics Series, https://doi.org/10.1007/978-3-319-95510-0_41

wide temperature range (−54C, 21C, and 71C). All of the testing was performed well below the glass transition temperature, T_g, of 125C so specimen aging wasn't a concern. For each of the bending experiments, ENF and MMB, the specimen was tested with both the CFRP and the GFRP on the compression side of the beam. This produced significantly different results as in one case the interfacial shear stress was working with the applied mechanical stress to produce fracture while in the other case the interfacial shear stress was working against the applied mechanical stress. For both the ENF and MMB tests, the crack was manually extended slightly beyond the Teflon precrack to ensure a sharp crack tip. In these tests, the crack growth may be unstable so a blunt crack tip may overestimate the strain energy release rate. All of the testing was conducted on an electromechanical test frame inside of an environmental chamber. The specimen temperature was measured on a dummy specimen that was placed within the chamber near the tested specimen that had an embedded type J thermocouple at the interface of interest. Each test allowed a sufficient soak time (approximately 5 min) at the test temperature before testing was initiated.

The calculation of the strain energy release rate varied by the type of test. The DCB experiments relied upon beam theory equations derived by Brown [6] that used the compliance of the specimen and the load and displacement measurements to calculate the strain energy release rate and current crack length for each data point. The ENF and MMB experiments relied upon separate measurements of the energy released and the area of the crack growth to determine a single value for the strain energy release rate for each specimen. The energy dissipation was measured by taking the integral of the load-displacement curve of each experiment. The crack growth used a combination of methods to take pre- and post-test measurements. Before each test, since the crack front could be seen through the translucent GFRP, an optical method was used to measure the initial crack area. A Keyence VHX-5000 digital microscope that has image stitching and a calibrated image measurement system obtained the initial crack area. This same method was attempted for the post-test measurements but it was unsuccessful. Since the interface of interest was typically tougher that adjacent ply interfaces, the crack front had a tendency to seed a new crack and grow it in addition to the initial crack. For this reason, a Mistras ultrasonic tank system was used to perform B- and C-scans of the specimen and calculate the total crack area. Since it may be difficult to pick up multiple cracks as the signal attenuates at each crack, each specimen was scanned from both faces.

The resulting strain energy release rate-mode mixity plots showed a dependence on temperature and specimen orientation. However any inherent shift in mode mixity due to residual stress changes was not very prominent. More testing is likely needed to reduce the amount of statistical variation. This simplifies the modeling required slightly. Current attempts to validate cohesive zone models which are used to predict delaminations have failed as the opposite trend of the effect that residual stress has on crack growth has been seen in the simulations. Future work can focus on accurately capturing the release of stored strain energy with less of a focus on mode mixity (Figs. 41.1, 41.2, and 41.3).

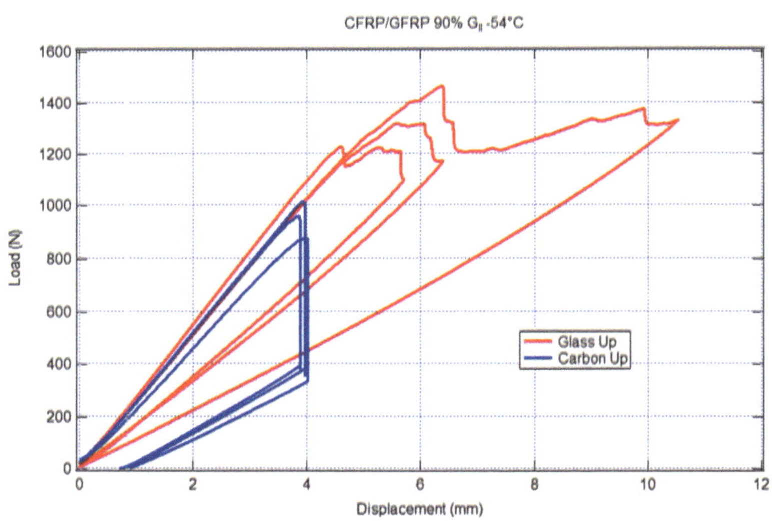

Fig. 41.1 Mixed mode bending of CFRP/GFRP composite with the CFRP and GFRP each on the compressive side of the beam

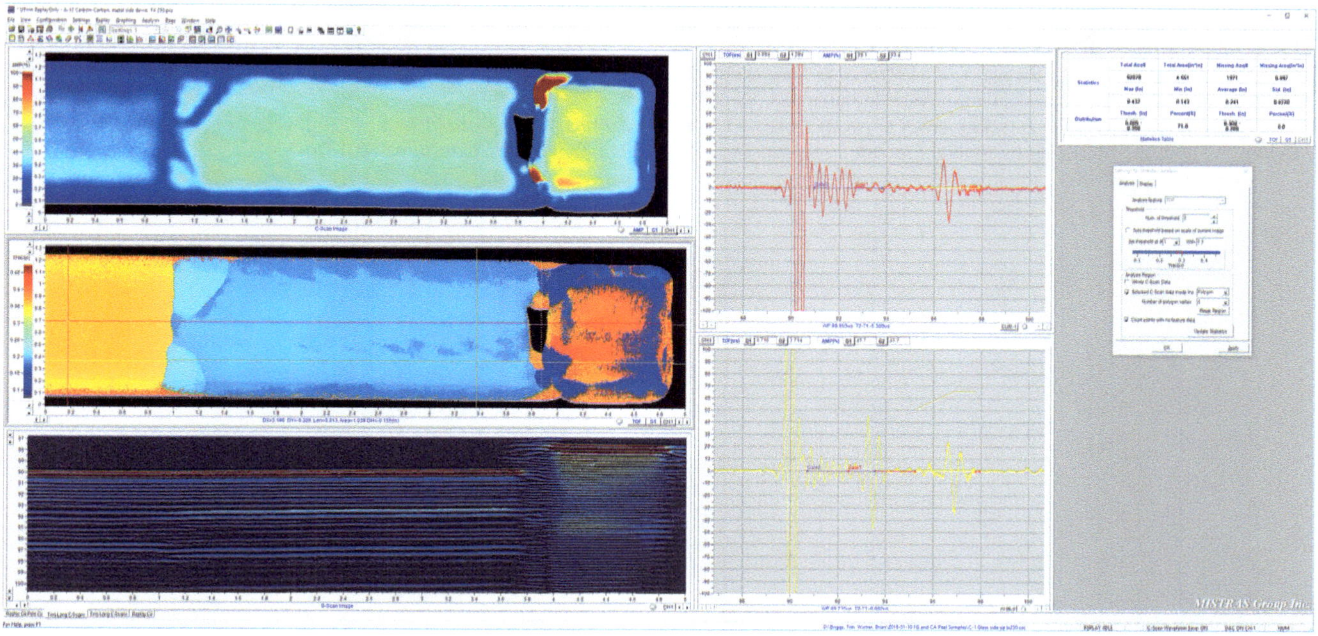

Fig. 41.2 Ultrasonic scan of tested specimen

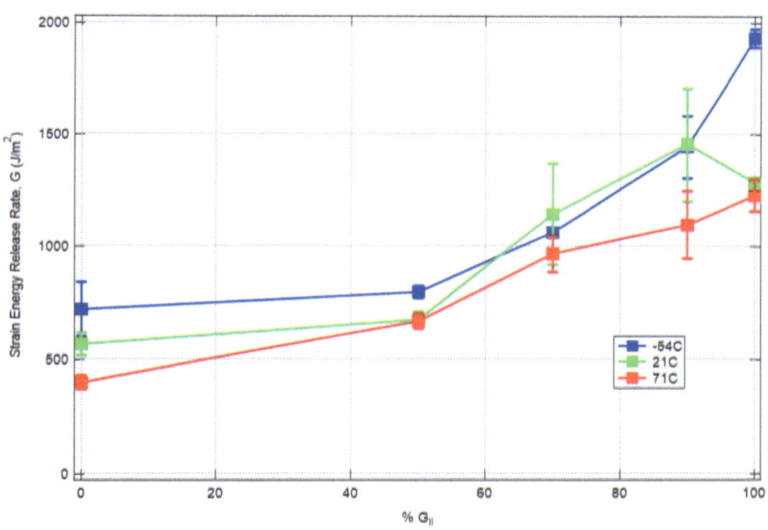

Fig. 41.3 Strain Energy Release Rate-Mode mixity plot for CFRP/GFRP hybrid composite with the GFRP on the compression side of the beam

Acknowledgement Sandia National Laboratories is a multimission laboratory managed and operated by National Technology and Engineering Solutions of Sandia LLC, a wholly owned subsidiary of Honeywell International Inc. for the U.S. Department of Energy's National Nuclear Security Administration under contract DE-NA0003525.

References

1. Yokozeki, T., Ogasawara, T., Aoki, T.: Correction method for evaluation of interfacial fracture toughness of DCB, ENF and MMB specimens with residual thermal stresses. Compos. Sci. Technol. **68**(3–4), 760–767 (2008)
2. Nelson, S.,Werner, B.: Experimental and computational investigations into the effect of process induced stresses on the mode I fracture toughness of composite materials. In: CAMX 2016. Anaheim, CA (2016)

3. Nairn, J.A.: Energy release rate analysis for adhesive and laminate double cantilever beam specimens emphasizing the effect of residual stresses. Int. J. Adhes. Adhes. **20**(1), 59–70 (2000)
4. Werner, B.T., Nelson, S.M., Briggs, T.M.: Effect of process induced stresses on measurement of FRP strain energy release rates. In: Society for Experimental Mechanics Annual Conference. Indianapolis, IN (2017)
5. Suo, Z.G., Hutchinson, J.W.: Interface crack between 2 elastic layers. Int. J. Fract. **43**(1), 1–18 (1990)
6. Brown, A.A., et al.: Modeling Interfacial Fracture in Sierra. Sandia National Laboratories, Livermore (2013)

Chapter 42
Analysis of Interfaces in AA7075/ Recycled WC Particles Composites Produced via Liquid Route

Marina Ferraz Viana and Maria Helena Robert

Abstract Casting of composites is a simple and low cost route to produce composites, which allows numerous possibilities in terms of reinforcement location in the cast part and high geometry flexibility of the products. However, interfaces between reinforcement and metal depend on the behaviour of the particles in liquid metal during processing. The work investigates the effects of parameters such as temperature and time of contact between WC particles and liquid aluminium, and reinforcement content, in the production of cast AA7075/WC composites. WC particles coming from recycled cutting tools are used as reinforcement for aluminium alloys AA 7075, widely used in aeronautical domain. Reinforcement contents of 15 and 30 wt %, temperatures of 700, 740 and 780C, and contact time of 1, 2, 3 and 4 h were the values of the investigated variables. Microstructure of cast composites produced under these conditions were analysed by OM, SEM and EDS microanalysis. A reaction layer was found in the interface between metal and WC, with composition and micro constituents depending on the studied parameters. A mechanism of formation of the reaction layer is proposed.

Keywords Composites · Aluminium · WC particles · Casting · Interface

42.1 Introduction

Casting of composites is one simple and low cost way of producing metal-matrix composites reinforced with ceramic particles. Such processes are not only flexible in terms of part geometry, reinforcement type and location in the product, but also favour adhesion between matrix and reinforcement, when compared to solid state processes. Generally, it is possible to incorporate up to 30% of particles by those methods. Depending on the nature of matrix and reinforcement materials, chemical reactions may occur while the reinforcement is in contact with the liquid metal – this is favoured by high temperatures and long contact times, and may be desirable when it leads to the formation of in situ reinforcement phases or chemical bonding between the matrix and the reinforcement [1–6].

The interface between matrix and reinforcement is a function of the nature of both materials and the behaviour of particles during processing [1, 2, 7]. In the specific case of particle reinforced composites produced by casting techniques, the ability of the liquid metal to wet the surface of the reinforcement and the chemical affinity between both materials are key in their integration [1, 7, 8]. Generally, ceramic phases whose chemical bonds have a strong metallic character, as WC and the TiC are more easily wet by the liquid metal [7, 8].

Poor adhesion between reinforcing particles and matrix hinders interface quality and is a common issue in the production of such composites. Low wettability and lack of chemical affinity are generally blamed as the root causes of such problems [1, 4, 8–10]. On the other hand, it has been observed that cemented carbides, such as the ones used for producing cutting tools, are not only exceptionally hard and wear resistant [11, 12] but also have great chemical affinity with aluminium [8, 13, 14]. Thus, those are an interesting reinforcement material for wear resistant aluminium composites.

Therefore, the aim of this work is to investigate the effects of parameters such as temperature, time of contact between liquid matrix and reinforcement and amount of reinforcement in the casting of aluminium matrix WC composites.

M. F. Viana
University of Campinas, School of Mechanical Engineering, Department of Materials and Manufacturig Engineering, Cidade Universitária, Barão Geraldo, Campinas, SP, Brazil

M. H. Robert (✉)
University of Campinas, Mechanical Engineering Faculty, Campinas, Brazil
e-mail: helena@fem.unicamp.br

P. R. Thakre et al. (eds.), *Mechanics of Composite, Hybrid and Multifunctional Materials, Volume 5*, Conference Proceedings of the Society for Experimental Mechanics Series, https://doi.org/10.1007/978-3-319-95510-0_42

42.2 Experimental Procedures

AA 7075 alloy (90.79% Al; 5.12% Zn; 2.19% Mg; 1.50% Cu; 0.22% Cr; 0.07% Si; 0.07% Fe; 0.02% Ti; 0.01% Mn) was used as matrix.

Cemented carbide granules, issued from recycled machining inserts, were used as reinforcement material. The granules, shown in Fig. 42.1a, are irregular and polygonal, and their dimensions range from 300 to 800 μm. Some of their surfaces have distinct colorations, which might be attributed to vestiges of coating materials. Fig. 42.1b shows the surface of one granule with greater magnification. It is possible to see the granules are comprised of WC microparticles whose dimensions range from 0.5 to 3 μm. The chemical composition of the granules is presented in Table 42.1.

The microstructure of one representative granule is shown in Fig. 42.2. As recycled material was used, there may be some variation between granules. Figure 42.2a shows three points where EDS micro analysis were effectuated; the numerical results of the analyses are presented in Table 42.2.

The microanalyses of points 2 and 3 show a high Co content in regions between the WC particles. Co is commonly used as a binder in the sintering of cemented carbides [11, 12]. Fe was also detected in those regions; its presence is probably due to contamination during the machining of ferrous materials. Machining inserts may contain different types of carbides in their composition, such as WC, TiC, TaC, NbC and CrC [11, 12]. Besides, they might be coated with TiCN, TiN, Al_2O_3, etc [11–13]. Indeed, there is a high content of Ti and Ta at Point 1. The high percentages of WC detected in all points refer to the neighbouring regions of the analyses points.

To produce the composites, the AA 7075 alloy was first melted in an electric furnace, in a ceramic crucible, and heated to a predetermined temperature, hereinafter called contact temperature. When the melt reached the contact temperature, it was removed from the oven, its surface was cleaned and a predetermined quantity of granules was added to it under manual agitation. Then the melt was returned to the oven at the contact temperature and kept there for 15 min., after which it was taken out again and agitated for 40 s. This operation was repeated until the contact time was completed; next, the melt was cooled down in air, resulting in a conic part.

Two series of composites were produced. The first one was fabricated as described above, as for the second one, a last was added to the procedure, which consisted of remelting the conic part in a cylindric steel die coated with alumina, at 740 °C for 45 min. The experimental conditions which were investigated are shown in Table 42.3.

All composites were sectioned, and prepared for metallographic analysis using polycrystalline 90, 45, 25, 9, 6, 3, 1, and 0.25 μm diamond abrasives and water as lubricant; optical microscopy, SEM and EDS microanalysis were used to analyse microstructures. Quantitative analysis of the area occupied by the carbide granules in the composites was performed in the second series, by analyzing OM images taken with a 50× magnification through free image analysis software.

Fig. 42.1 (**a**) General aspect of the carbide granules used as reinforcement; (**b**) magnified surface of one granule (SEM)

Table 42.1 Elemental composition of the cemented carbide granules used in the experiments (% wt)

W	Co	Ni	Ti	Ta	Fe	Zn	V	Cr	Ca	Cu	Al	Mn	P	Mo	Nb	Ag	Si
65.30	5.72	0.28	1.51	0.63	10.27	8.28	4.74	0.62	0.50	0.33	0.32	0.32	0.52	0.21	0.16	0.14	0.09

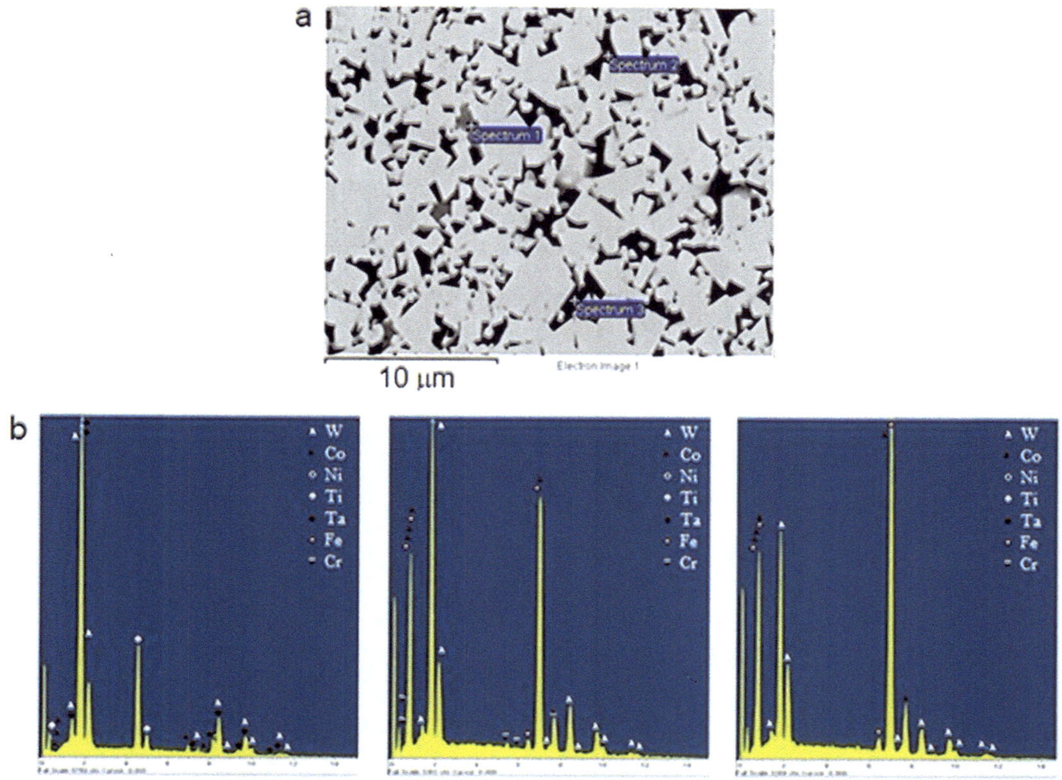

Fig. 42.2 (**a**) Microstructure of a representative granule with indication of EDS micro analysis points; (**b**) results from EDS microanalyses of points 1, 2 and 3, respectively (SEM)

Table 42.2 Average compositions of the regions indicated by Spectra 1, 2 and 3 in Fig. 42.2 obtained by EDS analyses (wt. %)

	W	Co	Ni	Ti	Ta	Fe
Spectrum 1	60.42	1.89	–	20.13	17.56	–
Spectrum 2	31.82	66.20	–	–	–	1.97
Spectrum 3	45.40	52.50	–	–	–	1.57

Table 42.3 Experimental conditions investigated in the production of AA 7075 reinforced with cemented carbide granules

	Reinforcement fraction (wt. %)	Contact time (h)	Contact temperature (°C)
First series	30	1	700
	30	2	700
	30	4	780
Second series	30	1	740
	30	3	740
	30	3	780
	15	1	740
	15	3	740
	15	3	780

42.3 Results and Discussions

42.3.1 Microstructure of the Composites.

Typical microstructures of composites produced in the first series of experiments, with the operational conditions described in Table 42.3, are shown in Fig. 42.3. These first composites were produced as a preliminary study of the behaviour of the WC particles in liquid aluminium, in order to allow of an informed choice on the time and temperatures of contact to be

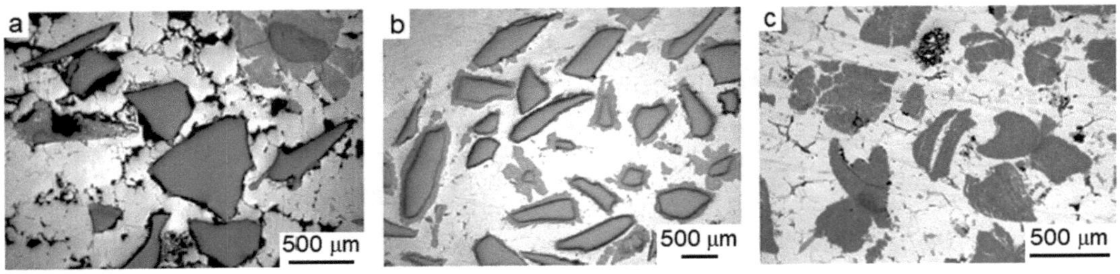

Fig. 42.3 Typical microstructures of composites produced in the first series of experiments, with contact times and temperatures of, respectively: (**a**) 1 h/ 700 °C; (**b**) 2 h/ 700 °C; and (**c**) 4 h/ 780 °C (OM)

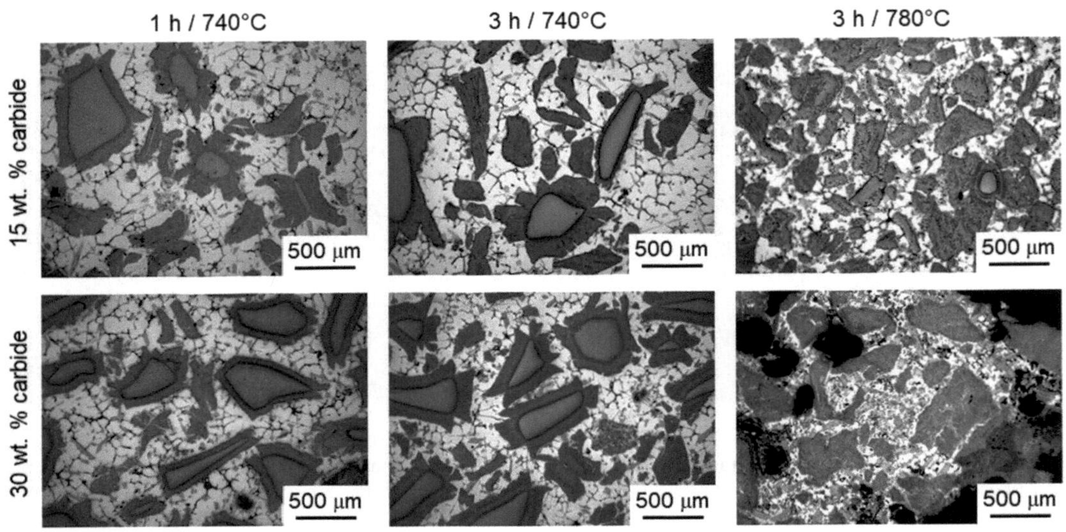

Fig. 42.4 Typical microstructures of the composites produced in the second series of experiments. The contact time (h), contact temperature (°C) and reinforcement fraction (% wt.) are indicated under each image (OM)

studied. Therefore, extreme conditions were chosen: temperatures of 700 and 780 °C, times between 1 and 4 h of contact, and 30%wt. reinforcement. At this scale, an even distribution of the reinforcement particles is observed. For the lowest temperature and contact time, voids are associated to the matrix/metal interface rather than any transition layers, as shown in Fig. 42.3a. When the contact time is increased in 1 h, the formation of a transition layer rather than voids are observed at the interfaces, as seen in Fig. 42.3b. Finally, for the maximum contact time and temperature tested (4 h and 780 °C), the interfaces between reinforcement and matrix become diffuse as the granules change morphology. Therefore, the conditions chosen for the second series of experiments where considered both intermediary in comparison to those tested in the first series, and propitious to the formation of a transition layer between reinforcement and matrix.

Typical microstructures of composites produced in the second series of experiments, with the operational conditions described in Table 42.3, are shown in Fig. 42.4. At this scale, an even distribution of the reinforcement particles and sound matrix/reinforcement interfaces were observed.

For the composites produced with the contact temperature of 740 °C, no significant difference with respect to the aspect of the carbide granules or the spacing between them was observed. In all composites, the presence of a reaction layer around the granules was observed. Some granules exhibited the same aspect of the reaction layer all over their surface, which suggests that the formation of the reaction layer might have consumed the granules. This last case is the usual situation in the composites produced with the contact time of 780 °C.

The estimated percent area fractions occupied by non-reacted granules, reinforcement layer and matrix are presented in Table 42.4. They were estimated by analysing OM images with the software Image J. These results give a better idea of how the reinforcement fraction and the contact time affect the formation of the reaction layer. For both reinforcement fractions, increasing the contact time from 1 to 3 h decreased the percent area occupied by the granules and increased the area fraction occupied by the reinforcement layer in such a way that the percent area occupied by the matrix remains constant. Thus, the formation of the reaction layer is favoured by the contact time. As the reinforcement content was increased from 15 to 30%

Table 42.4 Area fractions (%) occupied by reinforcement granules, reaction layer and matrix in composites produced with the contact temperature of 740 °C

% wt. reinforcement	Contact time (h)	Granules	Reaction layer	Granules + Reaction layer	Matrix
15	1	7	31	38	62
15	3	5	34	39	57
30	1	14	29	43	57
30	3	9	34	43	57

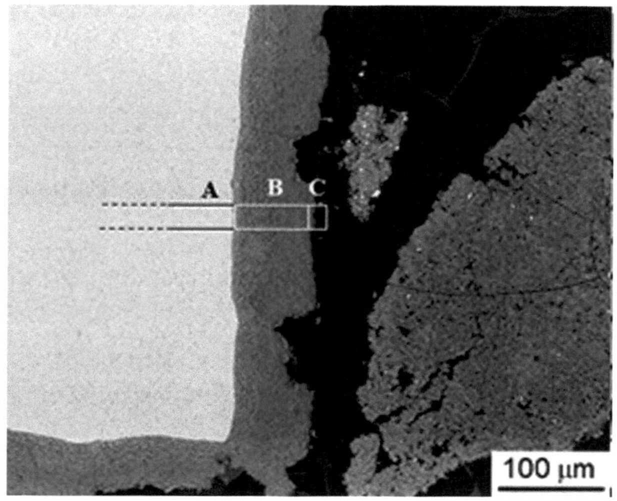

Fig. 42.5 Microstructure of composite produced with 1 h contact time and 740 °C contact temperature – regions A, B and C around the reaction layer are indicated (SEM)

wt., the percent area occupied by the non-reacted granules increased, but the area fraction occupied by the reaction layers remained virtually the same – that might indicate that the formation of the reaction layer is not strongly dependent on the weight fraction of reinforcement.

42.3.2 *SEM and EDS Microanalyses of the Composites*

Figure 42.5 shows a representative image of the microstructure observed in the composites fabricated with 1 h contact times and 740 °C contact temperatures. To facilitate the discussion, it was decided to label the three regions around reaction layer as follows: region A is the apparently unreacted carbide granule; region B is the reaction layer around the granules; and region C is the matrix close to the reaction layer.

Region A was analysed by SEM and EDS. The results were very similar to the ones presented in Fig. 42.2 and Table 42.2, and were thus omitted in this paper. However, we must say the portion of the original granules comprised by region A was little affected by the contact with the liquid matrix.

Region B might be divided in three sub regions, B1, B2 and B3, as shown in Figs. 42.6 and 42.7. Those differ both morphological and compositionally. Fig. 42.6 shows how the concentrations of the main elements in the alloy (Al, Zn, Mg and Cu) and in the granules (W and Co) vary across the thickness of the reaction layer. The Al and W concentrations vary especially near the borders of the reaction layer. The Al concentration is at a minimum level in region A, reaches an intermediate, constant, level all across region B and goes to a maximum in the matrix. The W concentration behaves in an opposite manner to the Al concentration. It is at a maximum in region B, at an intermediate level across region B and reaches a minimum in the matrix. As for the Co concentration, it reaches its highest levels in region A and decreases with the distance from the granule along regions B1 and B2. These observations suggest some mass movement from the granules to the matrix. Also, that the reaction layer might have intermediary Al and W concentrations when compared to the granules and the matrix. As for the other alloy elements, Zn and Mg are observed around the borders of the reaction layer, close to the matrix. The Cu concentration does not seem to vary much across the reaction layer.

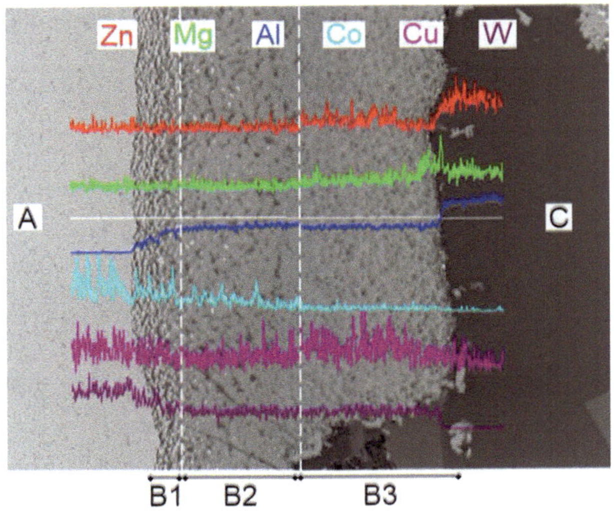

Fig. 42.6 Qualitative variation of the Zn, Mg, Al, Co, Cu and W concentrations across region B – distinct sub regions, B1, B2 and B3, are indicated, as well as regions and C (EDS – 112 μm line. EDS)

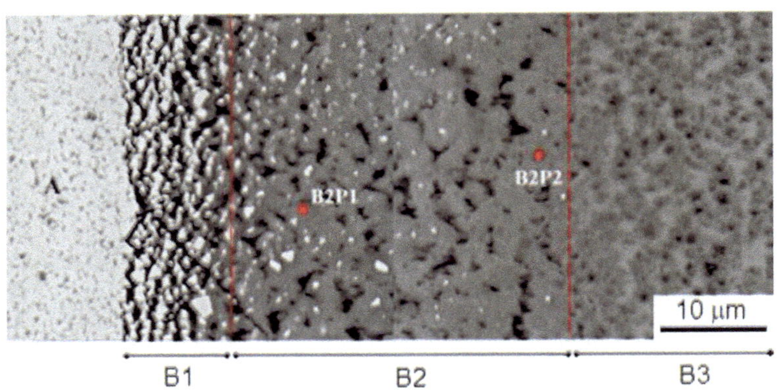

Fig. 42.7 (a) Microstructure of reaction layer with indication of EDS micro analysis points (numerical results from EDS microanalyses of points B2P1 and B2P2 are presented in Table 42.5 (SEM)

Table 42.5 Average compositions of the regions indicated by Points B2P1 and B2P2 in Fig. 42.7 obtained by EDS analyses (wt. %)

Point	Element (wt. %)				
	W	Co	Al	Zn	Mg
B2P1	53.42	2.38	43.57	0.42	0.22
B2P2	70.16	0.63	28.17	0.75	0.29

The sub regions B1, B2 and B3 are shown in more detail in Fig. 42.7. In the sub region B1, individual WC particles are observed – those particles seem to have turned themselves loose from the granule. Close to the granule, the WC particles seem to have the same dimensions as the particles inside it, but appear surrounded by the matrix. In sub region B2, both the quantity and the dimensions of the WC particles decrease as the distance from the granule grows. Also, a new, grey coloured phase appears and predominates as the distance from the granule increases. In region B3, more distant from the granule, no more WC particles are observed and the new phase seems thicker.

Results of the microanalyses points indicated in Fig. 42.7 are shown in Table 42.5. The analysis at point B2P1 allows evaluating the composition of the new phase that appears in B2; it is mainly composed of Al and W, contains a significant amount of Co and minor amounts of Zn and Mg. This result suggests that this new phase might have been formed by a reaction between the Al matrix and the WC particles; this reaction, apparently, consumed the WC particles. This new phase, for simplicity's sake called AlW phase from this point, seems to be partially continuous in region B2. At point B2P2, the EDS analysis indicated a high concentration of W at point, probably because there is a fragment of a WC particle at that point (the high concentration of Al and the presence of alloy elements Mg and Zn might be explained by the reduced size of the analysis region).

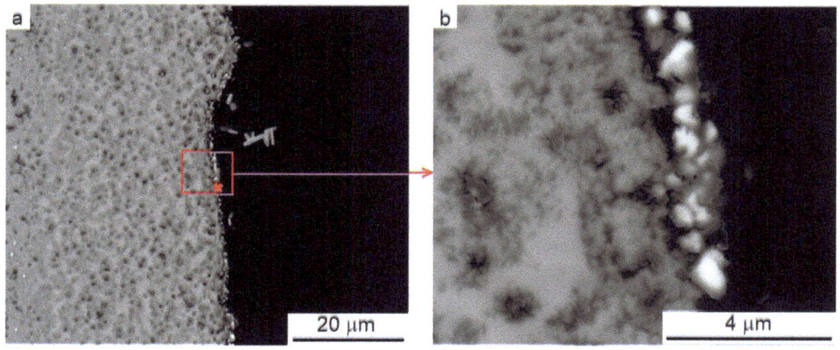

Fig. 42.8 (**a**) Microstructure of sub region B3 in the reaction layer with indication of a point where EDS micro analysis was performed. The analysis showed a composition of (in wt. %) 71.90% W; 028% Co; 25.22% Al; 1.98% Zn; 0.62% Mg. (**b**) Detail of the region indicated by a rectangle in **a** (SEM)

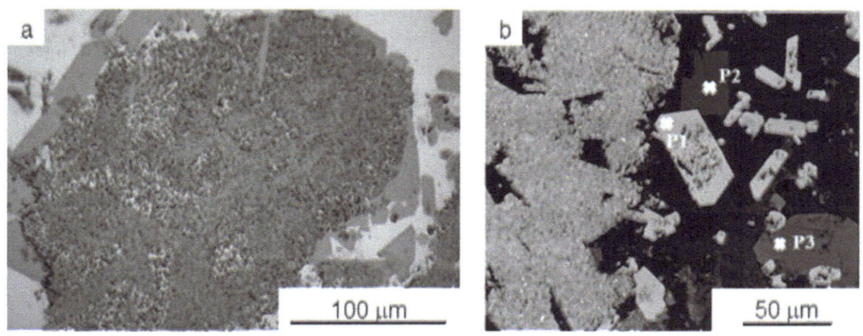

Fig. 42.9 (**a**) Typical aspect of the reinforcement in composites produced with 3 h contact time and 780 °C contact temperature (OM) (**b**) Region between the reinforcement and the matrix; EDS microanalyses where performed at points P1, P2 and P3, their numerical results are presented in Table 42.6 (SEM)

The microstructure of sub region B3 of the reaction layer, which is more distant from the granule, is presented with more detail in Fig. 42.8. In Fig. 42.8a, individual WC particles are observed just next to the matrix. The composition of that region, obtained by EDS microanalysis, is (in wt. %): 71.90% W; 028% Co; 25.22% Al; 1.98% Zn; 0.62% Mg. This confirms the particles are very similar to the particles that have turned themselves loose from the granule in sub region B1, as seem in Fig. 42.7, with the difference that the particles in Fig. 42.8 seem smaller and their contours seem less sharp. Their presence in the front of the reaction layer might be explained by the fact that they were not entirely consumed by reactions with the liquid Al, either because they were too large, or because of the presence of coating materials on their surfaces, which significantly reduce wettability [13].

Figure 42.9 presents typical aspects of the reinforcement phase in composites produced with 3 h contact time and 780 °C contact temperature. Most of the cemented carbide granules were consumed in reactions with the liquid metal. Instead of the compact granules surrounded by a reaction layer, discrete phases, composed of polygonal particles are seen, as in Fig. 42.9b. Those phases are present in the matrix as well. Generally, the increase in the contact temperature from 740 to 780 °C not only favoured more extensive reactions between the reinforcement granules and the matrix but also promoted the formation of other complex phases that were not observed in the composites processed at 740 °C.

In Fig. 42.9b, points where EDS microanalysis was performed in different phases are indicated. The numerical results of the analyses are shown in Table 42.6. Point P1 indicates a phase whose composition, as shown in Tables 42.5 and 42.6, is very similar to the composition of the AlW phase observed in the composites processed at 740 °C. Differently from the other phases indicated by points P2 and P3, that phase is ubiquitous all over the composite, its presence is much more extensive, and it has the same aspect as the previously observed AlW phase – therefore, it was concluded that it indeed that phase. However, it was able to develop more extensively, with a polygonal morphology, at higher temperature. The phase indicated by P2 presents high concentrations of Co and Fe in Al, while the phase indicated by point P3 presents a complex composition, where Al predominates but there are also high contents of W, Cr and some Fe and Ti. The Fe, Ti, and Cr might be from the cemented carbides, which contained those elements, as shown in Tables 42.1 and 42.2.

Table 42.6 Average compositions of the regions indicated by Points P1, P2, and P3 in Fig. 42.9 obtained by EDS analyses (wt. %)

Point	Element (wt. %)						
	W	Co	Ti	Fe	Al	Zn	Cr
P1	56.98	–	–	–	43.02	–	–
P2	–	20.35	–	11.63	68.01	–	–
P3	14.32	0.99	1.28	4.77	66.01	2.34	10.30

42.3.3 Phenomena Involved in the Formation of the Reaction Layer.

A mechanism of formation for the reaction layer is proposed based on the microstructures of the composites produced with contact temperature of 740 °C, and the observation of regions B1, B2 and B3 of the reaction layer. The sub region B1, which is adjacent to the granule, is characterized by the presence of disaggregated WC particles, which loosened themselves from the cemented carbide granule. They are morphologically similar to the particles found within the cemented carbide granule. In the transition between regions B1 and B2, the size of the carbide particles is progressively reduced as a new, platelike phase, which was called AlW, appears. In sub region B2, the carbide particles eventually disappear. That can be seen in Fig. 42.7. Finally, in the sub region B3, the AlW phase forms thicker agglomerates. In the interface between region B3 and the matrix, some remaining carbide particles that loosened themselves from the granules, but were either not entirely consumed or pushed away from the granule as the reaction layer was formed, are found; as shown in Fig. 42.8.

The disaggregation of the carbide particles from the granules might be explained by the high affinity between Al and Co [13, 14]. Al and Co form various intermetallic compounds even at low temperatures [15]. A solid solution of Al in Co, with Al concentrations around 1.5 wt. % Al, undergoes a eutectoid reaction around 350 °C, which is described in Eq. 42.1. The eutectoid product AlCo has a lower density (6.08 g/cc) than any of the allotropic forms of Co (8.84 and 8.79 g/cc for ε and α Co, respectively), and thus the transformation is accompanied by a volumetric expansion [14, 15]. Co is present as binder between the particles in machining inserts [11–14]. The Al would diffuse to the Co rich areas in the machining insert granules and form the phase AlCo, which is fragile and has a smaller density than any of the allotropic forms of the Co (and thus a greater specific volume) [14, 15]. The formation of that phase would cause local stresses to arise. As the AlCo is fragile, it would crack, separating the WC particles from the rest of the granule, as schematized in Fig. 42.10.

$$\alpha Co \rightleftarrows AlCo + \varepsilon Co \tag{42.1}$$

Still, AlCo might be formed even at very low Al concentrations in Co [15]. The authors studied failure of machining inserts during the cutting of Al. Their observations showed that when both Al and cemented carbides are heated together, the Al can diffuse up to 80 μm under the surface of the insert and form the AlCo phase. That renders the surface of the insert fragile and causes its failure – porosity and disaggregation of the cemented carbide are observed, which results in the loss of some particles. The microanalyses performed in this investigation did not specifically detect AlCo in sub region B1, but both the presence of Al and Co was detected by the EDS microanalyses in that region (see Fig. 42.6). Moreover, an AlCoFe rich phase was detected in regions near the reaction layer in the composites produced with greater contact temperature and time, as shown in Table 42.6.

Once they are loosened from the granule, the carbide particles are more closely surrounded by the liquid Al and reactions between the elements Al and W may form the AlW phase observed in sub region B2 (as shown in Fig. 42.7). For low W concentrations, two peritectic reactions, expressed by Eqs. 42.2 and 42.3, happen at 697 and 661 °C, respectively. EDS microanalyses in specific regions where the AlW phase is observed indicate W concentrations higher than 50% wt. and Al concentrations higher than 40% wt.. A comparison between this data and the approximate composition of the phases in Eqs. 42.2 and 42.3 indicated that, most probably, the phase observed in region B2 is Al_5W.

$$Al_{(l)} + Al_5W_{(s)} \rightarrow Al_{12}W_{(s)} \tag{42.2}$$

$$Al_{(s)} + Al_{12}W_{(s)} \rightarrow Al_{(s)} \tag{42.3}$$

Other authors have also observed reactions between W and liquid Al as WC particles were added to Al melts [8]. added WC particles to a Al melt at 830 °C and observed the formation of both $Al_{12}W$ and Al_5W. $Al_{12}W$ was formed with less frequency and appeared as polygonal plates. Al_5W was formed with more frequency and appeared as elongated and/or agglomerated plates. This behaviour is similar to the observed in Fig. 42.9 for the composite produced at higher temperatures. Also, the phase formed at lower temperatures exhibited, too, a strong tendency to agglomerate.

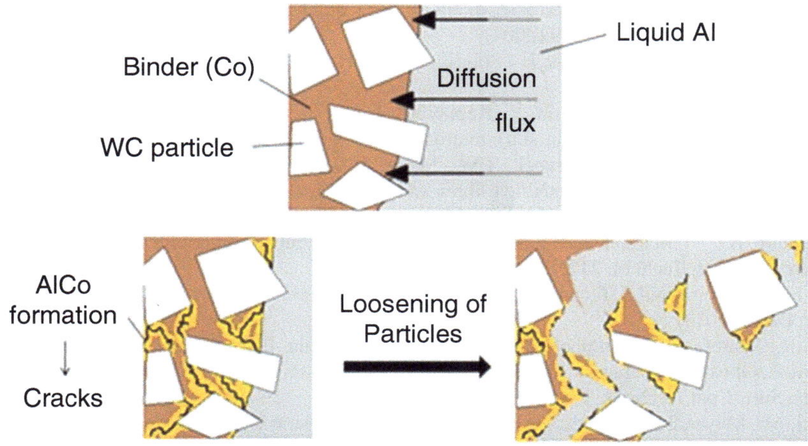

Fig. 42.10 Schema of the mechanism by which the WC particles would be separated from the granule

The formation of the reaction layer between the matrix and the granules might be desirable, as it proportionates a metallurgical bond between the matrix and the reinforcement and, possibly, a more gradual transition between the two materials. Other interesting result was that little difference was observed in the reaction layer of the composites fabricated with contact temperature of 740 °C – for that temperature, similar results are obtained after 1 and 3 h contact times. Therefore, if a reaction layer is desirable, it can be obtained with the blandest mixing conditions, of 1 h/ 740 °C.

42.4 Conclusions

This work investigated the possibility of producing, by casting, Al matrix composites reinforced with cemented carbide granules in such a way to promote metallurgical interaction between the matrix and the reinforcement.

The obtained results indicate that the contact between the liquid metal and the cemented carbide granules propitiates the formation of a reaction layer between granules and matrix at contact temperatures above 700 °C (for the time intervals studied). The formation of that reaction layer consumes the granules, at the same time it forms a metallurgical bond between matrix and reinforcement. For the contact temperature of 740 °C, the contact time (within the 1 and 3 h conditions studied) had little influence in the extension of the reaction layer formed. For the contact temperature of 780 °C, the formation of compact and/or complex phases in the material was observed. Moreover, the cemented carbide granules were mostly consumed by the formation of other phases at that temperature.

Acknowledgements The authors want to thank financial support from the Brazilian Government Agencies: CAPES and CNPq National Council for Scientific and Technological Development.

References

1. Kainer, K.U.: Basics of metal matrix composites. In: Metal Matrix Composites: Custom-Made Materials for Automotive and Aerospace Engineering, pp. 1–54. Wiley-VGH Verlag Gmbh & Co, Weinheim (2006. Cap.1)
2. Hunt Jr., W.H.: Metal matrix composites: Applications. In: Jürgen, K.H.B., Cahn, R.W., Flemings, M.C., Ilschner, B., Kramer, E.J., Mahajan, S., Veyssière, P. (eds.) Encyclopaedia of Materials: Science and Technology, 2nd edn, pp. 5442–5446. Elsevier, Oxford (2001)
3. Singh, S., Singh, I., Dvivedi, A., Davim, J.P.: SiCp-Reinforced Al6063 MMCs: Mechanical behavior and microstructural analysis. In: International Conference on Research and Innovations in Mechanical Engineering, 2014, Ludhiana/India. Proceedings… ed. Springer India, vol. 1, pp. 451–464 (2014)
4. Mazahery, A., Abdizadeh, H., Baharvandi, H.R.: Development of high-performance A356/nano-Al_2O_3 composites. Mater. Sci. Eng. A. **518**, 61–64 (2009)
5. Sukumaran, K., Ravikumar, K.K., Pillai, S.G.K., Rajan, T.P.D., Ravi, M., Pillai, R.M., Pai, B.C.: Studies on squeeze casting of Al2124 alloy and 2124 – 10% SiCp metal matrix composite. Mater. Sci. Eng. A. **490**(1–2), 235–241 (2008)

6. Naher, S., Brabazon, D., Looney, L.: Development and assessment of a new quick quench stir caster design for the production of metal matrix composites. J. Mater. Process. Technol. **166**, 430–439 (2004)
7. Karantzalis, A.E., Lekatou, A., Georgatis, E., Mavros, H.: Solidification behaviour of ceramic particle reinforced Al-alloy matrices. J. Mater. Sci. **45**(8), 2165–2173 (2010)
8. Lekatou, A., Karantzalis, A.E., Evangelou, A., Gousia, V., Kaptay, G., Gácsi, Z., Baumli, P., Simon, A.: Aluminium reinforced by WC and TiC nanoparticles (ex-situ) and aluminide particles (in-situ): microstructure, wear and corrosion behavior. Mater. Des. **65**, 1121–1135 (2015)
9. Wolbach, W.S., Bryan, S.R., Shoemaker, G.L., Krucek, T.W., Maier, R.D., Soni, K.K., Chabala, J.M., Mogilevsky, R., Levi-Setti, R.: Optimization of chemical reactions between alumina/silica fibres and aluminium-magnesium alloys during composite processing. J. Mater. Sci. **32**, 1953–1961 (1997)
10. Amirkhanlou, S., Niroumand, B.: Fabrication and characterization of Al356/SiCp semissolid composites by injecting SiCp containing composite powders. J. Mater. Process. Technol. **212**, 841–847 (2012)
11. Diniz, A.E., Marcondes, F.C., Coppini, N.L.: Tecnologia da usinagem dos materiais. In: Org (ed.) Materiais para ferramentas, pp. 77–104. Artliber Editora Ltda, São Paulo (2013)
12. Kalpakjian, S., Schmid, S.R.: Manufacturing Engineering and Technology, 6th edn. Pearson. Australia Group Pty Ltd, Melbourne (2009)
13. Chattopadhyay, A.K., Roy, P., Ghosh, A., Sarangi, S.K.: Wettability and machinability study of pure aluminium towards uncoated and coated carbide cutting tool inserts. Surf. Coat. Technol. **203**, 941–951 (2009)
14. Calatoru, V.D., Balazinski, M., Mayer, J.R.R., Paris, H., L'Espérance, G.: Diffusion wear mechanism during high-speed machining of 7475-T7351 aluminum alloy with carbide end mills. Wear. **265**(11–12), 1793–1800 (2008)
15. McAlister, A.J.: The Al-Co (Aluminium-Cobalt) system. Bull. Alloy Phase Diagr. **10**(6), 646–650 (1989)

Chapter 43
Investigation on Microstructure and Interfaces in Graded FE50007 / WC Composites Produced by Casting

Rodolfo Leibholz, Henrique Leibholz, Emin Bayraktar, and Maria Helena Robert

Abstract Graded composites of grade FE50007 nodular ductile iron reinforced with WC granules grinded from recycled cutting tools, were produced by casting, using lost foam technique. Reinforcement granules were forced to be located in a specific region of the part by controlling flow during pouring. Different casting conditions were investigated (reinforcing granules content and dimensions). Resulting microstructures were extensively analysed to investigate eventual phases modifications in the iron matrix among reinforcing particles and interface interactions or possible reactions. Results show that WC granules deteriorates in contact with liquid metal, by diffusion of binder elements, detaching WC individual particles, resulting in a smooth and continuous interface; any kind of reaction is observed in the interface. The presence of the ceramic particles modifies the alloy microstructure within the reinforced region due to thermal and chemical local features. Preliminary studies on wear behaviour for a specific cast product developed (an impact crusher) are also presented, indicated superior performance compared to commercial crushers.

Keywords Graded composites · Impact Crushers · Cast iron Hammers · Recycled WC · Fe-WC Composites

43.1 Introduction

Heavy impact hammers are extensively used in rotating devices in different sorts of industries to promote crushing of raw materials such as limestone, dolomite and others in sinter plants; ores in the mining industry; to shred sugar cane stalks in the sugar production plant, among others [1, 2]. During operation, such hammers are submitted to shock loads and severe wear in the region in contact with the hard material to be triturated - the beater head surface. As presented elsewhere [3] and also observed in previous work [4], the pattern of wear behaviour profile suffered during work in the striking surface of this product is not linear, presenting a increasing depth towards one of the edges, due to the uneven contact with hard materials.

Commercial hammers for the sugar industry are usually produced by forming of thick cast iron or steel plates (usually ordinary C-steels or harder steels like AISI A2 and others tool steels); the particular region submitted to severe wear - the beater head - is prepared by applying harder materials by means of welding layers. Typical work life of a impact hammer head can vary from 15 to 30 days, under continuous operation. Increasing the performance of these hammers by improving the ability of supporting the severe work conditions is an important issue; different approaches have been considered to produce surfaces with higher hardness and wearing behaviour, such as structure modifications through laser of plasma techniques, new heat treatments conditions for high alloyed steels in the striking surface of the head of the hammer to control hard micro precipitates, deposition of hard materials in pre-prepared surfaces by different techniques, etc [5–7]. Another approach is the addition of ceramic reinforcements in a layer in the surface of the hammer head, producing a composite material in the region. The thickness of such reinforced layer can be pre-determined according to the abrasion pattern resulted during work operation; even more, it can be tailored according to the required properties with a gradient content of the reinforcing element, in this case, a FGCM - functionally graded composite material can constitute the hammer head.

R. Leibholz · H. Leibholz
FEMAQ- Foundry Shop, Piracicaba, Brazil
e-mail: rodolfo_leibholz@femaq.com.br; henriqueleibholz@femaq.com.br

E. Bayraktar
Supmeca-Paris, Mechanical and Engineering School, Saint-Ouen, France
e-mail: emin.bayraktar@supmeca.fr

M. H. Robert (✉)
University of Campinas, Mechanical Engineering Faculty, Campinas, Brazil
e-mail: helena@fem.unicamp.br

P. R. Thakre et al. (eds.), *Mechanics of Composite, Hybrid and Multifunctional Materials, Volume 5*, Conference Proceedings of the Society for Experimental Mechanics Series, https://doi.org/10.1007/978-3-319-95510-0_43

Previous work [4] showed optimized conditions to produce impact hammers for shredding of sugar cane stalks in the sugar industry, by casting of FE50007 nodular iron containing a layer of recycled granules of WC from cutting tools as reinforcing material in the striking surface of the hammer head. This work analyses the effects of the presence of the reinforcing WC granules in the microstructure of the metal in the composite region and the interfaces between metal matrix / ceramic reinforcement. Additionally, preliminary studies on the wear behaviour of the of the reinforced layer are presented.

43.2 Experimental Procedures

The general geometry of the impact hammer produced in the experiments is presented in Fig. 43.1; total weight of cast iron product is around 33 kg. The hammer head with the striking surface submitted to severe abrasion conditions is highlighted. A layer of reinforced material was designed to be located in this region. Casting technique used lost foam pattern built in EPS (expanded poly-styrene); the metal was poured at 1480 °C.

The nodular cast iron employed was FE50007 (DIN 1693 – Deutsches Institut fur Normung), with ferrite / perlite matrix (70–80% of ferrite); C (3.2 wt%), Si (2.7 wt%) and Mn (0.46 wt%) are the main alloying elements; some residual Mg (0.035 wt%) may be present due to the graphite morphology modification treatment.

Recycled WC-based cutting tools from machining shops were grinded to dimensions ranging from 1 to 3 mm. General aspect and typical microstructure of WC granules are presented in Fig. 43.2. Chemical composition may change, as recycled material with different constitutions were employed. Tools are produced by sintering small particles of hard carbides, in this case predominantly WC, usually using Co or Co + Ni as binders. Most common commercial cutting tools contains binders content ranging from 6 to 16 wt%, depending on the properties required. Other elements can be present, such as Ti, Ta, Nb and Cr, which can be related to other carbides like TiC, TaC, NbC and CrC eventually participating in the composition of the tool, or to coatings like TiCN, TiN, Al_2O_3 [8].

Microstructure in Fig. 43.2 shows localized regions of high Co content used as binder element (3–12 wt% related to WC content is usual) of the individual WC particles, which present dimensions in the range 1–4 μm. EDS microanalysis

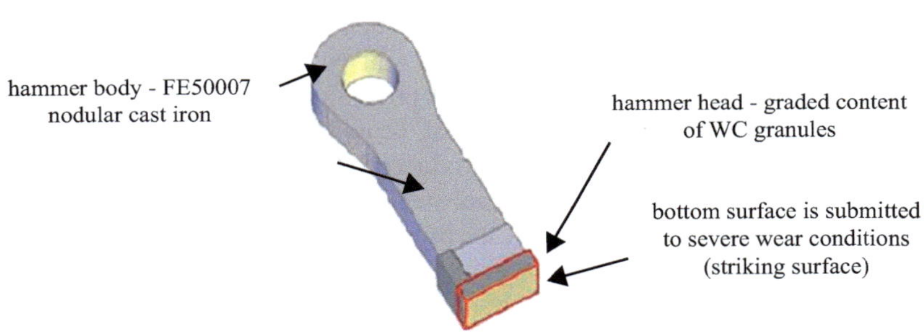

Fig. 43.1 Schematic representation of the impact crusher produced by casting, with a FGCM region in the hammer head

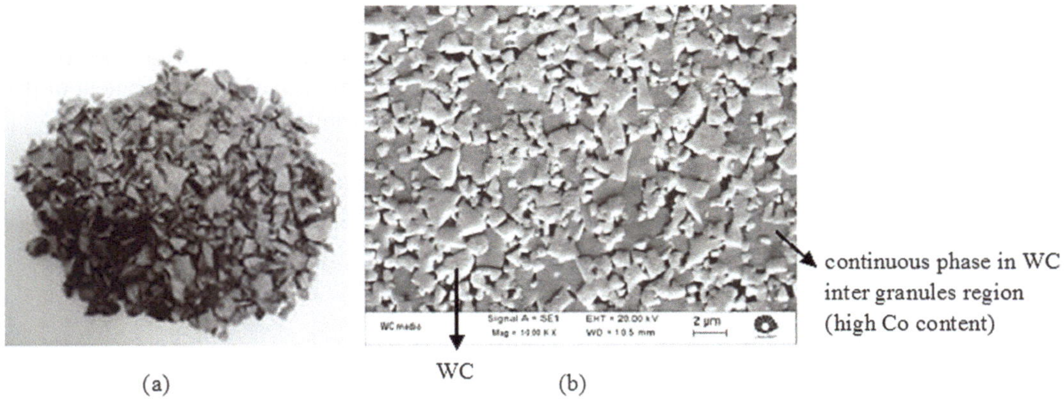

Fig. 43.2 (**a**) General aspect of WC granules grinded from recycled cutting tools; (**b**) typical microstructure (SEM/SE)

in different points of the continuous phase observed among WC particles indicated values up to 66 wt% of Co (regions as indicated in the microstructure as "rich Co content").

Cast hammers produced were sectioned and prepared by conventional metallographic procedures for microstructures analysis by means of OM, SEM, EDS and XRD. Additionally, the products were evaluated concerning wear behavior: micro wear by linear reciprocating tests were performed, as well as macro wear tests using rubber wheel method and finally, the general performance during work was observed. In the first case, tests conditions used a zirconia ball (2 mm diameter), 150,000 cycles (165 min) under 1.9 N and 15HZ. In the second case, tests were performed according to ASTM G65. In the third case, a hammer was assembled in a crusher rotor among commercial hammers, in a sugar cane plant, and submitted to work under normal conditions. Its general aspect was evaluated after crushing 234,000 ton of sugar cane stalks.

43.3 Results and Discussions

43.3.1 General Results

Typical cast hammers produced are presented in Fig. 43.3. General aspect of three hammers, with the head containing WC granules highlighted in the bottom part, is shown in (a); no evident casting defects are detected. Longitudinal section of the head, seen in (b1) shows a layer of WC-reinforced material varying from 8 to 15 mm of thickness, with a non-linear distribution profile, which increases towards one of the edges. This configuration follows the usual pattern of wear of the hammer during work. A homogeneous distribution of WC granules in the metallic matrix in the striking surface is observed in (c); in this surface, the relation of area occupied by the WC / metallic matrix is 87 / 13%.

43.3.2 Microstructure in the Hammer Body

Microstructures in different regions of the crusher produced were analysed. Figure 43.4 shows typical structure in the hammer body, circa 50 mm upper the limit of the reinforced layer in the hammer head. A ferrite / perlite structure with nodular graphite is observed throughout. Perlite / ferrite ratio is 70/40%. Graphite is classified as 75% of nodularization, and total area occupied by graphite is 12%. Such microstructure is typical for this grade of cast iron. Therefore, this region was not affected by the addition of WC particles to the material during casting.

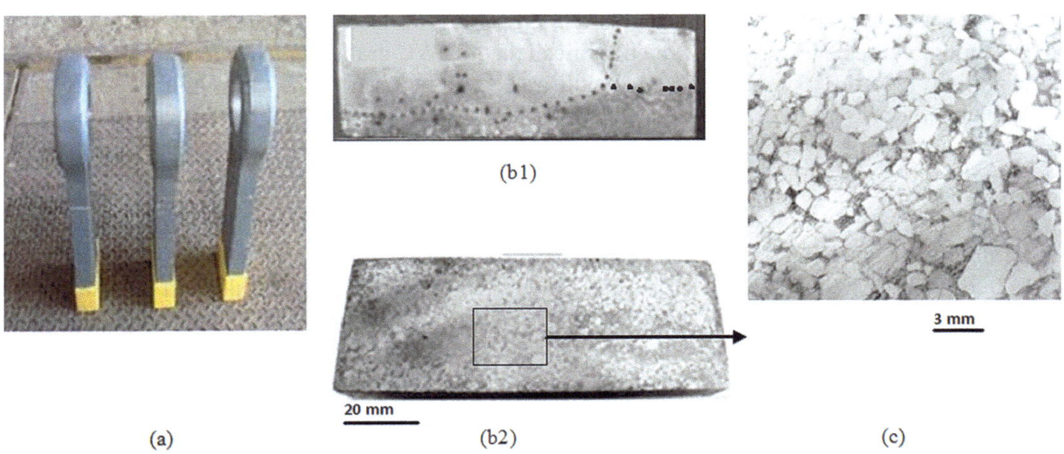

Fig. 43.3 (**a**) General aspect of cast products – bottom part is a MMC region; (**b1**) transversal section of the hammer head; (**b2**) striking surface of the hammer head; (**c**) typical microstructure in the striking surface (OM)

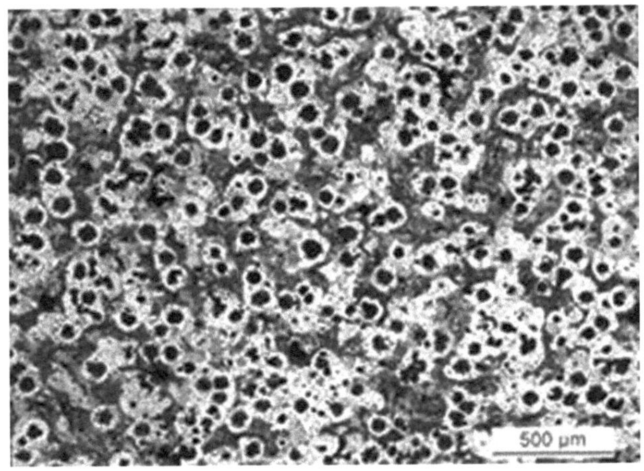

Fig. 43.4 Microstructure in the body region of the hammer produce in FE50007cast iron (OM)

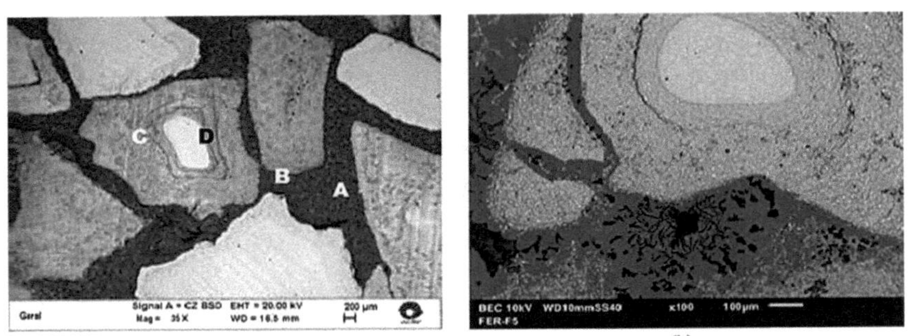

(a) (b)

Fig. 43.5 Microstructure in the striking surface of the cast hammer head with WC granules as reinforcement: (**a**) indication od distinct regions to be analysed; (**b**) detachment of parts of WC granules (MEV/BEC)

43.3.3 Microstructure in the Hammer Head: Striking Surface

Figure 43.5 presents general aspects of the microstructure in the bottom surface of the hammer head, it meaning the striking surface, where WC granules are concentrated. It can be noticed that the WC granules present different aspect when compared to the original condition: a porous external layer and some cracks can be observed, as shown in (b). Such cracks promotes detachment of granules pieces and even WC isolated particles. Four regions can be clearly defined and will be analysed separately: A – iron matrix among WC granules, B – matrix / reinforcement interface, C – external layer of the WC granule and D – internal region of the WC granule, as indicated in (a).

Region A – microstructure of metal matrix among WC granules
Different types of microstructures could be observed in the metallic matrix among WC granules in the composite region, as shown in Fig. 43.6. It is observed typical ferrite / perlite + nodular graphite phases for this kind of cast iron (similar to the structure in the body of the artefact), but also microstructures containing predominantly ferrite + globular graphite (b), or ferrite / perlite + degenerated graphite, or even low ferrite / perlite ratio + lamellar graphite (c).

These different microstructures can be consequence of the massive presence of the reinforcing WC granules in the region, which may have strong influence in the local solidification of the alloy, due mainly to two effects: thermal effect, as the granules are ceramic material, and so affect the solidification time; and chemical effect as the WC granules can interact with the liquid metal through chemical reactions or diffusion of elements, affecting solidification mode. The magnitude of such effects are certainly dependent on the distance among granules, which is uneven as the WC granules are randomly distributed in the liquid matrix. Therefore, metal matrix in the reinforced layer can be very heterogeneous when high WC content is added.

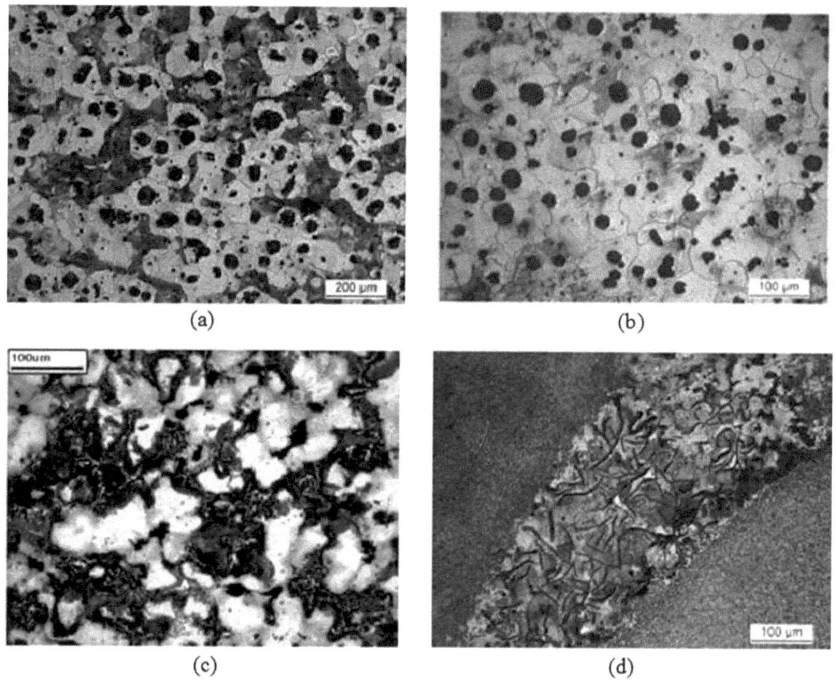

Fig. 43.6 Different structures observed in the metallic matrix (FE50007) among WC granules, in the composite layer in the head of the cast hammer: (**a**) ferrite / perlite + nodular graphite; (**b**) predominant ferrite + nodular graphite; (**c**) ferrite / perlite + degenerated graphite, (**d**) perlite / ferrite + lamellar graphite (OM)

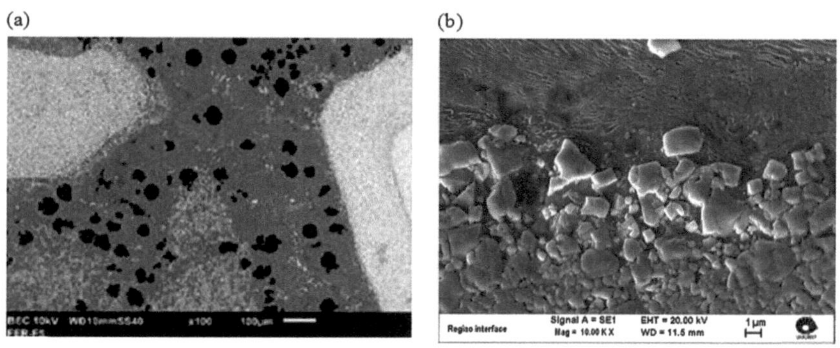

Fig. 43.7 General aspects of the metal matrix / WC granules interface, in the composite layer in the head of the cast hammer: (**a**) general aspect; (**b**) detailed characteristic. (SEM/BEC and SE)

Region B – microstructure in the metal matrix / WC granules interface

Typical aspects of metal matrix / WC granules interfaces, in the composite region of the hammer head produced are presented in Fig. 43.7. In all cases it can be observed smooth transition between the reinforcement and the metal matrix; neither apparent oxides layer or porosity, nor reaction zone in the transition region can be noticed. Indeed, it can be observed the ferrite or lamellar perlite phases directly in contact with the WC granule in the interface. As commented earlier, WC granules apparently suffer some kind of degeneration from the surface in contact with the hot liquid metal inwards, leading to the detachment of parts of the granules or even individual WC particles to the liquid metal. Liquid metal may penetrate in the surface of the degenerated granules, resulting in the smooth and continuous interfaces observed.

Microanalysis in different regions of the WC / metal interface helps to understand the interaction between liquid metal and reinforcing material. Figure 43.8 presents microanalysis results in the internal region of the WC granule, in its external, already detached region, and in the iron matrix circa of 200 μm far from the granule interface. It can be observed the presence of high Co and W contents in the metal (point 3 in the picture) and also high Fe content in the detached layer of the granule (point 2); some iron is also detected in the surface of the granule close to the detached layer of WC particles (point 1). Co

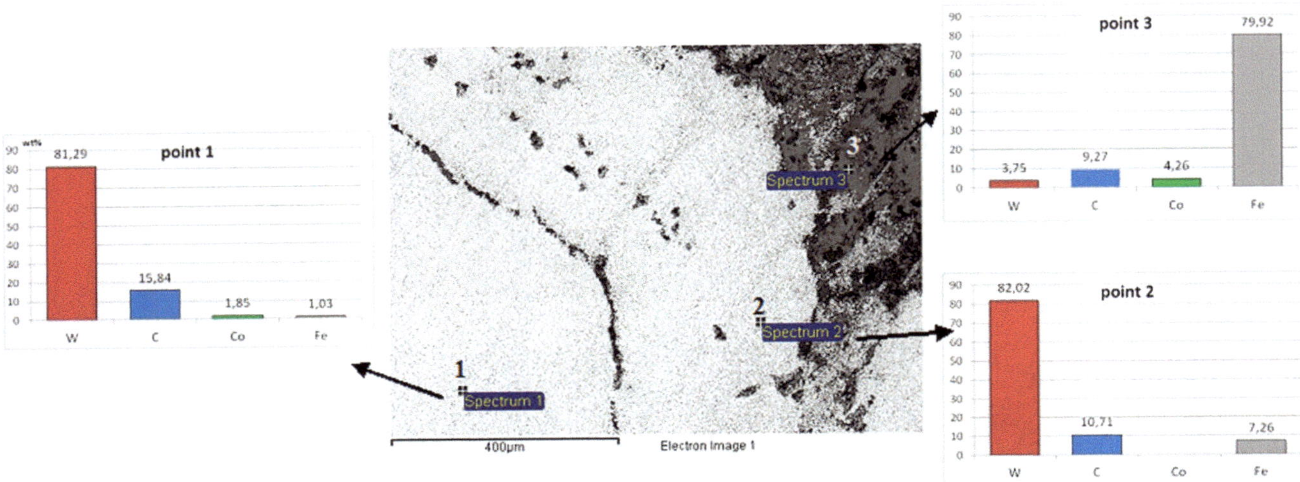

Fig. 43.8 WC granules / metal matrix interface in the composite layer in the head of the cast hammer (SEM/BEC) and results of local microanalysis in different regions as indicated in (a)

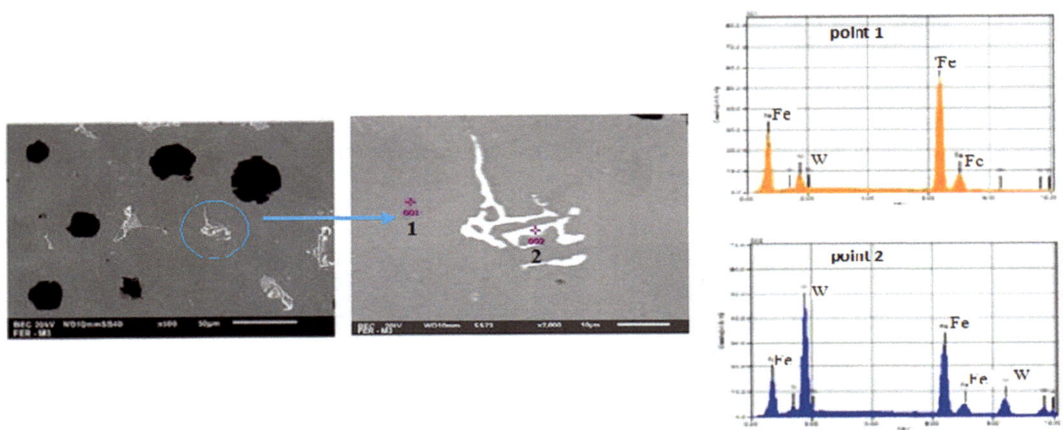

Fig. 43.9 FeCW rich phase in ferrite grain boundaries in the metal matrix among reinforcement, formed due to W diffusion from WC granules (SEM/XRD)

content is very low in the detached layer, according to other analysis in the same region, or absent, as shown here. These results suggest the diffusion of Co, used a binder in the WC granules, and W from the WC itself, to the liquid metal. On the other hand, Fe can penetrate in the porous, degenerated WC granules.

The diffusion of elements to the liquid metal can influence the metal microstructure formation, as mentioned earlier. For instance, Co can act as a ferrite stabilizer [9], while high W content in liquid cast iron can lead to the formation of complex Fe-C-W phases as observed in some regions in the matrix among WC granules, as presented in Fig. 43.9. This irregular shaped phase is present in ferrite grain boundaries, in a considerable frequency. XRD analysis indicate massive only the presence of Fe and W.

Regions C and D – microstructure of the WC granules

Figure 43.10 present the microstructure in two different regions of the WC granules used as reinforcement in the composite layer of the head of the cast hammer produced. As observed previously, the granules are degenetareted dusring contact with the liquid metal at high temperature; W, Co and Ni diffusion to the liquid was shown to be evident, resulting in the destruction of the bonding among individual WC particles in the granule of the cutting tool. Therefore, individual WC particles can be detached and suffer dissolution in the metal.

Analysing the external region the granules (region C), it can be noticed high porosity, highlighted in the back scattered electrons image, and irregular surface, with loose WC particles. On the other hand, region C, in the inner part of the granule, individual particles are still attached to their neighbours by the action of the binders, in spite of already presenting high

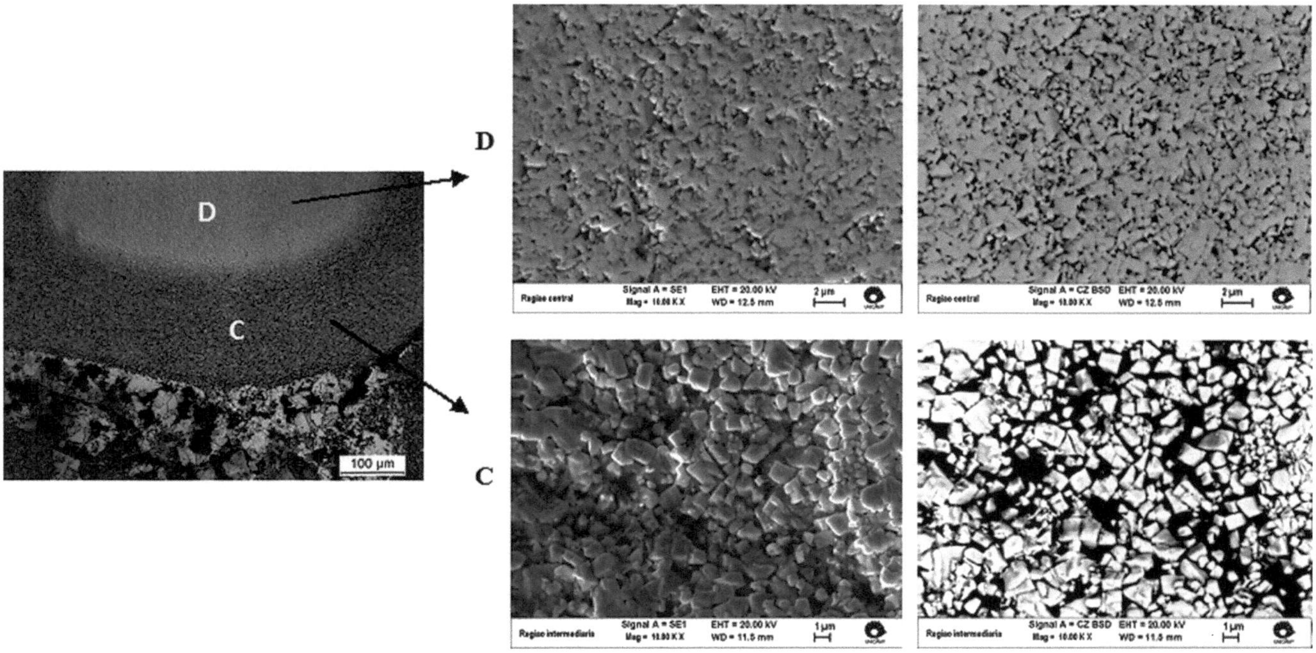

Fig. 43.10 Microstructure in two different regions of the WC granules in the in the composite layer in the head of the produced FE50007cast hammer: region C – outer surface; region D – inner region (OM, SEM)

porosity. Scanning different regions of the composite layer in the produced artefact, it could be observed some cases of completely degenerated WC granules; it occurs naturally with the smallest ones.

Results show that the utilization of WC granules as reinforcement in ferrous alloys by means involving liquid metal must take in account the degeneration of the reinforcing material. Resulting chemical alterations in the liquid can promote structure modifications during solidification, besides the eventual absence if the reinforcement itself in the structure. Therefore, appropriate production parameters must be used to avoid or minimize such events and to guarantee the presence of the required composite structure in the product.

43.3.4 Preliminary Evaluation of Wear Behaviour of the Developed Crusher

The cast hammer produced including a composite layer in its head was evaluated concerning wear behaviour in the striking surface considering different aspects: microwear, macrowear and overal performance during working in real conditions.

As it was observed a modification in the metal matrix microstructure in inter granules of the reinforcement region, micro erosion tests were prformed. Typical results is shown in Fig. 43.11: (a) presents in the wear profile path in the iron matrix between two WC granules, (b) the model of the cratere formed and (c) the pattern of colors used to measure the depht of the removed material. It can be noticed just a slight danification of the WC surface, while a deep cratere is formed in the soft matrix. Same experiment, under similar conditions (as indicated in item 2), was performed in the iron matrix in the body of the hammer, so, far from the composite layer, for comparative evaluation. Calculations resulted in the values of total volume loss/time of 72,876 μm^3/min in the metal between granules in the composite layer and of 490,710 μm^3/min in the metal in the body of the hammer, in the tsted conditions used. These preliminary results shows, therefore, an increase in the wear resistence in the magnitude of circa x7 in the FE50007 metal due to the structure modification in the intergranules regions in the reinforced layer.

As far as macro wear is concerning, besides the striking surface of the hammer head produced, also a sample of H13 tool steel was tested in the same conditions, to allow a comparative evaluation of the wear behaviour in the two types of materials. Results show that, in the specific conditions used in the tests, the composite FE 50007/WC presented 36–54 mm^3 of volume loss and around 0.50 g of mass loss, while the H13 steel presented 103–117 mm^3 and 0.86 g, respectively. These results show considerably superior performance of the nodular cast iron reinforced with WC particles compared to a hard tool steel.

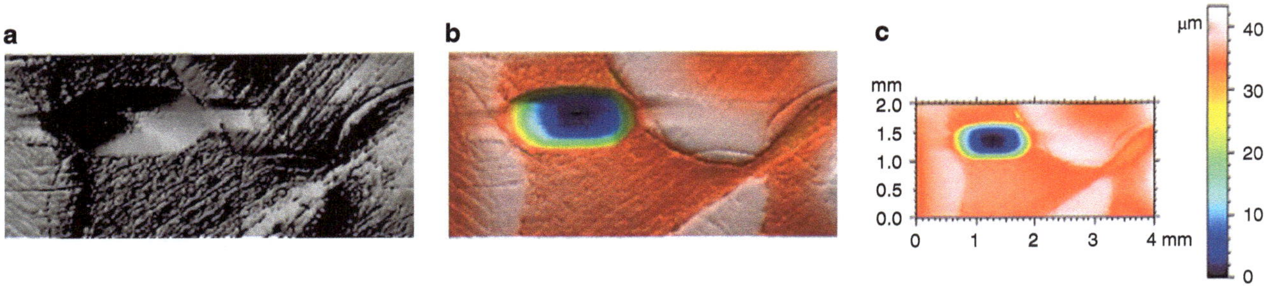

Fig. 43.11 (**a**) Typical damage by wear in the FE50007 matrix between two WC granules in the composite layer in the head of the produced crusher; (**b**, **c**) scanned topography

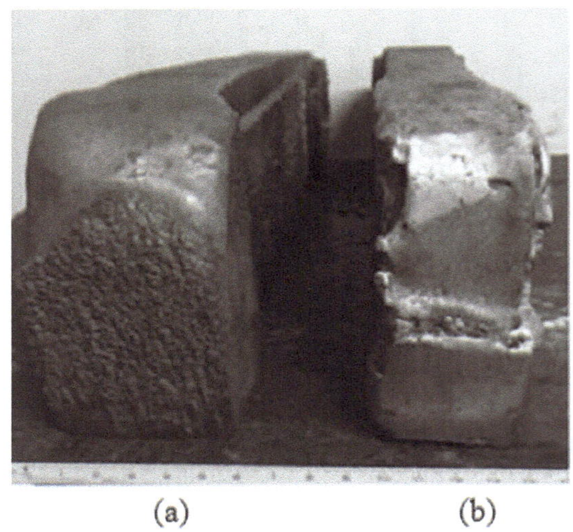

Fig. 43.12 General aspect of the striking surface of the hammer head of crushers submitted to the same work conditions in a sugar cane plant: (**a**) hammer head containing WC as reinforcement material; (**b**) commercial hammer, with surface coated by welding

Results of the evaluation of the crusher produced in this work, presenting a composite layer of WC reinforced material in the head surface, after submitted to normal work conditions in a sugar plant are presented in Fig. 43.12, comparatively to a commercial crusher (surface hardened by deposition of hard material by welding). After shredding 234,000 ton of sugar cane stalks, both hammers presented the damaged surfaces shown in the picture. However, it can be observed better conditions in the hammer with a composite surface. Mass loss was evaluated and results indicated values of 1.35 and 2.00 kg, over the total weight of 33 kg, respectively for the hammer with composite surface and commercial hammer with coated surface. Therefore, the composite material suffered mass loss around 32% lower than the weld coated material.

43.4 Conclusions

Results show that impact hammers can have their wear behaviour significantly improved, increasing also work life, when the striking surface of the hammer head submitted to the severe impact/wear conditions, is constituted by a composite material; instead a coated surface. WC granules recycled from cutting tools proved to be an appropriate reinforcement for ductile cast iron, as no deleterious reactions are present in the reinforcement / metal interfaces. Interfaces present a continuous and smooth transition. It was observed Co and W diffusion from the WC granules to the metal, leading to the detachment of WC particles to the liquid, where they can dissolve. Liquid metal can penetrate in the external surfaces of the deteriorated WC granules, promoting a smooth transition. These effect, added to the effect of the presence of WC material in the composite layer, promote microstructural changes in the metal matrix among WC granules. The presence of the hard WC particles, alongside with their effect in the microstructure result in significant superior wear quality of the hammer head containing WC as reinforcement when compared to head coated surfaces on commercial hammers.

Acknowledgements Authors want to thank financial support from Program French Catedra UNICAMP, Brazil/French Embassy for Prof. Bayraktar's visiting period at UNICAMP.

References

1. Mukhopadhyay, G., Palit, P., Bhattacharyya, S.: Development of AISI A2 Tool Steel Beater Head for an Impact Crusher in Sinter Plant – Metallography, Microstructure, and Analysis Application and Innovation for Metals, Alloys, and Engineering Materials. Springer Science + Business Media New York and ASM International. (2015). https://doi.org/10.1007s13632-015-0192-6
2. Umucu, Y., Deniz, V., Cayirli, A.: A new model for communition behaviour of different coats in impact crusher. Energy Sources A. **36**(13), 1406–1413 (2014)
3. CTC General Annuary, Centre for Sugar Technology, Brazilian Government, Piracicaba, SP, Brazil (2015)
4. Leibholz, R., Robert, H.H., Leibholz, H., Bayraktar, E.: Development of Functionally Graded Nodular Cast Iron Reinforced with Recycled WC Particles – Mechanics of Composite and Multi-functional Materials, (7), pp. 241–249 (2017)
5. Liu, A., Guo, M., Zhao, M., Wang, C.: Microstructures and wear resistance of large WC particles reinforced surface, metal matrix composites produced by plasma melt injection. Surf. Coat. Technol. **201**, 7978–7982 (2007)
6. Nascimento, A.M., Ocelík, V., Ierardi, M.C.F., De Hosson, J.T.M.: Wear resistance of WCp / duplex stainless steel metal matrix composite layers prepared by laser melt injection. Surf. Coat. Technol. **202**, 4758–4765 (2008)
7. Emamian, A., Alimardani, M., Khajepour, A.: Effect of cooling rate and laser process parameters on additive manufactured Fe–Ti–C metal matrix composites microstructure and carbide morphology. J. Manuf. Process. **16**, 511–517 (2014)
8. Sandvik.: Cemented Carbide, Sandvik New Developments and Applications – Sandvik Hard Materials, Materials Science and Engineering. pp. 1–12 (2005)
9. Hasse, S.: Duktiles Gusseisen – Handbuch fuer gusseirzeuger und gussverwender. p. 158 (1966)

Chapter 44
In-Situ Imaging of Flexure-Induced Fracture in Fiber-Reinforced Composites Using High-Resolution X-Ray Computed Tomography

Brian P. Wingate and Michael W. Czabaj

Abstract This work presents a new test method which allows for *in situ* high-resolution X-ray computed tomographic imaging of flexure-induced fracture in tape-laminate composites. Specimens with two distinct stacking sequences were tested to visualize, in 3D, the evolution of intralaminar ply cracks and delaminations as a function of the applied bending moments. The first laminate, which had small angle changes between adjacent plies, produced a fracture pattern that consisted solely of intralaminar cracks. The second laminate, having a more traditional quasi-isotropic stacking sequence, evolved a fracture surface that contained interacting intralaminar cracks and delaminations. The vastly different fracture surfaces obtained from both laminates provide invaluable, and previously unobtainable, 3D information for validation of existing and future progressive damage models.

Keywords Computed tomography · Damage evolution · Tape-laminate composites · Model validation

44.1 Introduction

Recent developments in finite-element (FE) based progressive-damage algorithms have enabled simulation of complex fracture patterns that can be observed in fiber-reinforced tape-laminate composites. FE codes based on the Floating Node Method (FNM) [1] or the Regularized eXtended-Finite Element Method (Rx-FEM) [2], are now capable of simulating initiation and evolution of intralaminar cracks and delaminations under a variety of loading conditions and laminate geometries. Despite these advances, there is a significant lack of experimental data that can be used to quantitatively evaluate the predictive nature of these codes. To date, most of the validation has been performed using in-plane uniaxial-tension static and fatigue tests, with progressive damage evolution imaged using 2D ultrasonic or X-ray radiography techniques. In this work, a new test is proposed which enables 3D imaging of flexure-induced progressive fracture in tape-laminate specimens using high-resolution X-ray computed tomography (CT).

44.2 Experimental Methodology

The test geometry proposed in this study consists of a loading fixture that applies bending to a vertical composite beam using two eccentric compressive loads as shown in Fig. 44.1a. The test fixture is designed to ensure that the central portion of the composite beam is always exposed to the incoming X-ray beam, thus allowing for uninterrupted imaging of the gage region during flexure. Moreover, the horizontal loading arms are sized to ensure sufficiently high bending moments in the gage region, while keeping the asymmetric stress distribution across the specimen height to a minimum. The entire flexure apparatus is designed to interface within the hot-cell experiment available at the Advanced Light Source beamline 8.3.2 (Lawrence Berkeley National Laboratory, Berkeley, CA), and if necessary, can be used at temperatures in excess of 1000° C [3].

The preliminary evaluation of the test fixture was performed by considering two composite beam designs, each with a distinct tape-laminate stacking sequence. The first design, shown in Fig. 44.1b, contained a stacking sequence with 15° angle changes between adjacent plies, which promoted extensive evolution of through-thickness transverse cracks. The second design, shown in Fig. 44.1c, contained a more traditional, quasi-isotropic stacking sequence with 45° and 90° angle changes

B. P. Wingate · M. W. Czabaj (✉)
Department of Mechanical Engineering, University of Utah, Salt Lake City, UT, USA
e-mail: m.czabaj@utah.edu

P. R. Thakre et al. (eds.), *Mechanics of Composite, Hybrid and Multifunctional Materials, Volume 5*, Conference Proceedings of the Society for Experimental Mechanics Series, https://doi.org/10.1007/978-3-319-95510-0_44

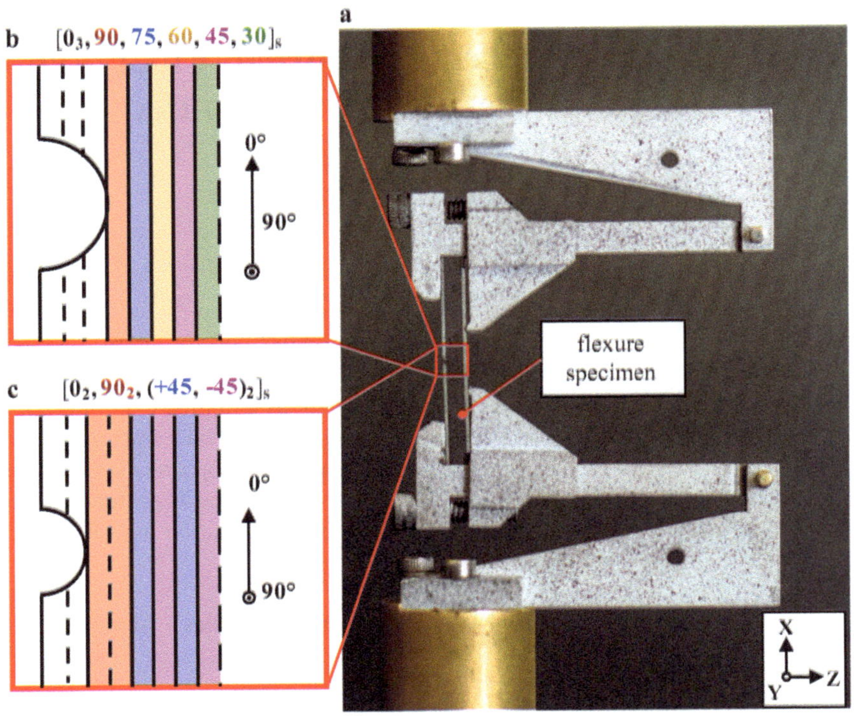

Fig. 44.1 (**a**) Vertical flexure fixture and specimen designed for in situ imaging of damage evolution (**b**, **c**) Detailed specimen geometry and stacking sequences for each tested laminate

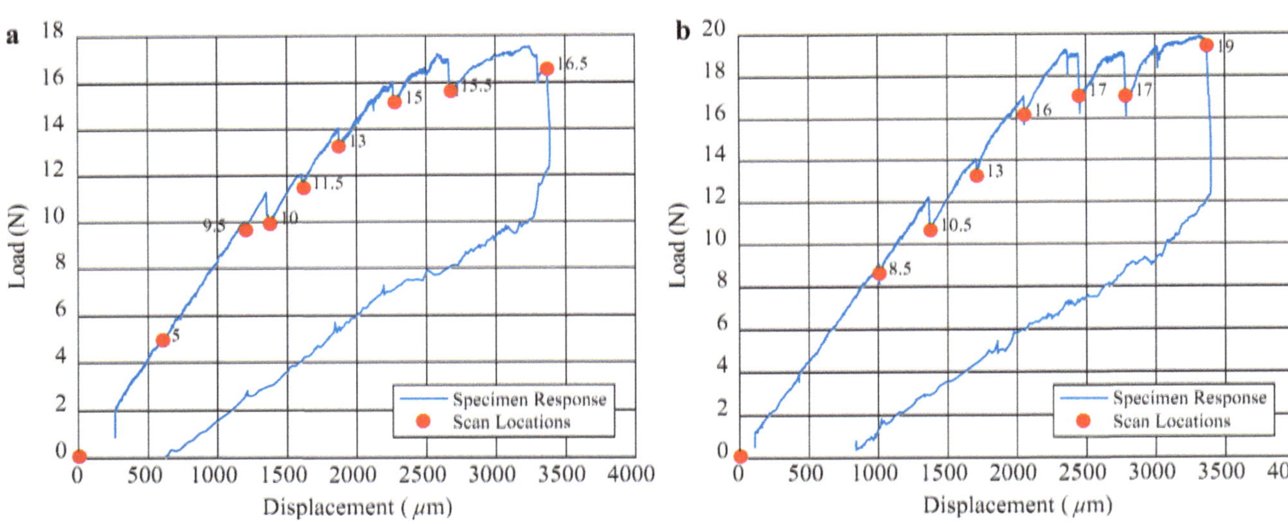

Fig. 44.2 Experimental response and scan locations for (**a**) 15° angle change specimen (**b**) 45° and 90° angle change specimen

between adjacent plies. This stacking sequence was designed to promote interaction and simultaneous growth of transverse cracks and delaminations. Both stacking sequences included a stack of 0° plies on the top and bottom surfaces to increase the overall flexural rigidity of the specimens. Prior to testing, a semi-circular notch was machined into each specimen to ensure initiation of fracture in the gauge region. The exact stacking sequences and machined notch geometry are presented in Fig. 44.1b, c.

X-ray CT *in situ* flexure testing was done in displacement control at a rate of approximately 0.5 mm/min. At each notable damage event, identified from live X-ray projections or by a drop in load, the test was paused and the specimen was scanned. This process was repeated several times until sufficient damage accumulated in the gage region. The force-displacement response for each laminate is shown in Fig. 44.2. In this figure, the red dots correspond to load levels at which the test was

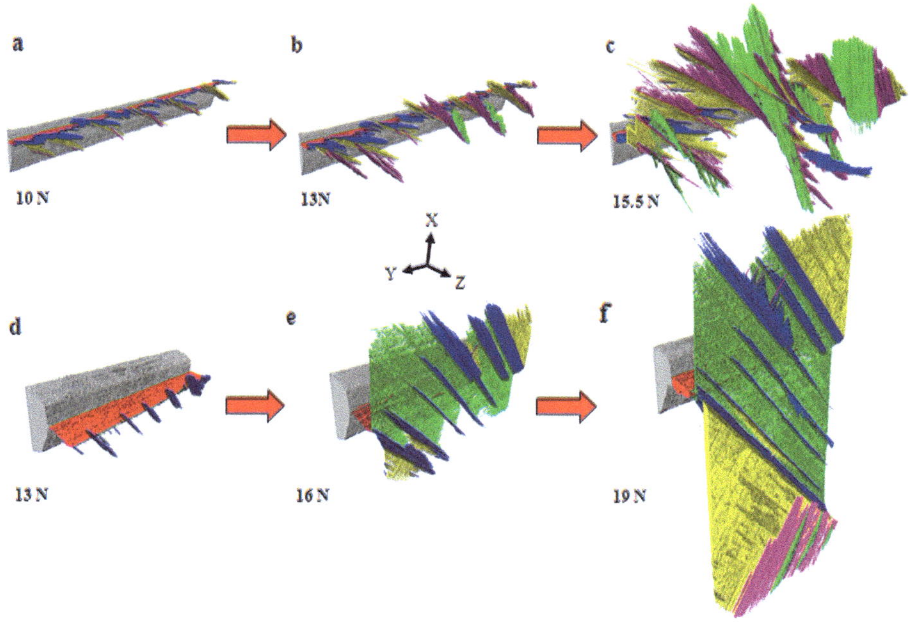

Fig. 44.3 Evolution of fracture surfaces at select load increments (**a–c**) Transverse-crack dominated fracture in 15° angle change stacking sequence (**d, f**) Delamination and transverse-crack interactions in 45° and 90° angle change stacking sequence

paused for X-ray CT imaging. Tomographs for each scan were reconstructed using the open-source Python package TomoPy [4], and the damage fracture surfaces were segmented using Avizo and ImageJ [5].

44.3 Results and Discussion

The resulting fracture surfaces for each specimen at three different load levels are shown in Fig. 44.3. The color scheme in each image corresponds to a different ply in the laminate as defined by the stacking sequences given in Fig. 44.2. Examining the image data for the first laminate (see Fig. 44.3a–c), the fracture process begins with width-wise growth of a transverse crack across the 90° (red) ply. As the load is increased, a series of nearly evenly-spaced transverse cracks extends into the 75° (blue) and 60° (yellow) plies. This process is repeated for the remaining plies. However, each new ply experiences crack coarsening with only two transverse cracks propagating past the midplane. Surprisingly, and contrary to conventional wisdom, most of the transverse cracks propagated across individual ply interfaces without forming delaminations (i.e. inter-ply cracks).

Examining the damage pattern in the second laminate (see Fig. 44.3d–f), the initial fracture also started with development of width-wise crack across the 90° (red) ply, which was followed by growth of evenly spaced transverse cracks across the +45° (blue) ply. As the load is increased, the specimen exhibits two distinct mechanisms of crack growth: (1) the transverse cracks in the +45° ply elongate along the fiber direction, and (2) delaminations initiate and extend along the 90°/+45° (green) and + 45°/−45° (yellow) interfaces. This pattern of growth continues as the load is increased, with the damage remaining below the laminate midplane until the final load of 20 N.

44.4 Conclusions

As described above, the two laminates exhibit vastly different evolution of fracture surfaces, which are invaluable for validation of existing and future progressive damage models. Work is currently underway to accurately quantify the evolution of damage in each case. The ongoing quantification includes measurements of the number, length and area of the evolving intra- and inter-laminar cracks as a function of the applied load. Moreover, a new method is being developed to accurately determine and segment the shape of individual plies, and the local variation in orientation of reinforcing fibers within each ply.

References

1. Pinho, S.T., et al.: A floating node method for the modelling of discontinuities in composites. Eng. Fract. Mech. **127**, 104–134 (2013)
2. Iarve, E.V.: Mesh independent modeling of cracks by using higher order shape functions. Int. J. Numer. Methods Eng. **56**, 869–882 (2011)
3. Bale, H.A., et al.: Real-time quantitative imaging of failure events in materials under load at temperatures above 1,600 °C. Nat. Mater. **12**, 40–46 (2013)
4. Gürsoy, D., et al.: Tomopy: A framework for the analysis of synchrotron tomographic data. J. Synchrotron Radiat. **21**(5), 1188–1193 (2014)
5. Schindelin, J., et al.: The ImageJ ecosystem: An open platform for biomedical image analysis. Mol. Reprod. Dev. **82**, 518–529 (2015)

MIX
Papier aus verantwortungsvollen Quellen
Paper from responsible sources
FSC® C105338
FSC
www.fsc.org

If you have any concerns about our products,
you can contact us on
ProductSafety@springernature.com

In case Publisher is established outside the EU,
the EU authorized representative is:
Springer Nature Customer Service Center GmbH
Europaplatz 3, 69115 Heidelberg, Germany

Printed by Libri Plureos GmbH
in Hamburg, Germany